P9-EAX-342

# SPACE TRAVEL

*A History*

# SPACE

*A History*

# TRAVEL

An Update of *History of Rocketry & Space Travel*

WERNHER VON BRAUN
FREDERICK I. ORDWAY III

*Revised in collaboration with*
DAVE DOOLING

*Foreword by*
FREDERICK C. DURANT III

*1817*
**HARPER & ROW, PUBLISHERS,** New York
Cambridge, Philadelphia, San Francisco,
London, Mexico City, São Paulo, Singapore, Sydney

This is the fourth edition of a book previously published under the title *History of Rocketry & Space Travel.*



*Original illustrations by Harry H-K Lange.*

Library of Congress Cataloging-in-Publication Data

Von Braun, Wernher, 1912–1977.
Space travel.

Updated ed. of: History of rocketry & space travel. 3rd rev. ed. 1975.
Bibliography: p.
Includes index.
1. Rocketry—History. 2. Astronautics—History. I. Ordway, Frederick Ira, 1927– . II. Dooling, David, Jr., 1950– . III. Von Braun, Wernher, 1912–1977. History of rocketry & space travel. IV. Title.
TL781.V6 1985 629.4'09 85-42596
ISBN 0-06-181898-4

85 86 87 88 89 MPC 10 9 8 7 6 5 4 3 2 1

# *Foreword to the Previous Editions*

The history of astronautics may be thought of as two separate histories—rocketry and space flight. These fascinating stories proceeded in parallel, unrelated fashion for hundreds of years until the last decades of the nineteenth century. At about that time, the first appreciation was gained of the necessity for rocket propulsion to achieve man's ancient, hitherto unattainable dream of flight into space.

The beginnings of rocketry date to the thirteenth century and the use of black-powder rockets. The Chinese are generally given credit for being the first to use the rocket as a propulsive device, although no hard factual reference to this exists. It is accepted by historians, however, that crude rockets were fired as military weapons in the thirteenth century in the Far East, the Near East, and in Europe. "Reaction propulsion" as an effect was known much earlier to the Greeks, but the principles were apparently not fully understood within the framework of the science of that period.

Authors Wernher von Braun and Frederick I. Ordway III are each eminently qualified to present this exciting and still widely unknown story. I have known both men for more than twenty years, during which period Von Braun has demonstrated his ability as the leading rocket engineer in the Western World. Ordway, through personal interest and initiative, has collected one of the largest private astronautical archives in the world and has been a major contributor to astronautical reference works. The interests of both men are broad—a necessity when studying a worldwide technological history of intermittent and sporadic progress spanning centuries. Both men have had a classic, as well as a scientific, education which enables them to recognize the impact of both the philosophical and technical aspects of the evolution of astronautics. From a historical standpoint, the significance of events and of individual efforts is not always evident, nor the relative impact easily seen in retrospect. In my opinion, this book represents an important contribution by recording the manifold contributions of many individuals from many different nations. It chronicles events which have led to perhaps man's greatest adventure—the exploration of space.

The history of rocketry and astronautics can, of course, never be completely told at any one time. "New" information and insights into the past will always remain to be discovered. Military archives, long-forgotten writings, and rare graphic presentations are continually coming to light. At the same time, new history is being written almost daily of space accomplishments. Another whole segment of this history has still to be written—that of the social and sociological implications and impact upon man resulting from the fruition of this investment of time, immense effort, money, and materials to achieve space mobility. This is quite apart from the impacts which will be felt from this new knowledge of the cosmos, as well as the new knowledge gained of man, himself.

At this instant, however, I know of no comparable reference work which documents rocket flight in such detail, contains as many references, or presents so many fine photographs and illustrations of past and current events. The authors have endeavored to encompass and present in a lucid fashion the hundreds of elements of this history and have succeeded admirably. Much is presented comprehensively for the first time in a single volume and the work promises to be an important standard reference.

It has always been surprising to me how little of the history of rocketry and space flight is known to the average intelligent person. This includes tens of thousands of persons professionally engaged in space programs. I would suspect that a significant percentage of the rocket engineers whom you might question would not know who Sir William Congreve was, let alone William Hale. Nor would they be likely to know that every major nation in Europe had rocket brigades in the nineteenth century; that rockets were used, albeit not widely, by the United States in the Mexican War in 1846–1847, and by both the Union and Confederate armies during the Civil War. Why should they know? One reason, I suppose, is that it can be both interesting and satisfying to know the history of one's profession. There is a cultural aspect, also, in being knowledgeable of the technical ancestry of modern space launch vehicles.

The study of astronautical history is exciting and

rewarding because one learns of the prescience and insight of brilliant intellects over the centuries. One learns also that improvements in technology and development of bold ideas were largely the result of study and research by individuals. These individuals, caught up in a dream, were driven relentlessly by the excitement of a concept and the possibility of contributing to its fulfillment. These stories make exciting history. Another aspect is that numerous and varied scientific disciplines and technologies are involved. Interdisciplinary study is required and for this reason it is sometimes difficult to make valid judgments of individual contributions.

Great strides in space accomplishment had to await modern technology and the team effort of thousands of minds simultaneously working on hundreds of problems in a tremendously complex system. Under the leadership of Major General Walter R. Dornberger and Dr. Von Braun, the opening of the German test station at Peenemünde in 1937 signaled the dawning of the systems approach necessary to achieve space flight. We know now that a great cooperative effort led by skilled administrators is required; a partnership of industry and government, and widespread technical communications. These, together with powerful tools such as computers, far-flung ground facilities for static-test and launch operations, and worldwide tracking stations make today's space technology possible. It is uniquely different from the individual creative effort, however brilliant, of forty years ago.

The human mind is still the common denominator of any team effort, however vast. The individual will still be the initiator of new concepts and inventions, and the solver of problems. Thus, the study of this history will reveal the creative mind in peace and in war over the centuries. Oftentimes, the name of the individual is unknown, but he is there and his contribution is evident.

I believe that this book will enlighten all who read it. It is hoped that reading it will encourage further interest and open new avenues of study of one or more of its facets, some of which have been barely touched upon because of space limitations.

Man's ancient drive to explore was limited to the Earth's surface until this century. Now the way to the Solar System is open. This is the story of how it began and how it became possible in our time.

FREDERICK C. DURANT III
Assistant Director, Astronautics
National Air and Space Museum

*Smithsonian Institution*
*Washington, D.C.*

*And to This Edition*

I am delighted to note the updating of this (now) classic history of rocketry and space flight. Authors Ordway and Dooling have added documentation of the significant events in astronautics of the past two decades. During this period mankind has established the capability for permanent presence in Earth orbit. Our solar system and our galaxy have been studied by space probes and powerful orbiting sensors. These tools have yielded more new knowledge of astronomy than all understanding in the hundreds of years before the dawn of the Space Age. Applications satellites utilizing their unique vantage points in space routinely provide Earth observation, navigation, and global communications of economic, social, and political benefit to all nations.

Now history, these momentous accomplishments are but small steps in mankind's emergence into the unlimited dimension of space.

FREDERICK C. DURANT III
Consultant

*Chevy Chase, Maryland*

# *Preface*

It has been a full decade since the third edition of this book appeared. During that decade, the Space Age matured to the point that both unmanned and manned activities became routine. Many twenty-fifth anniversaries were observed: in October 1982, that of the Soviet Union's Sputnik 1; in late January 1983, that of the first United States artificial satellite, Explorer 1; and in October of the same year, that of the National Aeronautics and Space Administration itself.

Following the 4 July 1982 Independence Day return of the space shuttle Columbia to Edwards Air Force Base in California, NASA declared the orbital flight test program completed and its Space Transportation System operational. At Edwards, President Ronald Reagan drew a parallel to the golden spike that was driven to celebrate the completion of the transcontinental railroad. Human beings could now travel to and from space almost at will and in space vehicles that were in large part recoverable. And they could stay for long periods in space aboard orbiting stations that could be revisited repeatedly. Thus, on 2 October 1984, three Russian cosmonauts returned from the repeatedly visited Salyut 7 after a record 237 days—nearly 8 months—in space.

The third edition of this book appeared in 1975 and carried this preface by von Braun and Ordway:

When the first edition of this book appeared in late 1966, the Space Age was nearly a decade old. The Americans had recently photographed a strip of Mars from Mariner 4 and the Soviets had impacted a Venera probe on Venus. In the manned space flight arena, after a slow start the United States had moved briskly ahead and had completed its one-man Mercury Earth orbital flights and was well into the two-man Gemini program. The Russians, for their part, had completed the Vostok and Voshkod series and, it was widely suspected, were on the verge of still more ambitious activities in space. The book looked forward, with great excitement, to America's Apollo Lunar landings scheduled to begin late in the 1960 decade. No estimate could be given of the target date of a supposed paralleled Soviet effort to reach the Moon, but it was speculated that it might be even earlier than Apollo.

Three years later, the Americans landed on the Moon, an event carefully covered in the second edition of this book. Although the Russians recorded important progress in their spacecraft probing of Venus, their manned flight activities were but modest. The "race to the Moon," it turned out, did not seem to develop into much of a race at all. From the vantage of mid-1969, we felt that it was too early to predict how the United States and other nations would react to Apollo 11, but the ebullient mood stirred up by the historic achievement seemed to presage, at the very least, a continuing interest in Lunar exploration leading to the establishment of bases on the Moon by the end of the 1970 decade. At the same time, plans for Earth and Lunar orbiting space stations and manned flight to Mars were unfolding in an atmosphere of cautious expectation.

Such far-reaching ambitions on the part of the United States clearly had to be coordinated under the umbrella of a long-range plan. For this purpose, President Richard M. Nixon established a Space Task Group chaired by Vice-President Spiro T. Agnew and consisting of NASA Administrator Thomas O. Paine, Air Force Secretary Robert C. Seamans, and the President's science advisor Lee A. DuBridge. Three alternative space goals were offered, all built on the premise of "a balanced manned and unmanned space program" with a manned Mars landing before the end of the twentieth century as the sought-for "focus" around which America's astronautical resources were to be brought into play. . . .

Unfortunately, the Space Task Group's recommendations went unheeded. No sooner had the enthusiasm created by Apollo 11's success abated than doubts as to the worth of America's space program began to be expressed. NASA, in the person of administrator Paine, immediately recognized the need "to find some focusing elements in the next decade or two." He found it hard to admit "that the United States, the wealthiest nation in the world, the nation that has the greatest capability that mankind has ever assembled, can't afford the really modest expenditures to continue an aggressive space program. To compare a decade of costs, in this time, we have spent a half-trillion dollars on our welfare programs. . . . I think if we indeed cripple the space program with the idea that this will help us move ahead on other fronts, it will have the contrary effect. I think that the nation will indeed become even more doubtful of its ability to carry out its responsibilities."

But, as a result of a complex mixture of events and attitudes (including the Vietnam war, urban and student unrest, racial disorders, increasing concern for the terrestrial environment, and growing shortages of fossil fuels and other resources) the space program wound down. In the late 1969 environment, no Apollo-type focus was to appear around which the next phase of the space effort could gel. The reusable space shuttle was the only major new program whose anchor held against the ebbing tide. The idea of backing a manned expedition to Mars was indefinitely deferred.

Despite the curtailment of the Apollo and follow-on programs and the failure of comparable Soviet programs to materialize, sound progress in astronautics was recorded during the five years between the appearance of the second and third editions. The Americans completed their manned Lunar landings with Apollo 17 in December 1972, and the following year orbited the Skylab embryonic space station. . . .

The authors wish to acknowledge the following for their help in preparing the third edition: C. Holley Taylor-Martlew, Cheltenham, England (research); Colonel Audrey E. Thomas, U.S. Air Force, Washington (military space programs); and Vladimir Belyakov, Embassy of the U.S.S.R., Washington (Soviet space program).

In this, the fourth edition, no solid, funded programs can be chronicled for the manned exploration of Mars—nor even for the resumption of manned flights to the Moon. But in recent years, the case for Mars has been argued in major symposia, and the feasibility of returning to the Moon and setting up permanent lunar bases was debated at a symposium at the National Academy of Sciences in Washington in late October 1984. And, of more immediate importance, in January 1984 President Reagan approved a program for the development of a space station. Many voices have been heard urging international cooperation in all these endeavors.

While large-scale manned programs failed to materialize during the decade, many important and exciting space events passed into history. In the unmanned arena, the Americans and Soviets continued to explore the worlds of the Solar System. Both probed cloud-covered Venus, unravelling with each flight more and more of her carefully guarded secrets. The twin United States Viking orbiters and landers added much to our knowledge of Mars, while Pioneers 10 and 11 and Voyagers 1 and 2 provided stunning views of, and a rich harvest of scientific information on, Jupiter, Saturn, and their moons. At the same time, the Europeans and Japanese intensified their research on solar-terrestrial relationships and other space phenomena.

The launching of scientific and applications satellites became so widespread during the decade as to barely warrant mention in the press. As launch vehicle technology increased in reliability, so did competition, as many nations and organizations vied for a share of the rapidly expanding launch services market.

Except for the short-lived United States–Soviet cooperative Apollo-Soyuz Test Project, carried out in mid-1975, the American presence in space evaporated until the advent of the space shuttle in the spring of 1981. The Russians, however, maintained an ever-expanding manned space effort throughout the decade, establishing a number of endurance records and other achievements.

The post-Apollo 1970's witnessed a gradual decline of the national space vision in the United States, as NASA struggled to keep its near-term options open within taut budgetary constraints. Yet the dreams of space did not wither away. Witness, for example, the almost cult-like popularity of proposals for space colonies; the extraordinary success of such films as *Star Wars* and its sequels; the simultaneous appearance on the best-seller lists of James A. Michener's *Space*, Arthur C. Clarke's *2010: Odyssey Two*, and Isaac Asimov's *Foundation's Edge;* and record-breaking crowds at the National Air and Space Museum in Washington, D.C., the Alabama Space and Rocket Center in Huntsville, and elsewhere.

Part of NASA's post-Apollo problem was its own record of success; it would not be an easy matter to top the lunar expeditions. In the aftermath of Apollo, many viewed the space shuttle as utilitarian, almost pedestrian. Its critics muttered that it had no grand mission to support, no challenging goal calling for a new surge into space. All the shuttle had, they complained, was a "hole in the middle."

It is puzzling that these same people did not try to fill this hole with their imaginations. By making access to and return from space routine, as von Braun and others had envisioned in the 1950's and 1960's, the shuttle would make possible not only permanent orbiting space stations but a return to the Moon. Now, the same shuttle critics are beginning to show grudging admiration for Spacelab and other missions, Solar Max retrieval and repair, the retrieval of Palapa and Westar satellites, and subsequent successes.

As the era of the shuttle matured in the mid-1980's, a subtle change began to be noticed. Space achievements were attracting diminished attention; the astonishing was becoming routine.

At the same time, a new generation of individuals was evolving who had not known a time when there was no space flight. Babies born the year of Sputnik 1 will celebrate their 30th birthdays in 1987. Soon thereafter, the destiny of the world will pass to the hands of a generation whose knowledge of pre–Space Age society will come largely from books. Such a generation is unlikely to turn its back on the stars.

DAVE DOOLING
*Huntsville, Alabama*

FREDERICK I. ORDWAY III
*Washington, D.C., and Huntsville, Alabama*

# *Acknowledgments*

This book is the result of many years of research and writing, and could never have been completed without the aid of scores of persons, libraries, and organizations all over the world. We are particularly grateful to the persons listed below, noting in parentheses the area of assistance rendered. We regret that space did not permit more of their material to be utilized, but it was invariably valuable to us in constructing a balanced picture of events. The titles given are those held at the time assistance was rendered. Many persons have since assumed higher posts in national space agencies, the aerospace industry, or elsewhere. Several have passed away.

David S. Akens, George C. Marshall Space Flight Center, NASA (United States rocketry and documentation); John Alden, Rare Book Collection, Boston Public Library (rare books); Kenneth H. Allen, North American Aviation, Inc. (postwar missiles); M. Almazov, Soviet News, U.S.S.R. Embassy, London (Soviet missile and space programs).

John Barbato, Space Systems Division, United States Air Force Systems Command (carrier vehicles and satellites); Charles Barr, Northrop Corporation (wartime United States rocket airplanes); Ingénieur Général J. J. Barré, Versailles (prewar French rocketry); C. W. Birnbaum, Douglas Missile and Space Systems Division (United States carrier vehicle information); R. Boccarossa, Engins Matra (French satellites, missiles); Aktiebolaget Bofors (Swedish missiles); Professor John A. Boyle, Department of Persian Studies, University of Manchester, England (Persian legends); British Museum, London (library research and reproductions); British Patent Office (rocket patents); Dr. William M. Bryant (translation of Latin texts); Werner Buedeler (German documentation); Francis A. Burnham, United States Air Force Systems Command, Space Systems Division (satellite programs).

L. J. Carter, British Interplanetary Society (documentation on British rocketry); Cinémathèque Française (film stills); A. V. Cleaver, Rolls-Royce, Ltd. (British rocketry); Contraves AG (Swiss missiles); Jean Coulomb, Centre National d'Études Spatiales (French space vehicles); R. Courcelle, Bibliothèque Royale de Belgique (research of French-language works on rocketry); Sir Alwyn D. Crow (British prewar and wartime rocketry); Peter Curtice, Central Office of Information, London (British rocketry).

A. J. Dalkin, Royal Artillery Institution, Rotunda Museum (early British and Indian rockets); Alain Danet, Paris-Match, Paris (French space developments); Melvin S. Day, NASA, Washington (technical documentation); Kurt H. Debus, John F. Kennedy Space Center, NASA (United States launch facilities); L. Decker, National Army Museum, U.K. (early British rocketry); A. P. DeWeese, New York Public Library (support in rare book and general research); Frank L. Dickey, Douglas Aircraft Company (postwar United States rocket and astronautic developments); Colonel John Joffre Driscoll (United States World War II rocketry); F. George Drobka, NASA, Washington (technical documentation); Frederick C. Durant III, National Air and Space Museum, Smithsonian Institution (broad support in information research, location of photographic materials).

Commander Burton I. Edelson, Office of Naval Research, London (space documentation); Krafft A. Ehricke, Autonetics Division, North American Aviation, Inc. (postwar United States space programs); Dr. Eugene M. Emme, NASA, Washington (postwar rocket and space history); Rolf Engel (prewar German rocketry); Colonel Denis Ewart-Evans, School of Artillery, Manorbier, Wales (British World War II rocketry).

Robert R. Finney, United States Army Missile Command (postwar Army missiles); Dr. William A. Fowler, California Institute of Technology (United States wartime rocketry); Captain Robert F. Freitag, USN-ret. (United States V-2 firings); Arnold W. Frutkin, NASA, Washington (postwar United States sounding rocket and international coöperation programs).

Les Gaver, NASA, Washington (spacecraft photography); Colonel Richard Gimbel, USAF-ret., Yale University Library (early space concepts); Colonel T. A. Glasgow, Aerospace Medical Division, Air Force Systems Command (United States space medical programs); Colonel C. V. Glines, Office, Assistant Secretary of Defense, Washington (United States postwar rocketry); Mrs. Robert H. Goddard (activities of Dr. Robert H. Goddard); M. G. J. Gollin (British liquid-fuel rocketry in World War II); Professor L. Carrington Goodrich, Association for Asian Studies, Columbia University (ancient Chinese rocketry); Grumman Aircraft Engineering Corp. (Apollo spaceship).

James W. Harford, American Institute of Aeronautics and Astronautics (early AIS/ARS rocketry); Gordon L. Harris, John F. Kennedy Space Center, NASA (United States launch facilities); Robert Hartwell, Department of Oriental Languages and Civilization, University of Chicago (ancient Chinese rocketry); T. A. Heathcote, National Army Museum, U.K. (early British rocketry); Dr. Heinrich Hertel, Technische Universität Berlin (World War II German rocket airplanes); Dr. C. N. Hickman (NDRC and prewar rocket research in the United States); D. N. Hoare, Bristol-Aerojet Ltd. (British sounding rockets); John M. Hughes, Aberdeen Proving Ground (United States Army missiles).

Imperial War Museum, London (World War I rocketry); George S. James, Aerojet-General Corporation (early United States JATO developments); L. L. Janssens, Société pour la Réalisation d'Engins Balistiques (French missiles); Joseph M. Jones, George C. Marshall Space Flight Center, NASA (carrier vehicle information); Jorge Salmon Jordan, *El Comercio*, Lima (Pedro E. Paulet work); Harry C. Jordon, Ballistic Systems Division, Air Force Systems Command (United States Air Force missiles) Brigadier Leonard Walter Jubb, British Defence Research Staff (prewar and World War II British rocketry).

Professor Väinö Kaukonen, Helsinki University (Finnish legends); Captain Richard K. King, Hq. Air Force Systems Command (United States Air Force ballistic missiles); Ernst Klee (World War II German rocketry); Zdenek Kopal, Department of Astronomy, University of Manchester (early lunar studies); J. Gary Kornmayer, General Dynamics Convair Division (Atlas missiles and carriers); Professor Edward A. Kracke, Jr., Department of Oriental Languages and Civilizations, University of Chicago (ancient Chinese rocketry); Krausskoff-Flugwelt Verlag (prewar German rocketry).

Fritz Lang (*Frau im Mond*—film); Lily Latté (*Frau im Mond*—film); Dr. Charles C. Lauritsen, California Institute of Technology (wartime United States rocketry); Willy Ley (prewar German activities); Library of Congress, Washington (support in rare book and general research); Robert E. Logan, American Museum of National History (early space flight concepts); Antonio Lulli, Peruvian Embassy, Washington (Pedro E. Paulet work).

Frank J. Malina, International Academy of Astronautics, Paris (wartime rocketry); Paul Mathias, Paris-Match, New York (French space developments); Metropolitan Museum of Art, New York (ancient astronomical and astronautical concepts); Bernard H. Mollberg (photographic analysis and selection); D. W. Morton, British Aircraft Corp., Ltd. (British missiles); Captain W. P. Murphy, USN, Office of Chief Polaris Executive, U.K. Ministry of Defence (postwar naval rocket experiments); H. J. Murray, Redstone Scientific Information Center (documentation); Professor Herbert Myron, Jr., Boston University (translation of medieval French material).

Brigadier F. S. Napier (British wartime rocketry); T. D. Nicholson, American Museum-Hayden Planetarium (astronomical information); Nord-Aviation (French rocketry); Novosti Press Agency, London (Soviet information, photographs); Colonel Frederick I. Ordway, Jr., USAF-ret. (World War I and War of 1812 research).

Geoffrey K. C. Pardoe, Hawker Siddeley Dynamics, Ltd. (British postwar rocketry, carrier vehicles); G. Edward Pendray (AIS/ARS prewar rocketry); Robert L. Perry, Rand Corporation (United States Rand satellite study); Dr. William H. Pickering, Jet Propulsion Laboratory (postwar United States spacecraft developments); M. F. Poffley, Ministry of Aviation, London (British World War II rocketry); George A. Pughe, Aerospace Technology Division, Library of Congress (Soviet translations; analyses of missile and space programs).

William C. Ragsdale (photographic analysis and selection); Donald L. Raymond, American Institute of Aeronautics and Astronautics (early AIS/ARS rocketry); G. Rear, Ministry of Defence—Royal Air Force (British missiles); Major General Ormand J. Ritland, USAF-ret. (United States ballistic missile and space systems); Royal Astronomical Society Library, London (literature research); Dr. Harry O. Ruppe, Technische Hochschule München (prewar German rocketry).

Max Salmon, Office National d'Études et de Recherches Aérospatiales (French space developments); Duncan Sandys, Member of Parliament (British World War II rocketry); Dr. Irene Sänger-Bredt (photographs and information concerning Eugen Sänger's prewar research); Major Francis N. Satterlee, Office of the Assistant Secretary of Defense, Washington (Defense Department missile and space programs); Andrew M. Sea, Manned Spacecraft Center, NASA (manned satellite photography, information); Mitchell R. Sharpe, Jr., George C. Marshall Space Flight Center, NASA (information search, location of sources); John Shesta (early AIS/ARS rocketry); Nathan Sivin, Harvard University (ancient Chinese rocketry); Colonel Leslie A. Skinner, USA-ret. (United States Army Ordnance prewar and wartime rocketry); Bart J. Slattery,

Jr., George C. Marshall Space Flight Center, NASA (carrier vehicle information); Brountislav J. Soshinsky (translation of historical Soviet material); Charles L. Stewart (information on surrender of German rocket scientists to American forces at close of World War II); F. D. Storrs, Short Brothers & Harland Ltd. (British missiles); Dr. Ernst Stuhlinger, George C. Marshall Space Flight Center (Saturn-launched satellites; Mars spaceship concept); Sud-Aviation (French rocketry); Svenska Aeroplan AB (Swedish missiles).

Major General H. N. Toftoy, USA-ret. (postwar United States rocket developments); Captain Robert C. Truax, USN-ret., Aerojet-General Corporation (Annapolis rocket experiments).

Edward G. Uhl, Fairchild-Stratos Corporation (bazooka developments); United States Intelligence Corps Agency, Office of the Chief (surrender of German rocket scientists to United States forces at end of World War II); University of Manchester Library, Manchester, U.K. (rare book research, reproduction of material).

Ivan Volkoff (ancient and medieval space flight concepts); Ruth von Saurma, George C. Marshall Space Flight Center, NASA (technical documentation).

Ronald C. Wakeford, Research Analysis Corporation (wartime British rocketry); Dr. Hellmuth Walter, Worthington Corp. (prewar and wartime German rocketry); Professor James R. Ware, Harvard University (ancient Chinese rocketry); Richard R. Wilford, Jet Propulsion Laboratory (spacecraft); William L. Worden, Boeing Company (missile and carrier vehicle information); Paul T. H. Yung, Library of Congress, Washington (ancient Chinese rocketry).

Accademia della Scienze di Torino (Italian nineteenth-century rocketry).

Beinecke Rare Book and Manuscript Library, Yale University, New Haven (rare book research); Biblioteca Casanatense, Rome (rare book research); Biblioteca Central, Diputación Provincial de Barcelona (rare book research); Biblioteca Nacional, Madrid (rare book research); Biblioteca Nazionale Centrale, Rome (rare book research); Bibliothèque Cantonale, Lausanne (Swiss military rockets); Bibliothèque de la Musée de l'Armée, Paris (military rockets); Bibliothèque Nationale, Paris (rare book and manuscript research); Bibliothèque Royale de Belgique (rare book research); Boston Public Library (rare book research); Dr. Bodo Bartocha, Office of Planning and Policy Studies, National Science Foundation, Washington (documentation); Gerald H. Bidlack, Communications Satellite Corporation, Washington (communications satellites).

Deutsches Museum, München (prewar and World War II German rocketry); Charles F. Ducander, Science and Astronautics Committee, U.S. House of Representatives, Washington (documentation).

Etablissements Ruggiere, Paris (documentation).

Free Library of Philadelphia (rare book research).

Massachusetts Institute of Technology Libraries, Cambridge (book research); Ministère de la Guerre, Paris (military rockets); Musée de l'Air–Service de Documentation, Paris (military rockets); Musée du Château de Rohan, Strasbourg (old fireworks); Musei Vaticani, Rome (early astronomical texts); Museo del Ejercito, Madrid (Spanish military rocketry); Museo di Armi Antiche della Pusteria di S. Ambrogio, Milano (old military rockets); Museo di Risorgimento, Palazzo Carignano, Torino (early fireworks); Museo Storico Nazionale Artiglieria, Torino (Italian nineteenth-century rocketry); Museu Militar, Lisbon (signal rockets).

National Army Museum, Sandhurst (early British rocketry); National Library of Ireland, Dublin (rare book research); National Maritime Museum, Greenwich (fleet-to-shore bombardment rockets); New York Public Library (rare book research).

Pierre Versins, Prilly, Switzerland (early astronautical concepts).

Royal Artillery Institution Library, London (early British rocketry); Royal Artillery Institution Rotunda Museum, London (early British and Indian rocketry); Rockefeller Library, Brown University, Providence (pyrotechnics).

Servicio Histórico Militar, Biblioteca Central, Madrid (Spanish military rocketry).

J. Gordon Vaeth, National Environmental Satellite Center, ESSA, Washington (meteorological satellites).

Widener Library, Harvard University, Cambridge (rare book research); Glen Wilson, Senate Committee on Aeronautical and Space Sciences, U.S. Senate, Washington (documentation).

*As Wernher von Braun would have wished, this book—now in its fourth edition—continues to be dedicated to the thousands of men and women all over the world whose efforts to explore the mysteries of space are inexorably turning into reality the dreams and aspirations of earlier generations.*

*It is also dedicated to his memory. Wernher von Braun died in Alexandria, Virginia, on 16 June 1977 at only 65 years of age. He was, as editorialized by* The Washington Post, *"without any doubt, a remarkable scientist, manager and dreamer. . . . it was his technical ability, experience and uncanny ability to organize others that made him the central figure in the space program. And it was his eloquent expression of that childhood dream of space travel that made him a national figure. . . . You can . . . think of him as he apparently thought of himself—as a man indentured only to a dream. He followed it where it led him. And, unlike most of us, he saw a large part of it come true."*

*In a White House statement President Jimmy Carter said: "To millions of Americans, Wernher von Braun's name was inextricably linked to our exploration of space and to the creative application of technology. He was not only a skillful engineer, but also a man of bold vision; his inspirational leadership helped mobilize and maintain the effort we needed to reach the moon and beyond."*

*At a memorial service for Wernher von Braun at the Washington Cathedral, Apollo 11 astronaut Michael Collins rhetorically asked about the man so vitally important to the nation's prowess in space. "Wernher von Braun was a study in contrasts," he answered. "He was, at the same time, a visionary and a pragmatist, a technologist and a humanist. . . . He was a master of the intricacies of his machines . . . yet he realized that his rockets could only be as successful as the people who made them, and he assembled an extraordinarily talented team, people who worked well with each other, and who were totally devoted to Wernher."*

*"Fortunately for the human race," eulogized former NASA administrator James C. Fletcher, "a few men arise in each century who 'see visions' and 'dream dreams' that give hope and spiritual nourishment to us all. . . . Such men cling to this vision despite all efforts to destroy it. Wernher von Braun was such a man . . . he clung to what seemed an impossible dream for his entire life, despite pressure of politics, bureaucratic entanglements, war, loss of fortune or even, especially, personal criticism."*

*"Life has been so beautiful. It surpassed even my wildest dreams. I will be forever grateful for all that was given to me." With these words, Wernher von Braun expressed his feelings when the last of the Apollo lunar expeditions returned safely to Earth. His memory, like his visions, will endure forever.*

# *Contents*

# 1 THE LURE OF OT

Before man could think about traveling to other worlds, he had to accept their existence. The idea is unquestioned today, but for thousands of years he thought of himself and the planet on which he lived as unique. The Earth was believed to be an unmatched pocket of life and change in a cold, dead, unalterable universe. From the beginning of thought, attempts were made to grapple with the mystery of the Earth's origin through myth and legend.

Nordic mythology created the giant Ymir, from whose body the land was born and whose sweat created the sea. From his skull was created the firmament, which remained unlighted until the gods brought forth the stars, harnessed the chariot of the Sun to the horse Arwaker, and hitched the Moon to Alswider, another mythical horse.

The Chinese had Tao, the "great original cause" that created a shaggy dwarf, P'an Ku. As the dwarf breathed, the winds began; as he opened his eyes, light was brought forth. When P'an Ku said the word *Sun* seven times, the Sun came into existence, followed by the Moon and stars. After P'an Ku died, his head became the mountains, his blood the rivers, his sweat the rains. His skin and hair turned into plants and trees, while the human race sprang from the insects on his body—perhaps the least flattering origin ever imagined for man.

The first steps toward a scientific view of the universe were taken in Babylonia and Egypt at least five thousand years ago. As early as 3000 B.C., Babylonian astrologer-astronomers were making methodical observations of the heavens. By the second millennium they had fitted the planets to the system of the zodiac, and by 1000 B.C. they apparently were keeping records of the movements of the brighter planets, as well as the Sun and Moon. Tables of the motion of Venus between the years 1921 and 1901 B.C. have been found by archeologists, and data on Mars and Jupiter were reduced for future reference.

In Egypt, astronomer-priests worked out a calendar of twelve 30-day months, with a five-day period to round out the year. Their celestial observations led to a star catalog that listed forty-three constellations by the thirteenth century B.C. The Egyptians knew that Mercury and Venus were closer to the Sun than the Earth, Mars, Jupiter, and Saturn. By 1000 B.C., the Sun-clock was in regular use in Egypt, an added sign of astronomical knowledge.

Great as these achievements were, they were just building blocks for the Greek astronomers in the city-states across the Mediterranean. Early in the sixth century B.C., Thales of Miletus, whom the Greeks credited with founding science, mathematics, and philosophy, traveled to Egypt to learn from its astronomers and returned to found what became known as the Ionian school of Greek astronomy. Thales was able to predict a Solar eclipse—its occurrence on 28 May 585 B.C. frightened the armies of Media and Lydia into making peace—and he measured the angular diameter of the Sun. Despite these advances, Thales clung to the traditional Egyptian view of the Earth as a circular disc floating on a great ocean inside a hemisphere of shining stars.

The members of the Ionian school questioned these traditional beliefs and proposed their own revolutionary theories. Anaximenes suggested that many bodies like the Earth exist in the heavens. Heraclitus, visualizing the sky as filled with pure fire, saw a universe of ceaseless change. Anaxagoras believed that the Moon was like the Earth and could support life. Anaximander pictured the sky as a sphere that revolved around the Earth—an idea that recurred frequently in history—with rings of universal fire burning around the sphere. The Sun, the stars, and the Moon were visualized as traveling within tubes of mist in the sphere. Anaximander knew that the Earth was not flat, but he pictured it as a cylinder, with the flat ends at the east and west.

A rival school, established by Pythagoras of Samos, flourished from the sixth to the fourth centuries B.C. One member of the school, Parmenides of Elea, was a strong proponent of a spherical Earth. He believed that the Earth, which he divided into five zones, was condensed from air, while the stars

# HER WORLDS

were of compressed fire. He looked upon a finite, motionless, spherical universe—whose apparent motions were illusory. The Sun, Moon, planets, and stars were thought to be arranged in bands around the Earth. Another thinker, Philolaus, went a step further and proposed that the Earth, the Sun, the Moon, the stars, and the planets revolved around a great sphere of fire, the center of the universe.

The Pythagoreans encountered opposition because they maintained that the Earth moved around the central sphere of fire, always keeping the same side toward the fire (Greece was on the shady side). There were two counterarguments: The gods were insulted by the idea of a moving Earth, and there was no evidence that the people who lived on the side of the Earth away from Greece could see the eternal fire.

There was no way of disputing the first argument, but to meet the practical question, the Pythagoreans invented a counter-Earth, or Antikhthon, which was placed so that it always protected our Earth from the eternal fire. They managed to avoid an explanation of why their Antikhthon was not visible from the Earth.

The idea of a spherical Earth became so imbedded in Greek thought that even Plato, an archconservative in political matters, accepted its truth. In the *Phaedo* he wrote:

> my persuasion as to the form of the Earth and the regions within it I need not hesitate to tell you . . . . I am convinced, then, that in the first place if the Earth, being a sphere, is in the middle of the heaven, it has no need either of air or of any other such force to keep it from falling, but that the uniformity of the substance of the heaven in all its parts, and the equilibrium of the Earth itself, suffice to hold it.

In Plato's cosmology, the planets moved with the heavenly sphere, with a circular movement in a direction opposite to their daily rotation. The Earth was at the center of the universe, with the Moon, the Sun, Venus, Mercury, Mars, Jupiter, and Saturn at increasing distances from it.

Plato believed that "weakness and sluggishness" prevented man from traveling upward through the air,

> for if anyone could reach the tip of it, or could get wings and fly up, then, just as fishes here, when they come up out of the sea, espy the things here, so he, having come up, would likewise descry the things there, and if his strength could endure the sight, would know that there is the true heaven, the true light, and the true Earth . . . the things beyond would appear to surpass even more the things here.

Aristotle, too, accepted the concept of a spherical Earth, writing that the opponents of the theory "fail to take account of the distance of the Sun from the Earth and the size of the Earth's circumference." He disputed the belief that the Earth floated in water, but he was convinced that the Earth was at the center of the universe. The stars, Aristotle said in his *De caelo,* were not only spherical and at a great distance, they were eternal occupants of a perfectly spherical universe centered on the Earth.

Since Plato and Aristotle were not primarily astronomers, their cosmological theories marked no great advances in thought. A lesser figure, Heraclides of Pontus (*c.* 388–315 B.C.), did produce some strikingly original ideas. He explained the daily rotation of the stars by assuming that the Earth turned on its axis, and he also discovered that Mercury and Venus revolved around the Sun rather than around the Earth.

These two observations paved the way for one of the great intellectual leaps forward in human history. Aristarchus of Samos (*c.* 310–230 B.C.) went the crucial step beyond Heraclides; he said that the Earth, too, revolved around the Sun. The idea was too revolutionary to be accepted by his contemporaries; indeed, it had to wait nearly two thousand years to be vindicated. Only Seleucus of Seleucia, a Chaldean who lived one hundred years later, accepted Aristarchus's theory as true.

Aristarchus also measured the distance from the Earth to the Sun, using geometrical methods. His

*In Babylonia, astrologer-priests began making methodical observations of the universe in about 3000* B.C. *This ancient seal shows a man about to take off for the Moon, lifted by a great bird.* (AMERICAN MUS. OF NATURAL HISTORY)

*In Egypt the Sun-clock was in regular use by 1000* B.C. *On the balustrade at Amarna figures representing Akh-en-Aton and Nefertiti hold offerings for the Sun god Aton.* (METROPOLITAN MUS. OF ART)

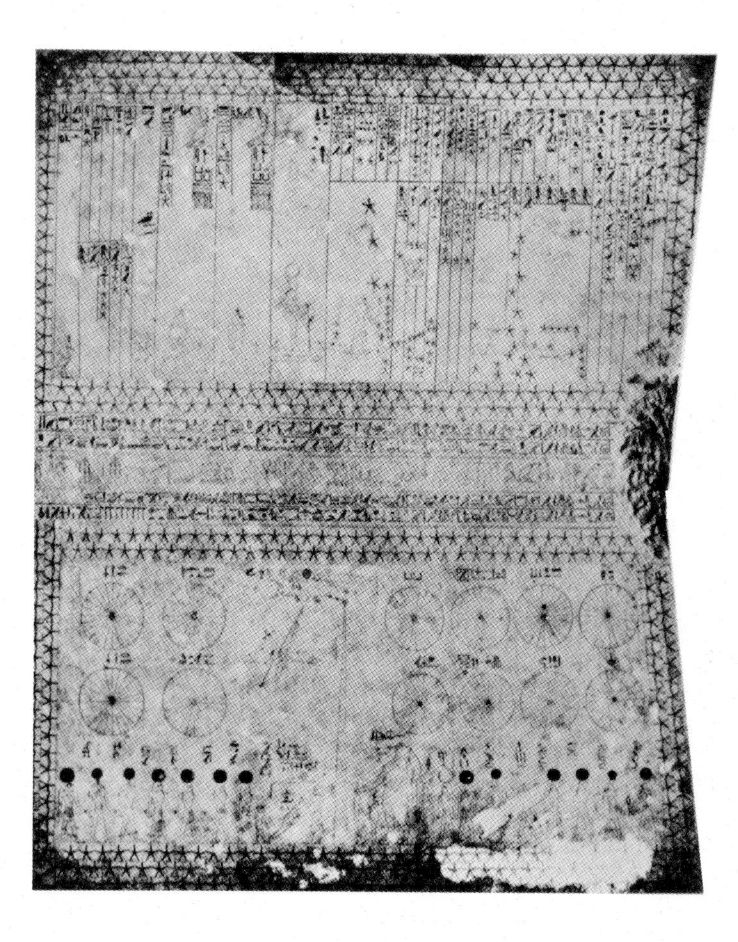

*Astronomical ceiling of the tomb of Sen-Mut, Thebes, dates from the Eighteenth Dynasty.* (METROPOLITAN MUS. OF ART)

figure was 4 to 5 million miles, twenty times too small but much closer to the truth than most contemporary estimates.

Later, a school of astronomers at Alexandria began to fill in blanks in astronomical knowledge. Eratosthenes (276–192 B.C.) made an accurate estimate of the circumference of the Earth, determined the angle of the ecliptic, and prepared a star catalog. Hipparchus, the greatest of all Greek astronomers, worked in Rhodes from 146 to 127 B.C. Hipparchus rejected the idea that the Sun was the center of the Solar System. But he did make accurate measurements of the size of the Sun and Moon, and he worked out a satisfactory theory to explain the motions of the planets. Hipparchus' star catalog, with 850 entries, was unequaled at the time and remained a standard reference for centuries.

The achievements of Greek astronomy were summed up in the work of Ptolemy, who worked in Alexandria from A.D. 127 to 141. Ptolemy's great work was the *Megiste Syntaxis* (*Great Collection*), known to us by its Arab name, *Almagest,* a mathematical and astronomical treatise that summarized all that was known about the universe. Ptolemy was more of a compiler than an original worker. He accepted Hipparchus's picture of the universe: The Earth was at the center, and was immovable, with the Moon, the Sun, and the planets orbiting in perfect circles. Ptolemy's treatise, as well as his tables of the Moon's motion, remained the supreme authority for more than twelve hundred years.

By Ptolemy's time, Greek science was looking backward, not forward. The light of Hellenic culture winked out as the Roman Empire fell. The torch passed, not to the barbarian hordes that swept over the empire, but to the Arabs who followed their Islamic faith in a tidal wave of conquest in the Near East. The Arabs translated the Greek and Latin texts, studied them, and added to their findings. In the ninth century, Muhammad al-Batani calculated the precession of the equinoxes; around 1000, Ibn Junis recorded both Solar and Lunar eclipses. But Arabic culture waned in the thirteenth century, and astronomical knowledge marked time for another three centuries.

*The Pythagoreans believed in a spherical Earth that revolved around a central sphere of fire. A counter Earth, between the two spheres, protected the Earth from the central fire.*

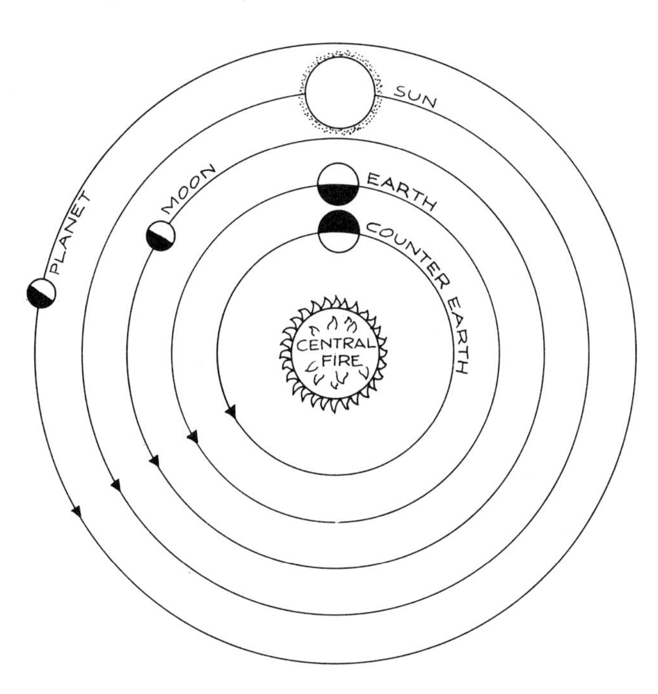


A single astronomical textbook, the *Tractatus de sphaera* (*The Sphere*) of Johannes Sacrobosco, contained most of what was taught in Europe during this period. Sacrobosco, who probably wrote in the thirteenth century, offered proofs that the Earth and heaven were spherical; among them were the observations that the Moon rose and set earlier in the east, and that eclipses occurred later in the Orient than in Europe. But Sacrobosco believed firmly that the Earth was the center of the universe, a belief he justified with this argument:

> To persons on the Earth's surface the stars appear of the same size whether they are in mid-sky or just rising or about to set, and this is because the Earth is equally distant from them. For if the Earth were nearer to the firmament in one direction than in another, a person at that point of the Earth's surface which was nearer to the firmament would not see half of the heavens. But this is contrary to Ptolemy and all the philosophers, who say that, wherever man lives, six signs rise and six signs set, and half of the heaven is always visible and half hid from him.

The same sort of argument was offered to prove that the Earth did not move.

> That the Earth is held immobile in the midst of all, although it is the heaviest, seems explicable thus. Every heavy thing tends toward the center. Now the center is a point in the middle of the firmament. Therefore, the Earth, since it is heaviest, naturally tends toward that point. Also, whatever is moved from the middle toward the circumference ascends. Therefore, if the Earth were moved from the middle toward the circumference, it would be ascending, which is impossible.

Much of the astronomical effort of the time went into the preparation of commentaries on the *Tractatus de sphaera.* One of the better commentaries, written

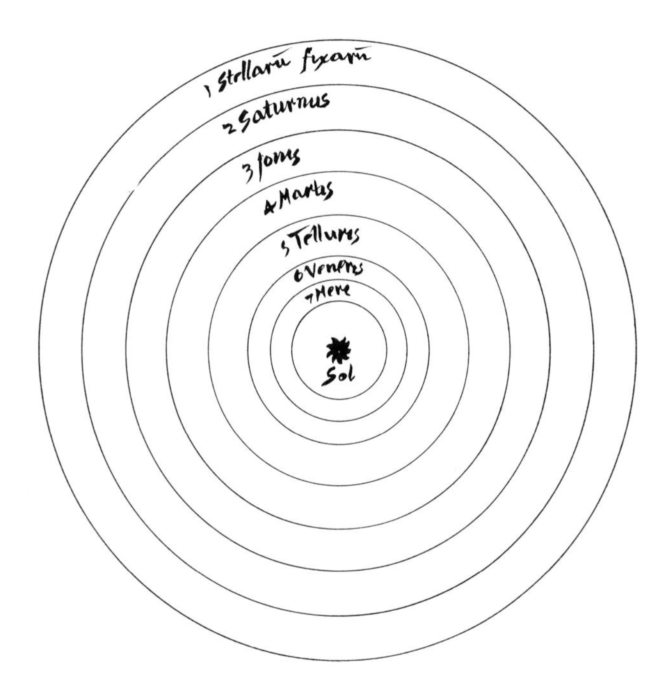


*By placing the planets in orbit around the Sun, Nicolaus Copernicus (1473–1543) revolutionized man's picture of the universe.*

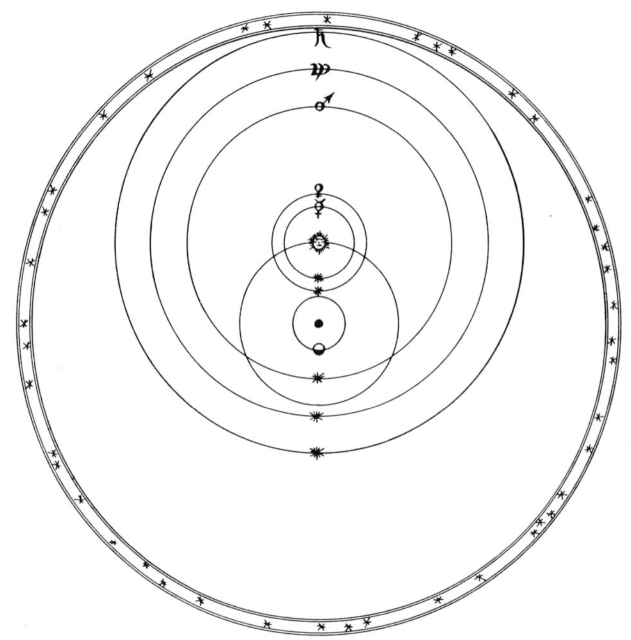

*Tycho Brahe (1546-1601) saw the Earth (center) as a stationary body, around which revolved the Sun, itself the center of motion of the known planets.*

by Robertus Anglicus in 1271, offered this description of the planets:

Saturn is of cold and dry nature; Jupiter of hot and moist nature; Mars of hot and dry nature which consumes by burning; but the Sun is of a hot and dry nature, which heat is life-giving; Venus of cold and humid nature with aerial humidity; Mercury hot with the hot, cold with the cold, following the nature of the planet with which it is in conjunction, the Moon of cold and humid nature with the humidity of water rather than that of air.

These fantasies dominated astronomical thought until the time of Mikolaj Kopernik (1473–1543), whom we know by the Latinized name of Nicolaus Copernicus. With Copernicus, man's picture of the universe changed irrevocably. In his monumental *De revolutionibus orbium coelestium* (*On the Revolutions of the Celestial Orbs*), published in 1543, Copernicus revived Aristarchus's heliocentric theory and established the true picture of the Solar System: the Sun in the center, with the Earth and the other planets orbiting the Sun. While Copernicus clung to the classical idea that the orbits of the planets were perfect circles, he accurately measured the distances of the planets from the Sun.

Where Copernicus was a towering theoretician, the greatest of his immediate successors, Tycho Brahe (1546–1601) was an outstanding observer, the best since Hipparchus. Brahe was able to measure the positions of the stars and planets with astonishing precision for pre-telescope times. Too conservative to accept the Copernican theory, he struck a compromise by proposing that all the planets but the Earth orbited the Sun, which orbited the Earth in turn. In 1582 he wrote:

I believe that it is absolutely and undoubtedly necessary to have the Earth fixed at the center of the universe, following the opinion of the ancients and the testimony of

*German astronomer Johannes Kepler (1571–1630) showed that the orbit of a planet can be represented by an ellipse, with the Sun at one of the two foci. A planet travels faster when it is nearer the Sun, and areas* A *and* B, *swept out during equal periods of time, are equal.*

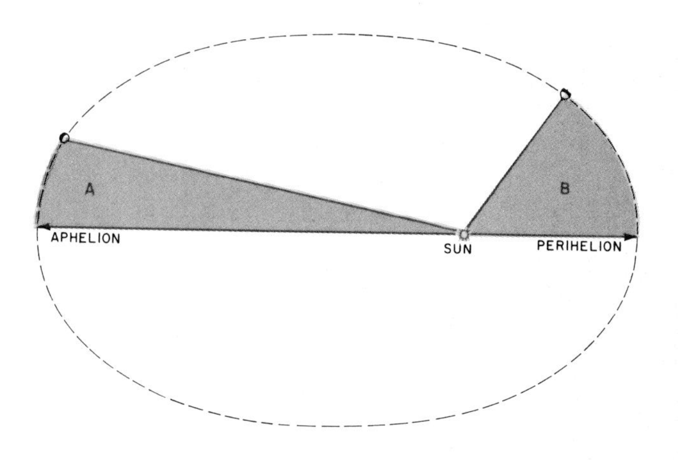

the Scriptures. I do not agree at all with Ptolemy in assuming that the Earth is the center of the orbits of the second mobile sphere; but I believe the celestial motions to be so arranged that only the Moon and the Sun, together with the eighth sphere which is the remotest of all and encloses the others, have the center of their motion in the Earth. The five other planets revolve around the Sun, which acts as their king and master, and the Sun will always be at the center of their orbits, being accompanied by them in its annual motion . . . . Thus the Sun is the regulator and terminus of all these revolutions, and, like Apollo surrounded by the Muses, he governs the harmony of the heavens.

Brahe's careful measurements were used by one of his pupils, Johannes Kepler, to develop a new set of laws about the planetary motions. Ceaselessly, for many years, Kepler tried theory after theory to fit the planetary orbits into a coherent explanation. Kepler finally arrived at his three great laws of planetary motion: The planets orbit the Sun in ellipses, not circles, with the Sun at one focus of the ellipse; lines drawn from the Sun to a planet will sweep over equal areas in equal periods of time; and the period of any planet's orbit is related to the planet's distance from the Sun. Kepler's findings, published in the *Astronomia nova* (*New Astronomy*) of 1609 and the *De harmonice mundi* (*On the Harmony of the World*) of 1619, set the stage for the next great advance.

This came when Galileo Galilei (1564–1642) first used a telescope to observe the skies. Galileo's observations destroyed forever the theory that the heavens were perfect and unchanging, far different from the crude and imperfect Earth. During a lifetime of observation Galileo discovered sunspots, from which he determined the Sun's rotational speed—he nearly ruined his eyesight while gazing at the Sun; he noticed many surface features on the Moon and worked out the height of mountains and the depth of craters; he discovered the four major satellites of Jupiter, which he called the Cosmian or Medicean stars, after Cosmo de' Medici II; he found that Venus, like the Moon, has phases; and he wrote that "the Galaxy is nothing else but a mass of innumerable stars planted together in clusters."

Galileo's discovery of the Cosmian stars proved that the Moon was not the only satellite in the Solar System. He told the world of his discovery in the *Sidereus nuncius* (*Sidereal Messenger*):

On the seventh day of January in the present year, 1610, in the first hour [sunset] of the following night, when I was viewing the constellations of the heavens through a telescope, the planet Jupiter presented itself to my view, and as I had prepared for myself a very excellent instrument, I noticed a circumstance which I had never been able to notice before, namely that three little stars, small but very bright, were near the planet.

He at first thought that they were fixed stars, but "when on January 8th, led by some fatality, I turned again to look at the same part of the heavens, I found a very different state of things . . . . I therefore concluded, and decided unhesitatingly, that there are three stars in the heavens moving about Jupiter, as Venus and Mercury around the Sun." Six days later he discovered the fourth satellite.

In his *Dialogo sopra i due massimi sistemi del mondo* (*Dialogue of the Two Chief Systems of the World*), published in 1632, Galileo defended the Copernican theory with brilliant arguments and biting satire, and brought the wrath of traditionalists on his head. Hauled before the Inquisition, he was forced to recant his views; he died, blind and under house arrest, in 1641.

Galileo's ideas lived on despite the Inquisition, and astronomers expanded his observations. Then, toward the end of the seventeenth century, Sir Isaac Newton tied all the observations together with his law of universal gravitation, which explained in pre-

*Isaac Newton (1642–1727) explained in mathematical terms practically every motion in the universe, from the fall of an apple to the orbits of the planets. This diagram illustrates his theory for a satellite: As the muzzle velocity of a cannon on the mountaintop is increased, the range of the shell also increases. If the velocity could be made high enough, the shell would not fall, but rather would remain in orbit.*

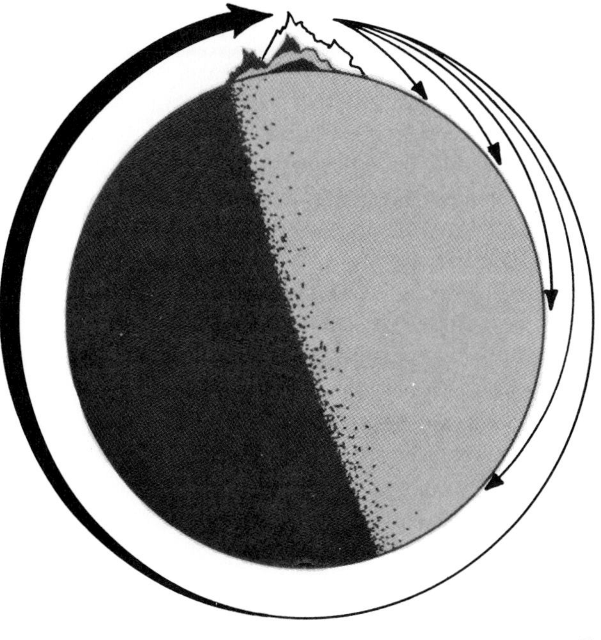

cise mathematical terms almost every motion of the universe, from the fall of an apple to the orbits of the planets.

Newton's great contributions to the dynamical nature of the universe were published in the famed *Philosophiae naturalis principia mathematica* (*Mathematical Principles of Natural Philosophy*) in 1687. He demonstrated in this work that Kepler's laws of planetary motion could be interpreted by his own universal laws of motion. In particular, he showed that the attraction of the Sun on a planet is directly proportional to the product of the Solar and planetary masses and inversely proportional to the square of the distance separating the two bodies.

"Newton was the greatest genius that ever existed and the most fortunate, for we cannot find more than once a system of the world to establish," said Joseph Lagrange (1736–1813). The history of astronomy for the next two centuries was largely the working out of the "system of the world" along Newtonian lines. A major effort was the *Mécanique céleste* (*Celestial Mechanics*) of Pierre Simon, Marquis de Laplace, a five-volume work published between 1799 and 1825, which worked out Newton's principles in enormous detail. Laplace demonstrated the stability of the Solar System; he and other astronomers made precise predictions of the future orbits of the planets. Variations from those orbits caused by the gravitational attraction of unseen planets led to the discovery of Neptune in 1846 by Urbain Leverrier and Pluto in 1930 by Clyde Tombaugh. Uranus had been discovered by Sir William Herschel in 1781, in the course of one of his methodical surveys of the sky.

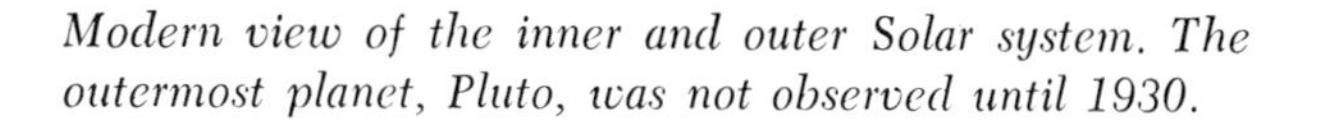

*Modern view of the inner and outer Solar system. The outermost planet, Pluto, was not observed until 1930.*

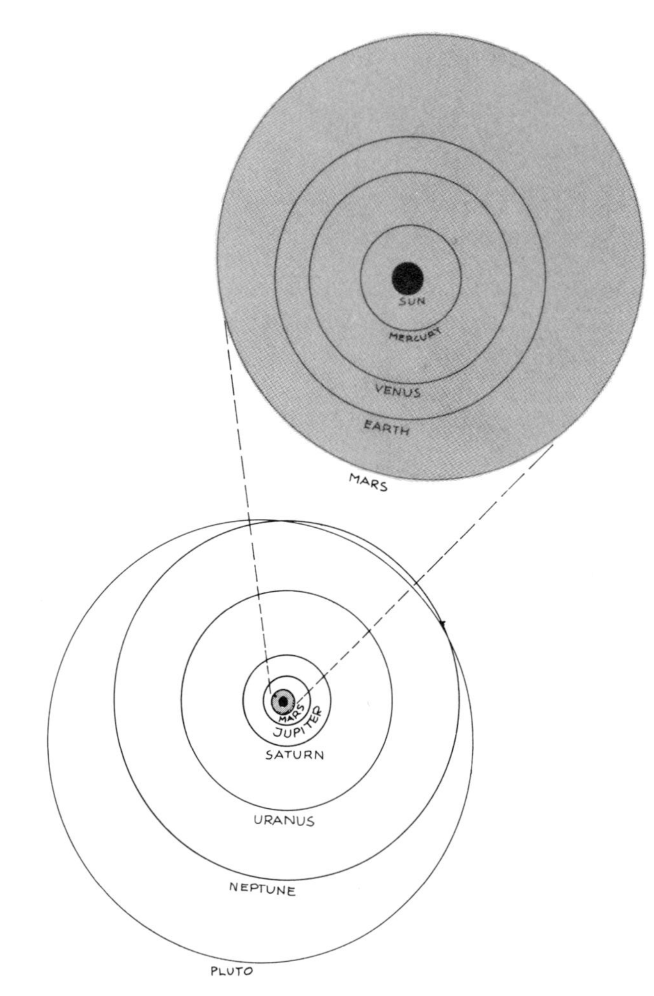


Starting in 1801, astronomers discovered that the large gap in the Solar System between the orbits of Mars and Jupiter actually was filled with thousands of small bodies, which they called asteroids. The first of these to be found, Ceres, has a diameter of 480 miles. It is still the largest known asteroid. More planetary satellites have been discovered—the total stands at thirty-one, including the Moon.

Astronomers now divide the planets into two groups. One group is made up of the four planets nearest the Sun: Mercury, Venus, Earth, and Mars, which are called the terrestrial planets because they resemble the Earth in size and density. The other group consists of the next four planets, Jupiter, Saturn, Uranus, and Neptune, which are relatively huge, but not nearly as dense as the Earth. They are believed to consist of metallic and solid hydrogen and helium cores surrounded by a layer of crystalline ammonia and finally by thick, turbulent hydrogen-rich atmospheres. Pluto, the outermost planet, remains in part a mystery; it is so different from the four giant planets that some astronomers believe it to be a satellite that somehow escaped the gravitational hold of Neptune.

The Solar System is over 7½ billion miles across, taking Pluto's orbit as the measure. This enormous distance is dwarfed by the space between the Sun and the nearest stars—a distance so great that light, traveling at 186,000 miles per second, travels for over four years from the nearest star to our Solar System.

It has taken civilized man some five thousand years to accumulate this store of knowledge. A trip to another world could not be imagined in anything approaching realistic terms until astronomy gave a true picture of the universe. As man's astronomical knowledge increased, however, the fictional space voyages devised by his restless imagination became correspondingly more sophisticated. This growing body of cosmic literature served both to express man's innate longing to reach other worlds and to stimulate further his desire to do so. Now that he at last knows the dimensions of the Solar System, he is ready to explore it. And when the first astronaut steps onto the Moon he will owe a debt to such writers as Jules Verne, who helped fire mankind's curiosity, as well as to the scientists like Sir Isaac Newton, who prepared the way for actually getting there.

No work of fiction dealing with cosmic travel predates the Christian era. The classical view of the Earth as a unique body in the center of an uninhabited, changeless universe made such speculations inconceivable. True, some thinkers, such as Plutarch (*c.* A.D. 46–120), believed that the Moon might resemble the Earth—in his *De facie in orbe lunae* (*On the Face on the Moon's Disk*) Plutarch contended that the Moon is a small Earth, inhabited by intelligent beings—but the standard view was given in the *Somnium scipionis* (*Scipio's Dream*) of Marcus Tullius Cicero, in which the dreaming hero realizes the uniqueness of the Earth:

> The Universe consists of nine circles, or rather of nine moving globes. The outer sphere is that of the heavens, which embraces all the others and under which the stars are fixed. Underneath this, seven globes rotate in the opposite direction from that of the heavens. The first circle carries the star known to men as Saturn; the second carries Jupiter, benevolent and propitious to humanity; then comes Mars, gleaming red and hateful; below this, occupying the middle region, shines the Sun, the chief, prince and regulator of the other celestial bodies, and the soul of the world which is illuminated and filled by the light of its immense globe. After it, like two companions, come Venus and Mercury. Finally, the lowest orb is occupied by the Moon, which borrows its light from the Sun. Below this last celestial circle there is nothing but what is mortal and perishable except for the minds granted by the gods to the human race. Above the Moon, all things are eternal. Our Earth, placed at the center of the world, and remote from the heavens on all sides, stays motionless and all heavy bodies are impelled towards it by their own weight.

This scheme, according to Cicero, was truly harmonious:

> The motion of the spheres creates a harmony formed out of their unequal but well-proportioned intervals, combining various bass and treble notes into a melodious concert. Such tremendous motions cannot take place in silence, and Nature has given a bass note to the lowest and slow orb of the Moon, and a treble note to the topmost and rapid orb of the starry firmament: between these two limits of the octave, the eight moving globes produce seven notes in different modes, and this number is the crux of all things. The ears of men are filled with this harmony and no longer capable of hearing it just as people living close to the cataracts of the Nile are no longer aware of the noise. The ear-splitting concert of the rapidly spinning Universe is so tremendous that your ears close up to keep its harmony out, just as you cannot bear to look at the fiery Sun because its piercing light dazzles and blinds you.

*One of the earliest fictional descriptions of space travel was written by Lucian of Samosata in the second century* A.D. *In* Vera Historia, *Lucian describes a sailing vessel that is lifted from the sea by a violent whirlwind and carried to the Moon.*

What is probably the first work of fiction describing what we now call space travel was the *Vera Historia* (*True History*) of Lucian of Samosata, a Greek sophist and satirist who wrote within a half century of Plutarch. Lucian provides all the necessary elements of space-travel fiction: a trip through space, a landing on another world, a description of that world, and a return.

It is unlikely that any other author has taken such pains to assure his readers that he is not telling the truth. Before beginning his tale, Lucian writes, "I shall at least say one thing true, when I tell you that I lie, and shall hope to escape the general censure, by acknowledging that I mean to speak not a word of truth throughout."

With that warning out of the way, Lucian describes a sailing vessel that is returning homeward when a great wind lifts it into the skies.

About noon . . . a most violent whirlwind arose, and carried the ship above three thousand stadia, lifting it up above the water, from whence it did not let us down again into the seas but kept us suspended in midair. In this manner we hung for seven days and nights, and on the eighth beheld a large tract of land, like an island, round, shining and remarkably full of light; we got on shore, and found on examination that it was cultivated and full of inhabitants, though we could not then see any of them.

The Lunar inhabitants are called Hippogypi, and they ride on three-headed vultures adorned with feathers "bigger than the mast of a ship."

As Marjorie Nicolson points out in her book *Voyages to the Moon,* "Lucian made his voyage . . . by mere chance; he suggested no previous idea of the possibility, no pondering upon means of conveyance, no plan or design." Nevertheless, the Lunar voyage was made, and the target world was explored, providing a model for tales of later centuries.

Lucian wrote a second story of space travel, the *Icaro-Menippus,* in which the trip to the Moon is carefully planned in advance. The hero, Menippus, uses birds' wings; hence the title, a reference to Icarus's brief flight.

In the story, Menippus strapped on a vulture's wing and an eagle's wing and, after making some test flights, reached the top of Mount Olympus. He then flew to the Moon, but that was not enough for this ambitious explorer, who kept flying to the stars and heaven itself. The trip took only three days. ("It shall be done, said I, and away I set out for heaven . . . in a little time the Earth was invisible, and the Moon appeared very small; and now, leaving the Sun on my right hand, I flew amongst the stars, and on the third day reached my journey's end.") The return trip was less adventuresome; Jupiter, angered by Menippus's intrusion, orders his wings cut off and directs Mercury to bear the mortal back to Earth.

Lucian's tale stood alone for centuries. The next surviving account of cosmic travel is found in the 60,000-verse epic poem *Shāh-Nāma,* published by the Persian poet Firdausī in 1010 following forty years of labor. Although it was written eight centuries after Lucian's *True History,* the Persian epic may contain the record of man's first imaginative venture into space, since it is a retelling of ancient legends.

The hero of the epic is Jamshíd, who reigned for seven hundred years over men, demons, birds, and fairies, and could transport himself on a demon-borne aerial throne into the heavens. The epic also tell of Kai-Kā'ūs, a mythical, headstrong king of Persia who was forever embarking on perilous adventures.

One day, Kai-Kā'ūs was persuaded by a dív, or demon, to attempt the conquest of heaven. After questioning wise men and astrologers, he decided on "crooked and ugly means." He sends men to steal young eagles from their nests, and has the birds fed on meat until they become "as strong as lions so they could pull down a mountain sheep." Then Kai-Kā'ūs builds a throne, with lances attached from which are hung legs of lamb. Binding four young eagles to the throne, he seats himself on it and is borne into the air as the eagles hurl themselves at the meat. The journey ends disastrously; the eagles become tired, fold their wings, and plunge "headlong from the black clouds, dragging down the king's throne and lances out of the air." The passage of the poem, translated for the authors by Professor John A. Boyle of the Department of Persian Studies, University of Manchester, England, follows:

The soul of that king was full of thought as to how he should rise into the air without wings.

He asked many questions of the learned as to how far it was from this Earth to the sphere of the Moon.

The astrologers spoke and the king listened, and he selected crooked and ugly means.

He ordered that during the night men should go to the nests of eagles,

Take a large number of their young ones and place one or two in every house.

He reared them for a year and a month on birds, roast meat and sometimes lambs.

When they had each of them become as strong as lions so that they could pull down a mountain sheep,

He constructed a throne of Qimāri (Cambodian) aloes-wood and strengthened the tops of the planks with gold.

He fastened long lances in the side and so made it ready.

He suspended legs of lamb from the lances, giving his whole mind to the matter.

Then he fetched four vigorous eagles and bound them firmly to the throne.

Kai-Kā'ūs seated himself on the throne having placed a goblet of wine in front of him.

When the swift eagles grew hungry they each of them hastened towards the meat.

They raised up the throne from the face of the Earth; they lifted it up from the plain into the clouds.

To the extent of the strength that was in them they directed their efforts towards the meat.

I have heard that Kai-Kā'ūs ascended to the firmament in order to pass beyond the angels.

Another said that he rose into the heavens in order to fight [them] with bow and arrow.

There are all sorts of traditions about this; only the Wise One knows the secret of it.

[The eagles] flew for a long time and then grew tired. So will be he that is seized by greed.

When no strength was left with the flying birds, they folded their wings according to their custom.

They plunged headlong from the black clouds dragging down the king's throne and lances out of the air.

They came towards such a forest as this; they alighted on the face of the Earth in Āmul.

Animal power is used in part of another great fantasy, the *Orlando Furioso* of Lodovico Ariosto, the first edition of which appeared in 1516. The hero of this tale, Astolpho, voyages to the Moon in a chariot drawn by four red horses in quest of the lost mind of Orlando. The mind turns up in a flask, but that is not all the Moon offers. Not only did the Moon, which was "swell'd like the Earth, and seem'd an Earth in size," possess most of the natural features of this planet, it also had cities, towns, and castles.

One of the most unusual works of early science fiction was written by Johannes Kepler. His *Somnium* (*Dream*) was not published until 1634, four years after Kepler's death. John Lear, who made a careful study of the work in his book, *Kepler's Dream*, published in 1965, suggested that the great astronomer was forced to present his ideas about the Moon in fictional form to avoid religious and political censure.

"The moon guide was designed for distribution to a restricted audience," Lear wrote. "Phrased in Latin, the international language of the learned, it was cast in the form of an allegory, the hidden meaning of which would be familiar only to scientists. As Kepler explained in a letter to his friend, Matthias Bernegger, the text of the geography had been deliberately strewn with 'almost as many problems as there are lines.'"

The *Somnium* is a fantastic tale, couched in supernatural terms, of a voyage from Earth—called Volva in the book—to the Moon, which Kepler called Levania. Kepler knew that there could be no dense atmosphere between the two worlds, and so he rejected animals and wings as means of transportation. Because he had no acceptable alternative, he dipped into the supernatural and chose demons to make the trip.

Kepler's demons abhor sunlight, but can travel at will during the night. Normally, it is impossible for them to pass between the two worlds, but from time to time when the shadow of the Earth intersects the Moon, they are able to cross. And, under certain rare conditions, they carry with them humans who have been given an anesthetic potion as protection against the ill effects of rarified air.

Kepler described the Moon in terms of the latest astronomical knowledge, but he added bizarre forms of life, vastly different from Earth's creatures. The Lunar beings envisioned by Kepler do not live in cities and towns; in fact, they possess no civilization at all. During the day they come out of caves and crevices to sun themselves briefly, only to seek the lengthening shadows and the coolness of the gloom again.

Kepler was followed by many writers who dealt with the theme of space travel in both poetry and prose. Of these the most famous is doubtless the "Speedy Messenger" by Domingo Gonsales, author of *The Man in the Moon: or a Discourse of a Voyage Thither*, published in 1638.

The name Domingo Gonsales was the pseudonym for an English ecclesiastic, Francis Godwin, Bishop of Hereford. The author mentions the rotation of worlds based on observations of their markings, something unknown before 1612, and he apparently understands that the attraction of a body varies in terms of mass and distance, a fact developed after

*In the sixteenth-century fantasy* Orlando Furioso, *the hero travels to the Moon in a chariot drawn by four red horses. According to the author, Lodovico Ariosto, the Moon "had most of the natural features of this planet, and cities, towns, and castle, too."*

1620. From these and other considerations, scholars believe the book was completed in 1630, three years before Godwin's death. It became quite popular, with some twenty-five editions in four languages published between 1638 and 1767.

The author states in the preface:

*That There should be* Antipodes *was once thought as great a* Paradox *as now that the* Moon *Should bee habitable. But the knowledge of this may seeme more properly reserv'd for this our discovering age: In which our Galilaeusses, can by advantage of their spectacles gaze the sunne into spots, & descry mountaines in the* Moon.

Domingo Gonsales, the protagonist and supposed author of the book, is a Spaniard of good but poor family, who has become ill while returning home from the East Indies, where he has made his fortune. Because of his illness, Gonsales and his servant, Diego, are taken off the ship and left on St. Helena Island. Seeking a way off the island, Gonsales trains some wild geese, called *gansas,* to carry a chairlike device ("engine").

I tooke some 30, or 40, young ones of them, and bred them up by hand partly for my recreation, partly also as having in my head some rudiments of that device, which afterward I put into practice . . . [and] began to cast in my head how I might doe to joyne a number of them together in bearing of some great burthen; which if I could bring to passe, I might enable a man to fly and be carried in the ayre, to some certaine place safe and without hurt.

To test his device, Domingo "fastened about every one of [his] *Gansas* a little pulley of Corke, and putting a string through it of meetly length, fastened the one end thereof unto a blocke almost eight Pound weight, unto the other end of the string [he] tied a poyse weighing some two Pound, which being done, and causing the signall to be erected, they presently rose all."

Following this initial success, Domingo sent a lamb aloft. Then came the time for the first manned flight. "I placed my selfe with all my trinckets, upon the top of a rocke at the Rivers mouth, and putting my selfe at full Sea upon an Engine . . . , my Birds presently arose, 25 in number, and carried me lustily to the other rocke on the other side, being about a Quarter of a league."

Gonsales and his birds eventually were picked up by a ship bound for Spain, but they were shipwrecked on one of the Canary Islands. Stranded among savages who continually "warre" against the Spaniards, Gonsales again mounted his "engine" and tried to fly to a nearby city.

But the birds took the "bitt between their teeth" and mounted upward, as if drawn "as the Loadstone draweth Iron." He alit temporarily on a strange land somewhere between Earth and the Moon. Soon the "*Gansa's* began to bestir themselues, still directing their course toward the Globe or body of the Moone." They sped along at about 85 miles per hour or, as Gonsales stated it, "Fifty Leagues in every hower." Since Godwin estimated the Moon to be about fifty thousand miles distant, at this rate the trip took some eleven days.

The traveler found that the Moon was like "another Earth," its surface covered by a "huge and mighty Sea," later revealed to cover "Three parts in Foure (if not more)." The gansas set him down atop a huge hill "where immediately were presented unto [his] eyes many strange and unwonted sights."

The trees were "three times as high as ours, and more then [*sic*] five times the breadth and thickness." And there were "herbes, Beastes and Birds; although to compare them with ours I know not well how, because I found not any thing there, any *species* either of *Beast* or *Bird* that resembled ours any thing at all, except *Swallowes, Nightingales, Cuckooes, Woodcockes, Batts,* and wild Fowle." The intelligent inhabitants of the Moon were "most divers but for the most part, twice the height of ours. Their colour never seen in an earthly world, and therefore neither to be described unto us by any, nor to be conceived of one that never saw it."

Although his stay on the Moon had been pleasant and instructive, the explorer became homesick, while his gansas grew restless "for want of their wonted migration." He therefore fastened himself to his "engine," signaled the birds, and took off, landing in China "in lesse than nine days." The quicker voyage homeward was explained by the greater attraction of the Earth. During the years that followed, many writers were to imitate Godwin and write about other inhabited worlds.

In 1638, the same year that Godwin's story was published, John Wilkins wrote the *Discovery of a New World; or, A Discourse tending to prove, that 'tis probable there may be another Habitable World in that planet.* Unlike the Godwin story, Wilkins's book was based on facts—at least, on the facts as they were known in the pre-Newtonian era. Wilkins's work inevitably contains many misconceptions and fundamental errors.

Wilkins was convinced—as many people still are, for no cogent reason—that the main problem in Moon travel is to lift the flier to the point between the Earth and the Moon where the Earth's influence ends. That point was believed by Wilkins to be not much farther "than that orb of thick vaporous air, that encompasseth the earth," or about twenty miles.

Once that altitude was attained—and Wilkins believed it could be quite easily—the rest of the voyage would be simple. And since "our bodies will . . . be devoid of gravity" no efforts would be exerted and no food would be required en route.

The next two major works dealing with travel beyond the Earth were written by Savinien de Cyrano de Bergerac, owner of the world's most famous nose. De Bergerac—wit, playwright, author, swordsman, philosopher, satirist, and part-time science-fiction addict—found time somehow to write *Voyage dans la Lune* (*Voyage to the Moon*) and *Histoire des États et Empires du Soleil* (*History of the States and Empires of the Sun*), which were published in 1649 and 1652, respectively. Since their author was well acquainted with the latest scientific studies, both are authoritative—if wildly fanciful—parodies on the theme of travel to other planets.

De Bergerac believed the Moon to be "a World like ours, to which this of ours serves likewise for a Moon." His first scheme for getting to the Moon was perhaps the most original ever devised. "I planted my selfe in the middle of a great many Glasses full of Dew, tied fast about me; upon which the Sun so violently darted his Rays, that the Heat, which attracted them, as it does the thickest Clouds, carried me up so high, that at length I found my selfe about the middle region of the Air."

Unfortunately, De Bergerac came down in Canada, not on the Moon. After a series of adventures in that country, Cyrano made another attempt at flight, this time in a "machine which I fancied might carry me up as high as I pleased . . . from the Top of a Rock [I] threw my self in the Air: But because I had not taken my measures aright, I fell with a sosh in the Valley below."

Returning to his flying machine, Cyrano found a group of soldiers busily tying firecrackers to it. Just as they lit the fuse, he leaped aboard. "Hardly were both my Feet within, when Whip, away went I up in a Cloud." Flames ignited tier after tier of rockets, lifting Cyrano ever higher. When the fireworks were exhausted, the machine fell "down again towards the Earth."

But Cyrano found himself drawn instead toward the Moon. He gives two reasons for this: His body was still greasy with marrow he had anointed himself with previously, and the Moon was "then in the Wain." He goes on to explain that since the Moon in that quarter would "suck up the Marrow of Animals, she drank up that wherewith I was anointed, with so much the more force that her Globe was nearer to me." Once he had landed on the Moon, Cyrano's imagination faltered; his return trip depended on supernatural means, and he came back to Earth as a spirit.

Not satisfied with the Lunar voyage, Cyrano built a new machine that would take him to the Sun and planets. He described it as a box 6 feet high and 3 feet square with holes below and in the cover, containing "a Vessel of Christal . . . made in a Globular Figure." It was "purposely made with many angles, and in the form of an Icosaedron, to the end that every Facet being convex and concave, my Boul might produce the effect of a Burning-Glass." A board within was provided for the pilot.

The operation of this unusual vehicle was somewhat complicated. First, it was necessary for sunlight to shine on the transparent Icosaedron

> which through its Facets received the Treasures of the Sun . . . . I foresaw very well, that the Vacuity that would happen in the Icosaedron, by reason of the Sunbeams, united by the concave Glasses, would, to fill up the space, attract a great abundance of Air, whereby my Box would be carried up: and that proportionable as I mounted, the rushing wind that should force it through the Hole, could not rise to the roof, but that furiously penetrating the Machine, it must needs force it on high.

As the machine mounts and the air becomes thinner, the device no longer serves, and Cyrano abandons it, continuing by a vague means of will power.

Most of the science-fiction works of the following years were variations on themes developed by writers like De Bergerac. A major event, in 1686, was the publication by Bernard de Fontenelle of a popular astronomy book called *Entretiens sur la Pluralité des Mondes* (*Discourses on the Plurality of Worlds*). A delightful work, it was read widely throughout Europe, partly because of its style but largely because of its fascinating speculations on the nature and habitability of other planets in the Solar System. Not only did De Fontenelle state that each known planet has its own race of people, but he went into detail about their appearance, civilization, customs, and habits. Oddly enough, De Fontenelle was not convinced of the Moon's habitability, believing that the air there was probably too rarefied. He gave relatively little attention to Mars, compared to such unlikely (to us) abodes of life as Mercury and Jupiter.

Four years later, Gabriel Daniel wrote *Voiage du Monde de Descartes* (*Voyage to the World of Descartes*), a novel that introduced the idea of soul or thought travel. The hero's soul separated from his body and soared out to the "Globe of the Moon" and the universe beyond, finding, among many other mysteries, the great master "Monsieur Descartes."

*Fictional space travel was accomplished in many ways. In David Russen's* Iter Lunare, *a giant spring was constructed on top of a mountain to catapult a man into space toward the Moon.*

The Moon was described in great detail, and was found to resemble the Earth. "One sees there fields, forests, seas and rivers. I see no animals, but I believe that, if they were transported there, one could nourish them, and perhaps they would multiply."

David Russen's *Iter Lunare: or Voyage to the Moon,* published in 1703, introduced the novel idea of using a spring catapult to propel a man into space from the top of a high mountain.

> Since Springiness is a cause of forcible motion, and a Spring will, when bended and let loose, extend itself to its length; could a Spring of well-tempered steel be framed, whose Basis being fastened to the Earth and on the other end placed a frame or Seat, wherein a Man, with other necessaries, could abide with safety, the Spring being with Cords, Pullies or other Engins bent and then let loose by degrees by those who manage the Pullies, the other end would reach the Moon.

Two years after *Iter Lunare* there appeared Daniel Defoe's *The Consolidator,* also a tale of Lunar travel. It tells how ancient peoples mastered the art of flying to and from the Moon, and how Mira-cho-cho-lasmo came to Earth to visit the emperor of China. Defoe reviewed many legends of flights to the Moon and several types of what today would be called spaceships.

Probably the most intriguing of these is the Consolidator, described as an engine "in the shape of a Chariot, on the backs of two vast Bodies with extended Wings, which spread about fifty yards in breadth, composed of feathers so nicely put together, that no air could pass; and as the Bodies were made of lunar Earth, which would bear the Fire, the Cavities were filled with an ambient Flame, which fed on a certain Spirit, deposited in a proper quantity to last out the Voyage; and this Fire so ordered as to move about such springs and wheels as kept the wings in most exact and regular Motion," described as "always ascendant." The story did not say what sort of propellant Defoe had in mind, but he appears closer to rocketry than most other writers of his time and, indeed, a long period afterward.

A quarter of a century later came Samuel Brunt's satiric *A Voyage to Cacklogallinia,* part of which concerns a Lunar voyage. After many varied adventures, the book's hero, Captain Brunt, was shipwrecked on Cacklogallinia, a strange land inhabited by bird people. The captain, befriended by Volatilio, quickly mastered their language and obtained a position in the government, where he could influence proposals made by the Cacklogallinians to reduce their government's burdensome debt.

One proposal involved an expedition to look for gold on the Moon. Brunt thought little of it and advanced many arguments against it, but to no avail; the idea had captured the imagination of the court, and the people as well. A great commercial enterprise unfolded, with shares sold as an investment in the gold everyone was certain existed on the Moon. Funds were promptly raised to finance the trip. The plan was to send two explorers to the Moon in a kind of palanquin, or streamlined flying chariot, propelled by birds. (The Cacklogallinians were birds themselves, but the palanquin was needed for the captain and Volatilio, who decided he, too, might want to ride in it.)

Before going to the Moon, the travelers investigated the environment away from the Earth and the ability of the birds to fly at high altitudes. The author knew that the air became less dense with increasing altitude, so the birds did not fly rapidly into space; instead, they rose slowly to accustom themselves to the changing atmospheric conditions.

The first test flight was made by Volatilio, who reported, "I ascended into the Mid-space, and found a vast alteration in the Air, which even here was very sensibly rarified." He explained that a "wet Spunge" provided welcome relief to the birds as they accustomed themselves to the thin air.

After several more flights, the time came for the last test before the flight to the Moon. "According to the Orders we receiv'd, *Volatilio* took his flight in an oblique Ascent, without a *Palanquin,* but wrapped up as warm as possible, accompanied by two servants." When Volatilio returned, he reported that he had passed the atmosphere "and, by Experience, had found my Conjecture true; for being out of the magnetick Power of the Earth, we rested in the Air, as on the solid Earth, and in an Air extremely temperate, and less subtle than what we breathe."

After some scheming by the courtiers to increase their profits from the enterprise, the voyage began. Even though it was made "with incredible swiftness," the trip was no overnight affair. "We were about a Month before we came into the Attraction of the Moon, in all which time none of us had the least inclination to Sleep, or Eat, or found our selves any way fatigued, nor, till we reach'd that Planet, did we close our Eyes." They floated in the weightless state until they descended to the Moon to begin their adventures. The Moon turned out to be inhabited by shades who lived quietly and peacefully without material wants or urges. The search for gold came to nothing—there was gold on the Moon, but the Lunarians did not allow the Cacklogallinians to take it.

After all this, Ralph Morris's 1751 novel, *A Narrative of the Life and Astonishing Adventures of John Daniel,* appears rather tame. With his son Jacob, the hero is stranded on a far-off island. They construct a device from materials salvaged from a shipwreck. It is made to fly by pump-operated calico wings supported by iron ribs—as the pump goes up and down, so do the wings. The machine is so efficient that the adventurers are carried not to a civilized country but—of course—to the Moon.

Hardly in the same class as these lightweight works is *Micromégas,* written by Voltaire in 1752 as a satire on man's pretensions to greatness. Voltaire does not start his tale on the Earth, or even in the Solar System, but on the gigantic star Sirius. The Sirians are proportionately large—the hero, Micromégas, is 120,000 royal feet high. A precocious lad, he had mastered geometry when he was only two hundred and fifty years old, and by the time he was four hundred and fifty he was busily studying—and writing about—the possibility of life on other worlds. Convicted of heretic beliefs, he was banished for a mere eight hundred years, a sentence Micromégas put to good use. He decided to explore the universe, using sunbeams, comets, and a sure knowledge of gravitation to make his trip pleasant and easy. Travelling across the Milky Way from star to star, he reaches the Solar System and settles down for a visit on Saturn, a puny planet to his way of thinking, whose inhabitants live only fifteen hundred years.

Micromégas strikes up a warm friendship with the secretary of the grand academy of Saturn, and the two argue about, and philosophize on, all manner of subjects. Finally, they visit the rest of the Solar System, flying first to the rings and moons of Saturn, then to Jupiter and Mars. At last they reach Earth, a tiny world the Sirian and Saturnian are convinced is uninhabited. But, peering idly through a diamond, Micromégas sights a whale swimming in an ocean he had assumed to be a mere puddle. Then a ship bearing explorers comes into view. Reluctantly, Micromégas acknowledges that even so tiny, so insignificant a world as the Earth can harbor rational beings.

Stories about space travel continued to appear at a steady rate. A few were original, but many were derivative, dull, and unimaginative. All underscored the persistent, romantic notion that there must be other worlds, inhabited by some kind of beings, and that somehow man can reach those worlds.

In a mid-eighteenth-century booklet, *Man in the Moon,* the hero, Israel Jobson, reaches the Moon by ladder; in a sequel, *The History of Israel Jobson, the Wandering Jew,* a chariot is used for the trip. Both books are believed to have been written by Miles Wilson, an English curate.

Not all speculative works were fictional. In 1698, Christian Huygens, a renowned scientist, wrote *Cosmotheoros,* or *Conjectures Concerning the Planetary Worlds,* in which he concluded that the planets were the abodes of rational beings. Emanuel Swedenborg, in 1758, wrote *Earths in our Solar System, which are called Planets, and Earths in the Starry Heavens,* which took much the same approach. In the book, Swedenborg's soul went out into the infinite. In the heavens, he wrote, are stars without end, around myriads of which are "thousands, yea, ten thousands of earths, all full of inhabitants." Swedenborg also gathered information from angels and spirits, which, he wrote, came to him from each of the Solar System's planets. Swedenborg mixed science, religion, and imagination to enrich the growing literature of supernatural voyages into the cosmos.

A 1775 work by Louis Guillaume de La Folie, *Le Philosophe Sans Prétention,* has some earmarks of the modern science-fiction novel. It tells of Ormisais, a Mercurian who arrives on Earth and tells his story to one Nadir, an Oriental. It seems that a Mercurian inventor, Scintilla, had created a marvelous electric flying chariot and had demonstrated it to his fellow scientists despite their scorn and ridicule. Ormisais was so certain that the chariot would not work that he casually vowed to fly it to the Earth. To his surprise, he was carried through space—his adventures are fairly standard—until he crashes on Earth and relates his story.

The fact that the space machine was an electric chariot is indicative of a change in science fiction. Readers now knew enough about science to insist on more realism from their fiction. But their insistence was limited to the means of travel; the wildest details about the nature of the planets and their imaginary inhabitants were still acceptable.

Joseph Atterlay's *A Voyage to the Moon with some Account of the Manners and Customs, Science and Philosophy of the People of Morosofia and other Lunarians,* published in 1827, is a good example of the new wave in science fiction. The author, whose real name was George Tucker, described a quite modern science-fiction device:

"The machine in which we proposed to embark, was a copper vessel, that could have been an exact cube of six feet, if the corners and edges had not been rounded off. It had an opening large enough to receive our bodies, which was closed by double sliding pannels, with quilted cloth between them." A metal called lunarium was used to "overcome the weight of the machine, as well as its contents, and take us to the moon." This antigravity concept became popular during the eighteenth and nineteenth centuries, although it had no more scientific credibility than geese or flying chariots. Still, it had an aura of science around it, and that was what counted most.

*In* Voyage to the Moon, *published in 1827, author Joseph Atterlay made an attempt at "science" fiction. Atterlay's spaceship, loaded with scientific equipment, rose to the Moon by virtue of an antigravity material called Lunarium.*

Eight years later, Edgar Allan Poe sent the hero of *Hans Pfaall—A Tale* (republished as *Lunar Discoveries, Extraordinary Aerial Voyage by Baron Hans Pfaall*) on a Lunar trip in a homemade balloon. Pfaall's reason for going to the Moon was hardly romantic; he was broke and heavily in debt, and the only way out was to flee the Earth.

Pfaall built a balloon, not forgetting "an apparatus for the condensation of the atmospheric air" to provide air for breathing as he passed through what he assumed was a rarified atmosphere extending to the Moon. The takeoff from Rotterdam at first seemed uneventful: "[I] was pleased to find that I shot upward with inconceivable rapidity (upon cutting the attachment cord), carrying with all ease one hundred and seventy-five pounds of leaden ballast." Presently, however, he had some trouble with kegs of gunpowder that gave the balloon an unexpectedly great jolt.

> a concussion . . . burst abruptly . . . and seemed to rip the very firmament asunder . . . . The balloon at first collapsed, then furiously expanded, then whirled round and round with sickening velocity, and finally, reeling and staggering like a drunken man, hurled over the rim of the car, and left me dangling, at a terrific height, with my head downward, and my face outward, by a piece of slender cord about three feet in length.

But Hans finally managed to pull himself back into the balloon.

As the balloon progressed Moonward, Hans's air condenser failed to work, but after experiencing some bad moments he was able to put it in working order. He averaged about one thousand miles a day, and by the eighth day he was so distant that "not even the outlines of the continents could be seen." Hans settled down to what had become a routine trip.

Seventeen days out of Rotterdam catastrophe struck. The balloon burst. "I was falling, falling with the most impetuous, the most unparalleled velocity!" Convinced he was falling back to Earth, Hans Pfaall braced himself for annihilation. But he found that the Earth "was over my head and completely hidden by the ballon, while the moon . . . the moon itself in all its glory—lay beneath me and at my feet."

Hans dumped all his ballast and even cut the balloon car loose, and presently settled gently down in the middle of a Lunar city populated by vast crowds of "ugly little people." After spending five years with the Lunarians, he sent one of them to Earth in his reconstructed balloon to ask forgiveness of his debts. Alas, the two-foot-high Lunarian was so afraid of the denizens of Earth that he did not wait for an answer. Hans was doomed to remain forever on the Moon, a debtor in exile.

The celebrated "Moon hoax" of 1835 showed just how ready most people were to accept the existence of life on other worlds. The hoax was the work of Richard Adams Locke, who presented a wild tale of Lunar creatures who had purportedly been seen through the telescope of Sir John Herschel. The report, under the imposing title of *Great Astronomical Discoveries Lately Made By Sir John Herschel, LL.D., F.R.S., etc, At the Cape of Good Hope,* appeared as a week-long serial in the New York *Sun* in September 1835. It gained credibility because it purported to record facts submitted by Sir John to the august but (unknown to most) defunct Edinburgh *Journal of Science.* The skill of the author and the tenor of an epoch when the public was ready to believe almost anything reported by science also had something to do with it. Locke's hoax, swallowed hook, line, and sinker at first, was exposed in short order—but not before many people had gorged themselves on descriptions of Lunarian civilization.

The thirty-year period between the Moon hoax and the publication, in 1865, of Jules Verne's immortal *De la terre à la lune* (*From the Earth to the Moon*) was a glorious era of science fiction. The public believed in the plurality of worlds, in inhabitants whose nature might not be known but who nevertheless existed, and in the certainty that the Moon and planets would soon be visited in reality as well as in the pages of fiction. While waiting for the actual physical contact to occur, many people thought it might be possible to communicate with the other worlds and thus learn more about them. Many schemes were hatched, including building huge fires, planting trees in geometrical patterns, and constructing mammoth mirrors.

Among the many authors of science fiction during this period was the Frenchman Achille Eyraud, whose short book, *Voyage à Venus,* contained a description of a spacecraft powered by the reaction principle. As pointed out by Alexandre Ananoff in his *L'Astronautique,* "If Cyrano first dreamt of using rockets to fly through the air, it is Achille Eyraud to whom goes the honor, in 1863, of having applied this principle to a true spaceship."

Eyraud was followed by Jules Verne, whose dominance in the field of science fiction still endures. The science in *De la terre à la lune* is nearly as accurate as the knowledge at the time permitted. Despite this, the voyage that Verne describes, while spectacular, is not feasible. Verne pictured a nine-hundred-foot-long cannon, pointing straight up into the sky. The cylindrical projectile, or rather the spacecraft which it fired, weighed some twenty thousand pounds, had a conical nose, and was fitted to accommodate three astronauts. Among the spaceship's interior features were walls "lined with a thick padding of leather," a middle section containing storage cupboards, and a lower area with a large, circumferential seat. Access was by means of a "narrow aperture" in the cone, secured by an aluminum plate. The travelers could look into space through four lens-shaped portholes with heavy metal lids. All manner of provisions were aboard, including chemically supplied air.

It is now known that the acceleration of a projectile fired from a huge cannon would destroy the voyagers, as would the heat generated as the spacecraft rushed through the atmosphere. Verne did not ignore the problems, but the solutions he provided were not adequate. The book ended with the firing of the cannon and the arrival of the spaceship into Lunar space, but the story was picked up again in Verne's *Autour de la lune* (*Around the Moon*).

After the takeoff, the travelers lost consciousness briefly because of the terrific acceleration. They survived because of a mechanism that absorbed the re-

*Hans Pfaall, hero of a story by Edgar Allan Poe, fled from the Earth in a homemade balloon because he was heavily in debt. He took off from Rotterdam, carrying an air condenser so he could breathe the rarefied air in space.*

coil. This consisted of a "bed of water, intended to support a watertight wooden disc, which worked easily within the walls of the projectile. It was upon this kind of raft that the travellers were to take their place. This body of water was divided by horizontal partitions, which the shock of departure would have to break in succession. Then each sheet of water, from the highest to the lowest, running off into escape tubes toward the top of the projectile, supplied with extremely powerful plugs, could not strike the lowest plate except after breaking successively the different partitions."

When they regained consciousness, the space voyagers felt uncomfortably warm. Their leader explained, "This stifling heat, penetrating through the partitions of the projectile, is produced by its friction on the atmospheric strata. It will soon diminish because we are already floating in Space, and after having been nearly stifled, we shall have to suffer intense cold."

Some 4,500 miles from Earth, a large meteor passed close enough to the spacecraft to throw it off course. The astronauts calculated that the ship, instead of landing on the Moon, would orbit it, giving them an excellent view of both sides.

The ship drew closer, until it seemed it might land after all. As a precaution, steps were taken to reduce the craft's velocity by firing "powerful fireworks" placed inside twenty steel-lined guns protruding from the hull. Before they were fired, the astronauts confirmed the earlier predictions that the spaceship would just barely miss the Moon. The closest approach was about twenty-nine miles, over the north polar regions.

A final attempt to land on the Moon was made as the travelers began to pull away, but the spaceship, instead of falling toward the Moon, plunged toward the Earth. It landed in the Pacific, where it was retrieved by an American corvette, the USS *Susquehanna.*

Verne's reasons for not permitting his space travelers to reach the Moon are obscure, but one can surmise what they may have been. Since he had resorted to a cannon to obtain the velocity necessary to depart from Earth, it would be hard for him (1) to arrange for the projectile to land on the Moon; and, just as important, (2) to devise a scheme for it to take off from the Lunar surface and return to Earth. Certainly there would be no cannon on the Moon waiting for the intrepid adventurers—unless they could be expected to construct one themselves once on the Moon (unlikely), or unless they could secure help from the local population (even less likely, since Verne, as opposed to most of his literary predecessors, postulated that the Moon was barren and lifeless).

Looking back over the hundred years since the appearance of Verne's two Moon tales, the debt modern astronauts owe him is apparent. His prodigious output brought scientific adventure to a reading public all over the world. His very name became a synonym for high adventure. And he was read with great respect by working scientists, so carefully did he do his scientific homework. Towering over the vast majority of later writers, Verne is still deservedly popular today.

As far as is known, the first fictional proposal for a manned space station appeared in Edward Everett Hale's story "The Brick Moon," published originally in *Atlantic Monthly* in 1869–1870 and later collected with other tales in the anthology *His Level Best and Other Stories.* Hale's theory was that a satellite placed in polar orbit would serve as a navigational aid for sailors, permitting them to determine longitude accurately and easily. He explained it thus:

> For you see that if, by good luck, there were a ring like Saturn's which stretched around the world, above Greenwich and the meridian of Greenwich, . . . anyone who wanted to measure his longitude or distance from Greenwich would look out of his window and see how high this ring was above his horizon. At New Orleans, which is quarter round the world from Greenwich, it would be just on his horizon . . . . So if we only had a ring like that . . . vertical to the plane of the equator, as the brass ring of an artificial globe goes, only far higher in proportion . . . we could calculate the longitude.

The Brick Moon was suggested as an alternative to this ring. In modern terms, the young Bostonians who are the heroes of the tale planned to put their artificial satellite into a polar orbit, where it would serve the purpose. They planned to use brick because "It must stand fire well, very well." An orbit 4,000 miles high is decided upon so that the Brick Moon can be seen "by a belt of observers six or eight thousand miles in diameter." The size of the Brick Moon is set at 200 feet in diameter, so that it can be seen from 4,000 miles away.

Plans had to be made for orbiting the moon, and, after discussions, the flywheel technique was decided upon. "We would build two huge fly-wheels, the diameter of each should be 'ever so great,' the circumference heavy beyond all precedent, and thundering strong, so that no temptation might burst it. They should revolve, their edges nearly touching, in opposite directions, for years, if it were necessary, to accumulate power, driven by some waterfall now wasted to the world."

At the proper time, the Brick Moon would roll

down "a gigantic groove" until it landed on both flywheels at the same time and was hurtled upward. "Upward; but the heavier wheel would have deflected it a little from the vertical. Upward and northward it would rise, therefore, until it had passed the axis of the world. It would, of course, feel the world's attraction at the time, which would bend its flight gently, but still it would leave the world more and more behind." Once in orbit, the Brick Moon would "forever revolve . . . the blessing of all sea-

*The first known proposal for a manned satellite appeared in a story by Edward Everett Hale about a Brick Moon. Its 37 inhabitants signaled the Earth in Morse code by jumping up and down on the outside of the satellite. People on Earth threw them books and other objects, some of which missed the Moon and went into orbit around it.*

*In Percy Greg's novel* Across the Zodiac, *the hero travels to Mars in a huge spaceship, propelled by an antigravity device called "apergy."*

men . . . the second cynosure of all lovers upon the waves, and of all girls left behind them."

The cost of the Brick Moon was calculated at $60,000—small change for modern times, but too much for the tale's heroes. Only after they have made their fortune can they consider the plan again; making further calculations, they come up with a total cost of $214,729, with each additional moon to cost $159,732. The money is raised, the wheels are built and put into operation, and plans are made for the launching.

But the Moon slipped prematurely down "upon these angry fly-wheels, and in an instant, with all our friends [construction workers and their visiting families], it hâd been hurled into the sky!" A year went by without news. Then a memorandum appeared in the *Astronomische Nachrichten;* the Brick Moon had been seen by Professor Karl Zitta of Breslau, who has named it Phoebe. Later, it is found to be orbiting 5,000 miles above the Earth's surface. Through opera glasses, it appears about the size and brilliance of Jupiter.

*Kurd Lasswitz, author of* Auf zwei Planeten (On Two Planets) *reasoned that if Martians were more intelligent, they—not Earthmen—would be the first to venture from planet to planet. Consequently, his travelers flew from Mars to Earth, where they set up a base at the North Pole.*

Observations of the wayward satellite showed signs of life. "Something is moving—coming, going. One, two, three, ten; there are more than thirty in all. They are men and women and their children!" The passengers had "survived that giddy flight through the ether, and were going and coming on the surface of their own little world, bound to it by its own attraction and living by its own laws."

It was soon learned that thirty-seven people were living on the Moon. Suspecting that they were being observed, they arranged themselves on the outside of the satellite and "at one moment, as by one signal, all . . . jumped into the air—high jumps. Again they did it, and again." It soon became apparent that they were signaling—short leaps and long leaps, dots and dashes—Morse code.

It turned out that there was plenty of air, food, and friends aboard: "What more can man require?" And it rained regularly, providing drinking water. The climate is good, even tropical, which helped form a soil, making it possible to grow palms, breadfruit, bananas, oats, maize, rice, wheat. Crops were harvested up to ten times a year.

Back on Earth, friends of the marooned space voyagers decide to send presents up to the Brick Moon by carefully wrapping the packages in many-layered wrappings. As the packages flew through the atmosphere, each layer would burn and disintegrate, but the inner portion often reached the Brick Moon unscathed. Some objects do get through and are retrieved, some are lost forever, and some take up orbit with the satellite. ("They had five volumes of the Congressional Globe whirling like bats within a hundred feet of their heads.")

Once the excitement wears off, life gets back to normal, both on Earth and on the Brick Moon. Communications continue back and forth, and occasionally presents are interchanged. Hale takes leave of his contented space heroes wondering if it can be possible that "all human sympathies can thrive, and all human powers be exercised, and all human joys increase, if we live with all our might with the thirty or forty people next to us, telegraphing kindly to all other people, to be sure? Can it be possible that our passion for large cities, and large parties, and large theatres, and large churches, develops no faith nor hope nor love which would not find aliment and exercise in a little 'world of our own?' "

From the latter part of the nineteenth century, fictional accounts of Moon voyages decline in importance and fall outside the mainstream of science fiction. There are relatively few exceptions to this trend. Not only was the Moon becoming recognized as a dead world, but it was just too close to the Earth to offer a sophisticated, science-oriented public the kind of romantic excitement associated with a Domingo Gonsales or a Cyrano de Bergerac. Interest was shifting to the planets.

A sign of the changing times came in 1880 with the appearance of Percy Greg's two-volume novel *Across the Zodiac.* Here a mysterious something called "apergy" is used to negate gravity, providing the means for a voyage to Mars. The spaceship, a huge thing with three-foot-thick walls, "resembled the form of an antique Dutch East-Indiaman." The deck and keel were "absolutely flat, and each one hundred feet in length and fifty in breadth, the height of the vessel being abour twenty feet." The apergy receptacle was placed above the generator in the center of the ship. From them "descended right through the floor a conducting bar in an antapergic sheet, so divided that without separating it from the upper portion the lower might revolve in any direction through an angle of twenty minutes." This sheath is used to direct "a stream of repulsive force" against the Sun or any other body.

The most noteworthy fact is that this ship is used to go to Mars, which had finally begun to assume the importance to science-fiction writers that it deserved. This was a result of increasing knowledge about the planet and the development of theories of the origin of the Solar System, which made Mars seem especially interesting. Greg wrote at the time of the discovery of Mars's "canals," and of the planet's two satellites (which he described in his novel). New data made Mars seem like an older Earth, and therefore a potential home for a more advanced race of intelligent beings.

Greg gives a haunting description of the planet. "The seas are not so much blue as grey. Masses of land reflected a light between yellow and orange indicating . . . that orange must be as much the predominant color of vegetation as green on Earth . . . . The sky, instead of the brilliant azure of a similar latitude on Earth, presented to my eye a vault of pale green . . . . The lower slopes [of a mountain] were entirely clothed with yellow or reddish foliage."

Kurd Lasswitz's *Auf zwei Planeten* (*On Two Planets*) carried the Mars theme several notches higher in the literary scale. Published in 1897, Lasswitz's book was based on the logical assumption that if Mars was the abode of a higher intelligence, the first space trip would be made from Mars to Earth. Accordingly, he had travelers from Mars flying to Earth and setting up a base at the North Pole. The method of space travel is a gravity-nullifying device, a material which, when formed into a shell-like structure, prohibits the passage of gravity and becomes weightless when the ports are closed.

H. G. Wells wrote his *War of the Worlds* as a magazine serial in 1897, and the story was published in book form the following year. It is the story of a Martian invasion of our planet in which the invaders score marked successes at first but then are vanquished by terrestrial diseases for which their bodies have no defenses. Some forty years later, a Welles named Orson terrorized the United States with an eerie retelling of the story in a radio program. But tales of an Earth expedition to Mars have not been lacking; Garrett P. Serviss's *Edison's Conquest of Mars,* which appeared soon after Wells's tale, told of a punitive expedition bound for the red planet.

The Moon was not entirely forgotten, however. A few years later, Wells published *The First Men in the Moon.* An antigravity material, Cavorite, was used to transport the spaceship the quarter million miles to the Moon. Wells populated the Moon with insectlike creatures living in underground caves and tunnels.

The appearance of *The First Men in the Moon* marks roughly the end of one era in science fiction and the beginning of another. The discovery was made in the closing years of the nineteenth century that the rocket reaction engine was the solution—the realistic, scientific solution—to the problem of space propulsion. Years of work lay ahead, but at last man knew how to go about the job. He need no longer invent implausible methods; nonfiction could begin to take over from the Cyranos.

The year 1900 did not signal the abrupt end of one epoch in space literature and the beginning of another. It is, however, a convenient point of departure from which to begin a new series of adventures based on science and technology instead of legend and fantasy.

The discovery of the importance of the rocket to space travel did not mark the end of science fiction; instead, it gave it a new and more mature outlook and greater popularity. Excellent works continue to be published, and they will appear long after man has landed on the Moon and the planets. But their role in stimulating the interplanetary venture inevitably declined and the practical work of going into space was stepped up. The debt to great speculators, from Lucian to Verne and Wells, was about to be paid.

# 2 A THOUSAND YI

While writers of science fiction used geese, bottles of dew, spirits, and chariots to carry their heroes to the Moon, the device that would finally make the trip possible was undergoing a slow but steady evolution, changing from a toy to a weapon, from a crude device to a relatively sophisticated machine. The time was to come when the rocket would take its place as the only possible means of space transportation.

The rocket is a reaction device, which works in accordance with Sir Isaac Newton's Third Law of Motion: For every action there is an equal and opposite reaction. A rocket can be compared to a continuously firing machine gun mounted on the rear of a rowboat. As the gun is fired to the rear, the recoil from the stream of bullets moves the boat forward. A rocket-motor's "bullets" are minute particles that are thrown out through a nozzle as a propellant is burned in a suitable chamber. The reaction to the discharge of these particles makes the rocket fly in the opposite direction.

Although a rocket is a reaction device, not all reaction devices are rockets. A rocket is a special case because it contains all the elements it needs to operate, including both fuel and oxidizer. A jet-airplane engine, by contrast, is a reaction device that uses the oxygen in the air to support the combustion of the fuel carried on board.

The reaction principle was known long before a true rocket was invented. In his *Noctes Atticae* (*Attic Nights*), Aulus Gellius describes the ancient pigeon of Archytas, dating back to about 360 B.C. Hanging from a string, the pigeon was made to move by steam blowing from small exhaust ports.

A more sophisticated device was the reaction wheel, or aeolopile, developed by Hero, a Greek resident of Alexandria. Hero lived at about the time of Christ, but his exact dates are unknown. Some scholars say the first century B.C., others the first century A.D., with the evidence pointing somewhat more strongly in favor of the latter, due to a reference in Hero's *Dioptra* to a Lunar eclipse that would have been seen in Alexandria in A.D. 62. The aeolopile, described in his *Pneumatica*, was, in effect, a primitive steam turbine, although apparently no practical use was made of the device. It was constructed of a hollow globe, which was pivoted to turn on two central trunnions. One trunnion also was hollow, permitting steam to pass through it into the globe. On opposite sides of the globe and at right angles to the axis of the pivots were two bent tubes, whose open ends were pointed in opposite directions. As the steam escaped through these tubes, the globe revolved.

Centuries later, the reaction to discharging steam was used to propel a model of a car designed by Jacob Willem Gravesande, a Dutch professor best known for his 1720–1721 two-volume *Physices elementa mathematica, experimentis confirmata sive introductio ad philosophiam newtonianam* (a textbook on Newtonian philosophy published in Leyden). Records also are available of schemes to build steam-reaction helicopters and even a man-carrying craft of unusual design.

The precise origin of the rocket itself is lost in the shadow of time. Almost all Western and Oriental histories credit the Chinese with the invention, but sources are seldom given for this claim, making it difficult to accept or reject. Part of the trouble in settling the question is caused by lack of clarity in descriptions of ancient weapons; often, it is difficult to tell from a vague description whether a projectile was powered by a rocket or merely carried powder or other material that burned.

There are many references, for example, to fire-arrows. These arrows were fitted with an inflammable material—pitch, bitumen, or resin. Launched by muscle power, they made flaming rocketlike arcs to their targets. And there was the fire-pot containing naphtha and other ingredients. Shot by hurling devices, the fire-pot spread its burning contents over a fairly wide area. In 305 B.C., in Greece, during the siege of Rhodes by Demetrius, the Rhodians are believed to have employed eight hundred flame-carriers in attempts to destroy the enemy's wooden siege en-

# ARS OF ROCKETRY

gines. A subsequent innovation was the shooting of fire-arrows and fire-pots by rocket devices, though just when they first were used is unclear (fire-pots themselves date from at least 1000 B.C.). Some scholars conclude from passages in the Byzantine princess Anna Comena's writings that rockets were used to carry fire-pots to the enemy in the ninth and tenth centuries within the Byzantine Empire, but this is generally regarded as unlikely. In 1450, Robertus Valturius, in *De re militari* (*Military Treatise*), wrote of what may have been rockets dating from Emperor Leo VI (886–911). It is said that the Byzantine ruler's warriors made use of "fire that is launched (or hurled)." The description, however, applies just as well to Greek fire slung by ballistas as to rockets.

Most authorities believe the invention of rockets is tied inextricably to the discovery of black powder which served as the first rocket propellant. The best available evidence, including both early Chinese documents and the writings of some of the first Europeans who visited China, indicates the Chinese certainly were the first to use black powder and, therefore, probably the first to use rockets as well.

The ingredients of black powder—technically, it should not be called gunpowder because guns came later, but it usually is—are charcoal, sulphur, and saltpeter. These have been known in China for perhaps two thousand years—charcoal since the very earliest times, and sulphur and saltpeter at least since the sixth century A.D. and probably as far back as the first century B.C. That the ingredient saltpeter is definitely of Chinese origin is indicated by the names given to this material by the Arabs, who called it "Chinese snow," and the Persians, who called it "salt from China." The three ingredients were known in China for many centuries, however, before they were combined into black powder.

The first firecrackers may have appeared in the Chin Dynasty (221–207 B.C.) or during the Han Dynasty (206 B.C.–A.D. 220). According to the Han work *Shen I Ching* (*Classic of Strange Spiritual Manifestations*), what was called Pao Chu, or "bursting bamboo," was "put into the fire," producing a noise that "frightened the spirit of the mountain." Earlier Chin works, *Ching Ch'u Sui Shih Chi* (*Annual Customs of Ching and Ch'u*) and the *Feng Su T'ung I* (*The Meaning of Popular Traditions and Customs*), speak of "bamboo bursting" and the "cracking of bamboo is like the roar of the wild animals." It is not known definitely that the bursting of the bamboo was caused by the explosion of black powder.

Other early Chinese writings also contain references to what was either black powder or a similar substance. A work on medicine written by Sun Saŭmiso, who died in A.D. 682, describes experiments on *fu huo fa* (calcination), including one in which the author combined equal amounts of saltpeter and sulphur, added some acacia seeds, and lit the resulting powder. Ch'ing Haü-Tzu (*c.* 809), an alchemist, reportedly set fire to saltpeter and sulphur mixed with a substance called *ma tou ling* that resembled black powder. During the Northern Sung Dynasty (A.D. 960–1126) the term Pao Chang was used to describe firecrackers, which are believed to have contained black powder. Another term, Yen Huo, meaning firework, is said to have originated during the reign of Yang Ti (A.D. 605–616). According to a 1947 article in *Isis* by Wang Ling, "On the Invention and Use of Gunpowder and Firearms in China," fireworks became popular during the T'ang Dynasty (A.D. 618–907) and were perfected by the time of the Northern Sung Dynasty.

By 1045, just twenty-one years before William the Conqueror invaded Saxon England, there is no doubt the Chinese were well acquainted with black powder. The *Wu-ching Tsung-yao* (*Complete Compendium of Military Classics*), published that year, contains many references to the subject. Written by a government official named Tseng Kung-Liang at a time when the Sung Dynasty under the Emperor Rjen Tsung was experiencing its first serious military threat, it is largely concerned with strategy. There are some statistics on the disposition of troops, a series of short notices on successful battles, and, most im-

*Jacob Willem Gravesande, a Dutch professor who studied Newton's laws of motion, designed a car that moved in reaction to discharging steam (see exhaust pipe, right).*

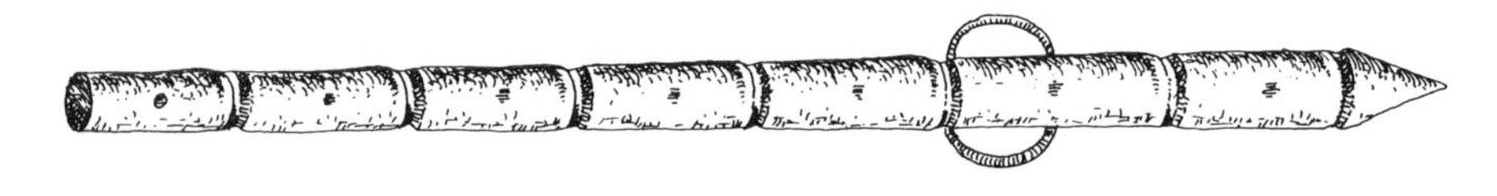

*Before William the Conqueror invaded England, the Chinese were experimenting with gunpowder. This fire-arrow, the* Huo yao pien chien, *was included in* The Complete Compendium of Military Classics, *published in 1045.*

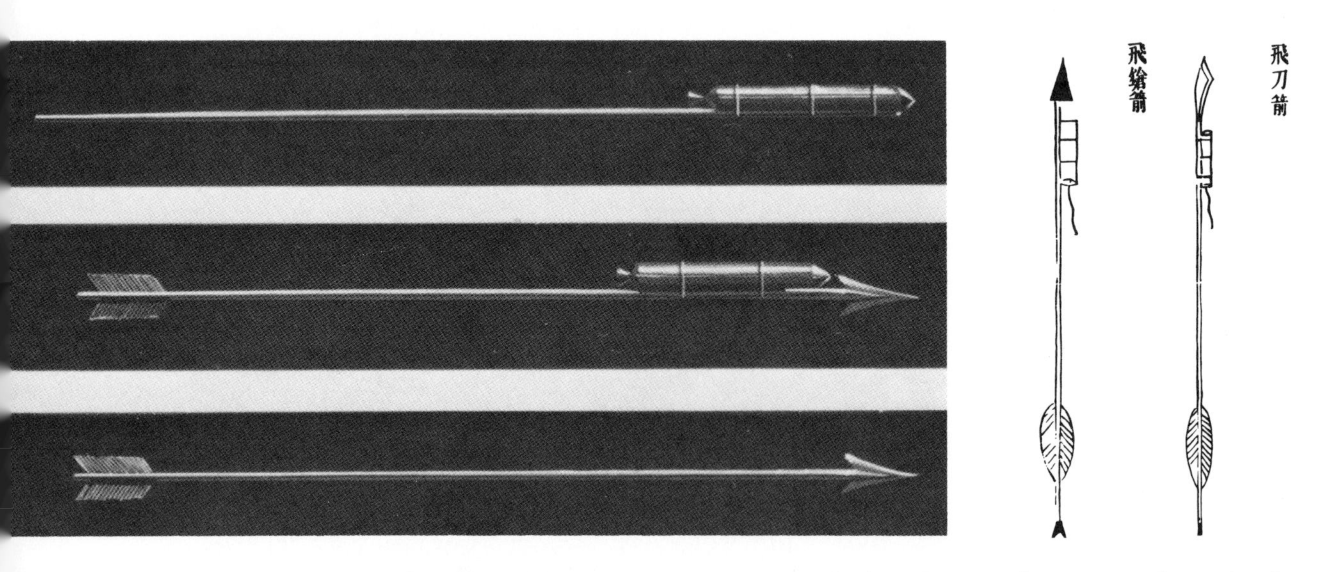


*Left, Chinese fire-arrows of the types described in* The Complete Compendium of Military Classics. *The two on top have explosives mounted on their sides. Right, actual illustrations from the* Wu Pei Chi. (RIGHT, PROFESSOR JAMES R. WARE)

portant, a section on weapons, including both text and many illustrations.

The *Complete Compendium* indicates that black powder, and possibly black-powder rockets, was used extensively during the Sung Dynasty, a period of brilliant cultural activity lasting from A.D. 960 to 1279. The book gives this formula for making gunpowder:

> 1 chin 14 ounces of sulphur, together with 2½ chin of saltpeter, 5 ounces of charcoal, 2½ ounces of pitch, and 2½ ounces of dried varnish are powdered and mixed. Next, 2 ounces of dried plant material, 5 ounces of tung oil, and 2½ ounces of wax are also mixed to form a paste. Then these ingredients are all mixed together, and slowly stirred. The mixture is then wrapped in a parcel with five layers of paper, which is fastened with hempen thread, and some melted pitch and wax and is put on the surface.

It is probable that the explosive properties of powder were unknown before the tenth century (unless the ninth-century work of Ch'ing Haü-Tzu is accepted as producing full-fledged explosive powder). Somewhere around the tenth or eleventh centuries, black powder was first compounded from its basic ingredients of saltpeter, charcoal, and sulphur, but this does not mean that it was used to propel rockets at that time. Some historians of technology mention that a rocket called San Kung Ch'uang Tzu Nu is described in the *Complete Compendium*, but close inspection of the eleventh-century handbook indicates the "rocket" is really a ballistalike device. Of course, it could have been used to launch rocket-powered fire-arrows as well as more conventional types, but the text accompanying the picture does not make this clear.

Fire-arrows are mentioned frequently in early Chinese sources, but—as in the case of the Byzantine writings mentioned earlier—it is difficult to determine whether the arrows were tipped with fire or whether they were propelled by rockets, which also would give them a fiery appearance.

A French missionary, Father Joseph Marie Amiot, in his *Mémoire concernant l'histoire, les sciences, les arts, etc. des Chinois* (Volume VIII) (*Memoirs on the History, Science, Arts, etc. of the Chinese*), observed that firearms were known in China from the beginning of the Christian era; Koung-Ming is reported to have used them around A.D. 200. But it is not clear what is meant by "firearms." Certainly fire was known, and could have been coupled with a weapon that was neither a gun nor any sort of a rocket.

Father Amiot also described a fire-arrow to which an early type of rocket may have been attached. According to Amiot: "The tube where the powder is placed must be extremely straight, should be only four inches long, and its end should be two inches from the fire. An arrow thus launched is equivalent to a very powerful gunshot." He does not say, however, when this fire-arrow was introduced.

Another type of fire-arrow was called the *ny-fung-yo*, or "powder that goes against the wind." W. F. Meyers, writing in the *Journal of the North China Branch, Royal Asiatic Society* (1871), described these as burning forward with "a sudden flame—so that no one durst approach them."

Some missionaries, such as Joseph P. G. Pauthier, who in 1821 wrote *Chine, ou description historique, géographique et littéraire de ce vaste empire d'après des documents Chinois* (*China, Historical, Geographic, and Literary Description of this Vast Empire According to Chinese Documents*) believed that the invention of the fire-arrow dated from the eighth century and that by the end of the eleventh century the Tartars had learned about it.

However, the French sinologists Joseph Toussaint Reinaud and Idelphonse Favé, in an indispensable work *Histoire de l'Artillerie: Feu grégeois, des feux de guerre, et des origines de la poudre à canon* (*History of Artillery: Greek Fire, Fireworks, and the Origins of Gunpowder*), published in 1845, and in an article, "Du Feu Grégeois, des Feux de Guerre, et des Origines de la Poudre à Canon chez les Arabes, les Persans, et les Chinois" ("On Greek Fire, Fireworks, and the Origins of Gunpowder by the Arabs, the Persians, and the Chinese"), which appeared in the October 1849 issue of the *Journal Asiatique*, state that, after years of research they discovered no documents leading them to believe that pyrotechnic devices existed prior to the thirteenth century. The French and other missionaries in China had access to vast quantities of documents; but, as Reinaud and Favé admit, the missionaries may not have known enough about weapons to interpret properly all that they read. Their reports, however, do provide additional evidence that incendiary compositions were known in China earlier than in the Arab countries and Western Europe.

The *Sung Shih Ping Chih* (*Military Memoirs of the Sung Dynasty*) refers to a new type of fire-arrow proposed by the general Fêng Chi-shang; it was subsequently made, and tested. Then, in the *Wu Li Hsiao Shih* (*Small Insights into the Principles of Things*), mention is made of a Yo I-Fang who, in A.D. 969, presented an improved fire-arrow to the emperor, T'ai Tsung, brother of the founder of the Sung dynasty, for which he was rewarded with a gift of silk. In 1002, T'ai Tsung's successor, Chen Tsung, used similar devices with ranges up to 1,000 feet.

Still, no direct mention is made during this period of powder being used to *propel* the arrows. But the Huo Yao Pien Chien type of fire-arrow may have been propelled by the force of the exploding black powder, for it is stated that five ounces of powder were placed at its end. And powder may have been used in conjunction with the San Kung Ch'uang Tzu Nu, described in the *Complete Compendium*.

An account of the use of these or very similar weapons is contained in the Sung history, which relates that "in the fifth year of Ch'un Hua [A.D. 994] an army of 100,000 men besieged the city of Tzu T'ung. A fierce attack was made, and the people in the city were greatly alarmed. Chang Yung ordered the hurling of stones by machines, and succeeded in pushing back the invaders. At the same time, fire-arrows were shot off, whereupon the enemy retreated." Other accounts say that fire-arrows were used by the Sung, Yuen, and Chin armies during the 1100's. In 1206, the second year of the K'ai Hsi, a Sung general, Chao Chun, fired the Huo Yao Pien Chien arrows "in order to burn down the wood, straw, and catapults of the enemy."

There is little doubt that powder-propelled fire-arrows were in fairly widespread use by the beginning of the thirteenth century. The Sung Dynasty, under continuous pressure from the north, had to rely more and more on technological developments to maintain its power and protect its civilization. Its ordnance experts introduced and improved incendiary projectiles of many types, explosive grenades, and possibly cannon. They also seem to have made good use of rocket fire-arrows at the battle of K'ai-fung-fu (then called Piang-king) in A.D. 1232, five years after the death of Genghis Khan. A description of the battle appears in Father Antonine Gaubil's *Histoire de Gentchiscan et de toute la dynastie de Mongous, ses successeurs, conquérants de la Chine* (*History of Genghis Khan and of the Mongolian Dynasty, its Successors, Conquerors of China*), published in Paris in 1739.

The town of K'ai-fung-fu, north of the Yellow River, was heavily besieged by Mongol hordes. The town's governor, Kiang-chin, took extraordinary measures which enabled the defenders to resist for many months the onslaught of at least thirty thousand invaders. Against Kiang-chin's defenses, the Mongols could do little; they withdrew, regrouped, changed generals, and unleashed a new offensive, which, according to Father Gaubil, the Chinese met with rockets.

The new weapon was devastating. "When it was lit, it made a noise that resembled thunder and extended 100 li [about five leagues]. The place where it fell was burned, and the fire extended more than two thousand feet (that is to say, it burned a circumference of two thousand feet) . . . . These iron nozzles, the flying powder halberds that were hurled, were what the Mongols feared most," said Gaubil.

Other writers described this weapon as "thunder that shakes the heavens." One source says that "an iron pot was used for that, which was filled with yo [the incendiary mixture]. As soon as it was lit, the pao [fire-projectile] rose and the fire exploded everywhere." This was some sort of explosive grenade, generally launched by catapult, but apparently on occasion by rocket.

The besieged townspeople also used a rocket-type weapon known as the *feï-ho-tsiang*. It was described as an arrow to which combustible material was attached. When lit, the arrow would take off rapidly and fly along a straight trajectory, spreading fire over a distance of ten paces upon landing.

Because of the difficulty of interpreting second- and third-hand reports by untrained observers, some authorities question whether true rockets were used at the siege of K'ai-fung-fu. In an article on "Early Chinese Military Pyrotechnics," which appeared in the November 1947 issue of *The Journal of Chemical Education*, Tenny L. Davis and James R. Ware conclude that the weapons described were flying spears rather than rocket-propelled arrows. These spears were "equipped with [a] fire tube which threw fire forward for a distance of about 30 feet . . . a reason-

*Probably invented in China, rockets were described by writers as "thunder that shakes the heavens." In the mid and late thirteenth century, the Mongolians used rockets, like those shown below, in battles ranging from Baghdad to Japan.*

able distance for fire to be thrown from a small tube, but . . . an unreasonably short trajectory for a rocket and one which would yield but little advantage."

Whether or not rockets played a part in this particular battle, there is substantial evidence that rockets were in general use at about this time. Rockets of one sort or another are mentioned regularly in accounts of battles following the siege of K'ai-fung-fu. For example, Constantine Mouradgea d'Ohsson, in the *Histoire des Mongols* (*History of the Mongolians*), published in 1834, wrote that rockets were used in the siege of Siang-yang-fu in 1271. They saw service in a battle between the Sung and the Yuan in 1274, and again in 1275. At about the same time the Mongols introduced them into Japan. According to the Japanese work *History of Japan's Humiliation,* in 1274, during the battle of Tsu Shima, fire-arrows were launched from Mongolian ships, while on land they were used by the army in attacking Iki Shima. Later, during the second invasion of Japan in 1281, rockets were launched in greater quantities with devastating effects. Once they learned about the properties of gunpowder, the Japanese began to develop fireworks; it is believed that both aerial and daylight types originated in Japan, a country that gave strong impetus to their development. Korea, Java, and India also adapted the Chinese invention by way of the Mongols.

Davis and Ware describe a variety of rocket-propelled arrows, their firing tubes, and several devices from which many arrows could be launched. These include "rocket-basket-arrows," which were fired from a cylinder of bamboo splints 4 feet long. Each cylinder contained from seventeen to twenty arrows on whose tips poison was smeared. Another contraption contained arrows "which will rush out on a solid front like 100 tigers"; they were fired from a frame, all 100 at a time, at targets up to 300 paces distance. There also was the "leopard-herd-rush-transversally" launcher, which could release forty arrows upon command, and the "long-snake-crush-enemy arrows," thirty of which were stored in a wooden box. These arrows were made of bamboo and were about 3 feet long. Each box of thirty weighed between 5 and 6 pounds, making this a highly mobile weapon.

Enemy armies, greatly impressed by the black powder and rockets of the Chinese, adopted them for their own use. Knowledge of these weapons was transmitted quickly to Europe, probably reaching the West by way of the Mongols and the Arabs.

The Mongols, for example, definitely used gunpowder at the Battle of Sejó, which preceded their capture of Budapest on Christmas Day in 1241. Contemporary accounts also indicate that this campaign featured what appears to have been the first gas attack on European soil. At the Battle of Sejó River, the Mongols set up on a pole "a long bearded head of horrible appearance" which emitted smoke with such a foul odor that the army of Heinrich von Schlesien was sent fleeing. Although effective as a gas, the real purpose of the smoke apparently was to serve as a screen for a Mongol attack.

Only seventeen years later, the Mongols are known to have used rockets in the Near East. According to the Arab writer Raschid-eddin, the Mongols employed fire-arrows while capturing the city of Baghdad on 15 February 1258. The rockets, containing black powder, were attached near the iron on the lance, with the wick placed on the opposite side. The Arabs subsequently referred to these arrows as "arrows from Cathay."

The exact date the Arabs adopted the rocket is unknown. Instructions for preparing black powder, the necessary propellant, are contained in a number of Arab works that date from the last half of the thirteenth century. Especially important are the writings of a Syrian military historian with the impressive name of al-Hasan al-Rammāh (the Lancer) Nedjm al-din (the star of religion) al-Ahdab (the hunchback). More simply referred to as al-Hasan al-Rammāh, his fame rested on his military writings.

His major work was the *Kitāb al-Furusīya wal munasab al-harbiya,* sometimes transliterated as *Ketab alferoussye ou al menassib alharbye* (*Treatise on Horsemanship and War Stratagems*). It is believed to have been written between 1285 and 1295, the year of al-Hasan al-Rammāh's death. In the foreword, the author announces that he will discuss, among other things, "The mixture of materials, the construction of machines, and sending of fire." He goes on to give a number of pyrotechnic recipes with careful instructions on how to prepare and purify the saltpeter portion.

The Arabs continued to be interested in rockets well after al-Hasan al-Rammāh wrote his book. Ibn Khaldūn (1332–1406) states in his *Kitāb al-'Ibar* (*Book of Wonders*), written in 1384, that the Arabs of North Africa were acquainted with the propulsive force of black powder in 1273. And the French historian Jean, Sire de Joinville, describes an odd, rocket-powered projectile used by the Arabs against the French under Louis IX during the Seventh Crusade.

Writing in 1268 about events that had occurred twenty years previously, De Joinville reports in his *Histoire de roy Saint Louis* (*History of King Saint Louis*) that the French first encountered this rocket while maneuvering along one of the eastern branches of the Nile in an attempt to take Damietta. The Arabs, who were on the other side of the river, launched "a projectile . . . which, when it had fallen on the bank [of the river], *came straight towards them, burning wildly;* it is doubtless *the egg that moves and burns.*" This interesting device was apparently a fairly flat object, filled with black powder and fitted with a tail to stabilize its path. Flames poured from little openings whose fuses were called Ikrikh. It was propelled by three rockets, "combined such that two of these rockets served as a guiding stick for the third."

Knowledge of explosive powder had penetrated Western Europe as far as England by the time De Joinville had come face to face with a rocket on the banks of the Nile. Roger Bacon (*c.* 1214–1220 to *c.* 1292) described the preparation of black powder before the middle of the thirteenth century, probably sometime in the late 1240's. He did not speculate on its use as a propellant for rockets, but he did describe what could have been rockets. The *Epistola Fratris Rog. Baconis, de secretis operibus artis et naturae et nullitate magiae* (*Epistle of Roger Bacon on the Secret Works of Art and of Nature and Also on the Nullity of Magic*) contains numerous references to saltpeter. Wrote Bacon:

> We can, with saltpeter and other substances, compose artificially a fire that can be launched over long distances. The light of lightning and the sound of thunder can also be perfectly imitated. By only using a very small quantity of this material much light can be created accompanied by a horrible fracas. It is possible with it to destroy a town or an army . . . . In order to produce this artificial lightning and thunder it is necessary to take saltpeter, sulfur, and *Luru Vopo Vir Can Utriet.*

The words *Luru Vopo Vir Can Utriet* form an anagram which hides the proportion of powdered charcoal to be added to the powder. With this mixture "you will make thunder and flashing, if you know the art."

Bacon's German counterpart, Albertus Magnus (1193–1280) also wrote about black powder and how to make it. In his *De mirabilibus mundi* (*On the Wonders of the World*), he gave this recipe: "*Flying fire:* Take one pound of sulfur, two pounds of coals of willow, six pounds of saltpeter; which three may be ground very finely in marble stone;—afterwards, a little later, at will, some may be placed in a skin of paper for flying or for making thunder."

A more explicit description of a rocket was given by Marchus Graecus, or Marc the Greek, in a work titled *Liber ignium ad comburendos hostes* (*Book of Fires and Burning the Enemy*). Probably written between 1225 and 1250, but perhaps as late as 1270, the *Liber ignium* goes into considerable length on the subject of gunpowder and provides many recipes of pyrotechnic devices of all ages, including Egyptian, Hellenistic, Byzantine, Arabic, and Latin. It gives instructions on combating enemies at long distances with rockets. The propellant, mixed in a marble mortar, contained 1 pound of sulfur, 2 pounds of charcoal, and 6 pounds of saltpeter. "A certain quantity of this powder" was to be placed in a "long narrow and well pressed casing." In order to "carry the fire the device must fly in the air."

Rather than use the word *rocket,* Marchus Graecus presents the terms *tunica ad volandum* or "casing destined to fly," and *ignis volatilis in aere,* "flying fire." He says that when lit, this rocket will fly immediately towards the desired destination (*evolat ad quemcunque locum volueris*). The casing "must be slender at both ends, wide in the middle, and filled with the powder under consideration. The covering that is to rise in the air can have several foldings [*plicaturas* in Latin]; the type used to produce a detonation can have many of them."

Other types of rockets are described. For example, "flying fire (*ignis volans in aere*) can be made with a mixture of saltpeter, sulfur, and linseed oil. After being mixed and placed in a tube or hollow cane (Latin, *canna*), and then lit, it will rise into the air. Still another type is composed of saltpeter, sulfur, and carbon provided with a wick made of papyrus (*in tenta de papyro facta positis*). Again, upon being lit it soars rapidly skyward.

Other pioneers of medieval rocketry were Muratori, who first used the word rocket in its Italian form, *rocchetta,* in 1379; Konrad Kyser von Eichstadt, whose *Bellifortis* (*War Fortifications*) of 1405 describes several types of rockets; Joanes de Fontana, author of a 1420 sketchbook, *Bellicorum instrumentorum liber* (*Book of War Instruments*), which contains suggestions for military rockets; and Jean Froissart (1338–*c.* 1410), in whose *Chronicles* the use of tube-fired rockets was proposed.

By the sixteenth century many Europeans were writing about, describing, and making proposals for all sorts of rockets, military and nonmilitary. An anonymous book published in Paris in 1561, the *Livre de cannonerie et artifice de feu* (*Book of Cannons and Fireworks*), tells how to make 3½-foot and 4-foot-long rockets. Similar instructions are found in the *Briefve instructions sur le fait de l'artillerie de France* (*Brief Instructions on Matters of French Artillery*) by Daniel Pavelourt (Paris, 1597) and *La Pyrotechnie* (*Pyrotechnics*) by Hanzelet Lorrain (Pont-à-Mousson, 1630).

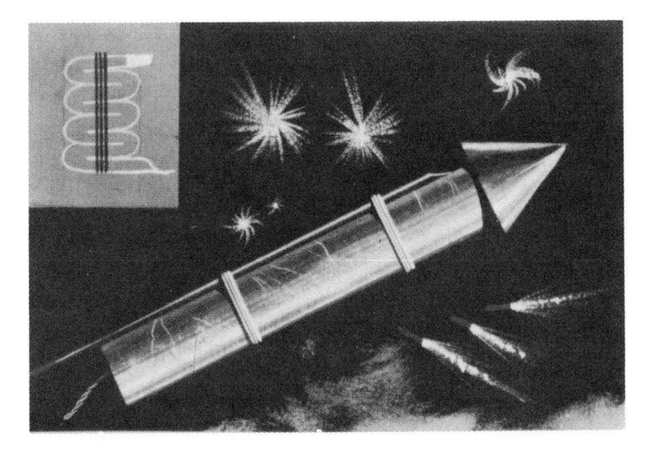

*French pyrotechnic expert Jean Appier Hanzelet Lorrain described this rocket in his 1630 treatise* La Pyrotechnie.

While rockets remained primarily military weapons, they also were used frequently for fireworks displays, and several books paid special attention to construction of this sort of rocket. The *De la pirotechnia* (*On Pyrotechnics*) of Vannoccio Biringuccio, published in Venice in 1540, while primarily a book on metallurgy, contains chapters on "Making Fireworks to be used in Warfare and for Festivals."

Nathaneal Nye's *The Art of Gunnery*, published in London in 1647, contains a 43-page section on rockets with the title *A Treatise of Artificial Fire-works for Warre and Recreation; Containing a Description to Make Sundry Kinds of Fire-works, both for Use and Pleasure, with lesse Labour and Cost than any Hath Heretofore been Published.* It includes instructions on how to make various pyrotechnic devices, has numerous illustrations of contemporary rockets, and tells the reader how to handle them. In order to fire a rocket "set your rockets mouth upon the edge of any piece of timber, battlement of a wall, top of the Gunners carriage, wheels, or any dry place whatsoever, where the rod or Twigge may hang perpendicular from it, then lay a Train of powder that may come under the mouth thereof, give fire thereunto, and you have done."

The Italians were the first Europeans to advance significantly the art of firework-making, with the Florentines and Sienese credited as being the first to place fireworks on wooden pedestals. Great fireworks displays were held regularly in many parts of Italy, which reigned supreme in pyrotechnics until the end of the seventeenth century when the French, under the influence of Louis XIV and Louis XV, began to take over the leadership. Frézier's *Traité des Feux d'Artifice* (Paris, 1747) is an excellent compendium of advances made in France by such men as Morel Torré and the brothers Ruggieri, as well as developments in the manufacture and use of fireworks in other countries.

The French military already had a tradition of rocketry. Rockets were used in the defense of Orléans in 1429, and again at the siege of Pont-Andemer in 1449. Rockets were used against Bordeaux in 1452, and a year later they were fired at Gand. The French had no monopoly on the subject; Kazimierz Siemienowicz, in his *Artis magnae artilleriae* (*Great Art of Artillery*), published in Amsterdam in 1650, wrote on several types of military rockets. And, in 1668, a German field artillery colonel, Christoph Friedrich von Geissler, experimented with rockets weighing from 55 to 120 pounds. By 1730 a series of successful flights had been made.

*Under kings Louis XIV and XV, the French led the world in the development of rockets. Such diverse styles as these were included in Frezier's* Traité des feux d'artifice, *published in Paris in 1747.*

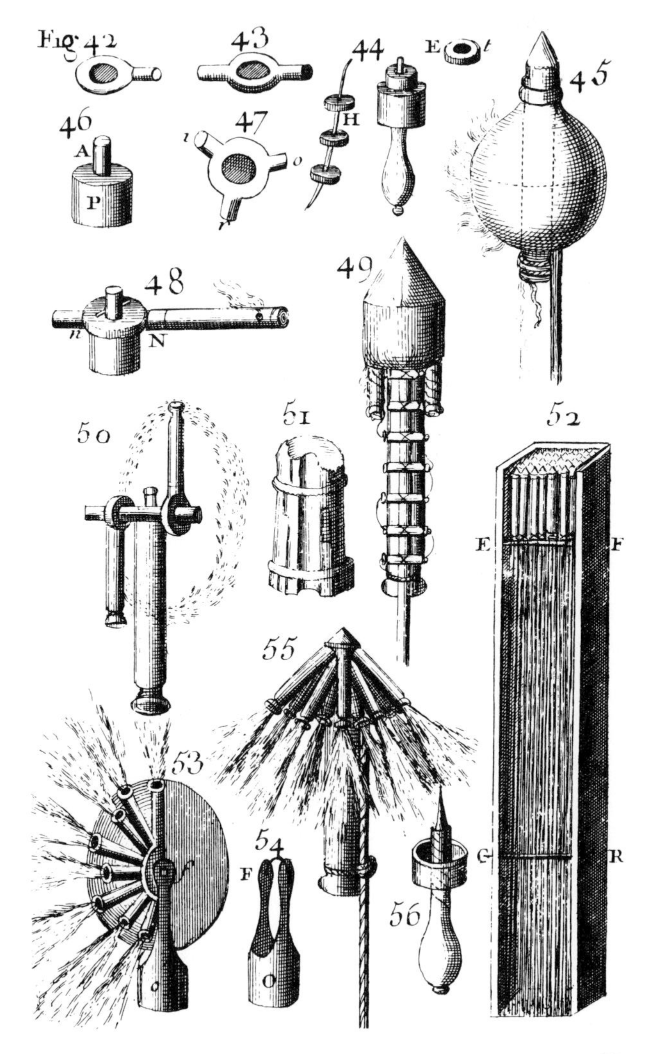

*Sir William Congreve (1772–1828) of Woolwich Arsenal, London. In the beginning of the nineteenth century, he developed war rockets that could be fired from land or sea. His rockets were used by the British in such diverse places as Copenhagen, Bologna, and the Potomac River.*

The eighteenth century was almost over, however, before Europeans became seriously interested in the military potential of the rocket—and then only because they suddenly found themselves on the receiving end of this weapon. The first major engagements with rockets that involved Europeans took place in India, where troops under Tippoo Sultaun of Mysore, fired them against the British during the two battles of Seringapatam in 1792 and 1799.

Two of Tippoo Sultaun's rockets are displayed in the Rotunda of the Royal Artillery Museum, Woolwich Arsenal, near London. One is a rude tube case, 10 inches long and 2.3 inches in exterior diameter, bound by strips of hide to a straight, 3-foot-4-inch-long sword blade. The iron tube of the second is 7.8 inches long and 1.5 inches in exterior diameter, secured by leather strips to a bamboo stick 6 feet 3 inches long.

Tippoo Sultaun's father, Hyder Ally, had built up a 1,200-man contingent of rocketeers by 1788. Subsequently, Tippoo Sultaun enlarged this corps to about 5,000. It is not known, however, how much of his strength he committed at either of the battles of Seringapatam and the accounts of British officers who took part in the campaigns against him differ as to the effectiveness of the rockets.

In the 1792 battle of Seringapatam, Tippoo Sultaun's army consisted of 36,131 men, including a rocket group of undisclosed size. One rocket unit, commanded by Cummer-dien Khan, had 120 men; another under Purneah, 131 men. On 22 April, twelve days before the main battle, rocketeers worked their way around to the rear of the British encampment, then "threw a great number of rockets at the same instant" to signal the beginning of an assault by 6,000 Indian infantry and a corps of Frenchmen, all directed by Mir Golam Hussain and Mahomed Hulleen Mir Mirans. The rockets had a range of about 1,000 yards. Some burst in the air like shells. Others, called ground rockets, on striking the ground, would rise again and bound along in a serpentine motion until their force was spent. According to one British observer:

> The rockets make a great noise, and exceedingly annoy the native cavalry in India, who move in great bodies; but are easily avoided, or seldom take the effect against our [British as opposed to Indian units attached to the British] troops, who are formed in lines of great extent, and no great depth.

The diary of a young English officer named Bayly gives a somewhat different picture of the rockets' effectiveness. "So pestered were we with the rocket boys that there was no moving without danger from the destructive missiles . . . ." He continued:

> The rockets and musketry from 20,000 of the enemy were incessant. No hail could be thicker. Every illumination of blue lights was accompanied by a shower of rockets, some of which entered the head of the column, passing through to the rear, causing death, wounds, and dreadful lacerations from the long bamboos of twenty or thirty feet, which are invariably attached to them.

Soon after their Indian experience, the British began developing rockets themselves. At the Royal Laboratory of Woolwich Arsenal, Colonel (later Sir) William Congreve initiated a series of experiments with incendiary barrage rockets. Congreve had been told that the British at Seringapatam had "suffered more from them [the rockets] than from the shells or any other weapon used by the enemy." In at least one instance, an eyewitness told Congreve, a single rocket had killed three men and badly wounded four others. It seemed to him that this might be a good weapon to use against the French. However, the weapon would have to be improved.

"In the year 1804," wrote Congreve in *A Concise Account on the Origin and Progress of the Rocket System*, "it first occurred to me, that, as the projectile force of the rocket is exerted without any re-action upon the point from which it is discharged, it might be successfully applied, both afloat and ashore, as a military engine . . . . I knew that rockets were used for military purposes in India; but that their magnitude was inconsiderable, and their range not exceeding 1000 yards." He then designed and built a 2,000-yard rocket, which he proposed be used in combat, as part of a "plan for the annoyance of Boulogne."

Congreve described his 32-pound rocket in clear, semitechnical terms. Its "carcass is the largest of the kind that has hitherto been constructed for use [apparently he did not know of von Geissler's 120-pounders]; it is completely cased in a stout iron cylinder, terminating in a conical head; it is 3 feet 6 inches in length, 4 inches in diameter, and weighs, when complete, 32 pounds . . . . The stick is 15 feet long, and 1½ inches in diameter, and is so constructed, that it may be firmly attached to the body of the rocket, by a simple and quick operation, at any required time." The rocket contained "about seven pounds of carcass composition," and cost one pound sterling. Congreve spoke of 13,109 rockets having been manufactured up to August 1806, and mentioned briefly of having experimented with 42-pounders with ranges of 4,000 to 5,000 yards.

Congreve's proposal to attack Boulogne was accepted by the British military. Ten launches were fitted with his incendiary rockets and, on 18 November 1805, they assembled off the city. The attack itself was scheduled for 21 November, but a sudden storm came up with such violence that the commander was "compelled to recall the vessels without a rocket having been fired." Five of the launches were swamped before they could retire from the bay.

The next attempt was planned for late spring or early summer of 1806, the flotilla to be outfitted with new 32-pound iron-case rockets capable of 3,000-yard ranges. To insure maximum accuracy, Congreve attached to each a 15-foot guiding stick. The attack was postponed until the fall when, on 8 October, eighteen boats with rockets aboard rowed into the bay. "In about half an hour above 2,000 rockets were discharged. The dismay and astonishment of the enemy were complete—not a shot was returned—and in less than ten minutes after the first discharge, the town was discovered to be on fire."

Even more spectacular than the attack on Boulogne was the barrage of some 25,000 Congreve rockets on Copenhagen in 1807. According to Baron Eben, who was in the city shortly after the bombardment, the "Danes were very much afraid of the rockets, and said they had burnt a great many houses, and besides, warehouses . . . ."

The British also used Congreve's rockets against the island of Aix at about the same time, then in 1809 against Callao, in 1810 against Cadiz, and in 1813 against Leipzig. From 1818 the British Army possessed an official rocket brigade, and other nations began to follow Britain's example. The Austrians formed a similar unit, supplied with rockets from a large factory at Wienerisch-Neustadt. The Russians also were active in war rocketry under the leadership of military engineers Alexander Zasyadko and Konstantin I. Konstantinov. Test fired in St. Petersburg in 1817, Zasyadko's rockets became the equipment of a special army unit. Subsequently, they were put into production at Russia's first rocket manufacturing plant, established in 1826 in St. Petersburg, and were used during the Russo-Turkish war from 1828 to 1829 and later in the Caucasus.

Congreve's rockets were employed frequently in the War of 1812 between Britain and the United States. Their best publicized moment came during a bombardment of Baltimore's Fort McHenry on the night of 13–14 September 1814, when a young lawyer named Francis Scott Key immortalized the spectacle of "the rocket's red glare" in a verse that later became the national anthem of the United States.

*In their first large-scale use of military rockets, the British fired 2,000 rockets, invented and built by William Congreve, on the city of Boulogne in 1807. Below are (left to right) 300-, 100-, 42-, 32-, 24-, and 18-pound Congreve rockets.* (ROTUNDA MUS. AND NATIONAL AIR AND SPACE MUS.)

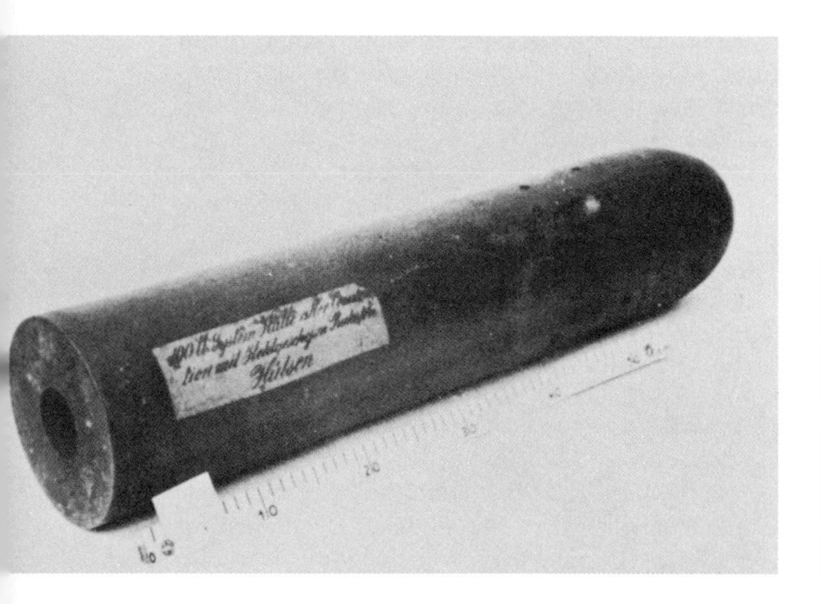

*The Austrians also developed military rockets, which they used during the mid-nineteenth century. Left, a 100-pound hollow-core Army Ordnance rocket. The length of its body is about 50 cm. Right, a painting in the Herres-museum, Vienna, shows Austrian troops firing stick rockets during battle.* (ROLF ENGEL COLLECTION, NATIONAL AIR AND SPACE MUS.)

Although dramatic, rockets were not particularly effective in this engagement. The bombardment was continued, more or less steadily, for twenty-five hours, but only four Americans were killed and twenty-four wounded.

The rockets witnessed weighed about 30 pounds and carried incendiary charges. They were fired from

*Immortalized in verse by Francis Scott Key, the "rocket's red glare" was seen frequently in the War of 1812. In the picture below, the 20-gun sloop* Erebus *(right), which had been converted by Congreve to a rocket ship, fires on the American Fort Washington on the Potomac River.* (NEWPORT NEWS MUS.)

the *Erebus*, a 20-gun sloop that had been converted under Congreve's direction into a rocket-firing bombardment vessel. The *Erebus* had some twenty long boxed frames extending from square openings, known as "scuttles," cut in the side of the ship. The boxes protected the interior of the ship from sparks and flames. Within them were large metal rocket-firing tubes. The tubes were fired by pulling lanyards.

Two basic Congreve designs were employed during the War of 1812: (1) case-shot rockets used as a substitute for, or as an auxiliary to, artillery; and (2) rockets loaded with inflammable materials whose purpose was to start fires.

The first, or case-shot, type contained carbine balls, which flew out like shrapnel when a charge of powder exploded. The rockets, when used with infantry, weighed from 3 to 12 pounds and were fired from a prone position. The rockets also could be fired from adjustable tripod stands, mounted on the decks of ships or in their rigging. Often they were fired from small boats, including those propelled solely by oars. They had a range of up to 3,000 yards.

First use of such rockets apparently was made by Rear Admiral Sir George Cockburn in the Chesapeake Bay area. Subsequently Lieutenant (later Sir) James Scott conducted an attack on shore targets from a boat propelled by oars. In describing the launching of the rockets he wrote:

> By good luck [for they were an uncertain weapon] in the first flight I let off, one of them fell directly into the

block-house and the other alighted in one of the batteries under it. Moving to the remainder of the boats, our gallant leader headed the attack and got possession of the batteries before the enemy could recover from the panic occasioned by the rockets.

Relating what it was like to be under rocket fire, one of the greatest American heroes of the war, Commodore Joshua Barney, wrote:

One of the enemy's rockets . . . fell on board one of our barges and after passing through one of the men, set the barge on fire and a barrel of gun-powder, and another of musket cartridges caught fire and exploded by which several of the men were blown into the water and one man very severely burned.

In another engagement on 8 June 1814, the British fired rockets against Barney's ships on the Patuxent River at the mouth of St. Leonard's Creek. The accuracy of the rockets was poor and they did little damage, but their range was greater than that of cannon and Barney's men were unnerved by the attack. On 10 June, the British did succeed in sinking two of his barges with rockets.

Some two months later, on 24 August 1814, during the Battle of Bladensburg, a seesaw fight between the British 85th Light Infantry Regiment and United States Attorney General William Pinkney's rifle battalion was quickly turned into an American rout when the British put their rockets into action. A special rocket squad, partially concealed in underbrush along the banks of a stream, fired the projectiles and caused such a panic among the Americans that they retreated headlong. "Never did men with arms in their hands make better use of their legs," wrote Lieutenant George R. Gleig, who commanded the British forces.

As the nineteenth century advanced, rocket designers concentrated on improving the weapon's accuracy. The normal way to control the flight direction of the early rockets was by stick. Congreve's incendiary 3.5-inch rockets, for example, were guided by a 15-foot stick attached to the case by hoops. Experiments in Britain, France, and the United States were next aimed at getting rid of the cumbersome stick by introducing a screw-shaped head. An American inventor named Court worked on the idea of constructing rockets so that the exhaust impinged on surfaces inclined to the main axis, producing spin. William Hale, an English inventor, gained fame for his further development of spin-stabilized rockets, which were subsequently used in Europe and Asia, and by the United States during the Mexican War in 1846–1848. Imparting spin to rockets was the first step since the introduction of the stick towards improving their accuracy.

Efforts also were made to increase range, but Congreve's standard rockets, which flew about 3,000 yards, remained pretty much in a class by themselves. The Swiss developed 6-pound rockets which were fairly accurate at 1,800 to 1,900 yards and, at 1,100 yards, could register three hits out of five attempts. American-made Hale-type rockets flew somewhat over 2,000 yards; 2.25-inch models weighed 6 pounds, while the larger 3.25-inch models weighed 16 pounds.

The American Army made limited use of rockets during the Mexican War.

On 19 November 1846, Major General Winfield Scott was selected to lead the United States expedition to Veracruz and then to Mexico City; his force included, among other elements, a brigade of rocketeers. By 4 December, recruiting posters were out urging "active, brave young men to serve with rocket and mountain howitzer batteries, now preparing by the Ordnance Department for immediate departure." Training took place at Fort Monroe, Virginia.

The battery, which included the rocketeers, was placed under the command of First Lieutenant George H. Talcott. Brevet Second Lieutenant Jesse Lee Reno commanded the rocketeer contingent. Its one hundred and fifty members and their equipment (including fifty 2¼-inch, 6-pound Hale rockets) sailed from Fort Monroe on 1 February 1847, on the bark *Saint Cloud.* It is believed that the rocketeers joined General Scott's forces at the island of Lobos some two hundred miles north of Veracruz toward the end of February. They sailed to Anton Lizardo and then to Sacrificios three miles southeast of Veracruz. On 9 March the landing took place, with sixty-seven surf boats each carrying between seventy and eighty men, among them the rocketeers. The troops quickly advanced to the city, which was placed under siege. Beginning on 24 March rockets were used against Veracruz's fortifications, contributing to their surrender on 29 March.

On 8 April the rocketeers moved inland, having been transferred from General William Scott Worth's to General David Twiggs's division, and advanced along a route discovered by Captain Robert E. Lee. The rocket battery was set up at La Atalaya after its occupation. Under Second Lieutenant Reno, thirty rockets plus forty rounds of spherical case-shot were fired in action, leading to the capture of El Telegrafo Hill on 18 April. Later, in August, rockets were again used in battles around Mexico City, particularly at Churubusco. And, during the storming of Chapultepec on 12 and 13 September, they proved their worth in softening up Mexican positions, keeping the defenders under a steady hail of fire. In 1848, the rocketeers were disbanded. Little is known of Mexican use of rockets during the engagement, though

Ordnance reports list Congreve rockets being in inventory with Santa Anna's forces.

During the period between the Mexican War and World War II Hale and Congreve rockets declined in importance, partly because of storage problems. When the Mexican War rockets were taken out of storage during the Civil War, it was found that their black powder charges had not maintained their bond with the cases. Rockets were used during the Civil War, but only sporadically and indecisively.

The Confederates under Jeb Stuart fired rockets at McClellan's troops at Harrison's Landing on 3 July 1862. Colonel James T. Kirk, 10th Pennsylvania Reserves, recalled that "on Thursday, the 3rd instant, while standing in the line of battle, I had one man wounded by a missile from a rocket fired from a rebel battery." The rockets were later reported to have been fired from "a sort of gun carriage." The Confederates also placed rocket batteries in service in Texas during 1863–1864. Both rockets and launchers were manufactured first at Galveston and later at Houston.

The first Union combat group to be given rockets was the New York Rocket Battalion. Organized by a British officer, Major Thomas W. Lion, it consisted of one hundred and sixty men. Their rockets were from 12 to 20 inches long and 2 to 3 inches in diameter; ranges were from a third of a mile to three miles. Accuracy was poor.

Light carriages with four wrought-iron tubes about 8 feet long could be used; or, alternately, 3¼-inch guiding rods bound together in an open framework. Sheet-iron launchers with 3-inch hollow tubes were also popular. The payloads of the Union rockets usually contained a highly inflammable compound, but occasionally musket balls were placed in a hollow head and exploded by time fuses. The New York soldiers were issued the equipment in March 1862, but never had the opportunity to use it in combat. However, in 1864, rockets were fired by Union troops under General Alexander Schimmelfennig in South Carolina, who found them "especially practical in driving the enemy's picket boats off the creek and, during the night, out of the harbor."

*In 1849, to celebrate the Peace of Aix-la-Chapelle, the British used rockets to light up the Thames.*

*In 1841, Charles Golightly was caricatured in the saddle of a flying machine that he designed, but never built or tested. He did receive considerable notice in the press—much of it satirical.*

Congreve's rockets did more than attract and occasionally inspire the military. In 1841, Charles Golightly received a British patent for a flying machine propelled by a steam rocket, a discovery that aroused great interest—much of it satirical. A model of the device was probably never constructed or tested. Like many inventors before him, Golightly was too far ahead of his time: almost ninety years would have to pass before man took to the air in a rocket-powered airplane.

Some modern writers refer to an even earlier "manned-rocket" concept. According to Nicolai A. Rynin (*Mezhplanetyne Soobshcheniya*, Vol. II, Part IV, Chap. 2, p. 10), a Chinese mandarin named Wan-Hoo, in about A.D. 1500, took "two large parallel horizontal stakes, which were tied together by a seat placed between them. Under this apparatus he placed forty-seven rockets which were fired simultaneously by forty-seven servants. However, the rockets under the mandarin's seat exploded irregularly and from the resulting fire unfortunately the inventor was consumed." This story, however, may be just a legend; neither Rynin nor anyone else who has mentioned the experiment has supplied any documentation for it.

Nineteenth-century experimenters also found new nonmilitary uses for rockets. In the early 1800's, rockets were developed to fire lifelines to stranded ships over which breeches buoys could be sent to rescue passengers and crew members. After some fifty years of line-carrying rocket history, Lieutenant Colonel E. M. Boxer of the Royal Laboratory in Britain developed, in 1855, a device consisting of two rocket cases so joined that when the first case had expended its propellant the second ignited. This tandem arrangement gave the rocket a much longer range than earlier models, and made it more effective in mercy missions. The Boxer rocket was kept in inventory by the British Board of Trade until well after World War I. Signal rockets also came to be a standard part of every ship's equipment. And whaling rockets came into use. The "California Whaling Rocket," for example, was made by Fletcher, Suits & Company of San Francisco and is described in the following terms:

> Our apparatus consists of a gun metal cylinder, filled with a peculiar composition made only by ourselves, to which is attached, in front, a bomb with a barbed point; inside the bomb is an explosive charge and a chain toggle, which is released by the bursting of the shell on entering the whale; an iron shaft is attached to the rear of the rocket, through which the whale line is spliced. There is absolutely no recoil . . . the hinged flange is thrown up by the rocket passing out, protecting the face from injury.

The manufacturer went on to boast that the device could kill whales at thirty fathoms (attested by a list of ten whaling captains "who recommend them to all parties interested in the whaling business").

Generally speaking, however, interest in rockets declined once the spur of war was removed. By the end of the nineteenth century, rocket research was being carried on by only a few experimenters.

Pedro A. Paulet, a Peruvian chemical engineer, is reported to have conducted experiments in Paris from 1895 to 1897 with a small, 200-pound-thrust rocket motor made of vanadium steel. He was forced to discontinue his work because of economic difficulties and his neighbors' complaints. For some unknown reason, however, Paulet did not report on his work until 7 October 1927, in Lima's *El Comercio*. A Russian engineer living in Germany, Alexander B. Scherschevsky, learned of the article which he subsequently summarized in his book *Die Rakete für Fahrt und Flug* (*The Rocket for Travel and Flight*), published in Berlin in 1929. According to Scherschevsky, the Paulet rocket's propellants were nitrogen peroxide and gasoline; ignition was by a spark gap in the combustion chamber, and tests were satisfactory. The motor "weighed a little over 5 pounds, producing its 200 pounds of thrust at 300 explosions per minute." Paulet claimed it could be operated for an hour without "suffering appreciable deformation."

If it had not been for Scherschevsky, Paulet would probably have gone unnoticed. As it was, the Peruvian's experiments caught the attentions of later German authors and subsequently those of writers on rocketry all over the world, with the result that he became widely accepted as the pioneer of liquid propellant rocketry. The 7 October 1927 "article" in *El Comercio* was a 2½-column letter written by Paulet from Rome in which he claimed "priority" for his invention. After drawing attention to the many plans for rocket airplanes and spaceships then current in Europe, Paulet said he had conceived such ideas "THIRTY YEARS AGO [*sic*] when I was a student at the Institute of Applied Chemistry at the University of Paris." He expressed the fear that his claims would not be believed and called upon his former student friends in the Latin Quarter to tell the world of his experiments, which were, nevertheless, "made, truly, without witnesses . . . ." Paulet died on 30 January 1945 with his claims still unconfirmed.

In Austria, Dr. Franz von Hoefft, of Vienna's

*An enterprising New Yorker found still another use for the rocket, as advertised in the* Whalemen's Shipping List and Merchants' Transcript.

CHANTS TRANSCRIPT. AUGUST 8, 1865.

PENJAMIN LINDSEY,
U. S. CONSUL, ST. CATHERINES. BRAZIL.
WILL furnish Beef, Vegetables, Groceries and other supplies for whaling vessels touching at that port, on the most favorable terms. Letters forwarded to his care will be received and delivered as addressed.
—REFERENCES—
Messrs. David R. Greene, & Co., Gideon Allen & Son, Swift & Perry, Henry Taber & Co., Thomas Knowles, & Co., Edward C. Jones, Jonathan Bourne, Jr., David B. Kempton. } New Bedford
jy28'63 tf

WRIGHT & CO.,
COMMISSION MERCHANTS,
RIO DE JANEIRO, BRAZIL.
REPRESENTED BY OUR AGENT,
JOHN S. WRIGHT, ESQ., No. 69 Wall Street, NEW YORK.

C. BREWER & CO.,
HONOLULU, S. I.
OFFER the services of their House, established in this place in 1828, as GENERAL SHIPPING AND COMMIS-

PATENT ROCKET HARPOONS AND GUNS.

FASTEN TO AND KILL INSTANTLY WHALES OF EVERY SPECIES.
WITH PROPER LINES AND BOATS,
SUCH AS WERE USED BY THE OFFICERS OF BARK REINDEER IN 1864,
ALL WHALES ARE SAVED.
N. B.—Two Months notice required to fill an Order for the Season of 1865.
—FOR SALE BY—
G. A. LILLIENDAHL, - - - - - - - - - - NEW YORK

*In 1906 Alfred Maul successfully took aerial photographs by attaching a camera to this rocket.*

Gesellschaft für Hohenforschung (Society for Altitude Research), proposed plans in 1928 for the development of rocket motors. And, in Germany, Wilhelm Gaedicke performed some preliminary studies of rocket-powered airplanes. At the turn of the century, rockets were being fired into clouds and exploded, hopefully to prevent hailstorms. Alfred Maul successfully lofted cameras to high altitudes with rockets and then took pictures of the Earth below, but his work was discontinued as the airplane arrived on the scene.

The ebb in interest in rockets that had begun in the nineteenth century continued into the first quarter of the twentieth, although the reasons for the decline changed. Lack of war as an incentive to weapons development was no longer a factor. Instead, it just seemed that the rocket had become obsolete. With the advent of radio, rockets lost importance as a method of signaling. Militarily, the rocket could no longer compete with artillery, where rifled barrels, breech loading, and other new techniques led to great increases in range and accuracy.

In the First World War, nevertheless, the Allies did make minor use of rockets, primarily for signaling and illuminating enemy positions. In an article on "The Use of Rockets and Illuminating Shells in the Present War," appearing in the July 1918 issue of the *Journal of Acetylene Lighting,* A. Bergman describes two kinds of rockets in common use at the front. One is the "ordinary, well known type that sails ahead of a tail of fire, and which finally bursts in a brilliant flash, illuminating a large area for a few seconds." The second type of rocket contained in its head a parachute to which a flare was attached. "Through this arrangement it is possible to keep a brilliant burning star suspended in the air for a comparatively long time, which is generally fixed to be about 30 or 35 seconds."

Both on land and at sea World War I rockets found service in laying smoke screens. As weapons of destruction they saw limited use, mostly in France.

*During World War I, Le Prieur rockets were sometimes fired from French and British biplanes or from the ground against German captive balloons. Otherwise, military rockets could not compete in range or accuracy with artillery of the day. Left, a Le Prieur rocket is fired from a BP12 in a ground test. Right, an H. Farman F40P has five such rockets on each side of the fuselage.* (IMPERIAL WAR MUS.)

*While the war was going on in Europe, Robert H. Goddard was beginning a long career of rocket research in the United States. As professor of physics at Clark University, Worcester, Mass., he conducted some early experiments on the campus lawn. Left, Professor Goddard beside a 1915 rocket chamber mounting in Worcester. Right, he loads a 1918 bazooka with a 3-inch projectile at the Mount Wilson Observatory, Calif.* (ESTHER C. GODDARD)

The French developed Le Prieur rockets (named after Naval Lieutenant Y. P. G. Le Prieur, who invented them) that were fired either from Nieuport (or other) airplanes or from the ground against German observation balloons. Normally, a biplane would carry four or five rockets mounted on each side of the fuselage.

In the United States, some work was done on short-range combat rockets. Dr. Robert H. Goddard, the father of modern rocketry, developed some rockets that were test fired just a few days before the war ended. His work was to bear fruit later, but in the frenzied atmosphere of World War I, it was generally overlooked.

At the same time, the first halting steps were taken toward development of guided missiles. There was no connection yet with rockets; it was something of a daring leap forward just to imagine that airplanes could be guided, without pilots, to a point where they would release their bomb loads on the enemy. The payoff was to come later, when the idea of a guided bombardment drone was wed to the propulsive force of the rocket.

During 1917, under the direction of Charles Kettering, the Delco and Sperry companies began to experiment with what appears certain to have been the first United States guided missile, a pilotless biplane known as the "Bug." Made largely of wood, the little plane weighed 600 pounds (including a 300-pound bomb payload) and was powered by a 40-horsepower Ford engine. Takeoff was from a four-wheel carriage running along a portable track. Flight direction was controlled by a small gyro, and altitude by an aneroid barometer. When target distance and wind conditions were determined, the "number of revolutions of the engine required to take the Bug to the target was calculated and a cam was set." Once the engine had propelled the missile the required distance, the cam dropped into position. The bolts that fastened the wings to the fuselage were pulled in and the wings detached, dumping the missile on its target. The Bug was tested successfully in 1918, before Army Air Corps observers in Dayton, Ohio.

A group led by Professor A. M. Low had already started a similar project in the United Kingdom. Low later recalled that the project was dubbed "A.T." so that people would think it was an "Aerial Target." The project was conceived in 1914 when Generals Caddell and Pitcher and Sir David Henderson, then Director-General of Military Aeronautics, proposed that radio could be used to direct a "flying bomb" to its target.

Low put together a team consisting of a captain (Poole) and two lieutenants (Bowen and Whitten).

*The first United States guided missile had nothing to do with rocketry. Rather, it was a pilotless plane, built mostly of wood and successfully tested in 1918. Powered by a 40-hp Ford engine, it took off from a four-wheel carriage running along a portable track. After the plane had flown a predetermined distance, the wings dropped off, dumping the missile on the target.* (U.S. AIR FORCE)

*The British experimented with unmanned airplanes that would be directed by radio to fly bombs to their targets. The plane at right, designed by De Havilland, was powered by a 35-hp engine designed by Granville Bradshaw. It was flight tested on 21 March 1917.* (IMPERIAL WAR MUS.)

*In 1927, the British developed the Larynx, a radio-controlled missile, which they flight tested from the HMS* Stronghold. *It could carry a 250-pound bomb at 200 mph to a target 100 miles away.* (ROYAL AIRCRAFT EST.)

*The Queen Bee (left), introduced in 1930, was followed by the Queen Wasp (right). Both were launched by catapult from naval vessels or landing installations. After completing their missions, they returned and landed on pontoons.* (ROYAL AIRCRAFT ESTABLISHMENT)

Experiments began at Brooklands and continued at Feltham, where the work force increased to forty. After much trial and error the radio equipment was developed and a monoplane constructed by De Havilland powered by "a beautiful opposed engine made by that outstanding designer Granville Bradshaw." Two tests were made in March 1917 at the Royal Flying Corps training school field at Upavon but "the first machine had engine failure on the runway and flopped ungracefully into the mud," according to a 1952 article by Low in *Flight.* (Major Gordon Bell, an observer of the "flight," was heard to say, "I could throw my bloody umbrella farther than that!") The next trial turned out better, Low continues, "for the machine took off, flew under control for a short time until, after a loop, the engine failed . . . [then] and with an appalling crash, the A.T. landed about three yards from where I was sitting and buried most of our beautiful work in the ground. But it had flown and it had worked."

Although work on the A.T. project terminated, British interest in missiles persisted until, in 1927, engineers at the Royal Aircraft Establishment developed a radio-controlled missile called the Larynx. A monoplane, it was flight tested both from the HMS *Stronghold* and from a testing ground in Egypt. At a speed of up to 200 miles per hour, it could carry its 250-pound bomb to targets as far as 100 miles away. In 1930, the Queen Bee was introduced, followed by the Queen Wasp, both launched by catapult from naval vessels or from landing installations. After undertaking their missions they would return, landing on pontoons.

During the Spanish Civil War, 1936–1939, the rocket staged a brief, and somewhat unusual, appearance. Converted sea-rescue rockets were placed into service for the purpose of transporting propaganda materials behind enemy lines. The nosecone was especially constructed so that it would burst open at a predetermined time and release its payload of propaganda leaflets, which were printed on very thin paper.

Although the Spanish Civil War was regarded as a testing ground for a coming world war, the peripheral role given to rockets provided a poor indication of the important role they were to play in the next few years. While practical use of rockets languished after World War I, a few men, working in obscurity and with limited funds, had laid the foundation for a theoretical and technical revolution. The pioneers of rocketry were about to receive the credit due them.

# 3 PIONEERS OF SPA

The idea gradually dawned around the turn of the twentieth century that the rocket was the key to space travel. Only a few individuals grasped this concept, and no one paid much attention to them at first. But this discovery was a landmark in human thought. At last man had the answer to a problem that had intrigued and baffled him for centuries. The discovery opened the universe to human exploration.

With the benefit of hindsight, it seems strange that the discovery did not come sooner. Men had known about military rockets for centuries, and the reaction principle was commonly employed in fireworks. But the fiction writers who dealt with space travel gave no indication, even on the rare occasions that they mentioned rockets in their tales, that there was a scientific basis for their use. The scientists, meanwhile, simply ignored the problem.

The potential of the rocket was realized independently by three different men, born in widely separated countries, who never even saw each other. Yet these men—Konstantin Eduardovitch Tsiolkovsky of Russia, Robert Hutchings Goddard of the United States, and Hermann Oberth, a Hungarian-born (Transylvanian) German—each came to the same conclusions about the future of space travel, conclusions that have become the basic working formulas of the space age.

Toward the end of the nineteenth century many technical advances were made that were destined to help translate the theories of Tsiolkovsky, Goddard, and Oberth into reality. Industrialization proceeded at a rapid pace. Enormous progress was recorded in metallurgy. Improved explosives became available and smokeless powder was invented. Scientists began to undertake searching investigations into heat engines (of which the rocket is an example) and learned how to liquefy gases that one day would be used as space vehicle propellants. Perhaps most important of all, the excitement of rapidly advancing frontiers of science and technology began to pervade the atmosphere, causing more and more young people to choose scientific and engineering careers, and leading universities to expand their scientific curricula.

The Russian scientist-schoolteacher Konstantin Eduardovitch Tsiolkovsky was the first to understand and develop the use of rockets in space travel. In his biography of the great pioneer, A. Kosmodemyansky writes that Tsiolkovsky "grasped the principle of obtaining motion by means of the reaction of ejected particles as early as 1883, but he created the mathematically precise theory of rocket propulsion only at the close of the century."

Tsiolkovsky really has only two potential rivals for the honor of being first—Nikolai Ivanovitch Kibalchich and Hermann Ganswindt. Considered more as historical sidelights, because neither contributed significantly to astronautical theory, these two men nevertheless deserve recognition for seeing an essential truth that others—including even Jules Verne—did not.

Kibalchich, born in 1853, became active in his twenties in anti-czarist circles, took part in the successful assassination plot in 1881 against Emperor Alexander II, was arrested shortly after the assassination, and sentenced to death. While spending his final days in jail, Kibalchich developed a scheme to propel a platform by rocket power. As the device flew, gunpowder cartridges would be fed continuously to the motor chamber. By changing the direction of the rocket motor's axis, the platform would change its flight path. Following the discovery of his proposal in police archives, excerpts were published in *Bylove* (*The Post*), 10 and 11, in 1918. Kibalchich wrote, "I am writing this project in prison, a few days before death. I believe in the practicability of my idea and this faith supports me in my desperate plight." This is the extent of his contribution. There are not even any indications, let alone records, of other studies Kibalchich may have carried out on rocketry and astronautics.

A little more is known about Ganswindt. He was a basement inventor who might be considered a German counterpart of Kibalchich. Around 1890, more than a decade before Tsiolkovsky's theories

# CE TRAVEL

were published, Ganswindt proposed a reaction-powered spaceship. Somehow he had stumbled across the idea that a reaction device would operate in space as well as on Earth; he could not explain why this should be, since he had no mathematical training. Ganswindt suggested that the spaceship be propelled by steel cartridges charged with dynamite (he failed to realize that combusting gases would have enough power). Each cartridge would be placed in a reaction chamber, one half being ejected by the force of the explosion and the other half striking against the top of the chamber to provide the reaction force. Below the chamber, suspended on springs, was the part of the spaceship that housed the crew. It had a center hole through which the "exhaust" of the motor passed.

Despite the credit given to Kibalchich and Ganswindt, Tsiolkovsky is universally regarded as the true pioneer of astronautical theory. As he himself expressed the development of his thoughts:

> For a long time I thought of the rocket as everybody else did—just as a means of diversion and of petty everyday uses. I do not remember exactly what prompted me to make calculations of its motions. Probably the first seeds of the idea were sown by that great fantastic author Jules Verne—he directed my thought along certain channels, then came a desire, and after that, the work of the mind.

Tsiolkovsky was born in September 1857 in the town of Izhevskoye, Spassk District, Ryazan Gubernia. Of humble origin, he showed an early great talent for science and invention and dedicated himself to the study of mathematics and physics. He read everything he could lay his hands on, and by his early teens, the germ of the idea of interplanetary travel had entered his restless mind. One day he conceived an admittedly impractical plan to send a vehicle into space which left him so "agitated, nay, shaken, that I could not sleep that night . . . . By morning I understood its futility, and the disillusionment was as great as the illusion had been."

But he persisted. He continued to read, teaching himself virtually everything he ever learned. In 1878 he became a "people's school teacher" and moved to Borovsk, in Kaluga Province. There he began to experiment in a home laboratory and to write reports on his findings. On the basis of his papers he soon was elected to the Society of Physics and Chemistry in St. Petersburg. And then, in 1883, he made the discovery that was to lead the world into the age of space flight.

An entry in his diary for 28 March 1883 shows a basic understanding of reaction flight.

> Consider a cask filled with a highly compressed gas. If we open one of its taps the gas will escape through it in a continuous flow, the elasticity of the gas pushing its particles into space will also continuously push the cask itself. The result will be a continuous change in the motion of the cask. Given a sufficient number of taps (say, six), we would be able to regulate the outflow of the gas we liked and the cask (or sphere) would describe any curved line in accordance with any law of velocities . . . . As a general rule, uniform motion along a curved line or rectilinear non-uniform motion in free space involves continuous loss of matter.

In the years that followed, he worked out the implications of his idea, refining it and putting it into scientifically acceptable form. "The old sheet of paper with the final formulae of a rocket device bears the date of 25 August 1898," he wrote long afterward. Five years later, in the journal *Naootchnoye Obozreniye* (*Scientific Review*) his first article on rocketry appeared. It was titled "Exploration of Space with Reactive Devices" ("Issledovanie Mirovykh Prostransty Reaktivnymi Priborami"). It had taken about two hundred years from the time Sir Isaac Newton expressed his law of action and reaction to the realization that a reaction device could enable man to escape from the planet Earth.

Tsiolkovsky worked alone with meager equipment and virtually no funds. His only assistance in pre-revolutionary times was in the form of a grant, totaling just 470 rubles, received in 1899 from the Academy of Science's Physics and Mathematics Department. He attempted no rocket-motor testing, concentrating on the theoretical aspects of reaction

*Russian scientist Konstantin Tsiolkovsky was the first to understand the use of rockets in space travel. Although he never built a rocket, he designed several and solved theoretically how reaction engines could escape from and re-enter the Earth's atmosphere.*

motion and interplanetary flight. He always kept his feet to the ground, despite his soaring thoughts. "I could never proceed without calculation. It was calculation that directed my thought and my imagination," he wrote.

*Konstantin Tsiolkovsky (left) in his home laboratory in Borovsk, Kaluga Province, Russia.*

*Tsiolkovsky's first rocket design, the 1903 spaceship, was powered by liquid hydrogen (H) and liquid oxygen (O). The explosives mix at* A, *producing heated gases, which expand and cool as they travel back through the tube, finally escaping at* B.

Tsiolkovsky not only solved theoretically such age-old questions as how to escape from the Earth's atmosphere and gravitational field, but he also described several rockets. The first, conceived in 1903, was to be powered by liquid oxygen and liquid hydrogen—a very modern propellant combination. "In a narrow part of the tube," he wrote, "the explosives mix, producing condensed and heated gases. At the other, wide, end of the tube the gases, rarefied and, consequently, cooled, escape through the nozzle with a very high relative velocity."

The Russian schoolteacher made another discovery—the *multistage rocket,* which he called the "rocket-train." Actually, this concept was not as new as Tsiolkovsky, who discovered it independently, thought; firework-makers had used the principle for at least 200 years. But Tsiolkovsky was the first to analyze the idea in a sophisticated manner. The multistage technique, he concluded, was the only feasible means by which a space vehicle could attain the velocity necessary to escape from the Earth's gravitational hold. His design for a "passenger rocket train of 2017" consisted of twenty single rockets, each with its own engines and propellants. More than 300 feet long, it was over 12 feet in diameter and was built in three layers of metal with quartz windows, a refractory material, and finally a highly refractory

*Tsiolkovsky's 1914 rocket spaceship was a further development of the 1903 model. The long curved tube led to the combustion chamber, which burned gaseous oxygen and liquid hydrogen.*

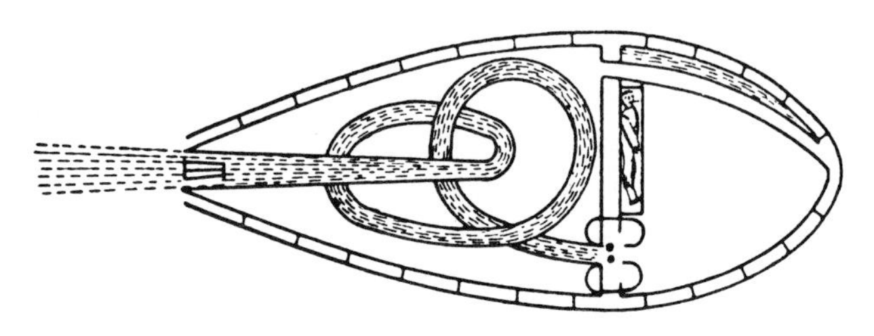

metal to protect the vehicle from the heat produced as it sped through the atmosphere. As each stage consumed its propellants it would be discarded, to keep the weight of the vehicle as low as possible. The next stage would take over the job of accelerating the spaceship, taking advantage of the velocity already given to it by the discarded first stage.

Tsiolkovsky described this important principle of rocketry in these words:

> If a single-stage rocket is to attain cosmic velocity it must carry an immense store of fuel. Thus, to reach the first cosmic velocity [his term for *orbital velocity*], 8 km/sec, the weight of the fuel must exceed that of the whole rocket (payload included) by at least four times. This will present considerable difficulties. The stage principle, on the other hand, enables us either to obtain high cosmic velocities, or to employ comparatively small amounts of propellant components.

Once he had worked out the basic principles of rocket dynamics, Tsiolkovsky devoted more and more time to speculations on space flight itself. It became very clear to him "that the device for moving in a void must be a kind of rocket, *i.e.*, be self-sufficient as regards both energy and the mass to thrust away from." He foresaw that a rocket could "navigate interplanetary space, interstellar space, visit planets or their satellites, rings, or any other celestial bodies, and then return to the Earth." He spent a great deal of time analyzing the components of rocket engines and calculating the energy contents of various propellant combinations, including liquid hydrogen, alcohol, kerosene, methane, and liquid oxygen.

As Tsiolkovsky's daring, yet carefully calculated, plans matured he was given increasing recognition. He was elected to the Socialist Academy (predecessor of the U.S.S.R. Academy of Science) in 1919, and later was granted a pension by the Soviet government. His writing continued unabated; from 1925 to 1932 alone some sixty works on astronautics, astronomy, mechanics, physics, and philosophy appeared. Tsiolkovsky died in Kaluga on 19 September 1935, two days after his seventy-eighth birthday.

*Tsiolkovsky's* Na Lune (On the Moon) *was published in Moscow in 1935. Shown here are the jacket of the book—it was bound with* Grezy o zemle i nebe, (Speculations on Earth and Heaven) *the only title on the cover—and the ship he described for flying through space.*

Although Tsiolkovsky's life spanned the entire period of the evolution of basic astronautical theory, he was not responsible for, and did not witness, the major practical developments in liquid-propellant rocket engines that took place in America and Germany in the 1920's and 1930's. There are many reasons why Russia did not capitalize on Tsiolkovsky's work, including political instability, lack of economic and technical resources, and failure of the military to appreciate the significance of his discoveries.

Tsiolkovsky's brilliant mind conceived theories and developed them, but did not lead him to perform practical experimental work. Ganswindt and Kibalchich also conceived an idea, but did virtually nothing to demonstrate its technical feasibility. But Robert H. Goddard combined theory and practice during an extraordinary, though often lonely, career that finally earned him the justified title "Father of Modern Rocketry."

Goddard clearly recognized that the entire sci-

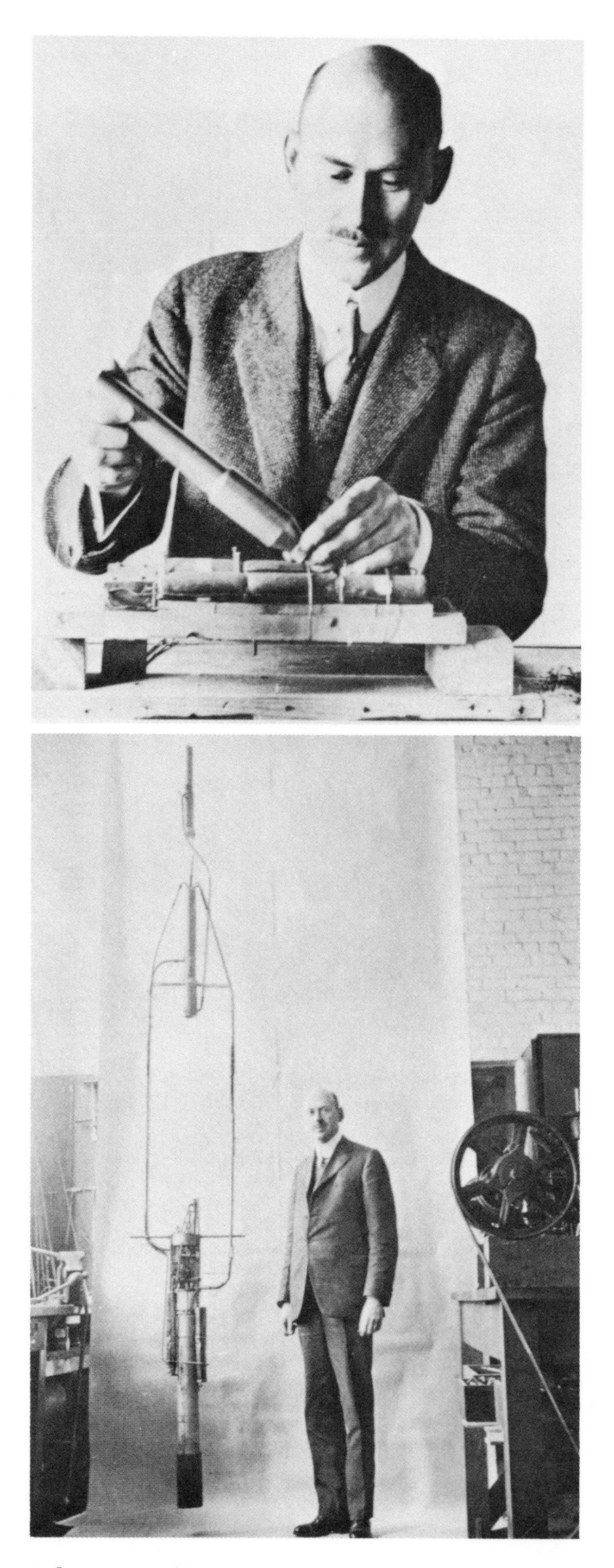

*Robert H. Goddard combined theory and practice in a long career of building and testing rockets. Here he is shown (top) with steel combustion chamber and nozzle for a 1915 rocket, and (bottom) standing beside a 1925 double-acting rocket in his experiment center in Worcester, Mass.* (ESTHER C. GODDARD)

ence of astronautics rested on the rocket propulsion system. Until the rocket was perfected there would be no trips through outer space, no landings on alien worlds. Goddard dedicated himself first and foremost to the rocket, although he never lost sight of its ultimate purpose.

Goddard was born almost exactly twenty-five years after Tsiolkovsky, on 5 October 1882, in Worcester, Massachusetts. He lived, studied, worked, and was buried there. During his life he was little known and less appreciated, yet his efforts had incalculable influence on the course of history. Only after his death did his country take notice of his genius.

As a boy Goddard showed an aptitude for science and engineering, with his interests channeled particularly into mathematics and physics. Along with science he found time to read such science-fiction classics as Wells's *War of the Worlds* and Verne's *From the Earth to the Moon;* and, like Tsiolkovsky, was inspired by them. In an autobiography written in 1927 (but only published in 1959 in the journal *Astronautics*), he acknowledged his debt to Wells and other science-fiction writers by recalling that they "gripped my imagination tremendously. Wells' wonderful true psychology made the thing very vivid, and possible ways and means of accomplishing the physical marvels set forth kept me busy thinking."

In 1902, while a student at South High School in Worcester, he submitted to *Popular Science News* an article titled "The Navigation of Space." It was not published. In a second article, he developed the scheme, as had Tsiolkovsky before him, of multistage spaceships. This article ended with the statement: "We may safely infer that space navigation is an impossibility at the present time. Yet it is difficult to predict the achievements of science in this direction in the distant future."

Cautious notes like this were to characterize Goddard's writings and statements on rocketry all his life. He had no doubt that the reaction principle underlying rocket motion eventually would permit man to explore the Solar System, "I began to realize that there might be something after all to Newton's Laws," he wrote. "[The Third Law] made me realize that if a way to navigate space were to be discovered or invented, it would be the result of a knowledge of physics and mathematics . . . ." But he was reluctant, as Tsiolkovsky and other European thinkers were not, to apply his full energies to the task of promoting space flight.

Meanwhile, Goddard's education continued. After graduation from Worcester Polytechnic Institute in 1908, he went on to Worcester's Clark University. He received a doctorate in 1911 and subsequently became professor of physics there. While a student and professor at Clark, Goddard accomplished an important portion of his work on rocketry. He began, in 1909, to make detailed studies and calculations of liquid-propellant engines, coming to the conclusion (again, like Tsiolkovsky) that liquid hydrogen and liquid oxygen would be an ideal combination. During a year at Princeton (1912–1913) he continued to work on the theory of rocket motion, further convincing himself that he was following a path that would one day reach to the stars.

Goddard kept detailed diaries of his activities, so that it is possible to follow closely the development of his rockets and to appreciate his method of thinking. His experiments and theories resulted in a succession of patents, most of which are basic to the operation of all modern rocket engines. For example, during July 1914, he was granted patents covering combustion chambers, nozzles, propellant feed systems, and multistage rockets ("a primary rocket, comprising a combustion chamber and a firing tube, a secondary rocket mounted in said firing tube, and means for firing said secondary rocket when the explosive in the primary rocket is substantially consumed").

As World War I approached, Goddard was involved in flight testing simple powder rockets near Worcester, some of which attained altitudes up to 500 feet; these tests soon suggested more elaborate experiments, ones that would remain out of the question unless he could find suitable financial support.

Attempting to obtain such support in September 1916, he wrote to the Smithsonian Institution outlining his work and his minimum financial requirements. After being requested to supply additional material to bolster his proposal, which involved a plan for making scientific measurements at high altitudes by means of rockets, he received on 5 January 1917 a grant of $5,000. Hardly a large sum, it was enough to allow him to begin in earnest his lifelong work. Soon, however, World War I caused a temporary change in his plans.

Instead of continuing with his high-altitude research, Goddard went to California to work on military rockets, including a forerunner of the World War II bazooka. In September 1918, Goddard showed two Signal Corps officers several rockets that were ready for production. One of them could be fired by a doughboy in the trenches; the largest could carry an 8-pound payload three quarters of a mile. Goddard's rockets were demonstrated at Aberdeen, Maryland, on 7 November 1918, just a few days before the Armistice. He had models weighing 5, 7½, and 50 pounds, which were fired from 2-inch and 3-inch tubes, 5½ feet long.

The end of the war meant an end of military interest in rockets; Goddard returned to Clark soon after the Armistice.

In 1919, and again in 1936, Goddard published two basic monographs, both appearing as Smithsonian Miscellaneous Collections. The first, titled *A Method of Reaching Extreme Altitudes*, was essentially the study Goddard had submitted to support his request for his first grant; the second, called *Liquid-Propellant Rocket Development,* was a report to the Daniel and Florence Guggenheim Foundation, which continued his financial support. His final major work, published posthumously in 1948, was entitled *Rocket Development: Liquid-Fuel Rocket Research, 1929–1941.*

The most famous of these three works is the first, a sober, learned exposition of the fundamentals of rocketry whose sections carry such titles as "Reduction of Equation to the Simplest Form," "Efficiency of Ordinary Rocket," "Calculations Based on Theory and Experiment," and "Calculation of Minimum Mass to Raise One Pound to Various Altitudes in the Atmosphere." It would have doubtless gone unnoticed by all but a tiny segment of the academic community were it not for the inclusion, at the end of the report, of the section: "Calculation of Minimum Mass Required to Raise One Pound to an 'Infinite' Altitude."

"Infinite altitude" meant space flight—escape from the Earth, a subject never far from Goddard's mind. And space flight in his day meant the Moon, a subject which he got around to after several paragraphs of preliminaries. Goddard broached the subject

*From this 7-foot frame (small structure, right, next to ladder) on a farm in Auburn, Mass., Goddard successfully tested the world's first liquid-fuel rocket on 16 March 1926. The rocket, which stood 10 feet tall, accelerated to a speed of 60 mph and flew 184 feet. The wooden frame at the left is a wind breaker.* (ESTHER C. GODDARD)

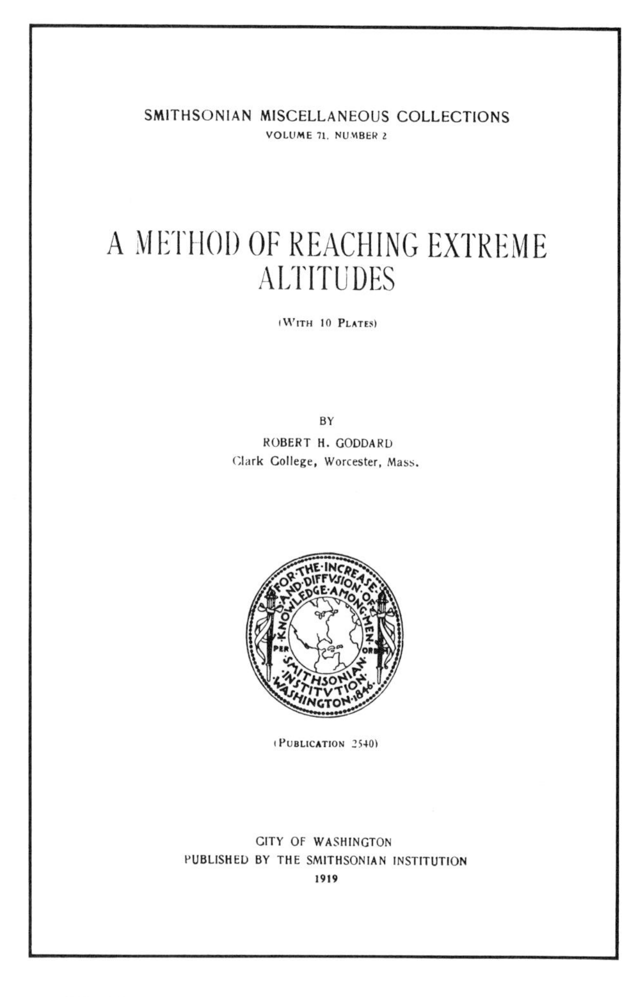

SMITHSONIAN MISCELLANEOUS COLLECTIONS
VOLUME 71, NUMBER 2

A METHOD OF REACHING EXTREME ALTITUDES

(WITH 10 PLATES)

BY
ROBERT H. GODDARD
Clark College, Worcester, Mass.

(PUBLICATION 2540)

CITY OF WASHINGTON
PUBLISHED BY THE SMITHSONIAN INSTITUTION
1919

*Goddard's most famous work was probably his least understood. A scholarly treatise on the fundamentals of rocketry, it contained a final section on how rockets could get to the Moon. The contents of the final section were sensationalized by newspaper reporters, who referred to Goddard as the "Moon man."*

obliquely, first noting that it would be interesting "to speculate upon the possibility of proving that such extreme altitudes had been reached even if they actually were attained." He realized that proving it would be "a difficult matter," even if a mass of flash powder were ignited at the peak of the trajectory, for "it would be difficult to foretell, even approximately, the direction in which it would be most likely to appear."

Then he came to his point: "The only reliable procedure would be to send the smallest mass of flash powder possible to the dark surface of the moon when in conjunction [*i.e.*, the 'new' moon], in such a way that it would be ignited on impact. The light would then be visible in a powerful telescope." He went on to calculate the amount of flash powder needed to be "just visible" and "strikingly visible" to a one-foot aperture telescope and to work out the total initial mass of the launching rocket.

Characteristically, he closed his report with words of caution:

This plan of sending a mass of flash powder to the surface of the moon, although a matter of much general interest, is not of obvious scientific importance. There are, however, *developments of the general method under discussion, which involve a number of important features not herein mentioned,* which could lead to results of much scientific interest. These developments involve many experimental difficulties, to be sure; but they depend upon nothing that is really impossible.

Goddard, so used to working quietly and alone, was completely unprepared for what occurred next. Newspaper editors seized on the statements in the paper's last section and sensationalized them, turning the professor from Massachusetts into the "moon man." Goddard, who had sought only the attention of fellow scientists, was profoundly irritated by the wave of publicity, some of which made him a butt of jokes. He finally decided that all he could do was remain silent until the newspapers lost interest, which they inevitably did.

The initial Smithsonian grant was used up by the summer of 1920, but Goddard's request for continued support brought the promise of another $3,500. At about the same time, arrangements were made for him to work for the United States Navy's Bureau of Ordnance-Indian Head Powder Factory in Maryland, where he remained from 1920 to 1923. There emphasis was placed on rocket depth-charges and rocket-boosted armor-piercing projectiles. Returning to Worcester, he carried out serious studies of liquid and solid propellants as well as of stabilization and guidance. From 1925 he concentrated on these all-important phases of rocket technology.

On 23 November 1929, occurred one of the most important events in Goddard's life—an unexpected visit from Colonel Charles A. Lindbergh, who had read of Goddard's work and was fascinated with its potentialities. Lindbergh, the world-famous aviator, subsequently arranged for a $50,000 grant to the rocket pioneer from the Daniel Guggenheim Fund for the Promotion of Aeronautics, to be paid through Clark University. A smaller grant from the Carnegie Institution was earmarked for test facilities.

Realizing that Massachusetts was too crowded for him to conduct the type of flights he now envisioned, Goddard went west to search for a suitable location—and found one at the Mescalero Ranch near Roswell, New Mexico. He, his wife, and four assistants set up shop there in 1930. From then until 1941, except for a break in 1932–1934, Goddard undertook one of the most amazing "lone-wolf" development programs in the history of technology.

To demonstrate the extent of his experimental efforts, a chronology of Goddard's major static and flight tests has been prepared. The material is derived from his second monograph, *Liquid-Propellant Rocket Development,* which covers activities from the time of his 1919 Smithsonian report to September 1935; *Rocket Development,* which brings his work up to 1941; and other sources. Mrs. Robert H. Goddard was kind enough to review this material and write a few paragraphs explaining it:

Despite the fact that the data for the accompanying list of milestones in the Goddard rocket researches were checked against several sources, all the figures may not be quite accurate. The measurement of pressures, flows, temperatures, and altitudes in the experiments of the early years was difficult. For example, the weight of liquid oxygen used in each test was hard to define, especially for flight tests, for some oxygen was used to cool the parts before the test, some evaporated in the tank during the final flight preparations, and occasionally some was left in the tank after the test. Further, the rocket was so small that it was not always possible, in the wide New Mexico sky, to ascertain its maximum height, with either the recording telescope or the moving picture camera.

Reliability of propulsion, stability in flight, and recovery were the primary aims in these early tests, rather than the attainment of high altitudes. Heights of one-half to one mile served our purposes. It should be borne in mind that no telemetering or other electronic instruments were then commercially available; our early measuring instruments, including barographs and recording telescopes, had to be made in our own shop. By July 17, 1941, however, the instruments had been sufficiently developed to give the details shown. My husband felt that his altitudes, speeds, and jet velocities were consistently on the low side in most of the tests, especially in the K series.

A few summaries of the number of static and flight tests have been made at points where the tests seemed to culminate, in order that the ratio of proving-stand tests to flight tests might be made clear; the addition of the average intervals between tests has been made because of frequent inquiries as to how often my husband ran his tests in New Mexico.

Descriptions of the pre-Roswell tests, and of the government work after we left Roswell, have been kept at a minimum. Only the New Mexico liquid-propellant period is fully tabulated, because only this Roswell work represents his full-time, self-directed experimental work.

*This Goddard rocket, tested in 1927, was equipped with a turntable for launching and a parachute for descending.* (ESTHER C. GODDARD)

*In this experiment, conducted in 1928, the test rocket was caught at the top of the tower.* (ESTHER C. GODDARD)

### Tests at Worcester, Massachusetts, 1915–1929

| | |
|---|---|
| 1915–1922 | Many tests were conducted with rockets using black and smokeless powders at Clark University and, during World War I, for the United States Signal Corps in California. In 1918, Goddard developed a prototype of the World War II bazooka rocket. Other work with solid-propellant rockets was undertaken for the United States Navy in 1920–1923. |
| 1920–1922 | Finding that solid propellants were inadequate for reaching the altitudes his theoretical studies had indicated were possible, Goddard turned to liquid oxygen and gasoline (he had mentioned using these liquids in his 1919 paper "A Method of Reaching Extreme Altitudes"). |
| 1923–1924 | A liquid-oxygen pump and engine were developed that, although they worked, were too small to be satisfactory. |
| 1924–1925 | Work progressed on a displacement cylinder model, two pumps and two engines, a powder igniter, and refractory-lined combustion chambers and nozzles. |
| 6 December 1925 | In order to reduce weight, Goddard returned to the simple pressure feeding of the liquids. At the Clark University physics laboratories he performed a 24-second test (thrust was great enough to raise the rocket during the last half of the run). |
| 30 December 1925 | Rocket trembled in its support for 8 seconds. |
| 3 January 1926 | Rocket again quivered for 17 seconds. |
| 20 January 1926 | The thrust produced raised the rocket motor the full distance it could travel; outdoor tests with a flight model commenced on 8 March 1926. |
| 16 March 1926 | *First flight* of a liquid-propellant rocket. Altitude: 41 feet; average velocity: 60 miles per hour; in air: 2.5 seconds. |
| 3 April 1926 | *Second flight* of a liquid-propellant rocket. Landed 50 feet from test stand after being in air 4.2 seconds. |
| 5 May 1926 | After several tests indicating the model was too small to permit refinements, it was decided to build a rocket twenty times larger. During 1926 a new tower was built and flow regulators, multiple liquid injection into large combustion chambers, means for measurement of pressure and lifting force, electrically fired igniter, and turntable for rotation were developed. |
| 18 January 1927 | The new large rocket was placed into the test tower for the first time. Rocket and turntable were lifted several times in the following months, but no flights took place with the large model. |
| 3 September 1927 | Construction began on a rocket one fifth the size of the previous one, *i.e.* four times larger than the 1926 flight models. Few, simple, easily replaced parts were used, and the fuel-injection system was improved greatly. |
| 18 July, 29 September, 10 and 20 October 1928 | Rocket started to rise, but it tipped and caught in the tower. |
| 26 December 1928 | *Third flight* of a liquid-propellant rocket. Rocket rose out of tower rapidly and tipped passing over observation shelter. Range was 204.5 feet, velocity over 60 miles per hour. This flight was followed by a series of static tests to develop liquid "curtain cooling" for inside of combustion-chamber wall, a regenerative cooling system, and better in-flight stability. |
| 17 July 1929 | *Fourth flight* of a liquid-propellant rocket. Started to lift at 13 seconds, rose at 14.5 seconds, reached top of trajectory at 17 seconds, hit ground at 18.5 seconds, landing 171 feet away. Flight was bright and noisy, attracting much public attention. |

### Tests at Camp Devens, Massachusetts, 1929–1930

| | |
|---|---|
| 3 December 1929– 30 June 1930 | Tests were carried out at an artillery range at Camp Devens, 25 miles from Worcester. Their object was to improve liquid-propellant rocket motor efficiencies, particularly the "curtain cooling" aspect. Sixteen static tests were made, but no flights. A few special tests were conducted with rocket-operated propellers. |

### Tests at Roswell, New Mexico, 1930–1932

1930–1932 — In July 1930 the project was moved to New Mexico under the auspices of Daniel Guggenheim of New York City, permitting Goddard to devote full time to rocket work. A series of thorough static tests in which the operating conditions were varied was first undertaken. For flight testing, a 5.75-inch diameter, 5-pound combustion chamber was used. In static testing, it produced a maximum thrust of 289 pounds over more than 20 seconds. Thrust was steady, exhaust velocity was over 5,000 feet per second. Owing to the Depression, the Guggenheim financial support ended in June 1932.

30 December 1930 — *Fifth flight* of a liquid-propellant rocket. Vehicle was 11 feet long, weighed (empty) 33.5 pounds. It rose to an altitude of 2,000 feet and a speed of 500 miles per hour. A gas pressure tank was employed to force the liquid oxygen and gasoline propellants into the combustion chamber.

29 September 1931 — Flight took place with jacket, streamline casing, and remote control. Rocket length was 9 feet 11 inches; diameter 12 inches; loaded weight 87.2 pounds, and empty weight 37 pounds. The rocket was in the air for 9.6 seconds, reached 180 feet, followed a trajectory described as "like a fish swimming."

13 October 1931 — Flight effected with simplified combustion chamber, parachute releasing. The 7.75-foot-long, 12-inch-diameter rocket reached more than 1,700-feet altitude, gave loud whistling noise on descent.

27 October 1931 — Flight took place with rocket using new gasoline shut-off valve. The dimensions of the rocket were similar to above, altitude was 1,330 feet, range 930 feet, total flight time 8.6 seconds.

23 November 1931 — Static test with modified oxygen-injection system used. Thrust was 270 pounds for 11 seconds; exhaust velocity 5,088 feet per second.

19 April 1932 — This flight rocket had pressure generated by liquid nitrogen, stabilization effected by gyro-controlled vanes. Length was 10 feet 9.5 inches; empty weight 19.5 pounds; altitude 135 feet; in air for 5 seconds.

### Tests at Clark University, Worcester, Massachusetts, 1932–1934

September 1932 — A grant from the Smithsonian Institution enabled Goddard, who resumed his full-time teaching at Clark University that fall, to carry out experiments that did not require flight testing.

September 1933 — A more extensive program was made possible in 1933–1934 by a grant from the newly founded Daniel and Florence Guggenheim Foundation. During these years, studies were made of insulators, welding methods for light metals, gyroscopic balancers, reciprocating and centrifugal pumps, jet pumps, and rocket chambers.

### Resumption of Tests at Roswell, New Mexico, 1934–1941

*A-series tests*

From September 1934 through October 1935 an A-series of tests was made with rockets using simple pressure-feed systems and stabilized by gyro-controlled blast vanes. The rockets were between 13.5 feet and 15 feet 3.25 inches long. Their empty weight varied from 58 to 85 pounds.

16 February 1935 — Flight test accomplished without automatic guiding device. Flight was short, rapid; parachute was released, checking the fall.

8 March 1935 — Flight test with equalizer (to prevent liquid-oxygen tank pressure from exceeding gasoline tank pressure), pendulum stabilizer, and 10-foot parachute. The motor fired for 12 seconds, producing small white flame; velocity of more than 700 miles per hour achieved (may have been supersonic). Rocket tilted to horizontal, landed 9,000 feet from tower.

28 March 1935 — Flight with improved gyro stabilization; rocket was 14 feet 9.75 inches long, weighed (empty) 78.5 pounds, reached altitude of 4,800 feet, range of 13,000 feet. Rocket corrected its path perfectly several times during 20-second flight, which was made at an average speed of 550 miles per hour.

**Resumption of Tests at Roswell, New Mexico, 1934–1941 (continued)**

31 May 1935 — Flight effected with new lift-indicator. Rocket length, 15 feet 1.5 inches; weight, 84 pounds; altitude, 7,500 feet; range, 5,500 feet. Stabilization excellent. Loud whistle produced on descent; rocket made 10-inch-deep hole on impact.

25 June 1935 — Flight test with new timing device for parachute and a cushioned gyro. Day was windy. Flight lasted 10 seconds. Rocket reached 120-foot altitude, tipping into wind as it left the launching tower.

12 July 1935 — Flight, with stronger and thicker air vanes. Motor fired 14 seconds; rocket reached 6,600-foot altitude with excellent correction up to 3,000 feet. Parachute was torn off.

29 October 1935 — New gasoline orifices used in this flight. Duration of thrust was 12 seconds, altitude reached was 4,000 feet, velocity was high. Rocket shot produced a wave of dirt, resembling a water wave, when it landed. Resulting hole was 6 inches deep.

*K-series tests*

From 22 November 1935 to 12 February 1936, Goddard worked on the K-series of tests, which consisted of ten proving-stand experiments designed to lead to the development of a more powerful 10-inch-diameter motor. Two of the most outstanding tests are listed below:

17 December 1935 — Rocket produced a thrust of 496 pounds for 14 seconds, with exhaust velocity of 4,470 feet per second. The liquid oxygen weighed 31 pounds, gasoline weighed 24 pounds, and the rocket weight was 225 pounds.

12 February 1936 — Attained thrust of 623.5 pounds for 4 seconds. Exhaust velocity was 4,340 feet per second. Liquid oxygen weighed 25.5 pounds; gasoline weighed 13.5 pounds.

*L-series tests*

This series involved thirty tests, divided into sections A, B, and C. The experiments were conducted from 11 May 1936 to 9 August 1938 and used nitrogen-pressurized rockets with 10-inch-diameter motors.

*Section A tests*

From 11 May 1936 to 7 November 1936, nitrogen-pressurized flight rockets were developed based on the 10-inch motors employed during the K-series of experiments. Rockets numbered L-1–L-7 were used, with lengths varying from 10 feet 11 inches to 13 feet 6.5 inches, diameters 18 inches, empty weight 120 to 202 pounds, loaded weights 295 to 360 pounds, liquid-oxygen weight about 78 pounds, gasoline 84 pounds, nitrogen 4 pounds.

31 July 1936 — Flight test. Altitude attained was 280 feet, duration 5 seconds, range 300 feet.

3 October 1936 — Flight rocket reached 200-foot altitude vertically in 5 seconds, at which time chamber burned through. Weight of liquid oxygen, 40 pounds; of gasoline, 46 pounds; and of nitrogen, 4 pounds.

7 November 1936 — Flight effected with cluster of four combustion chambers, each 5.75 inches in diameter. Rocket length: 13 feet 6.5 inches. It climbed to about 200 feet, fell to Earth near tower.

*Section B tests*

From 24 May 1936 to 19 May 1937; rockets L-8–L-15 were used in these tests. Experiments were made on 5.75-inch-diameter chambers with propellants of various volatilities. Other experiments involved the development of tilting cap parachute release, tests of various forms of exposed movable air vanes, tests of retractable air vanes, and tests of parachutes with heavy shroud lines.

18 December 1936 — Flight test with pressure storage tank used. Duration and altitude not recorded, but range was 2,000 feet, rocket landing with axis horizontal. Noise heard up to eight miles away, parts scattered over 300-foot area, most being recovered undamaged.

1 February 1937 — Flight with gyro, air, and blast vanes. Rocket was 16 feet 7.63 inches long; diameter was 9 inches. Firing time 20.5 seconds. Rocket reached altitude of 1,870 feet and corrected well. The ground behind the flame deflector turned green and was glazed by heat.

| | |
|---|---|
| 27 February 1937 | Rocket flight with new parachute release operated by gyro. It reached 1,500-foot altitude, and landed 3,000 feet from tower. Flight duration was 20 seconds, speed very high. |
| 26 March 1937 | Larger, movable air vanes used in this flight. Rocket soared to 8,000–9,000 feet (duration 22.3 seconds), corrected while propulsion lasted, then tilted. |
| 22 April 1937 | Flight test with larger movable air vanes; reinforced parachute. Rocket length: 17.75 feet; diameter: 9 inches; duration: 21.5 seconds. Rocket could not be followed to top of trajectory as it was nearly overhead. It landed about a mile from tower. |
| 19 May 1937 | Flight rocket had streamlined, retractable air vanes, wire-wound pressure storage tank (to reduce weight). Length was 17.67 feet; diameter, 9 inches; altitude achieved, 3,250 feet; and duration, 29.5 seconds. Stabilization was much improved. |

*Section C tests*

From 28 July 1937 to 9 August 1938; rockets L-16–L-30 featured light tank construction, movable tailpiece (*i.e.* gimbal) steering, catapult launching, and further developed liquid nitrogen tank-pressurization technique. The lengths of the rockets in this series varied from 17 feet 4.25 inches to 18 feet 5.75 inches, while diameters were 9 inches. Loaded weights were 170 pounds or more, empty weights 80 to 109 pounds. Static-test thrusts ranged from 228 to 477 pounds, exhaust velocities from 3,960 to 5,340 feet per second. The tests indicated extremely high temperatures for the exhaust. Pebbles of the cement gas deflector were fused and thrown out, starting fires more than 50 feet from the tower.

| | |
|---|---|
| 28 July 1937 | Flight of rocket with movable tailpiece steering. Wire-wound tanks used, barograph carried. Rocket length was 18 feet 5.5 inches, diameter 9 inches, loaded weight 162 pounds 5 ounces, empty weight 95 pounds 5 ounces. Carried 39 pounds of liquid oxygen, 28 pounds of gasoline. Flight lasted 28 seconds, rocket reached 2,055 feet. The parachute opened near the ground, checked speed. Rocket coasted ⅛ of ascent, landed 1,000 feet from tower. |
| 26 August 1937 | Flight had movable tailpiece steering; catapult launching. Length was 18 feet 5.5 inches, diameter 9 inches, loaded weight 162 pounds. Corrected well and strongly seven times during the flight, which took rocket to an altitude of more than 2,000 feet. |
| 24 November 1937 | In a short flight, rocket leaned after leaving tower, fell 100 feet away. Thrust was low. |
| 6 March 1938 | Rocket quickly left tower, producing little smoke; reached 500 feet before starting its coasting period. |
| 17 March 1938 | Rocket reached 2,170 feet in 15-second flight. Went vertical to 800 feet, then leaned to right, landing 3,000 feet from tower. Little smoke produced. |
| 20 April 1938 | Flight resulted in 4,215-foot altitude; duration of propulsion 25.3 seconds, landed 6,960 feet from tower. Rocket carried official barograph. Weight of liquid oxygen, 21.5 pounds; of gasoline, 34 pounds. |
| 26 May 1938 | Altitude of only 140 feet reached; on leaving tower rocket veered to right, landed 500–600 feet away. |
| 9 August 1938 | An altitude of 4,920 feet was recorded by telescopic observations, but according to the barograph carried in rocket only 3,294 feet reached. Rocket corrected well, and parachute opened at top of trajectory. Ground and telescope observers felt that rocket went considerably higher than indicated by the barograph. |

*Experiments leading towards development of propellant pumps*

From 17 October 1938 to 28 February 1939; models PT1–8 and P1–4 are associated with Goddard's developmental work on propellant pumps. After the successful flights of 9 August, Goddard turned again to the problem of pumps. He believed pumps were essential if very high altitudes were to be attained. Beginning on 17 October 1938 he made a thorough study (more than 20 proving-stand tests) of five models of small, high-speed centrifugal pumps, which had radically new features. This initial phase ended on 17 November 1938. Then, from 6 January through 28 February 1939 two pumps, called A and D, were selected for use in four proving-stand tests. From these tests it was concluded that a small chamber or gas generator, producing warm oxygen gas, should be developed to operate the turbines. Of these tests the following was the best:

**Resumption of Tests at Roswell, New Mexico, 1934–1941 (continued)**

7 February 1939 — With 29 pounds of liquid oxygen and 45 pounds of gasoline, a thrust of 671 pounds was produced for 12 seconds with an exhaust velocity of 4,820 feet per second. The liquid oxygen flowed at 2.15 pounds per second, the gasoline at 2.28 pounds per second; the mixture ratio was .94.

*Gas Generator tests*

From 24 March to 28 April 1939, 11 static tests were made near the shop of a new gas generator to drive turbines (models P5a–k). The best generator developed ran steadily for 10 seconds at 180 pounds per square inch pressure at greater than 250 pounds per square inch tank pressure, with liquid oxygen flow rate at 0.49 pounds per second. Later, from 18 May to 4 August 1939, this generator was used in eight static tests (P5–12) at the desert launching tower. The best two tests were:

17 July and 4 August 1939 — These runs gave thrusts of 700 pounds for about 15 seconds, with oxygen flow rates at 4 pounds per second and gasoline flow rates at 3 pounds per second. The exhaust velocities achieved were in excess of 3,200 feet per second.

*Static and flight tests with pump-driven rockets*

During the period 18 November 1939 to 10 October 1941, a series of 24 static and flight tests was made with rockets offering a large fuel capacity (models P13–P36). These vehicles used the rocket motors, pumps, and turbines that had been developed previously. They averaged 22 feet in length, were 18 inches in diameter, and weighed (empty) from 190 to 240 pounds. They carried some 140 pounds of liquid oxygen and 112 pounds of gasoline.

2 December 1939 — Static test at flight tower, steady 40-second run with thrust of 760 pounds.

15 May 1940 — Another static test, flame hot, thrust apparently high. Ground behind the flame deflector seemed to have melted.

11 June 1940 — A steady run of 43.5 seconds achieved in this static test, making it longest to date; red-hot stones seen to fly up out of cement gas deflector.

9 August 1940 — *First rocket flight with pumps:* rocket reached 300-foot altitude at very low velocity (10 to 15 miles per hour).

6 January 1941 — A static test made recording highest thrust to date: 985 pounds.

8 May 1941 — *Second rocket flight with pumps:* rocket reached 250 feet, then heeled away from tower.

17 July 1941 — In this static test, duration was 34 seconds; average thrust 825-plus pounds; exhaust velocity 4,060 feet per second; average thrust per pound of propellant per second 128; mechanical horsepower 3,040; weight of liquid oxygen, 131.7 pounds; weight of gasoline, 91.5 pounds; and ratio of oxidizer to fuel, 1:43.

**Summary of Goddard Static and Flight Tests**

| *Series and period* | *Static tests (number)* | *Flight tests (number)* | *Average interval between tests (days)* |
|---|---|---|---|
| First New Mexico series, 1930–1932 | 21 | 8 (5 left tower) | 20 |
| A-series, September 1934–October 1935 | 1 | 14 (7 left tower) | 28 |
| K-series, November 1935–February 1936 | 10 | 0 | 8 |
| L-series | | | |
| Section A, May–November 1936 | 4 | 3 (all left tower) | 25 |
| Section B, November 1936–May 1937 | 2 | 6 (all left tower) | 22 |
| Section C, July 1937–August 1938 | 7 | 8 (all left tower) | 25 |
| Pump tests, October 1938–February 1939 | > 24 | 0 | 5.5 |
| Gas generator tests, March–August 1939 | 19 | 0 | 7 |
| Pump-turbine tests, November 1939–October 1941 | 15 | 9 (2 left tower) | 28 |

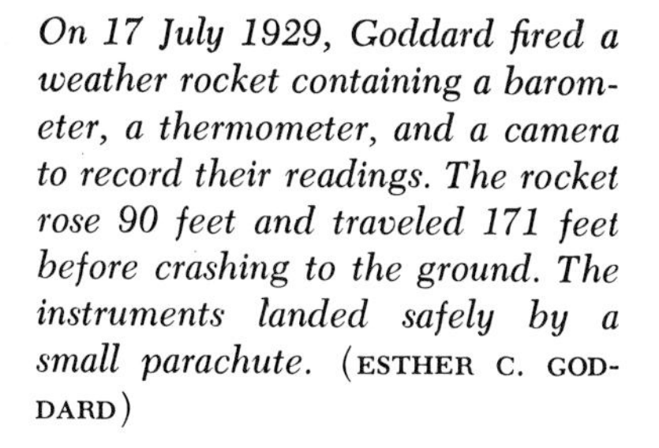

*On 17 July 1929, Goddard fired a weather rocket containing a barometer, a thermometer, and a camera to record their readings. The rocket rose 90 feet and traveled 171 feet before crashing to the ground. The instruments landed safely by a small parachute.* (ESTHER C. GODDARD)

In a letter to C. N. Hickman on 14 February 1937, Goddard emphasized how different working with liquid oxygen and gasoline propellants was from the old "multiple charge [solid propellant] days." He described his rockets thus: "the present rockets weigh between 90 and 100 lbs, have about 50 lbs of propellant, and fire steadily for 20 to 30 seconds. The present thrusts are from 200 to 300 lbs, and with some large chamber tests last month we had a thrust of about 800 lbs. The present chambers weigh about 6 lbs, which isn't bad, considering the above lift." He noted that his " 'staff' consists of Mrs. Goddard, and four men, besides me . . . ." He added that "It is, as you can imagine, a fascinating life. The drawback is that until there has been a great and spectacular height reached, no layman, and not many scientists, will concede that you have accomplished anything, and of course there is a vast amount of spade work, of much importance, that must be done first."

Several years later, in response to requests from L. T. E. Thompson of the Naval Proving Ground, Dahlgren, Virginia, Goddard addressed himself to the problem of rocket-accelerated all-purpose bombs. In a letter to Thompson on 24 February 1940 he described a combustion chamber operating under 350 pounds per square inch pressure that produced 700 pounds of thrust. The maximum diameter was 6 inches and length 2½ feet. Weight was but 6¾ pounds. Two high-speed centrifugal pumps supplied the propellants to the chamber from the liquid oxygen and gasoline tanks. He said that rapid starting

> was not needed, for flight tests. Nevertheless even with the design we are using at present, the rise to full working pressure is not much over a second. The appearance is rather striking. One moment everything is quiet and no flame is visible. The next moment there is a blast of flame and the entire tower shakes, which continues undiminished as long as the rocket is in the tower.

On 5 June 1940 Hickman wrote Goddard saying that he felt it his duty "to call the attention of our government to the possibilities of your rocket for defensive purposes." Goddard promptly replied, "Go ahead, and God bless you!" explaining that on a "recent trip East," the Guggenheim Foundation's attitude had changed and that he and Harry F. Guggenheim had met with a joint committee of Army and Navy officials in Washington on 28 May. Goddard presented information on both solid- and liquid-fuel rockets. The Army was "unsympathetic towards any long-range projectile," while the Navy had sev-

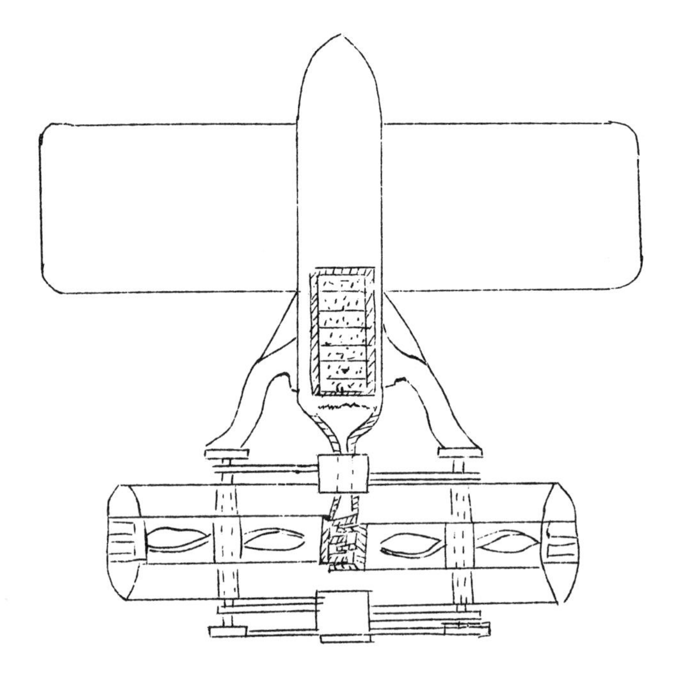

*A rocket-propelled airplane designed by Goddard and patented by him on 9 June 1931.*

eral ideas for applications of liquid rockets. The general attitude toward rocket research in Washington, however, seemed negative.

"Frankly," Goddard wrote, "I have been filled with disgust at the fact that no intensive fundamental work appears possible, and I suspect I have been hard to live with since my return. I am, however, endeavoring to make a few more attacks to see if it is not possible to carry on the work on the liquid fuel rockets, at least, on an intensive scale."

Following the meeting on 28 May, Goddard and Guggenheim met with Brigadier General George H. Brett, of the Air Corps Materiel Division. This session resulted in a proposal by Goddard on 27 July 1940 that his rocket technology be applied to the problem of assisted takeoff for bombers and other airplanes. A little over three months later, on 26 September, General Brett wrote that, while the Air Corps was "deeply interested in the research work being carried out by your organization under the auspices of the Guggenheim Foundation, it does not, at this time, feel justified in obligating further funds for basic jet propulsion research and experimentation." He added that when Goddard's experiments had "reached a point which indicates the probability of successful reduction to practice of a device, capable of being incorporated in or attached to an airplane to assist in accelerating takeoff and upon which an evaluation can be made in order to determine the feasibility and practicability of military application, the Air Corps will then entertain further proposals involving the actual construction, installation, and test of such device." Goddard was to produce on his own; then the Air Corps would become interested. Goddard later commented that "after trying to do a good piece of work over a period of years and actually getting flights before anyone else, it is discouraging to have the implication made that nothing of value has been accomplished."

In 1941, however, the mood of the government

*Goddard's later experiments, at a research center near Roswell, N.M., became increasingly more sophisticated. At left is a four-motor L-7 rocket that flew to a 200-foot altitude on 7 November 1936. Above, a pump-driven rocket tested on 1 August 1940.* (ESTHER C. GODDARD)

changed. Goddard's group began work in September under contracts with the Navy's Bureau of Aeronautics and with the previously uninterested Army Air Corps. In July 1942 the personnel and equipment were moved to the Naval Engineering Experiment Station at Annapolis, Maryland, where they continued until July 1945. During this period a liquid-propellant, jet-assisted takeoff unit for flying boats was developed and flight tested. A still more important assignment was the development of variable-thrust rocket motors, which required hundreds of proving-stand tests before a successful motor was finally achieved.

Recognition came late to Goddard, much of it after he had died in Baltimore on 10 August 1945. In 1959 he was honored by the United States Congress and received the first Louis W. Hill Space Transportation Award of the now defunct Institute of Aeronautical Sciences. A year later the Smithsonian Institution, under whose auspices he had worked for so many years, bestowed on him the coveted Langley Medal. One of the National Aeronautics and Space Administration's major facilities, the Goddard Space Flight Center, was named after him on 1 May 1959. And in 1960 the United States government awarded the Guggenheim Foundation and Mrs. Goddard $1,000,000 in settlement for government use of more than two hundred of the pioneer's patents.

*The first published work of German pioneer Hermann Oberth,* The Rocket into Planetary Space, *appeared in Munich in 1923. Scarcely larger than a pamphlet, the 92-page text explained theoretically how rockets can launch payloads into orbit around the Earth. The book's cover is shown here.*

The third great pioneer of rocketry and astronautics, Hermann Oberth, was born in the town of Hermannstadt, Transylvania (which had been incorporated into Hungary), on 25 June 1894. The son of a German-speaking family, Hermann Oberth is considered German rather than Hungarian; he studied in German, writes in German, later moved to Germany, and in his adult life became a German citizen.

At the time of Oberth's birth, Tsiolkovsky was in his middle thirties while Goddard had not yet reached his teens. For a long time the three remained ignorant of one another, partly because of differences in age, partly because of the distances between them, partly because they wrote in different languages.

Like his predecessors, Oberth's interest in space flight received its initial stimulus from the great science-fiction writers of the nineteenth century, particularly Verne. In his autobiography, he wrote that "At the age of eleven, I received from my mother as a gift the famous [Lunar] books . . . by Jules Verne, which I had read at least five or six times, and, finally, knew by heart." Even at this age he realized that Verne's scheme of firing the astronauts to the Moon by cannon was not feasible, for "the travelers inside the missile would have been crushed without pity by the enormous acceleration." He started to look for alternative means of reaching the satellite.

Oberth's first ideas were far-fetched—a magnetic acceleration device in a long tunnel from which the air had been evacuated, an airplane with silk propellers, a large wheel developing high centrifugal forces—but he persisted until he came upon the reaction principle. Again, he acknowledged a debt to Jules Verne, who in *Around the Moon* had suggested using rockets to reduce the fall of the spaceship toward the Moon and to permit maneuvering in space. But he wrote, "I should tell a lie in stating that I was delighted at this discovery. I was not pleased at all with the enormous fuel consumption, the hazards of rockets containing solid fuels, the difficulty of handling liquid fuels, the high costs of the chemicals, etc."

Shortly before World War I, Oberth became interested in combat rockets. In 1917, he proposed to the German War Department the development of a liquid-propelled, long-range bombardment missile. After the war, in 1922, he made similar proposals and added speculations on the feasibility of space flight.

Oberth learned of the publication of Goddard's 1919 report through newspaper accounts. Unable to obtain a copy in Germany, he wrote Goddard for one in 1922, adding, "I think that only by the common work of scholars of all nations can be solved this great problem . . . to pass over the atmosphere of our earth by means of a rocket."

Oberth's first book, the now-celebrated *Die Rakete zu den Planetenräumen* (*The Rocket into Planetary Space*), was published in Munich the following year. Oberth added a three-page supplement to assure his readers that if they compared Goddard's 1919 publication with his own "one can easily see that I have worked completely independently." More than thirty-five years later, Oberth was still bothered by the thought that he might be accused of having been inspired by Goddard, emphasizing in a 1959 article that "I had carried on my investigations completely independently of Goddard's work."

*Die Rakete zu den Planetenräumen* is a small book—with but 92 pages of text, scarcely larger than a pamphlet. But it is a thorough discussion of almost every phase of rocket travel, including the abnormal effects of pressure on the human body.

Oberth's little book demonstrated many of the truths that today are taken for granted—that a rocket can operate in the void, and that it can move faster than the velocity of its own exhaust gases. He realized that it would be possible for a rocket to launch a payload into orbit around the Earth if the required velocity could be generated, a thought which led him, like Goddard and Tsiolkovsky, to investigate many propellant combinations. He also described in detail the design of a rocket, which he called the Modell B, that he felt could be used to explore the upper atmosphere; discussed the merits of alcohol and hydrogen as rocket propellants; and included a section on applications of rocket techniques.

The book rapidly went through its first edition, and Oberth began work on his *Wege zur Raumschiffahrt* (*The Road to Space Travel*), a 423-page expansion of the first book which was published in Munich in 1929. These two books were important not only for the many genuinely new thoughts they contained about the problems of space flight, but also for the inspiration they gave other scientists to work on rockets, a drive that has led directly to today's space achievements.

Unlike Goddard, Oberth did everything he could

*Above, a scene from Fritz Lang's film* Frau in Mond *showing the arrival of the spaceship on the Moon.* (ARFOR ARCHIVES)

*Below, an interior scene of the spaceship en route to the Moon, showing the effect of weightlessness.* (CINÉMATHEQUE FRANCAISE)

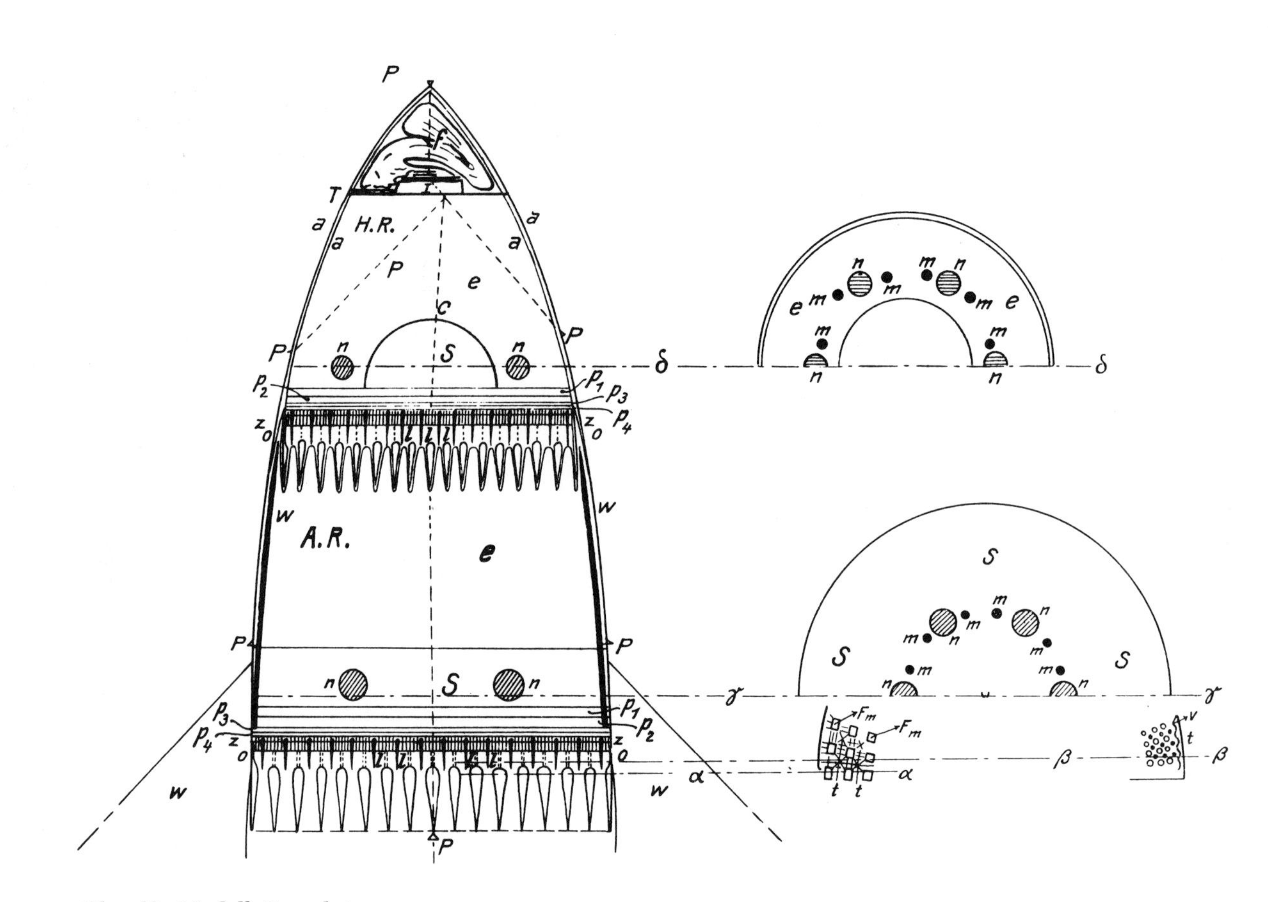


*Oberth's Modell E rocket*

to publicize rocketry in general and his own work in particular. He became a technical advisor to the Ufa Film Company and director Fritz Lang, who was filming a movie called *Frau im Mond* (*Girl in the Moon*). As a stunt dreamed up by publicity agents, Oberth, with Rudolf Nebel and Alexander B. Scherschevsky as his assistants, was to design, build, and fly a rocket. More a theoretician than an engineer, Oberth had little success. The rocket was actually designed and constructed, but it quickly became apparent that many basic component changes would have to be made if the rocket were ever to fly. A small test motor built by Oberth, the Kegeldüse, was successfully static-tested on 23 July 1930 in a suburb of Berlin. In September, Ufa's test funds ran out and Oberth returned to Transylvania to teach again.

Oberth was always active in movements to forward the use of rockets, and encouraged others to work, to lecture, and to write on rocket and space flight. He joined (and in 1929 became president of) the Verein für Raumschiffahrt, Germany's Society for Space Travel. Although to many almost a legend, he is still alive; since World War II he has written additional books, worked for a few years in the United States with his erstwhile student and helper Von Braun at the Army Ballistic Missile Agency, and returned to West Germany where he now lives in semiretirement, still consulting and writing on space flight.

The period spanned by the lives of the three great pioneers, two in Europe and one in America, is over one hundred years—from Tsiolkovsky's birth in 1857 to the present. Their significant contributions cluster between the mid-1880's, when Tsiolkovsky first developed the relationship between rocketry and astronautics, and the 1930's and the early 1940's—a period exceeding half a century. By the time he was in his early twenties, each man had made his basic discoveries and had laid down the foundations of his future activities—the two Europeans concentrating to varying extents on theoretical astronautics and the American on the practical development of liquid-propellant rocket motors.

What these three pioneers started, others were eager to finish. The impetus provided by their discoveries attracted dozens of eager engineers and scientists into rocketry and astronautical research. Major technical barriers still had to be overcome, but for the first time a sizable group of men had begun to take space travel seriously.

Professor Oberth entzündet die Kaltgas-Düse.

Mit mehr als 3 m hoher Flamme zischt die Riesenfackel der Düse gen Himmel, um ihre Kräfte auf den Hebel des Meßinstrumentes wirken zu lassen

## Vorarbeiten für die Weltraumrakete

Neue Versuche Professor Oberth's

Der Raketenforscher Professor Oberth hat jetzt gemeinsam mit dem Verein für Raumschiffahrt Flüssigkeitsraketen konstruiert, die sich auf dem Versuchsstand der Chemisch-Technischen Reichsanstalt bereits recht gut bewährt haben. Allerdings sind bisher lediglich Düsenbrennversuche durchgeführt worden, bei denen der durch den Rückstoß der mit einer Geschwindigkeit von 1700 bis 2000 m in der Sekunde ausströmenden Gase entstehende Druck experimentell gemessen wurde. Die aus der Düse ausströmenden Gase werden durch die Verbrennung von Benzin in reinem Sauerstoff erzeugt. Der Vorteil der Flüssigkeitsraketen besteht darin, daß sie längere Zeit betrieben werden können, während die bisher benutzten Pulverraketen schon nach wenigen Sekunden abgebrannt waren. Gleichzeitig kann man die Brennstoffzufuhr bei den Flüssigkeitsraketen in gewissen Grenzen regulieren. Bei den Versuchen in der Chemisch-Technischen Reichsanstalt wurden vorgeführt eine Kaltgas-Düse und eine Spaltdüse.

Die Spaltdüse wird montiert, um die beim Brennen entstehenden Kräfte zu messen

*Oberth failed in his attempt to construct a rocket for the German movie* Frau im Mond, *but he did build a motor, which he tested in a Berlin suburb. The above report on the test appeared in the newspaper* Boten aus den Riesengebirge *on 10 August 1930. At left, Klaus Riedel and Rudolf Nebel watch Oberth make last-minute adjustments. Lower right: 18-year-old Wernher von Braun tightens fuel connections on the rocket motor.*

*Hermann Oberth at the Ufa Film Company working on demonstration rocket motor for the movie* Frau im Mond.

*Hermann Oberth as he appeared in 1958 while living in the United States. He has since returned to Germany, where he consults and writes on space flight.*

# 4 THE LEGACY OF

The modern rocket took shape during the two decades that followed World War I. As the writings of Tsiolkovsky, Goddard, and Oberth attained wider circulation, a growing number of rocket enthusiasts began the work that led directly to space flight. By the end of the 1930's, on the eve of another war, military work on rockets was being carried on vigorously in several nations.

Public interest in rocketry and support of rocket development varied immensely. In Germany and Russia, and to a lesser degree in Great Britain and the United States, popularizers digested the concepts of the pioneers and transmitted to the public not only their spirit but the more easily understood—and often more spectacular—proposals they had made. The Americans seemed to be the least receptive, remaining skeptical about space flight right up to the orbiting of the first artificial satellites in the late 1950's. Outside of a dedicated handful of experimenters, talk of rockets and space travel was viewed as crackpot by the public and as unscientific by most scientists.

The first flight of a liquid-propellant rocket in Europe took place on 14 March 1931, when an 11-pound methane-oxygen rocket soared to an altitude of nearly 1,000 feet near Dessau, in Germany. This achievement was symbolic of German leadership in rocket flight between the two wars, but it did not mean that Germany stood alone. Significant work was performed in the United States, Russia, Britain, France, and—to a lesser extent—in Italy, and there was ample cross-fertilization between rocket researchers of different nations.

During the 1920's and 1930's, rocket and interplanetary societies were established in the major western European nations and in the United States. Most of the theoretical and experimental work was conducted in these nations and—with the notable exception of Goddard's—was reported regularly in rocket and space journals that were distributed to society members and to selected subscribers.

Communications between the early rocket and space societies were frequent during these years. Letters and reports circulated among the European nations and crossed the Atlantic, and the rocketeers themselves visited each other. In January 1931, France's pioneer space-flight scientist Robert Esnault-Pelterie came to the United States to deliver a lecture to the American Interplanetary Society. And, in the same year, G. Edward Pendray, vice-president of the American society, traveled to Berlin to meet members of the German space-flight society and observe tests at their proving grounds. Nikolai Alexsevitch Rynin of the Soviet Union regularly sent out communications on Russian activities, and from Germany Willy Ley and others provided a constant stream of information.

To further cement the camaraderie among the early rocketeers, most of the inner circle became members of several societies. To give one example, in 1934 alone the British Interplanetary Society named as fellows Guido von Pirquet of Austria, Robert Esnault-Pelterie of France, Jakov Isidorovitch Perelman of Russia, Willy Ley and Ilse Kühnel of Germany, and Edward Pendray and Ernst Loebell of the United States.

Virtually all of the civilian experimenters and promoters were much more interested in what the rocket could do, theoretically at least, to further their dreams of visiting the Moon and planets, than in rocketry itself. For the first time in history, they believed, man was in a position to do something about his ingrained desire to visit other worlds. He possessed the rocket, primitive though it might be, and a technological base that made the future appear infinitely "promising."

In Russia, Tsiolkovsky's work provided impetus for additional rocket research. In 1924, the Soviets created a Central Bureau for the Study of the Problems of Rockets (TsBIRP) and an All-Union Society for the Study of Interplanetary Flight (OIMS), both of which staged an exhibition on rocket technology in 1927. These two groups were responsible for the "second generation" of Soviet rocket re-

# HE PIONEERS

search, in which the work started by Tsiolkovsky in the early 1900's was brought up to date and expanded.

The leading exponent of rocketry in this second generation was Fridrikh Arturovitch Tsander, who began to develop liquid-propelled rocket engines during the 1920's. By 1930, tests were possible. His two rocket motors, called OR-1 and OR-2, were propelled, respectively, by gasoline and air and gasoline and liquid oxygen. Successful static tests were conducted in 1930 and again in 1933, the first developing over 10 pounds of thrust and the second over 100 pounds of thrust. In the same period Valentin Petrovitch Glushko developed rocket engines at the Gas Dynamics Laboratory in Leningrad and later at the Reaction Scientific Research Institute. In 1932, the year before Tsander's death, he published a significant book entitled *Problems of Flight by Means of Reactive Devices* (sometimes translated as *Problems of Reactive Flight*), while Glushko, with G. E. Langemak, wrote *Rockets, Their Construction and Utilization* in 1935.

After Joseph Stalin became master of the country, the rocket groups created in 1924 were replaced by the better known Len-GIRD, or Group for the Study of Reaction Motion (Gruppa Isutcheniya Reaktivnovo Dvisheniya) in Leningrad, and Mos-GIRD, a similar organization in Moscow. These eventually became the State Reaction Scientific Research Institute.

There is some confusion as to when these organizations were formed. G. A. Tokaty, former chief of the Aerodynamics Laboratory of the Zhukovsky Academy of Aerodynamics of the Soviet Air Force, has said that Len-GIRD was established in 1931, but most other sources give the year as 1929. It may have been set up unofficially in 1929 and not recognized officially until two years later. Whatever the date, the Leningrad group was founded by Nikolai Alexsevitch Rynin and Jakov Isidorovitch Perelman, while the Moscow branch was founded by I. P. Fortikov. These GIRD groups were affiliated with the larger Society for the Promotion of Defense and Aero-Chemical Development (Osoaviakhim).

The members of these groups published scientific reports of their own, translated foreign literature —particularly German—and organized several congresses. One example was the All-Union Conference on the Study of the Stratosphere held March–April 1934, whose proceedings were published by the U.S.S.R. Academy of Sciences in 1935. Special conference papers on *Rocket Technology* and *Jet Propulsion* were released by the Union of Scientific Technical Publishing Houses in 1935–1936.

Beginning as early as 1928 and extending through 1932, an even more monumental publishing effort had taken place: Rynin's great *Mezhplanetyne Soobshcheniya* (*Interplanetary Communications*). (The Soviets used *interplanetary communications* to mean "space travel" or "space flight." Their term *cosmonautics* is equivalent to the preferred Western term astronautics.)

*Interplanetary Communications* was a nine-volume encyclopedia that offered abundant proof of Soviet interest and activity in astronautics. All but the last two volumes were published before a single book on interplanetary flight had appeared in the United States or Great Britain. During the 1930's only three English-language books were published, compared with several dozen in Russia. Perelman's *Mezhplanetyne Puteshestviya* (*Interplanetary Travels*) went through ten editions and sold 150,000 copies, far more than similar books in any language. Rynin's work was printed in a much smaller edition. Only 1,000 to 2,000 copies of most of the volumes were run off, although the third volume had a printing of 15,000 copies.

The first two volumes of the encyclopedia were a history of science fiction, starting with legends and early fantasies and running up to the works of Verne and Wells. The third volume was a study of proposals for communicating with other worlds; it gave special attention to the energy sources needed for interplanetary signaling, including some far-out

**Some Early Russian Liquid-Propellant Rocket Engines**

| *Engine Designation*[a] | *Thrust (pounds)* | *Propellant combination* | *Organization* | *Designer* | *Year* | *Remarks* |
|---|---|---|---|---|---|---|
| OR-1 | 11 | gasoline and gaseous air | TsGIRD[b] | Tsander | 1929–1932 | Combustion tests, of which over 50 were made, proved basic design concept, led to OR-2. |
| ORM-1 | 44 | toluene and nitrogen tetroxide | GDL[c] | Glushko | 1930–1931 | Basic combustion and ignition research. |
| ORM-4 thru ORM-23 | varied | toluene and nitrogen tetroxide | GDL | Glushko | 1932 | Gradual upgrading of performance. |
| OR-2 | 110 | gasoline and liquid oxygen | MosGIRD[d] | Tsander | 1931–1933 | Combustion chamber cooled by the oxidizer, nozzle cooled by water circulating through closed-circuit system. An all-wood glider made by C. I. Cheranovskii designed to be powered by engine. Uprated to 150 pounds of thrust to power GIRD-X rocket. |
| ORM-24 thru ORM-49 | — | kerosene and nitric acid | GDL | Glushko | 1933 | Water-cooled, extended duration testing. |
| ORM-50 | 330 | kerosene and nitric acid | GDL | Glushko | 1933 | Built to power antiaircraft missile designed by M. K. Tikhonravov. |
| ORM-52 | 660 | kerosene and nitric acid | RNII[e] | Glushko | 1933 | Regenerative cooling by fuel introduced. |
| ORM-53 thru ORM-64 | 1300 | kerosene and nitric acid | RNII | Glushko | 1934–1936 | Further testing of engine parameters at higher thrust. |
| ORM-65 | 110–390 | kerosene and nitric acid | RNII | Glushko | 1936 | For use in flying bomb and in rocket airplane designed by S. P. Korolev. |
| ORM-67 thru ORM-70 | 660 | kerosene and nitric acid | RNII | Glushko | 1937 | Engine testing; no known application. |
| ORM-101 | 175 | kerosene and tetranitromethane | RNII | Glushko | 1937 | Combustion research with new combination. |
| ORM-102 | 220 | kerosene and tetranitromethane | RNII | Glushko | 1937 | Combustion research and engine performance at higher thrust rating. |

[a] The OR designations refer to *opytnaya raketa,* or experimental rocket. The addition of the *M* for motor used by Glushko. Occasionally, the designation *OP* is given, taken from the first two letters of the word *opytnaya* alone.
[b] Group for the Study of Reaction Motion under Tsander within the Central Council of the Osoaviakhim.
[c] The Gas Dynamics Laboratory in Leningrad.
[d] In April 1932, the GIRD group under Tsander was established in Moscow and became known as MosGIRD; it was also under the Osoaviakhim.
[e] State Reaction Scientific Research Institute, resulting from a merger in 1934 of MosGIRD and GDL.

ideas such as altering the paths of comets and planets.

The next three books were a history of rocketry into the twentieth century, a mathematical study of jet propulsion, and a volume on airplanes and long-range artillery. Tsiolkovsky got a volume to himself, including an autobiographical sketch.

The eighth volume, whose 350 pages made it the longest in the series, examined the astronautical writings of the major world figures—Oberth, Goddard (whose *A Method of Reaching Extreme Altitudes* was translated in full), Esnault-Pelterie, Scherschevsky, Ley, and many others.

The final volume, entitled *Astronavigation,* gave descriptions and photographs of the Moon and planets and their orbits. It included an exceptionally detailed chronological bibliography of work on space flight, both fiction and nonfiction.

As *Interplanetary Communications* and a host of other books were published, the Russians continued their experimental work. Groups led by Tsander, M. K. Tikhonravov, and Glushko built and flight tested liquid-propellant rockets during the early and mid 1930's. One, the GIRD-X, weighed 65 pounds, was about 8½ feet in length, and slightly more than 6 inches in diameter. It reached an altitude of more than 3 miles on 25 November 1933. Another was fired

МЕЖПЛАНЕТНЫЕ СООБЩЕНИЯ

Н.А. РЫНИН

ЛУЧИСТАЯ ЭНЕРГИЯ

1 9 3 1

*Cover for volume 3 of Rynin's* Interplanetary Communications, *a nine-volume encyclopedia published between 1928 and 1932. Volume 3, the biggest seller, was a study on communications with other worlds.*

the next year and by 1936 an Aviavnito sounding rocket had exceeded an altitude of 3.5 miles. It was over 10 feet long, just under a foot in diameter, and weighed 213 pounds.

During the mid-1930's work began on solid-propellant military rockets. This effort eventually resulted in the Katyusha rockets that were fired in great quantities on the German troops in World War II. Another group developed rocket engines for airplanes. The Russians also developed plans for a guided missile; this advanced research project had to be canceled, however, when Germany invaded Russia at the beginning of World War II.

Great as was the progress in Russia, the German rocket effort far outstripped it. The quick sale of the

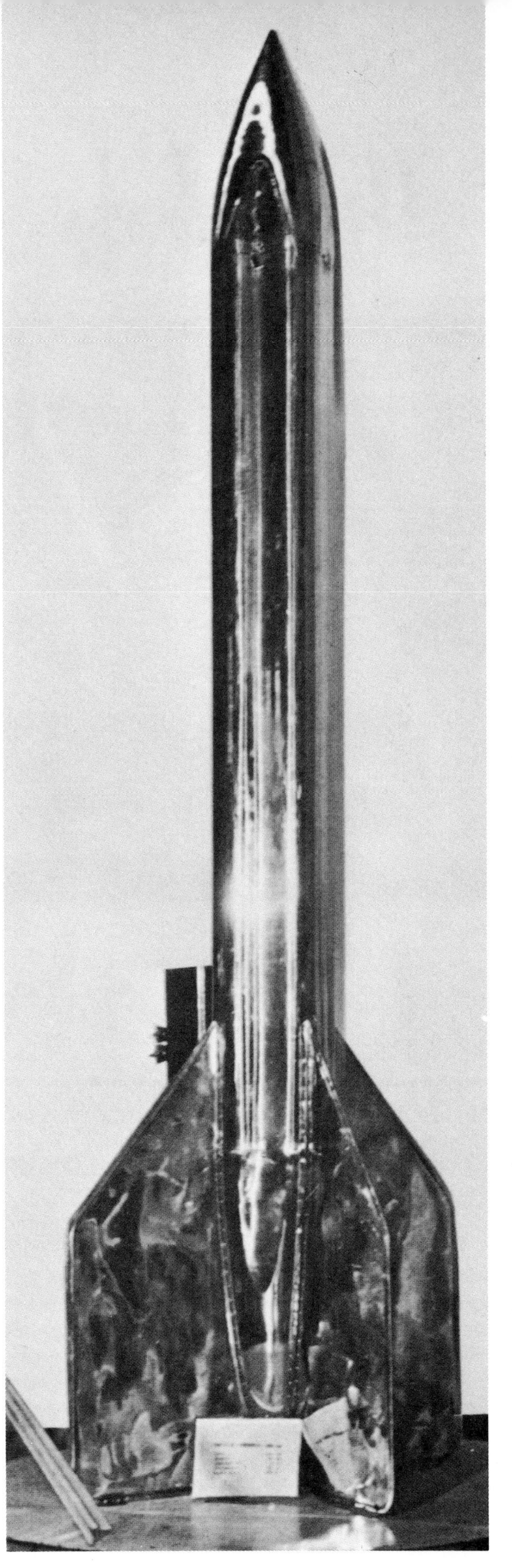

*The Aviavnito sounding rocket was one of several liquid-propellant rockets built in the Soviet Union in the mid-1930's. Ten feet in length, it reached an altitude of 3.5 miles in a 1936 test.* (U.S.S.R. ACADEMY OF SCIENCES)

*Cover of a January-June 1927 summary issue of the monthly journal* Die Rakete, *published by the German Society for Space Travel.*

first edition of Oberth's *Die Rakete zu den Planetenräumen* (*The Rocket into Planetary Space*) was an indication of how much Germans in the early 1920's were fascinated with the idea of space flight.

In early June 1927, rocket and space enthusiasts in Germany founded the Verein für Raumschiffahrt (Society for Space Travel). Its membership rose rapidly to around five hundred, enough to support a journal, *Die Rakete* (*The Rocket*). Between the time Oberth's first book was published in 1923 and the appearance of his *Wege zur Raumschiffahrt* (*Road to Space Travel*) in 1929, a number of other works appeared, including Walter Hohmann's *Die Erreichbarkeit der Himmelskörper* (*The Attainability of Celestial Bodies*) in 1925, Willy Ley's *Die Fahrt ins Weltall* (*The Flight into Space*) and *Die Möglichkeit der Weltraumfahrt* (*The Possibility of Space Travel*) in 1928, and Max Valier's *Der Vorstoss in den Weltenraum* (*The Advance into Space*) in 1924, which, in 1930 (after five printings), was enlarged and retitled *Raketenfahrt* (*Rocket Flight*). Of these, Hohmann's mathematical treatment of orbital and interplanetary flight mechanics was the most advanced. The book is still referred to today.

The VfR circle did more than write books. In early 1928, space flight popularizer Max Valier, who wanted to publicize the capabilities of the rocket, got together with automobile tycoon Fritz von Opel, who wanted publicity for his automobile, and a manufacturer of powder rockets named Friedrich Wilhelm Sander. Their combined talents produced the world's first rocket-powered automobile.

The car's first test—corresponding to a flight test of a more conventional rocket—took place on 15 March 1928 at Opel's track near Rüsselsheim, with Kurt C. Volkhart at the wheel. The results were not spectacular, though the little Opel car did move. The next test was not much better; the car got up to just 45 miles per hour. The experimenters then decided to replace the standard car they had been using with a vehicle specially designed for 400-pound Sander rockets.

The new car, called the Opel-Rak 1, was a converted racer whose conventional engine was replaced first by six Sander rockets and then by eight. On 12 April 1928 successful runs were made at speeds up to 55 miles per hour over 2,000- to 3,000-foot distances at Rüsselsheim. Then a twelve-rocket experiment was made and, though five rockets failed to ignite, the little car reached more than 70 miles per hour. The

*The second generation of rocketry was international in spirit. Goddard's 17 July 1929 firing in Worcester, Mass., was reported in the July 1929 issue of* Die Rakete *(incorrectly stated as 18 July).* (ARFOR ARCHIVES)

**Extra-Blatt**

der Zeitschrift „DIE RAKETE" :: Breslau, Juli 1929
Offizielles Organ des Vereins für Raumschiffahrt E.V. in Deutschland

In letzter Minute geht die Nachricht ein, daß Professor G o d d a r d in Worchester, Massachusets, am 18. Juli 1929 eine Versuchsrakete von 3 m Länge und 70 cm Durchmesser abgeschossen hat, die eine enorme Höhe erreichen sollte. Die Rakete startete ordnungsgemäß, explodierte jedoch bereits in geringer Höhe mit mächtigem Knall, so daß die Fensterscheiben in der Umgegend zertrümmert wurden.

Es ist dies seit längerer Zeit ein neues Lebenszeichen von G o d d a r d, das deutlich zeigt, wie auch drüben am Raumfahrtproblem intensiv gearbeitet wird.

Opel-Rak 1 was followed by the more streamlined Opel-Rak 2, which was fitted with short, inverse wings to hold it to the ground as it sped along. Opel-Rak 2, outfitted with twenty-four rockets and with Opel himself at the wheel, sped down a track in Berlin on 23 May 1928 at a maximum speed of nearly 125 miles per hour.

These tests were followed by railway-car experiments in June and October 1928. The first, conducted between Celle and Burgwedel, were powered by twenty-four large powder rockets that accelerated the car to well over 100 miles per hour. The second series of tests was held between Blankenburg and Halberstadt. A brief résumé in the VfR journal, *Die Rakete,* of 15 October 1928 said that the car was built jointly by Valier and the J. F. Eisfeld firm. One run was successful; the other, with a larger and heavier payload, was not.

Rocket-sled experiments were also made; and, on 30 September 1928, a rocket-glider flight took place at Rebstock, near Frankfurt. Often mistakenly hailed as the first rocket airplane, the glider was powered by sixteen Sander rockets, each producing 50 pounds of thrust. Piloted by Fritz von Opel, it reached 95 miles per hour.

The first rocket-powered airplane had been flown just three months earlier at the Wasserkuppe mountain by the Rhön-Rossitten Gesellschaft, a glider group. On 11 June 1928, a sailplane named *Ente* (Duck) powered by two Sander slow-burning rockets, covered just over three quarters of a mile in slightly more than sixty seconds. These pioneering flights were ahead of their time, and soon were discontinued.

At about this time, Oberth had his brief adventure with the Ufa Film Company—probably the oddest source of research and development funds in history. When Oberth went back to teaching, some of his helpers and fellow VfR members, among them Rudolf Nebel, Klaus Riedel, Willy Ley, and Wernher von Braun (then a young student), obtained permission to use an abandoned military ammunitions dump in Reinickendorf, a suburb of Berlin, to test their rockets. Dubbed *Raketenflugplatz,* or rocket airfield, it became famous as the site of the early experiments of several of the men who later played major roles in the German army's V-2 missile. The engineering genius of the group was Riedel, an exceptionally talented, self-made man with no formal technical education. He had been involved in rocketry since the early VfR days and participated in many experiments. He died in an automobile accident toward the end of World War II.

*Opel-Rak 1, the first successful rocket car, was a converted racer whose engine was replaced by commercial sea-rescue rockets. In tests at Opel's track near Russelsheim, it reached speeds of more than 70 mph. Above, Opel-Rak 1, with its driver, Kurt C. Volkhart.* (ROLF ENGEL)

*Max Valier, a member of the German VfR (Society for Space Travel), became interested in rocket cars as a means of publicizing the capabilities of the rocket. He is shown here in his liquid-propellant rocket car.*

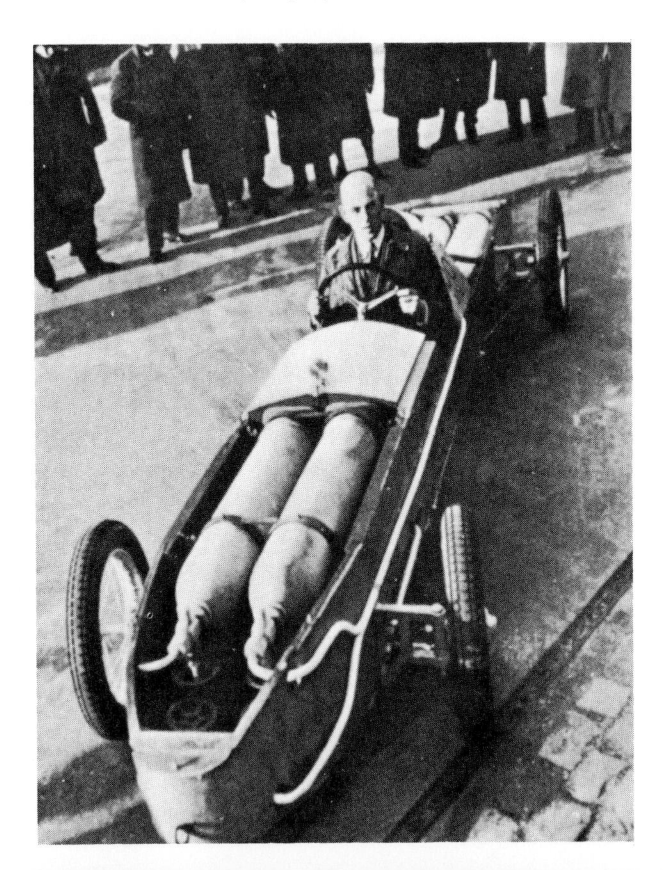

*Valier's rocket sled, RS-1, glided along the snow on pontoons. In tests on 22 January 1929 it reached 65 mph.*

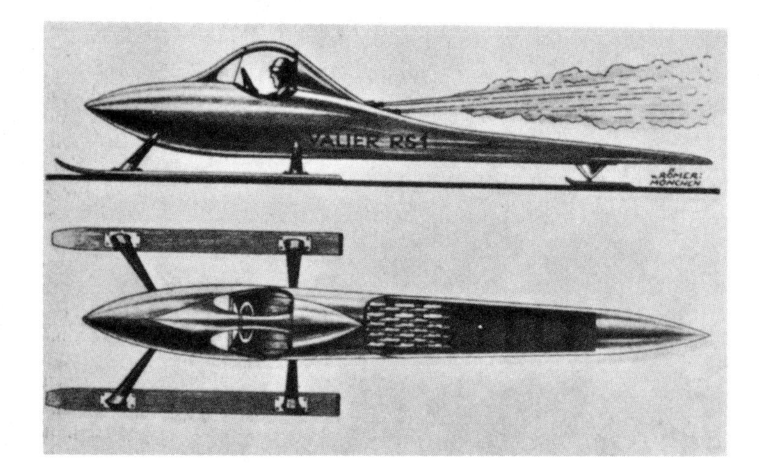

*The first rocket-powered airplane was a sailplane named "Ente" (Duck). Powered by two slow-burning Sander rockets, it traveled three quarters of a mile in slightly more than 60 seconds in tests on 11 June 1928.*

*Automobile manufacturer Fritz von Opel piloted his own rocket glider in tests near Frankfurt on 30 September 1928. Its 16 rockets, each producing 50 pounds of thrust, were built by Friedrich Sander.* (THOMPSON RAMO WOOLDRIDGE)

*In 1929 Hermann Oberth, shown here with moustache and necktie, attempted to build a spaceship for the Ufa Film Company movie* Frau im Mond. *Flanking him are co-workers at Ufa.*

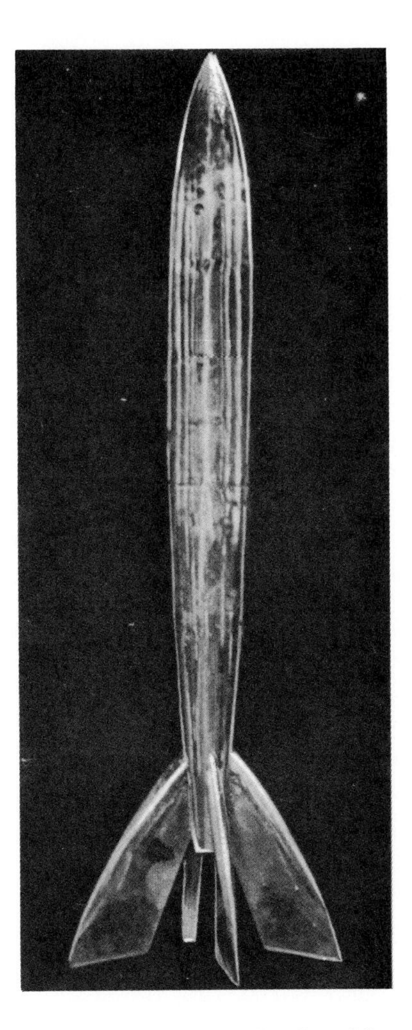

*Demonstration rocket developed by Oberth for* Frau im Mond.

*After Oberth left the film studios, some of his helpers moved their testing equipment into an abandoned ammunitions dump near Berlin. Their "Raketenflugplatz" (rocket airfield) became famous as the site of the early experiments of several of the men who later developed the German war missiles. The above picture shows (left to right) Rudolf Nebel, Wernher von Braun, and Kurt Heinisch at the Raketenflugplatz.*

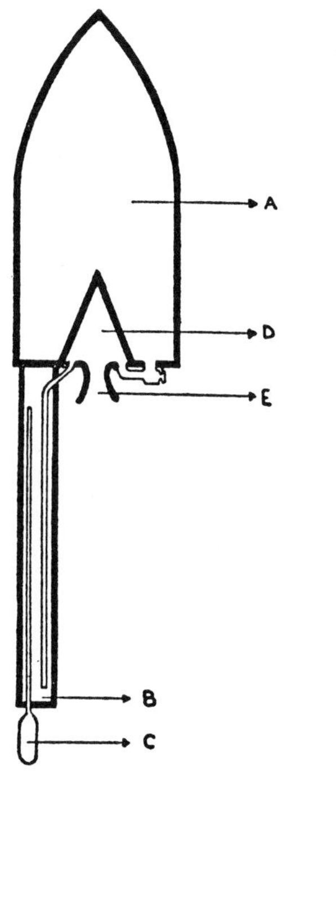


*At left, Mirak 1, a rocket developed by the Raketenflugplatz group, had a 1-foot-long body and a 3-foot tail section. Liquid oxygen was stored in the head (A); gasoline, in the tail (B). The carbon dioxide charger was at (C), combustion chamber at (D), and exhaust at (E).*

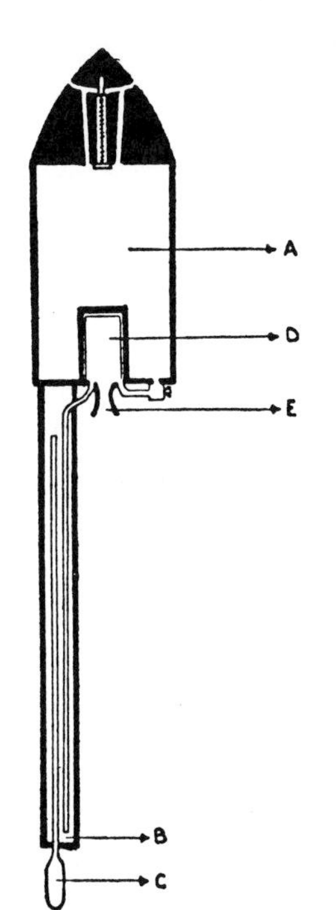


*Mirak 2, built with a modified cooling system, developed 70-pound thrusts during the spring of 1931. The letters in the diagram are the same as for Mirak 1.*

For more than four thousand years, man's mind has slowly opened to the universe around him. From the beginning he saw stars and planets, the Sun and Moon, and an occasional comet and meteor. But what the heavens meant to precivilized man, how he interpreted the lights in the sky, and his first glimmerings of understanding remain largely unknown to us.

To the ancient Babylonians, astronomy was inextricably tied up with astrology and magic. As early as 3000 B.C. they were observing the motions of Venus, and by 2000 B.C. they had recognized the movement of the planets across the zodiac. Astrologer-astronomers developed the notion of a 360-day year composed of twelve 30-day months, and could calculate in advance the positions of the Sun, Moon, and planets.

The Egyptians added five days to the year, for a total of 365. They developed sun clocks, refined the concept of the zodiac, and by 1300 B.C. identified at least forty-three constellations and the five planets visible to the naked eye: Mercury, Venus, Mars, Jupiter, and Saturn.

Neither the Babylonians nor the Egyptians understood the nature and form of heaven and earth. Egyptian papyri relate that in the beginning the universe contained a mass of water called Nu. From it germinated the Sun, a flat Earth, and, floating in the waters above, the stars—many if not all of which were thought to be gods. Babylonian ideas were quite similar, as were those of the early Hebrews. According to the Bible, God said: "Let there be a firmament in the midst of the waters . . . and let the waters under the heaven be gathered together unto one place, and let the dry land appear."

Across the Mediterranean, Thales of Miletus, who lived approximately 624 to 547 B.C., integrated elements of Babylonian and Egyptian astronomy into a culture destined to dominate Western thought for centuries. Thales and his followers envisioned a disklike Earth floating on water.

Later Greek astronomers improved upon this. Pythagoras, born in 572 B.C., correctly believed in a spherical Earth, and implied that the universe might be spherical too—an idea that appeals to many modern astronomers. Anaxagoras theorized that the Moon does not itself shine, but rather reflects sunlight. And Heraclides of Pontus (*c.* 388–315 B.C.) hypothesized, accurately, that the Earth rotates on its axis and that Mercury and Venus revolve around the Sun. But only Aristarchus of Samos, who died in 230 B.C., placed this planet in its proper orbit around the Sun. That the Earth was the center of the universe remained the accepted view until the time of Copernicus (1473–1543).

As man's knowledge expanded, his imagination also began to encompass new worlds. In the second century A.D., author Lucian of Samosata wrote two tales describing voyages to the Moon. From then until the appearance of the *Shāh-Nāmā* in Persia in the eleventh century and of *Orlando Furioso* in the early sixteenth century, the "space-travel" theme lay dormant in the minds of men. (If it did occur to someone, there is no surviving reference to it.) Only following the discoveries of Nicolaus Copernicus in the sixteenth century was the approximate character of the worlds beyond the Earth finally explained. It took another three and a half centuries for man to figure out scientifically how he could voyage to the Moon and planets.

The key to space travel was the rocket, the only known propulsion system that will operate in a vacuum. Invented by the Chinese in the thirteenth century (and perhaps even earlier), it was used alternately for war and for amusement over a period of hundreds of years. Powder rockets of many shapes and sizes were fired with varying degrees of success by the Chinese, Mongolians, Arabs, Indians, Italians, British, French, Austrians, Americans, Russians, and others. So slow was their development that, as late as World War I, rockets could hardly be considered more than minor adjuncts to the conventional weapons then in use.

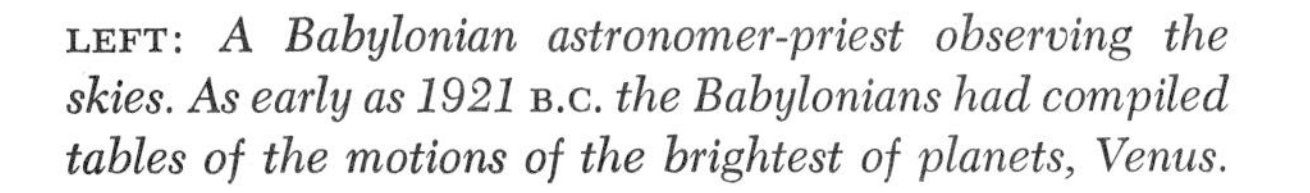

LEFT: *A Babylonian astronomer-priest observing the skies. As early as 1921* B.C. *the Babylonians had compiled tables of the motions of the brightest of planets, Venus.*

BELOW: *Egyptian Sun and Moon gods. The ancient Egyptians, who developed Sun clocks and Solar calendars, were never able to distinguish the gods from the planets, the stars, the Moon, and the Sun. Amon-Ra, the all-powerful Sun god, was widely revered.*

ABOVE: *In the sixth century* B.C., *Ionian philosopher Thales predicted a Solar eclipse and measured the angular diameter of the Sun. In spite of these advances, he clung to the traditional view of a flat, disk-like Earth, floating in a great ocean.*

RIGHT: *When Kai-Kā'ūs, mythical king of Persia, wanted to conquer the heavens, four specially trained eagles lifted him on his elaborate throne into the firmament. According to Firdausi's epic* Shāh-Nāma *(*A.D. *1010), the eagles "raised up the throne from the face of the earth, they lifted it up from the plain into the clouds."*

Harry Hans Kurt Lange

RIGHT: *The first device known to have operated on the reaction principle was developed by Archytas of Tarentum during the first half of the fourth century* B.C. *A founder of theoretical mechanics, he built a pigeonlike device that hung by string and moved from one position to another in reaction to the force produced by jets of discharging steam. Behind the pigeon is the aeolopile, a steam-driven wheel described by Hero of Alexandria in the first century* A.D.*—the second known application of the reaction principle.*

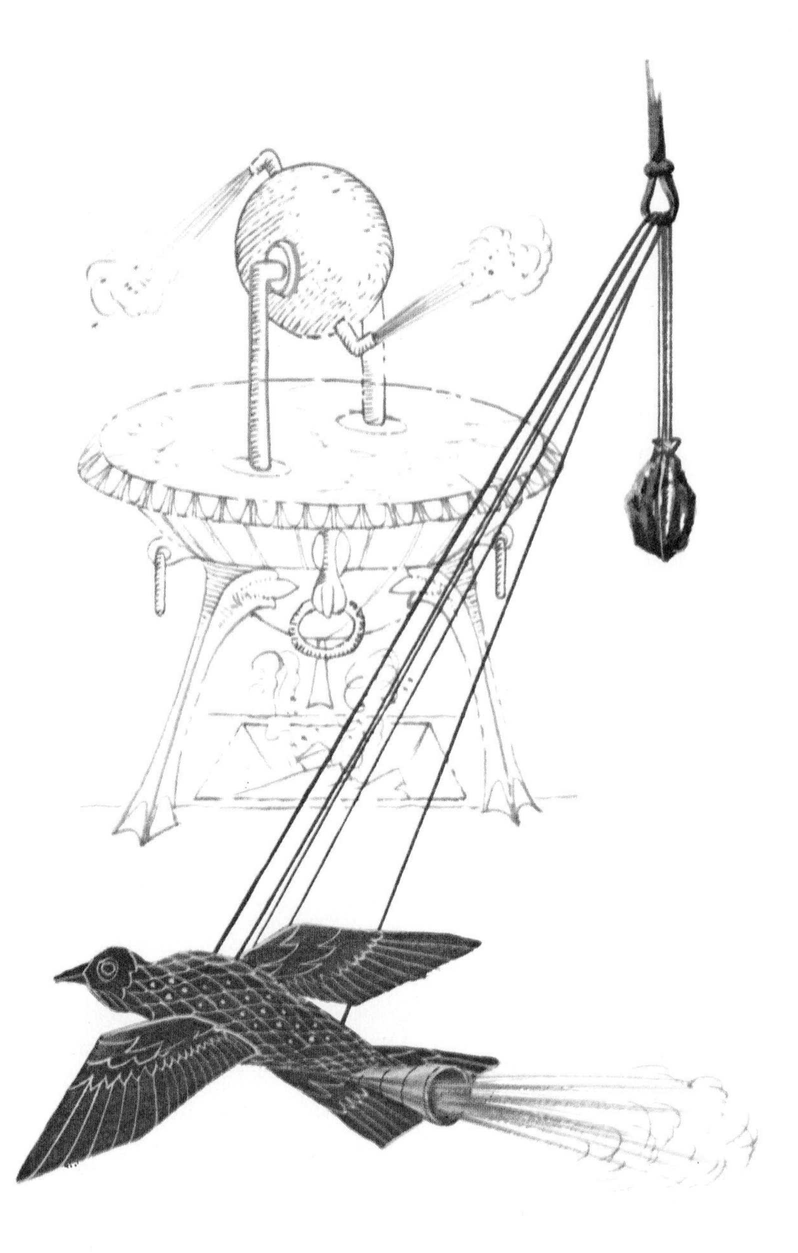

BELOW: *Rockets, probably invented in China, spread quickly to other countries. The Arabs used the rocket-powered "egg that moves and burns" against French invaders in the Seventh Crusade. It was described by Syrian historian al-Hasan al-Rammāh in a late-thirteenth-century* Treatise on Horsemanship and War Stratagems.

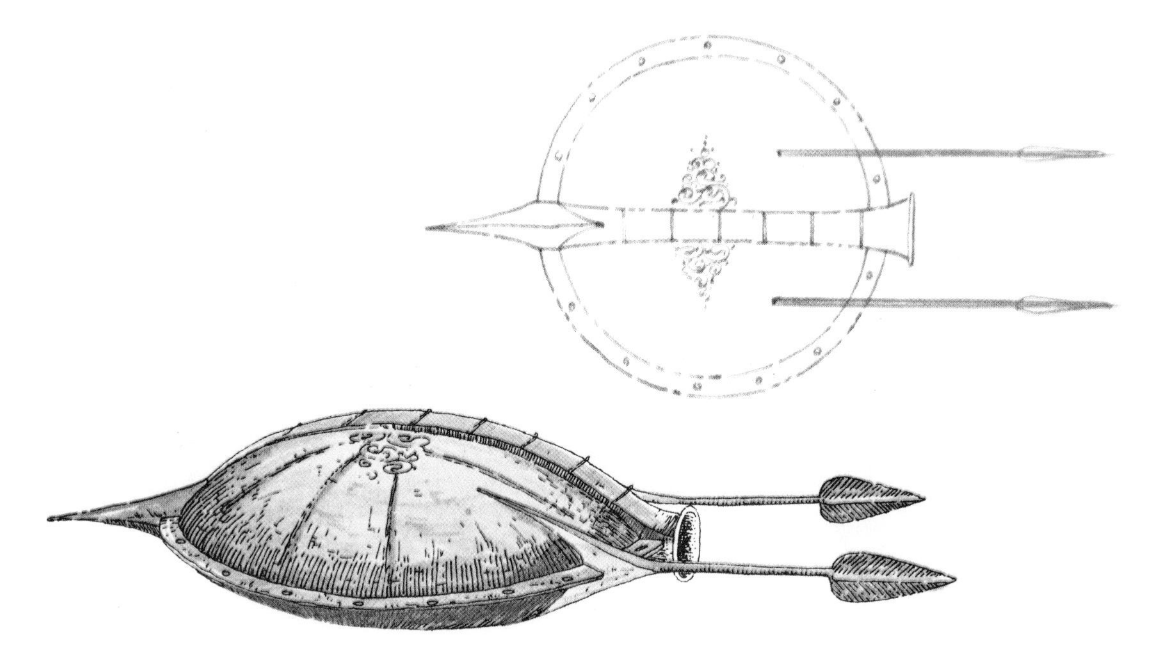

*Chinese armies in the thirteenth and fourteenth centuries employed a variety of rocket-type devices against the Mongolians. Carrying such descriptions as "long snake crush enemy" and "leopard herd rush transversally," these "fire-arrows" were fired over ranges of hundreds of feet. But their accuracy was not great, and consequently they were not very effective.*

*Domingo Gonsales, hero of Bishop Francis Godwin's* The Man in the Moone *(1638), had no intentions of going to the Moon in this novel device driven by goose-power; he only wanted to escape from an uninhabited island on which he was stranded. To do this he trained a flock of birds to fly him back to civilization. But by the time they were trained their migrating season had begun. And, to Domingo's dismay, their other home was not on our fair planet but—where else?—on the Moon. Thus, a Lunar voyage was made quite by accident.*

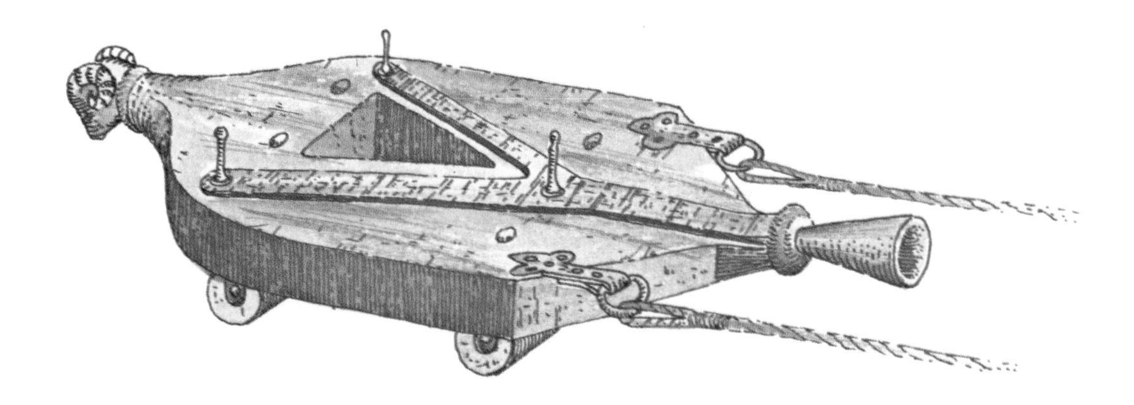

*In his* Bellicorum instrumentorum liber, *Joanes de Fontana described such novel devices as a rocket-propelled cart designed to break through enemy strongpoints, and a rocket-propelled torpedo made of wood.*

*Fiery as they may have been, nineteenth-century rockets were so inaccurate that they were all but replaced by new and improved artillery. With a few exceptions, such as the Civil War rocket launcher, above, they virtually disappeared by the second half of the century. In the First World War, French Le Prieur rockets (below, about to be launched against a German observation balloon) were occasionally fired from Nieuport biplanes. Still, they were indecisive. Another great war had to come and go before the rocket finally became an important military weapon.*

Beginning in August 1930 the VfR-Raketenflugplatz experimenters tested a series of rockets called *Mirak,* a contraction of "Minimum Rakete." Mirak had a body about a foot long and a tail section about 3 feet long. The tail held the tank for the gasoline that fueled the rocket, while the oxidizer (liquid oxygen, or LOX) was stored in the head. The tiny copper motor resembled Oberth's Kegeldüse, but it was immersed in the liquid oxygen, which was expected to furnish more than enough cooling. Mirak 1 had a successful, if brief, stationary test in August in Bernstadt, Saxony. But during the next test, held in September, the oxygen tank burst and the rocket was destroyed. Mirak 2 was built and tested, with a substantially modified cooling system. By the spring of 1931, Mirak 2 was regularly developing thrusts of around 70 pounds in 10-second tests. Finally, it too exploded. The theory was sound, but the technology was just too primitive.

Most of the early difficulties stemmed from cooling, or rather the lack of it. The combustion chambers usually burned through after a short firing period. Their materials were unable to withstand the heat of the burning gases, and the heat transfer to the coolant was inadequate. Clearly the experimenters had to develop better means of cooling their rockets before they could proceed to flight testing.

A motor was finally built that had aluminum walls cooled by water, rather than the liquid oxygen of the earlier Miraks. The motor was incorporated not in Mirak 3 but in a new rocket that Willy Ley nicknamed "Repulsor." Repulsor 1 soared to a height of about 200 feet on 14 May 1931; Repulsor 2 reached about the same altitude and achieved a long range for that time of 2,000 feet on 23 May. The third Repulsor did better; in early June 1931 it reached an altitude of at least 2,000 feet and crash-landed a similar distance away. The experimenters had hoped to recover it by parachute, which, to their annoyance, was torn off during the ascent.

Several more No. 3 models were flown before a redesign that resulted in Repulsor 4, a rocket that used a single stick for stability. On its maiden flight in August 1931, Repulsor 4 flew up approximately 3,300 feet, where its parachute opened and floated it gently back to Earth. Other flights of Repulsor-4 models attained altitudes of one mile—no mean accomplishment in those days.

But the Repulsors had just missed out on the honor of being the first successful European rockets. A few months earlier, Johannes Winkler, with the financial support of Hugo A. Hückel, had built a 2-foot-long, 12-inch-diameter rocket fueled by liquid oxygen and liquid methane. Near Dessau, at Gross Kühnau, the Hückel-Winkler 1 was launched on 14 March 1931, to an altitude of perhaps 1,000 feet. The next in the series, Hückel-Winkler 2, did not do as well. After taking off near Pillau in East Prussia on 6 October 1932, it caught fire and crashed to the ground. Maximum altitude: 10 feet.

The day before the first Hückel-Winkler shot, Karl Poggensee, in a test near Berlin, had fired an experimental solid-fuel rocket to a 1,500-foot altitude. Fitted with an altimeter, cameras, and a device to measure velocity, it was recovered by parachute at the end of the flight. Other rocket pioneers active at about this time included Reinhold Tiling, Gerhard Zucker, and, in Austria, Friedrich Schmiedl.

Reinhold Tiling built six solid-propellant rockets, four of which he fired from Osnabrück in April 1931. One exploded at 500 feet, two climbed to from 1,500 to 2,500 feet, and one reached 6,600 feet, burning for 11 seconds and reaching a maximum speed of 700 miles per hour. Tiling launched his last two rockets, of more advanced design, from Wangerooge, one of the East Frisian Islands. One of them may have climbed as high as 32,000 feet. About 5 feet long, they consumed 14.3 pounds of propellant and carried a payload of 11 pounds.

Starting in February 1931, Friedrich Schmiedl of Austria fired solid-propellant rockets with mail payloads, mostly between Schöckel and Radegund and Schöckel and Kumberg, for several years. The rockets delivered hundreds of letters. Tiling tried the same idea in 1931; a few years later, so did Gerhard Zucker, who envisioned regular rocket mail service across the English Channel. Like Schmiedl and Tiling, Zucker used powder rockets, but his attempts were spectacular only for the explosions they produced. The longest shot he attempted was from Harris to Scarp, in the Western Isles of Scotland, on 31 July 1934. The rocket destroyed itself in an explosion.

Another sidelight to prewar rocketry was the so-called Magdeburg project. Rudolf Nebel had conceived the idea of building a manned rocket, but before embarking on such an ambitious project, first set out to accumulate needed flight data with a small test rocket. Accordingly, with the help of Herbert Schaefer, he built a rocket which was subsequently fired on 9 June 1933, at Wolmirstedt near Magdeburg. The initial test met with scant success: the motor did not provide enough thrust to permit the rocket even to fly free of the 30-foot launching tower. Several later attempts produced the same results until, on 29 June, one finally managed to leave the tower, fly horizontally about 1,000 feet, and land only slightly damaged. It was later rebuilt into

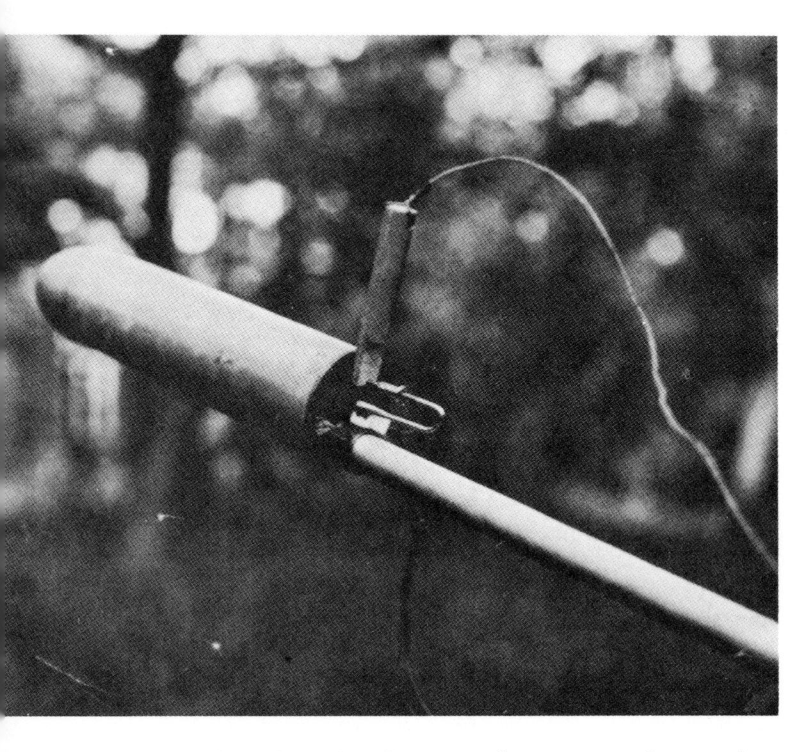

*Mirak 2 at the Raketenflugplatz. Liquid oxygen is in large tank at left; fuel, in long, thin tank at right.*

*The VfR experimenters designed an improved cooling system, aluminum cooled by water, for the Repulsor series. Repulsor 2, shown here on the test stand at Raketenflugplatz, was fired 2,000 feet on 23 May 1931.*

a Repulsor-type rocket and launched from Lindwerder Island in Tegeler Lake near Berlin, reaching an altitude of some 3,000 feet before crashing 300 feet from its tower. Additional tests were conducted from a boat on Schwielow Lake in August 1933 before the project was abandoned altogether.

The early successes of the VfR experimenters were not enough to sustain it during the economic depression that gripped Germany. The VfR program fell on hard times, along with the rest of the world, during the 1930's. By 1932, membership in the rocket society began to drop severely. The problems of the rocket pioneers were compounded by increasing objections of the Berlin police to rocket flights within the city limits. In desperation, the VfR group sought support from the German army. After a demonstration flight of a Repulsor at the army proving grounds at nearby Kummersdorf in the summer of 1932, the army invited Wernher von Braun to do the experimental work for his doctor's thesis on rocket combustion phenomena at Kummersdorf.

*Shortly after this picture was taken of its launching in August 1931, Repulsor 4 rose to a height of 3,300 feet, where its parachute opened and floated it back to Earth.*

Work at the Raketenflugplatz ended during the winter of 1933–1934. The VfR was in shaky financial condition and could not pay its accumulated bills. In January 1934, a year after Hitler's rise to power, the Raketenflugplatz was turned back into an ammunition dump by the military. The international ties of the VfR and the Raketenflugplatz irritated the new regime, and it was not long before the Gestapo would make its presence felt. In this stifling atmosphere private interest in rocketry all but vanished.

At Kummersdorf, meanwhile, the army continued experiments under the direction of Captain Walter Dornberger and Von Braun. The Heereswaffenamt-Prüfwesen (Army Ordnance Research and Development Department) gradually developed the Versuchsstelle Kummersdorf–West as a site for developing and static-testing rockets. Between November 1932, when Von Braun began work on his doctor's thesis with the help of a single mechanic, and 1937, when the Kummersdorf–West staff had grown to about eighty, the group worked on several small rockets. This work laid the groundwork for the V-2.

The main source of support within the army for the Kummersdorf project was Artillery General Karl Becker, who, with his mentor Carl Cranz of the Prussian Academy for Military Engineering, had long championed the rocket as an artillery weapon. The Cranz-Becker textbook, *Ballistics,* contained a chapter on the subject, ample evidence of the authors' thinking on the potential of the rocket. It was Becker who encouraged and sponsored Von Braun's doctoral thesis and, after his graduation, put him to work in the army under Major von Horstig and his assistant, Captain Dornberger. And it was Becker who, in 1935, proposed in vain to Hitler a program to develop a long-range bombardment rocket.

The first new rocket to be developed at Kummersdorf was the A-1, an abbreviation for Aggregate 1. Powered by an alcohol–liquid oxygen regeneratively cooled engine developing 660 pounds of thrust, the A-1 was a relatively simple rocket. The alcohol-cooled engine was in the bottom bulkhead of the single pressurized metal tank which was half filled with alcohol. While the alcohol fuel filled the lower half of the metal tank, liquid oxygen was loaded into a thin-walled, open fiberglass container inserted into the upper half of the metal tank. A small compressed-nitrogen flask, pressurizing the single metal tank, thus provided identical feed pressure for both propellants. The payload was a 70-pound gyroscopic flywheel in the rocket's nose, which provided stability during the 16-second powered flight.

After a few successful static tests, the first A-1 was destroyed by an explosion caused by delayed ignition. Model tests had shown, meanwhile, that the flywheel had to be closer to the rocket's center to be effective. Since the location of the open oxygen container within the fuel tank posed the danger of a flashback explosion after a powered flight, the arrangement was changed. The fuel tank was shortened, a separate metal tank was provided for the liquid oxygen, and the gyrowheel was placed between the two tanks. The new configuration was called A-2. In December 1934 on the North Sea island of Borkum, two A-2 rockets, "Max" and "Moritz"—named after characters in a humorous German book—made two successful straight-up flights to altitudes of about 6,500 feet.

The Borkum success was a shot in the arm for

*Above, Reinhold Tiling (left, with side to camera) standing beside black-powder rocket he launched from Osnabrück in April 1931. Below, the Tiling rocket, in flight. It rose to a height of 2.5 miles.*

*Gerhard Zucker envisioned rocket mail service across the English Channel. The longest shot he attempted was from Harris to Scarp, in western Scotland, on 31 July 1934. But the rocket blew up before takeoff.* (AIAA)

*The Nazis, who disliked the VfR's international ties, turned the Raketenflugplatz back into an ammunitions dump. Rocket research was brought under the army's control at Kummersdorf. Below, Hitler watches a Kummersdorf demonstration in March 1939.*

Kummersdorf. The staff was increased and allotted more funds. The Dornberger–Von Braun organization grew, as did the power of the rocket engines. Soon both were too large for the restricted confines of Kummersdorf. A move to another location became necessary in April of 1937. The site chosen was Peenemünde on the Baltic coast, not far from the small town of Wolgast. The first task of the Heeresversuchsstelle Peenemünde (Army Experimental Station Peenemünde), of which Von Braun became technical director, was to develop a new liquid-fuel rocket with adequate growth potential. The second assignment: Prepare the A-3, now virtually complete, for a test firing out over the Baltic Sea.

While the move into the vast new experimental station was accomplished easily, the launching in the fall of 1937 of three A-3's was a different matter. All three test models of the 21-foot-long, 1,650-pound rocket failed. The 3,300-pound-thrust alcohol–liquid oxygen engines behaved well enough, but a new and quite sophisticated inertial guidance system did not. The problem was solved—but only after nearly two years of labor. Meanwhile, larger, more powerful, longer range rockets were on the drawing boards.

The pace of activity at Peenemünde quickened as war drew nearer. In March 1938, German troops invaded Austria. At the end of September, the Munich pact won the Sudetenland for Hitler. Only six months later all of Czechoslovakia capitulated. As these events unfolded, the Ordnance Department demanded with new urgency that the experimenters at Peenemünde prove their worth by developing practical weapons.

The Ordnance Department wanted a long-range missile, capable of flying 150 to perhaps 200 miles with a one-ton warhead. The rocket had to be transportable by railroad and truck, and hence of a size compatible with Germany's existing rail and road network—tunnel widths and heights, road curves, and the like. Above all the rocket had to be reliable.

The response to these criteria was the A-4 rocket, which became known as the V-2, or second weapon of retaliation, when it was used near the end of the war. But the jump from the problem-beset A-3 to a reliable A-4 could not be made without an interim vehicle to test the guidance system, which may have been a luxury for the A-3 but would be a vital necessity for the larger missile. Hence, the A-5 was born.

The A-5 used the same propulsion system as the A-3 and was outwardly identical to the new A-4, although much smaller in size. The A-5 had a greatly improved structure and an improved, yet simplified, guidance and control system. Its first flight, in the fall of 1938 from the nearby Greifswalder Oie island—without the new guidance system—proved the soundness of the A-5's mechanical design. A second series of launches, this time with the new guidance, came off perfectly. The first vertical flight, in the summer of 1939, reached an altitude of 7.5 miles. Other flight models did as well. Some were launched vertically, others along slanting trajectories to simulate the operational trajectory of the forthcoming A-4. Several A-5's were recovered by parachute and flown again.

While most of the work on military rockets in Germany was performed by the Dornberger–Von Braun team at Peenemünde and its associated industrial and university contractors, there were other efforts before and during the war, particularly those sponsored by the Luftwaffe. In 1936 the Luftwaffe had invited the Austrian Dr. Eugen Sänger to build a rocket research establishment at Trauen near Hannover. It was ready when the war started. Sänger's experience had been gained in liquid-propellant rocket engine testing at the University of Vienna, where he had fired Diesel oil–oxygen gas motors developing a little over 50 pounds of thrust for 20 or 30 *minutes,* incredibly long burning times. Sänger visualized many applications for his rocket engines, some of which were designed to use liquid oxygen and conventional fuels, others liquid ozone and metallic fuels (aluminum powder suspended in fuel oil). But his work at Trauen was ended in 1942, and Sänger was transferred to another research institute, where he turned his interest to air-breathing jet engines and worked on conceptional studies of an intercontinental rocket-boosted bomber.

Other German research workers had been trying to mate rockets and airplanes. Early in the 1930's, Hellmuth Walter, a research chemist at the Hanseatische Apparatebau Gesellschaft in Kiel, suggested the use of hydrogen peroxide for propelling torpedos. When the idea failed to work out, Walter turned to aeronautical applications. At the Chemical State Institute in Berlin he developed a rocket engine operating on 80-percent-strength hydrogen peroxide. It was so successful that in 1935 he established his own firm, Hellmuth Walter Kommanditgesellschaft.

The Air Ministry became interested in a proposal to install a Walter rocket engine on a Heinkel Kadett airplane, which was done in February 1937. Although the rocket produced only 220 pounds of thrust, the test was successful. It was the first rocket-assisted takeoff in history. According to Walter "later in that year and during 1938, a great number of flight tests were made with ATO's [assisted takeoff units] at 300 to 500 kg thrust with land and sea planes without any accidents whatsoever." Rocket-powered depth

*The 21-foot A-3 on the test stand at Peenemünde, 1937. A-3's engine, which ran on a combination of alcohol and liquid oxygen, was the basis for the wartime V-2.*

*In Vienna, Dr. Eugen Sänger developed liquid-propellant rockets that burned for 20 to 30 minutes. Here Dr. Sänger (right) is shown pouring liquid oxygen into a container at his test station in 1934.*

charges and mines designed to be dropped from aircraft were also developed. Walter engines had many wartime applications.

The Bayrische Motoren Werke (Bavarian Motors Works) at Munich also developed liquid rockets for airplanes. Starting in 1938, under the direction of Helmut von Zborowsky, the company investigated a myriad of propellant combinations before settling on nitric acid doped with 10 percent sulfuric acid to reduce corrosion and a "visol" fuel that consisted of vinyl isobutyl ether and aniline. Rheinmetall-Borsig in Berlin also worked with nitric acid at about this time, while Schmidding und Dynamit A.G. in Bodenbach developed solid-propellant rockets to assist airplane takeoffs.

Peenemünde also had done some work on rocket engines for airplanes. In January 1935 Major von Richthofen, head of Luftwaffe airplane development, visited Von Braun at Kummersdorf to inquire about the feasibility of adapting a liquid-propellant rocket engine to power military aircraft. An engine developing 2,200 pounds of thrust on liquid oxygen and alcohol propellants was soon available. With the help of the airplane manufacturer Ernst Heinkel, it was installed by Von Braun and his associates in a Heinkel He-112. Static testing began during the summer of 1935 and flight testing in the spring of 1937, at about the same time the Kummersdorf group had transferred to Peenemünde. While the He-112 never became operational, it did provide a stepping stone to later rocket airplane developments and helped inspire Heinkel's work on JATO (jet assisted takeoff) devices.

In neighboring France the rise of rocketry was less spectacular than in Germany. Like Germany, Russia, and the United States, France had a leading pioneer, Robert Esnault-Pelterie. Though his contributions to astronautical knowledge were less important than those of Tsiolkovsky, Goddard, and Oberth, he was a well-known airplane inventor, a brilliant individual, and a member of the French Academy of Sciences. As early as 1907 he began thinking seriously about astronautics. Like Tsiolkovsky and Oberth, he was a highly effective publicist for the idea of interplanetary travel and rocketry.

Esnault-Pelterie launched his public campaign in February 1912 when he delivered a major lecture in St. Petersburg, repeated back home in Paris to the Société Française de Physique in November. The Société was the most prestigious organization to provide a forum up to that time for what many people still considered to be a fantastic subject. Only Esnault-Pelterie's reputation in other fields permitted him to lecture, albeit cautiously, on the "Considera-

*Hellmuth Walter, a research chemist from Kiel, built rocket engines fueled by hydrogen peroxide. The peroxide was manufactured in a plant in Lautenberg in the Harz Mountains.*

*Robert Esnault-Pelterie began theorizing on space flight as early as 1907, and subsequently published basic works on the subject. This picture was taken in 1929.* (ARFOR ARCHIVES)

tions sur les résultats d'un allègement indéfini des moteurs" ("Considerations on the Results of an Unlimited Lightening of Motors"—a title referring to the fact that as a rocket uses up its propellants it becomes progressively lighter).

Fifteen years later, on 8 June 1927, he appeared before the Société Astronomique de France to tell astronomers about the results of his further theoretical research in astronautics. The Society subsequently published the written text as a 98-page book, *L'exploration par fusées de la très haute atmosphère et la possibilité des voyages interplanétaires* (*Rocket Explosion of the Very High Atmosphere and the Possibility of Interplanetary Travel*). This title, more audacious than the 1912 delivery, combines Goddard's conservatism and Oberth's optimism. At the same meeting Esnault-Pelterie revealed that he and a banker friend, André Louis-Hirsch, had established a 5,000-franc prize to be awarded annually to the author of the most outstanding work on astronautics. Called the Prix REP-Hirsch (REP-Hirsch Award), it was awarded by the Astronautical Committee of the Société Astronomique de France. The first recipient in 1928 was Hermann Oberth, who was so highly esteemed that the prize was doubled to 10,000 francs.

Esnault-Pelterie's greatest contribution was the publication in 1930 of a book entitled *L'Astronautique* (*Astronautics*), which, together with its 1934 supplement, *L'Astronautique-Complément*, covered virtually all that was then known of rocketry and space flight.

Although Esnault-Pelterie's major interest was theoretical astronautics, he was well aware of the military implications of rocketry. On 20 May 1928, he proposed to French army general Ferrié a plan for the development of ballistic bombardment missiles against which he could imagine no defense. He wrote that such weapons could deliver "over several hundreds of kilometers . . . thousands of tons" of destructive payload, all within a few hours. (He was obviously thinking in terms of salvo firings like the World War II V-1 and V-2 offensives.) "Moreover," he added, "the necessary ground installations would not entail great expense and would doubtless be infinitely less burdensome than if it were a question of delivering the same load by aeroplanes."

His proposal resulted in the appointment of ingénieur général J. J. Barré to his laboratories in 1931, which in turn led to work approved by the Commission des Poudres de Guerre at Versailles first on liquid-oxygen–gasoline motors, then on nitrogen peroxide–benzene motors, and one powered by liquid oxygen and tetranitromethane. In October 1931 tests

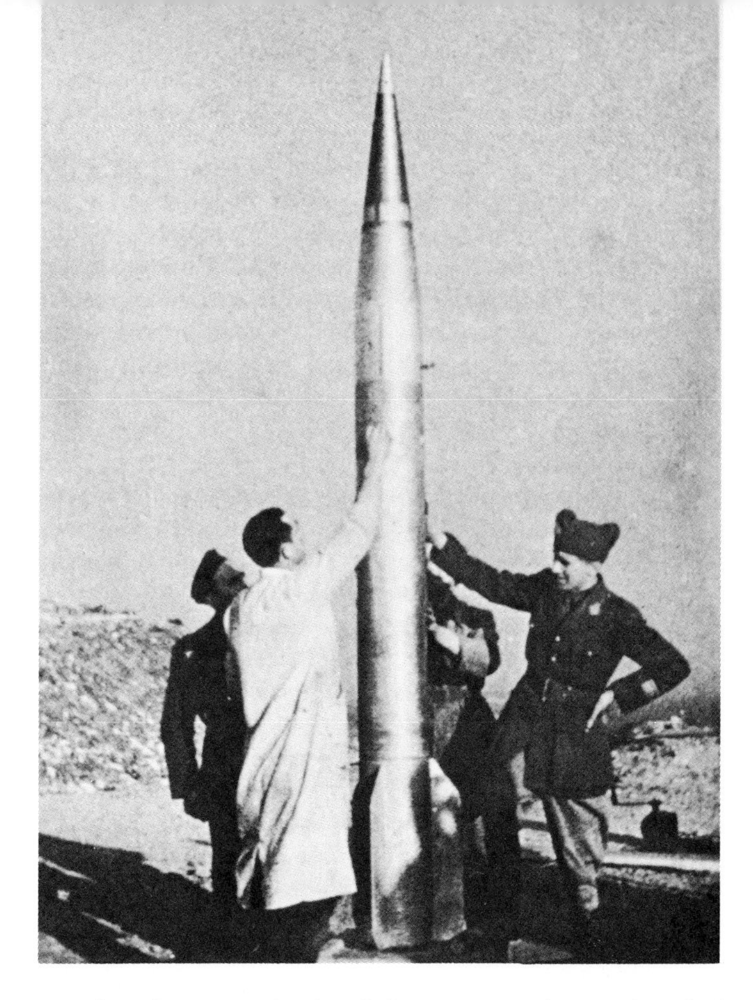

*French rocketeer J. J. Barré with the experimental EA-41 launched near Toulon.*

of the last, an accident occurred, causing Esnault-Pelterie to lose four fingers.

In 1934 a study contract was let to Esnault-Pelterie by the Direction des Études et Fabrications d'Armement under the general supervision of ingénieur général Desmazières. There, in addition to liquid rocket work, 80-mm solid-fuel rockets were developed whose application was to have been to accelerate bombs. Elsewhere, the Services de l'Armement Français studied, in 1939, the use of 1,000-pound-thrust JATO units for assisting heavy bombers to take off. Air Liquide, a private concern, worked for a short period on a 100-pound-thrust test motor under Air Ministry contract at Champigny and at Seyne. French rocketry continued sporadically, and without conclusive results, until the outbreak of war.

Experimental work in rocketry also was conducted by Henri F. Melot who, beginning in 1916, spent nearly three decades developing airplane boost motors at the Institut des Arts et Métiers in Paris. Black-powder rockets were studied from 1932 to 1935 by Louis Damblanc, a civil engineer, at the Institut Aérotechnique at Saint Cyr and some tests were made—the aim being to realize more reliable signal, Coast

Guard, and anti-hail devices. Similar experiments were made at the Établissements Ruggieri and at the École Centrale de Pyrotechnie in Bourges. In the field of astronautics, the great 1937 Paris exhibition at the Palais de la Découverte featured an impressive space-flight exhibit, including contributions from other European countries and from America.

Incredible as it may seem, there was some independent French rocket activity during the German occupation. According to Barré, General Arnaud and Colonel Gentil of the Sous-direction du Service de l'Artillerie directed that the Section Technique de l'Artillerie should secretly operate, in Lyon, under the completely false name Service Central des Marchés et de Surveillance des Approvisionnements (Central Market Service and Supplies Supervision)! Directed by Colonel (later General) Joseph Dubouloz, studies were initiated by Barré to develop an experimental motor operating on liquid oxygen and gasoline ether called the EA-1941 (or EA-41). Static tests began on 15 November 1941 at Larzac, and within a year, seven tests had been successfully conducted at both Larzac and at Vancia near Lyon—"unknown to the occupiers, of course." Barré describes the rocket as the first conceived and built in France, that is, the first airframe powered by a liquid-propellant rocket motor—previously only motors alone had been constructed. The EA-41 motor could be regulated in terms of its propellant consumption, was cooled by its own fuel, and was pressurized by nitrogen gas.

The clandestine group planned to fire their test rockets in 1942 from a field at Beni-Ounif de Figuig in Algeria, but the Allied landings prevented them from doing so. They awaited the Liberation; and, in March 1945, at the Etablissement d'Expériences Techniques at La Renardière near Toulon, the long-awaited firings began. During the course of the first set of two firings in March 1941 all went relatively well, but during the second set, one rocket landed about 20 miles from La Renardière and could never be located.

Across the Channel in Great Britain, prewar rocketry suffered from two crippling handicaps. The British had no great man similar to Oberth or Tsiolkovsky to rally around. And the British government was stubbornly hostile to rocketry. The Explosives Act of 1875 forbade rocket testing, and the government enforced the law to the letter. Rocketry in Britain literally never got off the ground.

British rocket enthusiasts got around these two barriers as best they could, concentrating on public education and on providing a literary forum for astronautical scientists and engineers. In these endeavors, they were more successful.

In 1933 the British Interplanetary Society was founded. Unlike other rocket societies, this group has kept its name and original objectives intact to the present. Apparently the only attempt to make the BIS adopt a more conservative name occurred in 1934, a year after it was founded, when it was suggested that the word "Interplanetary" be dropped from the society's name. The society's president scorned the proposal:

> Are we to pander to public opinion . . . an opinion which held to ridicule the votaries of heavier-than-air flight, and which refused to credit the marvel of wireless telegraphy to such good effect that the inventor died heartbroken, deserted even by his friends, who also deemed him mad? . . . it seems to me that a change in name, regardless of the reason for it, would be universally misconstrued as an admission of doubt, as a confession that the interplanetary idea only belongs to the realm of extravagant fiction.

In 1934 the first issue of the BIS *Journal* appeared. It became universally recognized as the most distinguished publication devoted to space flight in the world. The importance the British attach to their *Journal* was evident from its very beginning. P. E. Cleator, the first president of the BIS, editorialized in the *Journal's* second issue, published in April 1934:

> The tremendous importance of a substantial and interesting Journal was brought home to me during my conversations with the German experimenters. In the year 1929, the old Germany rocket society . . . ceased the publication of their Journal, *Die Rakete*. The immediate result was the loss of over 600 members! It happened like this: there came a time when the Society had to choose between publishing the Journal and carrying out certain important and costly experiments. Eventually, it was decided to sacrifice the Journal. Now the new programme was all very well for those members who happened to live in Berlin, for they could take part in, or witness, the experiments. But not so for the majority of members, who were scattered throughout the country. With the loss of the Journal, they were deprived of their only real link with the Society. The moral is clear. The Journal of a Society constitutes a vital connecting link . . . . The Journal must come before experimental work.

But the British knew the importance of experimental work. As Cleator later said, "No one is more eager than I am to organize and to begin our share of actual experimental work," but until finances were more secure and conditions more propitious he was unwilling to promote testing. Attempts to get government support were completely unsuccessful. As late as 1934 the Air Ministry was reported to have "evinced not the slightest interest" in liquid-propellant rockets, though engineers at the Royal Aircraft

Establishment had done some investigations of rocket motors in the mid-1920's.

The first British book on astronautics was Cleator's *Rockets Through Space: The Dawn of Interplanetary Travel,* which appeared in 1936. Written at a popular level, it did not contain any profound insights into the new science. What Cleator set out to do, however, he did well. His volume is the best written of the first group of books on rocketry. Since he was close to the center of communication provided by the *Journal,* Cleator was able to include reports on the many activities taking place on both sides of the Atlantic.

In December 1934 Alwyn Douglas Crow (later Sir), Director of Ballistic Research at the Woolwich Royal Arsenal, proposed to Sir Hugh Elles, Master of General Ordnance, that the British begin to investigate the possibilities of developing rocket weapons powered by smokeless cordite powder of the unrestricted burning type. Crow later wrote that

> in April 1935 the Research Department at Woolwich Arsenal was asked to put forward a program for the specific investigation of the possibility of developing rockets, utilizing cordite as the propelling agency. Preliminary work was started . . . in May 1935, and by the summer of 1936 encouraging advances in technique had been achieved.

With this encouragement, the Committee of Imperial Defense's Subcommittee on Air Defense Research directed Crow and his team to investigate the use of rockets for antiaircraft defense, long-range attack, air-to-air combat, and assisted takeoff for heavy aircraft. During 1936 and 1937, small motors based on 2-inch- and 3-inch-diameter solid cordite charges were perfected. Larger motors with 11 cordite sticks were also studied. Test firings were conducted in a covered range in England during July and August 1936, outdoors in November, and again during the second half of 1937, before they were transferred to the more favorable climate of Jamaica. Some 2,500 3-inch rockets were fired in Jamaica during the first few months of 1939 under the direction of Brigadier Allan Younger. According to Brigadier Leonard Walter Jubb, a participant, the Jamaica trials were devised "primarily to obtain information on the external ballistics of the rockets." Simultaneous tests were conducted in Britain to prove out the design of a launcher. Both single and twin launchers ultimately were developed for Army and Navy deployment.

Crow, who was responsible for starting this series of tests, ultimately headed Britain's wartime rocket program. In 1938 he left his post as Director of Ballistic Research at the Woolwich Royal Arsenal, a position he had held since 1919, and moved his rocket research team to Fort Halstead, Kent, where he set up what was to be known as the Projectile Development Establishment for the U.P. (unrotated projectile) under the War Office. In 1939, Sir Alwyn became Chief Superintendent of Projectile Development. Shortly after Britain entered the war, his title became Controller of Projectile Development in the Ministry of Supply.

Some notable prewar rocket research also was performed by individual scientists and engineers in Italy, although a strong national development program never was sustained.

The Italian General Staff sponsored a rocket research program beginning in 1927 under the direction of General G. A. Crocco with the assistance of his son Luigi and the collaboration of the Bombrini Parodi Delfino chemical firm. According to an article by Crocco on "Instruction and Research in Jet Propulsion" (*Journal of the American Rocket Society,* March 1950), tests first were conducted with powder rockets with some success, but by 1929 the Italian General Staff canceled the program because dispersion was too great and velocity too low (rockets launched during the experiments had reached a velocity of only about 1,000 feet per second).

Beginning in 1929 Crocco and his associates then began to investigate liquid propellants, selecting gasoline as the fuel and nitrogen dioxide as the

*British rockets being prepared for launch at Port Royal, Jamaica, in February 1939.* (BRIG. LEONARD WALTER JUBB)

oxidizer. Assisting in this work were Corrado Lani of the Establishment of Constructions, Italian Air Ministry; scientists at the Chemical Institute of the University of Rome; and Riccardo M. Corelli. By 1930 a combustion chamber had been developed which operated for up to ten minutes at 140-pounds-per-square-inch pressures. Crocco later reported that "By the end of 1930, the available funds were exhausted, and despite the promising results obtained, the General Staff did not renew its contract." A familiar end to many early government-sponsored rocket research programs.

Two years later Crocco and Corelli obtained support from the Aviation Ministry for a study of propellants, concentrating first on a mixture of 70 percent trinitroglycerin and 30 percent methyl alcohol, and later on, nitromethane. Again, although preliminary results warranted continuing the study, the program was phased out in 1935.

In the United States, the most crucial influence was the aloofness of Robert H. Goddard from other American rocketeers. Goddard did not join in when rocket enthusiasts met in New York City on 21 March 1930 to found the American Interplanetary Society. He did not publicize his advanced ideas, nor did he help other rocket researchers. Intent on his own work, he allowed others to make the public speeches. The great American rocket pioneer did very little to nurture a climate in which large-scale rocket research might flourish.

The society's announced aims were to promote "interest in and the experimentation toward interplanetary expeditions and travel . . . and the raising of funds for research and experimentation." David Lasser, the first president, was a technically trained individual then serving as editor of *Wonder Stories*, a science-fiction magazine. Other founders included a chemist, William Lemkin; several writers and newspapermen, C. W. van Devander, G. Edward Pendray, C. P. Mason, Laurence Manning, Fletcher Pratt, and Nathan Schachner; and an engineer, Clyde J. Fitch. The society's first activity was a ceremony in which explorer Sir Hubert Wilkins presented it with a copy of *Discovery of a New World*, a volume published in 1638 by John Wilkins, Bishop of Chester, Sir Hubert's ancestor.

The AIS attracted a number of engineers and scientists into its ranks. The four-page first issue of its *Bulletin*, which appeared in June 1930, reported news of the society and events in the United States and abroad. It maintained fairly regular publication, though it probably changed its name more frequently than any other journal in history.

The society itself remained almost continually on the defensive in its dealings with the public. Interplanetary travel was the target of such ridicule in the United States that in 1934 the American Interplanetary Society changed its name to the less frightening American Rocket Society. It kept this name until 1963 when, after merging with the Institute of Aerospace Sciences, it became the American Institute of Aeronautics and Astronautics.

Goddard, while not a member, did contact the society occasionally, if only to correct an error. In the June–July 1931 *Bulletin* he provided an accurate account of his 17 July 1929 rocket firing, which had been reported incorrectly in the preceding issue. He ended his letter with these words:

> The article in the *Bulletin* also speaks of my reticence in giving my results to the public. It happens, as I have explained to Mr. Lasser, that so many of my ideas and suggestions have been copied abroad without the acknowledgment usual in scientific circles that I have been forced to take this attitude. Further, I do not think it desirable to publish results of the long series of experiments I have undertaken until I feel that I have made a significant further contribution to the problem.

About a year after the founding of the American Interplanetary Society, Pendray traveled to Europe, visiting the German experimenters at the Raketenflugplatz in April 1931. In an article on "The German Rockets" in the May 1931 issue of the *Bulletin*, he reviewed his trip and supplied details and drawings of the Mirak series of rockets. Pendray's trip helped stimulate experimenters in the United States to conduct tests themselves. A research committee quickly was established by the society, and in the summer of 1931, plans for testing were drawn up. The projected tests were described in the September issue of the *Bulletin* as providing a means "of studying the operating qualities of rocket motors, in order to determine what power they might develop under various conditions."

The first of the society's rocket designs, according to a report in the March 1932 *Bulletin*, was "patterned . . . in general after one of the successful German liquid-fuel rockets . . . the type known as the two-stick Repulsor but containing . . . so many changes and inventions of our own that it must stand or fall upon its own merits." It was to use gasoline and liquid oxygen and be constructed largely of a special aluminum alloy "cast by the Aluminum Company of America from a pattern made upon specifications designed by ourselves."

The motor of the test rocket developed by AIS members (under the direction of H. Franklin Pierce) was somewhat larger and heavier than the German Mirak motors and boasted a number of apparent propellant-feed improvements. Mounted on a shackle at the front end of two 5.5-foot-long, 1.5-inch-diameter cylindrical tanks, containing the liquid oxygen and gasoline propellants, the water-cooled, 6-inch-long, 3-inch-diameter motor was made of cast aluminum. At the rear were four aluminum guiding fins, and at the opposite end, a cone-shaped nose bay with parachute.

Although Rocket No. 1 was completed in January of 1932, tests could not begin until a suitable firing location was found. This turned out to be a New Jersey farm near Stockton, some one hundred miles from New York City. Work began there in August 1932 to prepare two dugouts, one for close observation and the other for controls. A proving stand was built which, according to Pendray,

> consisted of two rounded up-right members of wood, held in place by a stout framework of planks. The up-right members, constructed to guide the rocket in the event of an actual flight, were fourteen feet high. The span of the supporting framework was a little under twelve feet. The whole apparatus stood on the ground as an independent unit, the lower parts protected from flame by sandbags, flat stones and banks of earth.

In order to engage the upright members of the proving stand, guides were placed along the sides of the rocket; the uprights were "copiously soaped" to insure movement of the guides. During static tests a 12-foot-long pole held down the rocket; it was "pivoted to an upright post at its outer end, and held down at its inner end by a coil spring which was in turn fastened to the proving stand."

Many difficulties faced the early experimenters. They found it exceedingly uncomfortable to load liquid oxygen into the relatively long, narrow oxidizer tanks ("it requires sometimes as long as fifteen minutes to empty a quart, most of which either evaporates from the furiously boiling supply in the funnel, or spatters down the sleeves of the operator, inflicting innumerable tiny frostbites, like little scalded spots"). The valves, which had to be activated by fuse-wire connections from the dugout, gave them trouble. And the remote firing control was "a complete failure," leading the rocketeers to consider going forth with a "box of matches and a gasoline soaked torch . . . and light the fuse between the tanks of the rocket."

The first static test, held on 12 November 1932, was a success. Recalling the event a short while afterward, Pendray, still in a state of near euphoria, said:

> It is impossible adequately to describe that sight, or to convey the feeling it gave us. I suppose we were excited; but there was a certain majesty about the sound and sight which made it impossible for the moment to feel excite-

*The American Interplanetary Society static tested its first rocket between two upright "guides" on a New Jersey farm in 1932. At right, AIS president David Lasser, fore-*

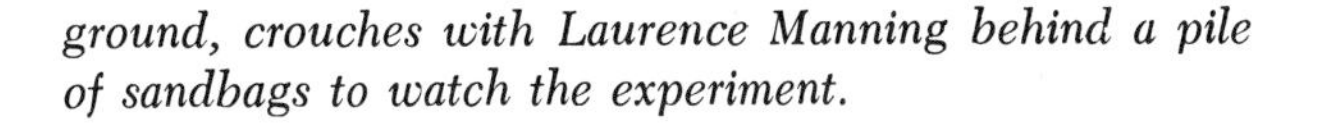
*ground, crouches with Laurence Manning behind a pile of sandbags to watch the experiment.*

*Max Kraus, secretary of the American Interplanetary Society (left), G. Edward Pendray, chairman of the experiment committee, and Bernard Smith, "valveman," ready Rocket No. 2 for firing at Marine Park, Great Kills, Staten Island, N.Y. on the morning of 14 May 1933.* (G. EDWARD PENDRAY)

ment as such. We forgot to remain behind the shelter of our earthworks. Moreover, we forgot to count the seconds as they passed in that downward pouring cascade of fire.

The exhaust pattern was "clear and clean, of a bluish-white color, and quite steady." About the only flaw in the test was the depletion of the oxidizer before the fuel, with the result that the excess gasoline "came spurting out, throwing a shower of fire all around the foot of the rocket and proving stand."

A flight test of Rocket No. 1, planned for the next morning, had to be canceled because of bad weather. Based on the 60 pounds of thrust developed during the test, the experimenters calculated that the 15-pound rocket might have attained an altitude of 6 miles.

The rocket was overhauled, and became the society's Rocket No. 2. The tanks were placed closer together, the parachute bay was removed, valving was improved, and balsa-wood vanes replaced the aluminum fins. A new streamlined nosecone had a hole to admit cooling air, instead of the cooling jacket of Rocket No. 1. The launching rack, designed by Laurence Manning and Alfred H. Best, was made of two 2-inch pine poles, 15 feet long, supported below by 4-by-4 timber and above by wooden shackles. It was tilted 5 degrees to insure that the rocket would fly out over the water, where it would do no damage upon landing and could be recovered for post-flight inspection.

At 11:20 in the morning of 14 May 1933, the flight was made from a new testing grounds at Great Kills, Staten Island, under the direction of Pendray and his assistants, who included Manning, Best, Bernard Smith, Carl Ahrens, and Alfred Africano. The motor, instead of burning for 20 to 30 seconds as it had in the November 1932 static test, shut down at an altitude of about 250 feet after burning only 2 seconds; the liquid-oxygen tank had burst, the fins separated from the tanks and the motor, which continued in flight, landing about 400 feet offshore (where they were recovered). The heat of the motor's

*ARS Rocket No. 3 being prepared for static tests on Staten Island in September 1934. Shown here are (left to right) John Shesta, G. Edward Pendray, and Bernard Smith.* (G. EDWARD PENDRAY)

exhaust caused a rapid pressure buildup that ruptured the tank.

During the rest of 1933 the society continued to hold lectures, publish its journal (now called *Astronautics*), and develop rockets. Three different designs were produced almost simultaneously—Rockets No. 3, 4, and 5. The first of these, designed by Africano, Pendray, and Smith, was about 5.5. feet high and 8 inches in diameter, operated on one quart of gasoline and four quarts of liquid oxygen, and weighed 20 pounds. Rocket No. 3's 60-pound-thrust motors produced a 2-g acceleration.

Rocket No. 4 was somewhat longer (7.5 feet over-all) and narrower (3 inches), and had a smaller propellant capacity—only a pint of gasoline and a quart of liquid oxygen. It was shaped to keep air resistance to a minimum and to permit the motor to ride above the main body, with the exhaust gases deflected outwards by four canted nozzles. Four blades were in the rocket's nose "so arranged that during flight they were held in a collapsed position along the axis of the rocket by air resistance. When the speed slackens away, the blades are opened by springs, and the rocket descends tail-foremost, revolving as the whirling blades break the speed of fall" (*Astronautics,* March 1934). The inventors probably had an eye on Reinhold Tiling's activities in Germany when they devised this scheme. Designed by Ahrens, Best, Manning, and project leader John Shesta, No. 4 had its first static firing, on 10 June 1934, at Staten Island. The test ended with a motor burnout.

The motor was replaced, and changes were made in the rocket. Larger exhaust nozzles were installed, a water-cooling jacket was added, wider propellant inlets were substituted, and the rotor blades were removed in favor of a parachute recovery system. These modifications occupied the design team during July and August of 1934; by September the rocket was ready for another test. Early in the morning of 9 September, again at Great Kills, the flight took place. This time the results were good. After takeoff from the society's new steel launching rack, the motor burned for approximately fifteen seconds, enough to lift the rocket along a trajectory that was described as "excellent." The maximum altitude was 382 feet, the range 1,338 feet. Although not accurately measurable, the velocity appeared to exceed 600 miles per hour, possibly approaching 700 miles per hour. The only disappointment was the failure of the chute to open.

After the flight of Rocket No. 4, the experimenters decided not to attempt to fly Rocket No. 3 because of its motor inadequacies. Attention shifted from flight testing of liquid rockets toward development and static testing of more reliable and more powerful rocket engines. Rocket No. 5, because of its radical design, was never built, though some of its components were developed and used in subsequent models.

The results of the new rocket motor tests, conducted on 21 April, were reported by John Shesta a few months later in the June 1935 issue of *Astronautics.* Short and long nozzles were used, both operating alternately at pressures of 150 and 300 pounds per square inch. At the high pressures the motors with short nozzles developed 59 pounds of thrust at 430 seconds total impulse, compared with 46 pounds at 380 seconds for those with long nozzles. Thrusts and total impulses at the lower pressure were, respectively, 25 and 17, and 280 and 180. These tests were conducted on a special new proving stand at Crestwood, New Jersey.

*Rocket No. 4, only three inches wide, was shaped to minimize air resistance. Exhaust gases were deflected outward through four nozzles.*

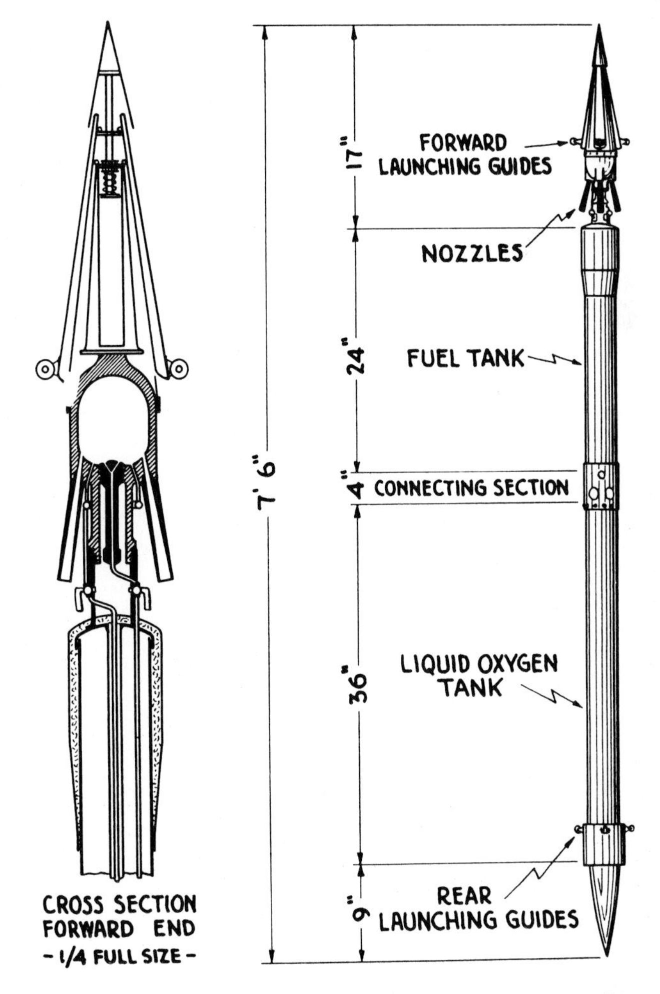

Additional tests were undertaken in July and August and again in October 1935. These continued into 1939, conclusively demonstrating that cooling combustion chambers with water was not a satisfactory solution to the problem of dissipating excess heat, a finding that opened the way for important advances in the development of cooling techniques.

The idea of using the fuel instead of water or air to cool the combustion chamber—regenerative cooling—had been considered previously by experimenters in both Europe and America. Eugen Sänger had achieved successful test results in 1933 with a Diesel fuel oil–oxygen gas engine. And Harry W. Bull had done preliminary work in the same year at Syracuse University, cooling the nozzle area only. But it was not until James H. Wyld built what he called a "self-cooled tubular regenerative motor" in 1938 that sustained work began in the United States on full regenerative cooling. The fuel was circulated in a cooling jacket surrounding the combustion chamber and nozzle, not only keeping the motor cool, but preheating the fuel and thus improving its combustion characteristics. The first tests of the 2-pound motor were made under the direction of the American Rocket Society's Experimental Committee on 10 December 1938, with satisfactory results.

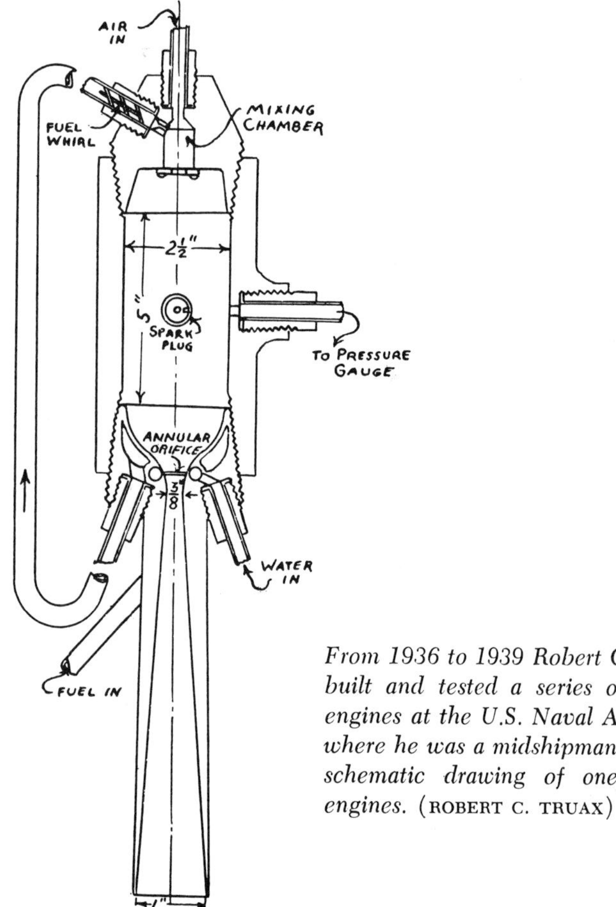


*From 1936 to 1939 Robert C. Truax built and tested a series of rocket engines at the U.S. Naval Academy, where he was a midshipman. Left, a schematic drawing of one of his engines.* (ROBERT C. TRUAX)

Other organizations were active in rocketry in the United States during the 1930's. The Cleveland Rocket Society had a rocket-testing site some 12 miles east of the city. The site included a 12-foot rocket-motor test stand, a roofed-over control trench, and a laboratory. Members built two types of motor: one made of chrome-nickel steel with water cooling and the other of a light alloy cooled by its liquid-oxygen and gasoline propellants. Experiments were under the technical direction of Ernst Loebell.

At Syracuse, Bull carried out a series of detailed investigations on the key component of the rocket motor, the combustion chamber, with the aim of bringing the reliability of the rocket up to that of more conventional motors. He and his associates varied every element of the chamber, including length, diameter, fuel-injection system, nozzle diameter, cooling system, and more. Most of the experimental chambers consisted of four sections of 2-inch tool-steel stock, produced about 2 pounds of thrust, operated on gasoline and liquid oxygen, and fired from around 20 seconds to over 100 seconds. The final design resulting from these studies, which were carried out in the early 1930's, had four gasoline and eight liquid-oxygen inlets. It was made of tool steel, and had what Bull termed "cooling fins" welded to the sides.

During the mid-1930's, some experiments with small rocket planes also took place in the United States. At Greenwood Lake, New York, F. W. Kessler sponsored the flight of two 15-foot-wingspan, aluminum airplanes powered by liquid oxygen–alcohol rocket motors developing between 40 and 65 pounds of thrust for up to 35 seconds. The two planes, carrying cargoes of mail (4,323 covers and 1,823 postcards), took off from the frozen lake on 23 February 1936, after 10-second takeoff runs. One reached an altitude of about 1,000 feet before the combustion chamber burned out, producing a side thrust that sent it spinning to the ground. The wings of the second plane tore off at about 15 seconds. The airplanes and motors were built by Nathan Carver, and flight operations were supervised by Willy Ley.

From 1936 until 1939 Robert C. Truax, a midshipman at the United States Naval Academy, developed, built, and tested small liquid-propellant rocket engines. Truax's work was to lead to the wartime Navy-sponsored rocket research program at Annapolis. Truax was a spare-time experimenter who had to scrounge help—a trait common to most amateur rocketeers, past and present. He later recalled: "I selected as the main body of the thrust chamber a nickel steel pinion gear. The hub of this gear appeared to be of proper thickness and quality to withstand almost any pressures which might be generated." He spent eight months at a lathe, working on his first test combustion chamber. As he put it, "When my masterpiece was completed, I took it to the head of the Marine Engineering Department and requested permission to set it up in the foundry and fire it. In perhaps justifiable concern over the future of Isherwood Hall [where the foundry was located], permission was denied." Truax persisted, and finally persuaded officials at the Experiment Station across the Severn River to allow him to test his rocket engine and to furnish him with additional working materials.

With the help of a welder, he built a test stand and propellant tanks, the latter from old pipe. "We then made closures for the tanks by burning circles out of boiler plates, welding them in, and providing them with gussets which appeared . . . to be about adequate in thickness and strength." Three tanks were built, one each for the oxidizer and fuel and one for cooling water.

Instrumentation was characteristically simple and direct, involving the use of the Bourdon tube pressure gages, an Eastman Kodak timer, and best of all, a stock room scale on which the thrust chamber was mounted in a nozzle-up position . . . . The instruments were then photographed with a Boy Scout camera at intervals determined primarily by the time required to wind the film . . . . The fuel consumption was measured by means of a boiler gage glass.

Since liquid oxygen was not available, he chose compressed air as the oxidizer; gasoline served as the fuel. "The first tests were run during lunch hours when the workmen from the shop would . . . gather around the rocket and amuse themselves by throwing stones into the jet to see how high they would be hurled." Some 25 pounds of thrust was generated by the engine in a December 1937 test. At first Truax reported, the motor ran "like a motorcycle engine." Then, "with extremely careful adjustment of valves, there came a short, smooth roar, and then again an infernal popping." Despite some success, the motor was redesigned with a fuel-cooled nozzle and chamber and reverse fuel injection.

By September 1938 several new chambers were ready, using refractory-lined nozzles in place of the earlier water-cooled ones. Among the materials that Truax tried were silicon dioxide, silicon carbide, aluminum oxide, and tungsten carbide. The new engine operated satisfactorily under constant chamber-pressure conditions for 10 to 45 seconds; thrust varied from 6 to 25 pounds in a series of seven runs. In December 1938 Truax obtained permission from the Annapolis authorities—who had kept an eye on his activities—for tests using oxygen in place of air. In the February 1939 *Astronautics* he was able to report on the motor, which was constructed of stainless steel, and was 14 inches long, 3 inches in diameter, and weighed slightly more than 2 pounds.

Between 1932 and 1938 the United States Army Ordnance Department conducted sporadic research on rockets, principally under the direction and impetus of Captain (later Colonel) Leslie A. Skinner at the Aberdeen Proving Ground in Maryland. A variety of single- and double-base powders were static-tested until a double-base powder of German manufacture was selected for continued develop-

*The Army Ordnance Department, under the direction of Leslie A. Skinner, concentrated on air-to-air rockets that could be fired from one plane to another. Below, Skinner's diagram for a liquid-propellant engine operating on gasoline and nitrous oxide.*

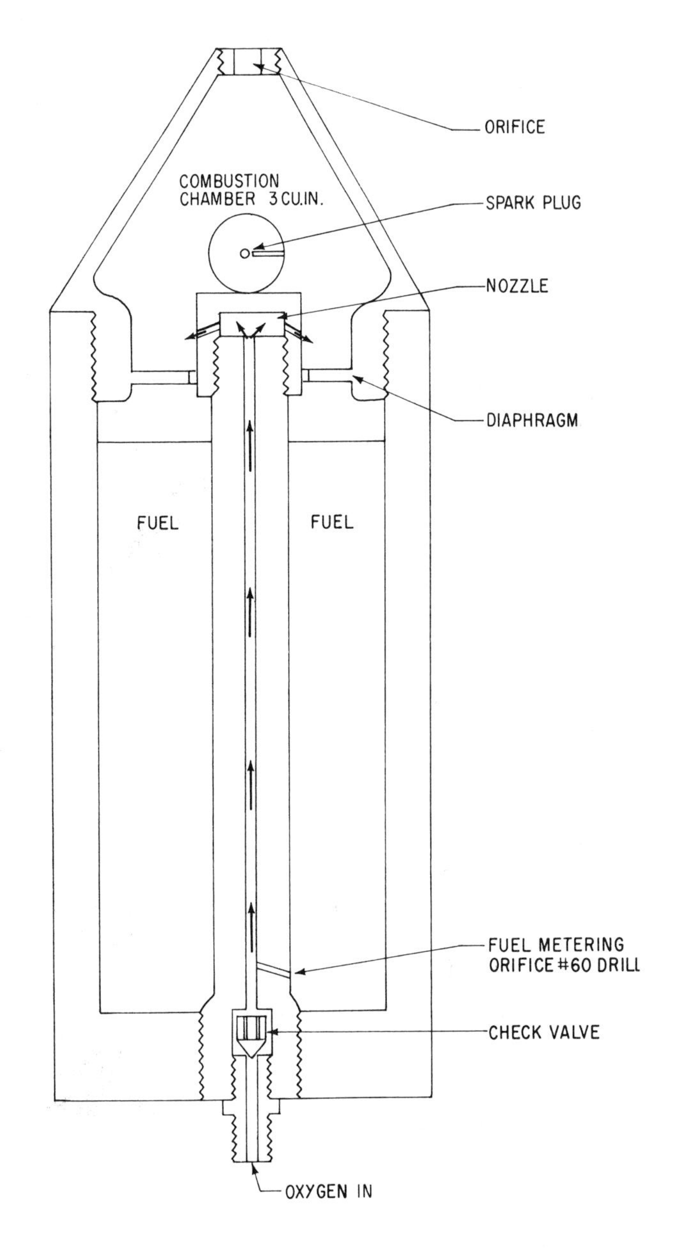

ment. Skinner's primary goal was to develop a rocket that could be fired from one airplane against another —in modern terminology, an air-to-air rocket. Skinner modified 81-mm trench mortar ammunition, loaded the new propellant, and fired the resulting rocket from a simple pipe launcher. The rocket's range was fairly good, but Skinner described its accuracy as "extremely poor."

When Skinner was transferred to Hawaii in 1938, rocket work was stopped at Aberdeen. He returned to the United States in 1940 and was assigned to the Bomb and Pyrotechnic Section of the Chief of Ordnance. His official title was Director of Army Projects of the newly formed Rocket Ordnance Section (Section H) of the National Defense Research Committee. NDRC's rocket group was under the direction of Clarence N. Hickman, who had worked with Goddard in California during World War I, and later became a research director at Bell Telephone Laboratories. A letter written by Hickman in June 1940 to Frank B. Jewett, president of the National Academy of Sciences, was instrumental in calling the government's attention to the importance of Goddard's work. Supporting Skinner and his associates from his position within NDRC, Hickman was able to push the Navy into developing rocket-accelerated bombs and the Army into developing both the antitank weapon that became the bazooka, and the 4.5-inch aircraft and artillery rockets.

The ARS and other groups of experimenters in America and in Europe concentrated on liquid-propellant motors during the 1930's, primarily because liquids contained greater energy than solids; that is, they produced more pounds of thrust per pound of propellant burned per second. Being more powerful, they appeared more attractive for eventual space-flight applications. But liquid fuels were difficult to handle, and being relatively new, little was known about them.

Solid propellants also had both advantages and disadvantages. Since they had been used for centuries, much more was known about them. Solid propellants could be stored for relatively long periods of time; they were comparatively simple to work with, being less hazardous and corrosive than many liquids; they formed an integral part of the motor case, eliminating the need for complicated loading procedures, "plumbing" hardware, and injection apparatus to feed the fuel into a combustion chamber. Finally, the solid propellants were more reliable and less expensive than the liquids.

On the minus side, the solid propellants were less powerful. Once a solid propellant had begun to burn, it could not be shut off and then restarted by closing and opening a valve. And since the solid propellant was part of the motor case, it could not provide the directional control obtained by swiveling the combustion chambers of a liquid-propellant rocket. When such factors as thrust control, noise, vibration, and acceleration were considered, solid propellants also were at a disadvantage.

Many of the problems inherent in both solids and liquids eventually were solved by the early rocketeers, who continued experiments on both types of fuel. While researchers in the United States spent most of their time developing liquid propellants, the solid-propellant motor was not ignored.

Beginning in September 1937 in Pawling, New York, the ARS Experimental Committee test-fired seven small rockets powered by dry-fuel cartridges, including a two-stage model looking something like the German Repulsors, and designs developed by ARS experimenters Shesta, Pierce, Africano, Goodman, and Wyld. They ranged in weight from 1.19 to 2.28 pounds, took from 3 to 7 seconds to make their ascents, and reached altitudes of 100 to 1,500 feet. More solid-fuel rocket tests were conducted by Peter van Dresser and Alfred Africano during the summer of 1935 in Danbury, Connecticut, the main purpose being to determine where the motor should be placed in the rocket.

In September 1939, the ARS Experimental Committee conducted another series of flights at Mountainville, New Jersey. The maximum altitude achieved in 23 shots was 1,930 feet, reached during a 24-second flight of a 39-pound rocket; none of the other rockets got as high as 800 feet. A set of twelve firings was made on 19 November 1939 at the same location, using Unexcelled Fireworks Company 2-, 3-, 4-, and 6-pound commercial powder rockets. While none reached the peak altitude of the September series, they averaged considerably higher—none under 500 feet, five over 800 feet, and one over 1,000 feet.

A considerable amount of amateur rocketry during the 1930's was organized by university groups under the supervision of professors of mechanical and other branches of engineering. Scattered nonuniversity experimenters were active, some forming local rocket clubs. Most of these closed down when World War II started, but they did serve as training grounds for many rocketeers.

One group that stayed in business after Pearl Harbor was the Galcit Rocket Research Group in California. The Guggenheim Aeronautical Laboratory of the California Institute of Technology (hence the name Galcit), had started a program of rocket research and development in 1936. Operating under a fund established by Weld Arnold, the group made thorough studies of reaction-engine technology in general and rockets in particular, both liquid- and solid-propellant based. The experimental group consisted of Arnold himself, Frank J. Malina, Hsue-shen Tsien, Edward S. Forman, John W. Parsons, and A. M. O. Smith. Theodore von Kármán, Galcit's director, became more and more active as the project matured. With the help of data gathered by ARS experimenters, Galcit progressed rapidly to the rocket-motor test phase. A variable-pressure combustion chamber (100 to 1,000 pounds per square inch) was developed, operating on methyl alcohol and gaseous oxygen. By 1938 the Army Air Corps took notice of Galcit; in a modest way, what had happened in Germany was about to take place in the United States. That December General Henry H. Arnold, commanding the Air Corps, asked the National Academy of Sciences' Committee for Air Corps Research to sponsor a development program for rocket units to help heavily loaded planes take off from short runways. On 1 July 1939, the Academy sponsored the establishment of the Jet Propulsion Research Project at the California Institute of Technology. Von Kármán was the director, and three members of the Galcit group, Malina, Forman, and Parsons, were his assistants. A year later the Army Air Corps took over the project.

The rocket-boosted takeoff project, which became known as JATO, built on experience accumulated since 1936. Both liquid- and solid-propellant motors were studied. The final production units, firing for 10 to 30 seconds, could lift an airplane high enough to continue its flight unassisted. Since only rapid-burning solid propellants were known then, much time was spent on developing slow, or restricted, burning propellants that would provide a constant thrust. By 1941, Galcit 27, which delivered 28 pounds of thrust for 12 seconds, was developed. Tests of motors burning this solid propellant began a few months before the United States entered World War II.

Scientists and engineers working on liquid-propellant JATO units discarded liquid oxygen, the conventional oxidizer, because JATO units had to be mobile. Liquid oxygen was ruled out because it cannot be stored for long periods and is not easy to transport for long distances. As an alternative, the experimenters began investigating the possibility of using red fuming nitric acid usually referred to as RFNA. Although corrosive and poisonous, RFNA can be stored under certain conditions. Whether it would decompose efficiently with a fuel was not known, but tests during the fall of 1939 showed that it mixed with, and supported the combustion of, gasoline and benzene. A test stand was constructed and an experimental motor was built. In May 1940 the first tests were held, and in July the program was given additional support by the Army Air Corps. The immediate objective was development of a 1,000-pound-thrust motor, which was to become the prototype of a new type of rocket widely used during World War II.

As World War II approached, the future of rocketry had begun to take definite shape. Germany, which had recognized more clearly than other countries the military potential of the rocket, was clearly ahead in the field, but no nation was without its group of rocketeers who would be given their opportunity when military needs for weapons arose. The rocket would soon be placed into combat on a large scale—simple, unguided weapons at first, then complex and terrifying missiles. Although they would not be decisive, they would herald the dawn of a new concept of war.

Rocketry had struggled through its infancy. Just ahead was an accelerated adolescence. The transition between the struggling experimenters on the vacant lot and the massive missile complexes of today was about to be made.

# 5 THE ROCKET RE

World War II abruptly changed the course of rocket history. Until then there had been little continuity of development throughout the world. During the 1920's and 1930's progress was made by fits and starts and goals were unclear. But, with the coming of war, rocketry suddenly flowered.

Every major combatant nation had a rocket program. Under the impetus of war, rockets increased dramatically in size, range, and accuracy. Many types of rockets proliferated. A basic distinction was made between the rocket *per se,* which traveled over a preordained and fixed trajectory, and the missile, a device which could be guided while in flight toward its target. Separate families of rockets and missiles also evolved. Depending upon their primary missions, the weapons were grouped under such headings as surface-to-surface, surface-to-air, air-to-air, and air-to-surface. The war saw the development of every basic type of rocket and missile used today.

A tight curtain of secrecy was wrapped around rocket development during hostilities. When peace came, information about the research programs of the Western allies and the Axis powers became public knowledge. The Soviet Union, however, has maintained its policy of secrecy until this day. Some technical and manufacturing details of rockets twenty or more years old still are not available.

What is known about Soviet World War II rockets has been gleaned painstakingly from the few official articles that have been published. The best known of all Russian rockets was the Katyusha, a solid-propellant infantry support weapon that was manufactured and fired in enormous quantities. The standard model was 6 feet long, 5.1 inches in diameter, and weighed 92.5 pounds, of which more than half—48.2 pounds—was payload. Katyusha was fired from ground- or truck-mounted racks, and had a range of over 3 miles. Much of the double-base powder that propelled Katyusha came from the United States through lend-lease.

Other versions of the rocket were fired from the "Stalin Organ," a mobile launcher that could fire a broadside of 30 to 48 rockets. The smallest rocket was 3.3 inches in diameter, weighed 17.5 pounds, and had a range of less than 3 miles. A 30-pound model had twice this range.

The unguided solid-propellant rockets used against airplanes by the Russians are believed to have been variations of the Katyusha, but details about them have not been revealed. Russian air-to-surface rockets, fired by airplanes against troops and vehicles, were introduced late in 1941. The most widely used model carried an explosive charge of 2.2 pounds, which slammed into the target at up to 1,150 miles per hour. The rocket had a solid propellant, was nearly 2 feet long and 3.2 inches in diameter, and weighed a total of 13.2 pounds.

"Rocket bombs" weighing 56 and 220 pounds were used against tanks and other armored vehicles. A Stormovik Il-2 fighter would carry eight 56-pounders on special wing racks; the German troops had a healthy respect for these weapons.

Little can be said about the Soviet World War II rocket program because of Russian reticence. The story of the Japanese rocket effort is almost as brief, but for another reason. With little interest in rocketry, the Japanese military had the most backward rocket program of all the major belligerents. The barrage and close-support rockets used by Japanese troops during the war were often improvised at the front. The weapons varied widely in description, with diameters ranging from 3.2 to 18 inches and weights from 12 to 1,500 pounds. Typically, ranges were short—from 300 to 500 feet. The Japanese did introduce a bazooka-type antitank rocket, but this was more or less a copy of captured American specimens.

The Japanese also experimented with some more sophisticated rockets and missiles. In an effort to combat United States air attacks, the Japanese Naval Technical Research Institute developed solid-propelled surface-to-air barrage rockets weighing from 13 to 53 pounds and began research on a series of four Funryu guided missiles. Airframes for the Funryu

# JRNS TO WAR

series were built at the naval dockyard at Yokosuka and the engines at the Naval Powder Arsenal and at Mitsubishi Heavy Industries, Ltd. The two completed missiles in this series, Funryu 2 powered by solid propellant, and Funryu 4 powered by liquids, could reach altitudes of 3 to 20 miles, respectively, and were guided by radio commands from the ground. Funryu 1, which never left the experimental stage, was to have been command-guided and directed against surface vessels, while Funryu 3 was to have been a liquid version of Funryu 2.

In the air-to-surface category, two unguided rocket-propelled bombs, weighing 224 and 815 pounds and each with a range of 3 miles, were introduced during the war, while three guided missiles were still being tested when the war ended. The air-to-surface missiles, developed by the Army Bureau of Aeronautics, were models 1A, 1B, and 1C of the I-go series. 1A and 1B, made of wood and powered by hydrogen-peroxide motors, were manufactured by Mitsubishi Heavy Industries and the Kawasaki Aircraft Industry Company. 1A's motor, firing for 75 seconds at a maximum thrust of 530 pounds, was to be guided to the target by radio command. The 1B model had a slightly more powerful motor that gave it a 5-mile range. The 1C model, designed to home on the shock waves produced by naval guns, was canceled in 1945.

One sinister curiosity of the Japanese rocket program was a wooden, rocket-powered suicide plane that saw action in 1945. The Japanese had a fling with their own version of the German Me-163, but the only one of the two rocket planes built that ever flew cracked up on its maiden flight from Oppama on 7 July 1945. Their Ohka (sometimes called Marudai) Kamikaze plane was more successful—unfortunately for its pilots.

Ohka was conceived in the spring of 1943 and developed in the following year. The standard model was a wooden monoplane with two fins attached to a 2,645-pound bomb, which accounted for more than half the plane's total weight. Ohka was 20 feet long, with a 16.4-foot wingspread. It was carried to the vicinity of its target by a mother plane. The suicide pilot would then glide for about 50 miles, making a beautiful target at the rather slow speed of 230 miles per hour, before igniting the craft's three 1,700-pound rocket engines. These propelled the plane into its target—or the sea—at a speed of 600 miles per hour after their 10-second burn. The Ohka was first used in combat in April 1945. The American nickname for the airplane was *Baka,* the Japanese word for "fool."

Unlike the Russians, who have not told much about their wartime rockets, and unlike the Japanese, who did not have much to talk about, the British have released information about a wide range of effective surface-to-air antiaircraft rockets which were developed to protect troops and ships.

Appreciating the threat of German air raids to her cities, Britain began work on these defensive weapons before the war and initial tests were highly encouraging. The first surface-to-air rocket placed in production was a 2-inch barrage model whose per-

**Funryu Surface-to-Air Missiles**

| *Missile* | *Length (feet)* | *Diameter (inches)* | *Weight (pounds)* | *Altitude (miles)* | *Velocity (mph)* | *Propulsion* |
|---|---|---|---|---|---|---|
| Funryu 2 | 7.9 | 12 | 815 | 3 | 525 | solid |
| Funryu 4 | 13.1 | 24 | 4,190 | 20 | 650 | liquid |

**I-go Air-to-Surface Missiles**

| *Missile* | *Length (feet)* | *Span (feet)* | *Weight (pounds)* | *Range (miles)* | *Velocity (mph)* | *Propulsion* |
|---|---|---|---|---|---|---|
| I-go (1A) | 18.9 | 11.8 | 3,085 | 6–7 | 340 | liquid |
| I-go (1B) | 13.4 | 8.5 | 1,500 | 5 | 340 | liquid |

formance paralleled that of the British 3-inch anti-aircraft gun. The 2-incher was propelled by a tube of cordite, which was placed in the low-carbon sheet-steel case in such a way that all the exposed surfaces burned when the electrical ignition signal was given. The motor walls were protected from the high-combustion temperatures by a spray composed of sodium silicate solution containing a suspension of finely ground alumina. A shell ring, secured by spring pins, enclosed the motor at the front end. The motor burned for 2.8 seconds and produced 17 pounds of thrust.

The 2-incher was designed for defense against low-flying bombers. Soon a more powerful 3-incher was demanded. Most of the technical problems of producing the bigger rocket were overcome by the summer of 1940, and by the end of the year, the 3-incher was being used to supplement heavy gun defenses. A salvo of up to 128 rockets could be fired from a battery of twin-barreled launchers (which the British called "projectors").

The development of these projectors was spurred by a meeting held on 20 May 1940 in the private room of a small hotel in Wales, near the Aberporth test grounds, at the request of the Director of Naval Ordnance. The War Cabinet had directed that rocket weapons should be used "forthwith" to protect ships, factories, and other targets against air attack.

*During World War II, the Soviet Union fired Katyusha rockets from the ground or truck mounted racks. The standard Katyushas were 6 feet long, weighed 92.5 pounds, and had a range of 3 miles. Below, they are being fired against the Germans in the battle for Sevastopol.* (U.S.S.R. ACAD. OF SCIENCES)

By July, the firm of G. A. Harbey of Greenwich had designed and manufactured the first ten projector units; by September, one thousand had come off the assembly line. In October, Major Duncan Sandys took command of the experimental Z-rocket battery. The following March, Sandys, by then a lieutenant colonel, organized a regiment of 3-inch rocket batteries to defend Cardiff. The unit downed its first German plane on 7 April 1941.

Sandys, who had been evacuated in June from the British Norwegian expedition, had at first been given orders to take the existing Z batteries from East Hamstead Park in Berkshire to combat German dive bombers in Malta, but quickly found the units inadequately trained, poorly officered, and without sufficient reserves of rocket ammunition. Accordingly, he took immediate steps to select a new group of officers, including Captain (now Colonel) Kenneth Post as his second in command, and request additional time to bring the batteries up to combat proficiency.

In September he moved to Aberporth in Wales, persuading Sir William R. Cook, Crow's number 2 man, that the UP-3 projectile was as applicable against high-flying enemy bombers as against Stuka dive bombers—which were only rarely employed against British targets. Sandys and Post tested the UP-3's with new fuzes against Queen Bee drones with such success that they were able to convince General Sir Frederick Pile, head of the AA Command, that three Z batteries could promptly be activated, one to be kept at Aberporth for training and development and two to be placed into operational combat service in Cardiff, also in Wales. Sandys set up the various radar elements, projector batteries, and command equipment (including the bomber order-of-battle predictor from which individual batteries received information as to bearing, elevation, and fuze settings) in Cardiff's Penarth Golf Course.

From January 1941 into the spring Sandys literally commuted between Cardiff and Aberporth, remaining in the former city until midnight or 0100 when the German raids would normally terminate, then driving to the test and training center where he would supervise the next day's activities. On Good Friday his chauffeur fell asleep at the wheel during the trip, causing an accident that so injured Sandys that he had to leave active military service. Subsequently, he became Junior Minister in the War Office, heading up an intelligence investigation on the capabilities of German long-range bombardment rocketry.

Both rocket and projector were refined later. Two types of 3-inchers were used, a finned version and the UP-3 unrotating projectile model. The finned version was 6 feet long and weighed 56 pounds; the UP-3 was 4 feet long and weighed 110 pounds. Time fuzes were used for targets at altitudes up to 4 miles and photoelectric fuzes above that height. The UP-3 had a lethal radius of 65 feet.

The single-rocket projectors were replaced by twin "Pillar Box" types, which more than doubled firing rates, and about 100 3-inch antiaircraft gun mountings were adapted into 9-barrel projectors, which were used in Great Britain and North Africa.

By December 1942, 91 batteries, each with 64 twin-barrel projectors, had fired a total of 65,000 3-inch rockets against enemy aircraft. The use of rockets would have been even greater if the factory that produced their fuzes had not been destroyed twice in air raids.

Since the rockets were always used with conventional antiaircraft weapons, their effect is difficult to assess. However, an official report dated 2 November 1944 said: "There is no doubt that the deterrent effect of rockets is considerable. Few pilots will fly straight if they believe their aircraft to be the target for a rocket salvo. AA [antiaircraft] rocket fire, therefore, fulfills the important function of discouraging accurate bombing." The report said that pilots were able to see and avoid the rockets, but that their evasive action disrupted bombing runs.

A tribute came from eyewitness Captain H. Spencer, whose rocket-armed ship, *City of Lincoln*, was attacked by German planes in June 1942. "I had many opportunities for using my P.A.C.'s [the rockets] during the action and have the highest opinion of their value," Spencer wrote. "[They] kept the aircraft up all the time and on several occasions I saw them swerve violently just before dropping their bombs. It did not strike me that this was due to anything but the P.A.C. rocket. Personally I should never like to be without my P.A.C.'s in any aircraft action."

In a typical combat operation in North Africa, Z Battery 124, under air attack near Bône, fired 800 rockets in 10 salvos, hitting a number of the 50 raiding airplanes. The battery itself suffered no damage from the enemy.

An interesting variation on this theme was the

*The Japanese Funryu 2, a solid-fuel guided missile, could reach altitudes of 20 miles and was guided by radio commands from the ground. The midsection of the Funryu is shown here.* (MITSUBISHI INDUSTRIES)

*Japan's famous suicide plane, the Ohka Kamikaze, was propelled by three 1,700-pound rockets and carried a 2,645-pound bomb. The Kamikaze plane shown here was found on an airfield in Okinawa two days after the American landing.* (U.S. ARMY)

*Sir Alwyn D. Crow (right), head of the British missile effort, shows Prime Minister Winston Churchill a 2-inch rocket in Shoeburyness, Essex, in 1941. The 2-incher was the first in a series of surface-to-air rockets developed by the British to protect their cities from low-flying German bombers.*

Snare project, designed to down low-flying enemy planes by entangling their propellers in wires. Officially called Antiaircraft Parachute and Cable Rocket Projector, the Snare was generally installed on the flying bridge of a ship. A lanyard fired a 3-inch rocket, which released a parachute at 550 feet. Connected to the parachute was a 200-foot wire with another parachute at its opposite end. A 90-foot trip wire, attached to the ship, opened the second parachute. As the two parachutes floated down, the wire provided a nasty obstacle for enemy bombers. Snares were generally fired when the enemy airplane was 2,000 feet from the ship at a height of 500 feet or less. The first kill, a German dive-bomber, was chalked up in July 1940.

Toward the end of the war, the British produced a new winged antiaircraft missile to combat Kamikaze attacks. The missile, called the Stooge, was 10.5 feet long and weighed 740 pounds, of which 220 pounds was warhead. It was launched from an adjustable ramp by four solid-propellant rockets that fired for 1.6 seconds and developed 5,600 pounds of thrust; then a 760-pound-thrust sustainer engine boosted the Stooge to its top speed of 500 miles per hour. With a range of 8 miles, the Stooge was guided by radio commands.

British air-to-surface missiles got off to a later start. Not until July 1941 did the Ordnance Board establish a requirement for a preliminary feasibility study. In August, General Mason MacFarlane saw Russian airplanes launch their rockets; the next month Sir Alwyn D. Crow, head of the British rocket effort, met with officials at the Ministry of Aircraft Production to outline a development program.

The first test took place in October, when a Hurricane fighter fired a 3-inch rocket at a ground target. By May 1942, flight trials had progressed to the point where Hurricanes were carrying eight rockets, four under each wing, firing them from rail launchers.

However, it was decided that tanks were too small and maneuverable to serve as targets for air-launched 3-inchers. The belief that airplanes would have to fly dangerously close to the ground to use the rockets led to the decision to fly rocket missions only against naval targets, including submarines.

After the Air Ministry published in December 1942 a detailed report on air-to-surface rocket trials at the Pendine Range and the Boscombe Down Experimental Aircraft Station, the Fleet Air Arm accepted the 3-incher. In April 1943, a Swordfish from HMS *Archer* sank a German U-boat with rockets—a first in sea warfare. The rocket could punch through a submarine's hull after traveling 50 feet underwater, and was equally effective against surface ships; rockets later sank the Italian liner *Rex.*

The Air Ministry had second thoughts about the

*Toward the end of the war, the British produced the Stooge missile to combat Kamikaze attacks. Stooge had a range of 8 miles and was guided by radio control.* (MINISTRY OF AVIATION, LONDON)

3-inchers, which were eventually used against tanks. The airborne version, called the RP-3, was a spin-stabilized rotating projectile. It was 5.35 feet long, weighed 60 pounds, had a range of about a mile and traveled at a top speed of 1,000 miles per hour. The rockets were usually fired in pairs from airplanes.

The surface-to-surface missile entered the British inventory even later. While consideration had been given to development of a 60-pound, 3,500-yard-range missile for use by demolition troops and sappers (engineers) as early as 1939, and a prototype was ready by June 1940, progress was made slowly.

By January 1943, a 5-inch-diameter rocket was virtually ready for production. It was to be launched either from a two-legged, two-rail launcher that was 6 feet long, weighed 40 pounds, and could be handled by a single soldier, or from a truck-mounted launcher with six pairs of rails.

But at the last minute the Army decided against using the weapon. The reasons boiled down to doubts about the rocket's range and reliability, a belief that it was not needed in the field, and a shortage of sapper personnel to handle the weapon.

If the Army didn't want the 5-incher, the Navy did. A few months before the 1943 Army turndown, a review of the Dieppe Commando raid, conducted at Combined Operations Headquarters, had concluded that more firepower should have been brought to bear on the landing area before the assault. The 5-inch rocket was suggested as an ideal weapon to make up the lack in future amphibious assaults, especially if it could be fired in heavy salvos from specially designed landing craft.

That led to the outfitting of a Landing Craft Tank (Rocket) with sextuple "Mattress" projectors a short time later. The 5-incher itself was equipped with a new high-explosive shell and an improved cordite motor. The end result was a craft that could fire a devastating salvo of from 800 to 1,000 rockets in less than 45 seconds at ranges up to 3,000 yards. After trials early in 1943, six LCT(R)'s were dispatched to the Mediterranean to support the landings in Sicily and Italy that were to take place during July.

After the Sicily landing, an eyewitness, Lieutenant Commander M. Mulleneux, wrote, "Whatever the destructive effect of the barrage, the effect on morale is shattering . . . ." C. F. Bruce, a headquarters operations planner who also witnessed the landings, said that

> all officers to whom I spoke out there were unanimous in their praise of these craft. On the sea when the LCT(R) is firing it presents a most impressive spectacle with its sheets of flame followed by the rushing of rockets overhead and the colossal "clumps" as they land on the target. One LCT(R) fired quite close to the craft I was in, and the sight and sound of it greatly cheered all the soldiers and helped to make them forget how cold and wet they were.

*A Beaufighter MK-1 launches two 3-inch rockets against a German ship off Norway. Britain's first air-to-surface weapon, the 3-incher was used primarily against naval targets, including submarines.* (SIR ALWYN D. CROW)

The Navy's success caused the Army to take a new look at land applications for the weapon that it originally developed. Demonstrations for War Office experts, held at the Army's Sennybridge Range in Wales, produced no immediate results, but development work continued. By the end of the year,

*British troops load rocket launcher near Reichwald, Germany. This "Land Mattress" launcher could be mounted on truck-trailers or self-propelled vehicles.*

Crow's Projectile Development Establishment offered the Army a spiral rail launcher that gave the 5-incher better accuracy, smaller fins, less weight, and a warhead of larger diameter. A proposal by Lieutenant Colonel Michael Wardell in March 1944 led to the development of a "Land Mattress" launcher that could be mounted on truck-trailers or self-propelled vehicles. A prototype, with 32 spiral rail barrels, was ready in May, and was demonstrated to observers from the War Office and Canadian Army Headquarters in June and again in July. The Canadians were impressed enough to order a dozen 30-barrel projectors, which they used that fall when their troops crossed the Rhine and Scheldt rivers.

While the 5-incher was relatively accurate at long ranges, it ran into difficulties at shorter distances. Brigadier A. F. S. Napier, military advisor to Crow, wrote that "at low angles of elevation the range dispersion is very great so a new method of varying range had to be devised; this consisted in placing over the nose of the shell flat discs of varying diameter known as spoilers; these have the effect of increasing air resistance at high angles of elevation."

The 5-incher had a quick-burning, multiple-stick cordite propellant. Eleven sticks of cordite were in a space between two tubular walls, with perforations allowing a free flow of gas.

The British undertook no major wartime programs involving liquid rocket motors. A. V. Cleaver, chief engineer for rocket propulsion at the Aero Engine Division of Rolls-Royce Ltd., one of Britain's leading postwar liquid-propellant rocket researchers, made this comment to the authors:

> I can assure you that British developments on liquid propulsion were quite negligible up to the end of the Second World War. In fact, literally the only work of this nature which was undertaken here was the project to develop a liquid oxygen–petrol rocket engine during the closing years of the war. This was done by a man [now dead] called Dr. Isaac Lubbock, who worked for the British Shell organisation, and is quite well known for his other work on developing fuel injectors for the early gas turbine jet engines. . . . The work was generally done in association with the Royal Aircraft Establishment at Farnborough, and eventually developed into a project for a small vehicle called "LOP/GAP." This stood for "Liquid Oxygen Petrol/Ground-to-Air Projectile," and this in turn evolved into an improved vehicle known as RTV.1 [Rocket Test Vehicle 1] which was used in some quantities during the early post-war years at Woomera [Australia] for research on early British anti-aircraft guided weapon systems.

Lubbock's wartime research on liquid-propellant rockets was done under a cooperative contract between the Asiatic Petroleum Company (which later became Shell Petroleum Company, Ltd.) and the Ministry of Supply, whereby the government paid for materials and the company underwrote labor costs. Lubbock had served during World War I in France, had graduated from Cambridge University in mathematics and engineering, and had lectured at the Royal Military Academy at Woolwich prior to becoming head of the company's fuel oil department in 1926. Following the evacuation of British forces at Dunkirk in the Second World War, he began considering the possibility of bombarding the Germans from England across the channel by rocket-powered weapons.

He first investigated solid-propellant rockets, submitting proposals to both Lord Cherwell (Professor F. A. Lindemann, Churchill's scientific advisor) and Sir Alwyn D. Crow in January 1941. Crow replied that solid-propellant work was well in hand but suggested that Lubbock look into the feasibility of using liquid-propellant rockets to assist heavily loaded Wellington bombers to become airborne. With virtually no information on liquids available to them, Lubbock, M. G. J. Gollin, and a few associates started theoretical investigations of the performance of gasoline with liquid oxygen before embarking on experimental research. According to Gollin, among the unknowns they had to deal with were:

> how to insulate a liquid oxygen vessel, how to purge the propellant lines and chamber, how to reduce nitrogen gas pressure, how to handle the excess heat produced during combustion, how to initiate combustion, how to measure thrust, how to calculate the specific impulse of the engine, and how to cool the nozzle.

The solutions to these and many other problems had to be worked out from scratch since Lubbock's group did not have access to German prewar work and was not aware of the liquid-propellant rocket experiments of Goddard and the American Rocket Society across the ocean. During the first year, Lubbock's staff totalled eight, including one boy, and his budget a mere £10,000.

The first cold flow tests were made during February and March 1941 at the Fuel Oil Technical Experimental Station at Fulham, followed by hot combustion tests between May and September at Cox Lane, Chessington. In a series of pilot runs code-named Plow (for petrol–liquid oxygen with water as a temperature moderator), a thrust level of 60 pounds was achieved. Later, full-scale engine tests developing 1,600 to 1,900 pounds of thrust for 20- to 35-seconds duration were made at the Langhurst Flame Warfare Station near Horsham in Sussex. The first run was made on 15 August 1942 for 5 seconds; at the end of September, the experimenters fired for 23 seconds at a maximum thrust of 1,750 pounds.

Lubbock was important to the British in more

ways than one. British Intelligence found his knowledge of liquid propellants extremely useful, and he was called in frequently to help interpret the probable characteristics of German liquid-propellant rocket weapons. Lubbock was unquestionably Britain's leading authority on the subject.

Despite Lubbock's work on liquid-propelled ATO units, the major British effort in this area was with solid-fueled rockets. Some creditable, if not spectacular, progress was made, with solid-propellant motors for lifting airplanes from ships. The first model to be developed, called CRC for cordite rocket catapult, was used on merchant ships. It consisted of a trolley to which were attached two solid-propellant rockets. When enemy planes attacked, the rockets would be ignited and the trolley would hurl the attached airplanes aloft. When escort carriers were attached to convoys, an ATO system was developed to insure that many airplanes would get into the air quickly. The ATO unit was dropped by the airplane after burnout.

The United States drew on British rocket knowledge after Pearl Harbor. The United States had very little experience in high-energy solid propellants of the type neeeded for extended-range, high-speed weapons. The only commercially available product was ballistite, a double-base smokeless powder derived from British trench mortar powder. A visit by Sir Henry Tizard to the United States as the head of a British scientific mission, a trip by Charles C. Lauritsen of the National Defense Research Committee to Great Britain in the summer of 1941 to study their production facilities, and a demonstration of British antiaircraft rockets at the Aberdeen Proving Ground early in the war helped bring British know-how to bear on the problem. A pilot plant was obtained from Britain, and Lauritsen and William Fowler at the California Institute of Technology led the effort that ended with a practical production technique. Within a few years, suitable solid propellants were being produced at the Army's Radford Ordnance Works, the Navy's Indian Head Powder Factory, and the Sunflower Ordnance Works, operated by the Hercules Powder Company.

This effort was part of an over-all United States rocket program which, although modest in comparison to German wartime development, was much larger than commonly believed. The United States conducted some experiments with virtually every type of rocket and missile. While relatively few kinds of rockets actually were used by United States combat forces during the war, those that were produced were made in huge quantities. By 1945, the Army was spending $150 million a year on rockets, and the Navy was spending $1.2 billion. The Navy alone had 1,200 plants making rockets or their components.

The United States rocket program was under the direction of Division A (later 3) of the National Defense Research Committee, a coordinating agency established by President Franklin D. Roosevelt on 27 June 1940. The head of NDRC was Vannevar Bush. Its Division A (Armor and Ordnance) was directed by Richard C. Tolman and was divided in turn into two sections of civilian scientists and engineers who worked on rockets and missiles. Section H (named for its leader, Clarence N. Hickman, whose letter urging a rocket development program was a seminal event in American rocket history), and Section L (for Lauritsen, who also was vice-chairman of Division A).

Hickman described in detail to the authors the genesis of Section H, which was to play a vital role in the development of United States rocketry during World War II. In 1917–1918, while studying for his master's degree under A. G. Webster at Clark University, Hickman had met Goddard—at the time working under a small Smithsonian grant to develop a multiple-charge rocket. Acting upon the suggestion of L. T. E. Thompson, Webster's assistant, Goddard sought Hickman's aid in solving some mechanical problems, with the result that the two worked together during the First World War in California, and later at the Aberdeen Proving Ground in Maryland. After the war Goddard, Thompson, and Hickman continued their association. By the time NDRC was created, Thompson was in charge of research at the Navy's Dahlgren Proving Ground, so it was rather easily arranged for Hickman's Section H to begin its testing program at that Virginia site. According to Hickman:

> It was not long before Dahlgren became crowded with other work and they wanted to get rid of my testing there. Dr. Thompson suggested that I could use the old valley part of Indian Head [Naval Powder Factory, Indian Head, Maryland] where the Navy had done the testing of big guns. . . . Dr. Tolman, Dr. Thompson, and I paid a visit to Indian Head and decided that it would be a good move, so the Indian Head Propulsion Laboratory was organized with me as director. I still retained my position as chairman of Section H, Division A.

As work expanded, a contract was made with the George Washington University for personnel and services. A team of consultants, including Goddard, was built up, and before long much of the administrative work was taken over by R. E. Gibson, appointed vice-chairman of Section H. Much of the manufacturing work was handled by the Budd Wheel Company of Detroit.

As expansion continued, Hickman again had to look for new research and testing facilities. "Dr. Van Evera, who was head of the George Washington con-

*The bazooka, one of the most popular and effective weapons of World War II, was developed late in 1941 by an Army team of Leslie Skinner, C. N. Hickman, and Edward G. Uhl. Above, the original bazooka, autographed by its developers.* (C. N. HICKMAN)

tract, and I paid a visit to Cumberland [in early 1944] to see an Army Ordnance factory that had been built for manufacturing small arms ammunition. They had closed the plant because they found they did not need it. We thought it was just the place for us so we took it over." Still later, the Section H group moved to the Allegany Ballistics Laboratory at nearby Pinto, West Virginia, where they worked closely with Army Ordnance, the Chemical Warfare Service, and the Air Corps.

On the West Coast, meanwhile, Hickman assisted in the establishment of Section L under Lauritsen, which "took over the work at Cal Tech on rockets." Outlining the role of his section in Pasadena, Lauritsen told the authors:

> During the years 1941 to 1945, we designed and developed all the rockets that were used by the U.S. Navy during the war. We established and operated the Naval Ordnance Test Station, China Lake, California, for testing rockets, training Navy personnel, and for pilot production.

There were many facilities involved in the overall rocket program, both Army and Navy: the Picatinny Arsenal, the Aberdeen Proving Ground, Wright Field in Ohio, the Dover (Delaware) Army Air Force Base, the Navy Proving Ground at Dahlgren (Virginia), and the Naval Ordnance Test Station at Inyokern (California). Many of these activities and those under NDRC were coordinated by an *ad hoc* Committee on Controlled Missiles established in June 1942 by the Joint Committee on New Weapons and Equipment. This was replaced in January 1945 by the Guided Missiles Subcommittee, which was organized by the Joint Chiefs of Staff, placed under the chairmanship of Bradley Dewar, and kept under the control of the same Joint Committee. Still later, the subcommittee's functions were taken over by the Joint Research and Development Board.

The first company created solely to produce rockets was Reaction Motors, Inc., of Pompton Plains, New Jersey, which is now a division of the Thiokol Chemical Company. Reaction Motors was founded in late 1941 by members of the American Rocket Society. Their work on regeneratively cooled liquid-propellant rocket engines eventually led to development of JATO units and test missiles during the war.

A second pioneering company, Aerojet Engineering Corporation, was organized in Azusa, California, by a group associated with the California Institute of Technology and interested in producing the JATO units developed by Galcit. Aerojet JATO units were used in large numbers during the war; the company was absorbed by the General Tire and Rubber Company in 1944.

Of the various American rocket weapons developed during the war, undoubtedly the best known was the bazooka, a rocket-propelled grenade that was employed with great success on all fronts, European, African, and Pacific.

The basic idea for this weapon was not new; Goddard had done work along the same line during World War I. The events that led to the development of the bazooka started in December 1940, when Colonel Leslie Skinner of Army Ordnance presented to Ordnance Colonel Gregory J. Kessenich a tentative design for a bazookalike weapon. At about the same time, Henry H. Mohaupt, a Swiss engineer, offered the Army Ordnance Technical Staff a concept for an armor-piercing projectile. A grenade fitted with "the special Mohaupt head" was tested and found to be satisfactory—except that it could not be fired from the shoulder, a basic requirement, because of its recoil.

Skinner then applied the techniques of rocketry to solve the recoil problem. Technical advice made available by Hickman during the early research and development program helped lead to a quick solution. Skinner, with the help of Lieutenant Edward G. Uhl, soon had his one-man tank killer close to production.

Under the supervision of Lieutenant Colonel W. T. Moore, the first launcher and rocket parts were produced at the Frankford Arsenal in the spring of 1942. Uhl, now a captain, fired the first bazookas from the shoulder, initially at NDRC's test ground, then

*Built to be fired from the shoulder, the original bazooka was 2.36 inches in diameter and about 7 feet long. Its rocket grenade, right, was a converted mortar shell with added fins, nozzle, and method for holding the rocket propellant.* (COL. LESLIE A. SKINNER)

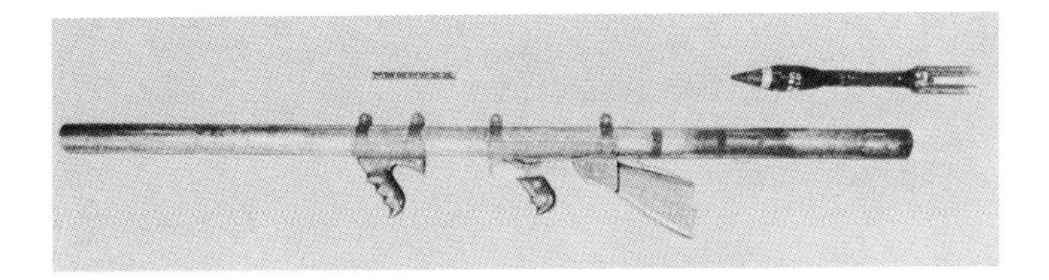

at the Aberdeen Proving Ground. According to Hickman, during one of the earliest tests of the tube at Aberdeen, no sight for the weapon was available. So,

> when one of the officers asked him [Lieutenant Uhl] to demonstrate the recoilless gun, he improvised a rear and front sight using wire. He hit a moving tank nine times out of ten shots and they said, "This is what we want." A major, who was present, asked what that thing was. They told him it was a recoilless gun. He said: "It looks like Bob Burns' Bazooka to me." Then and there the name Bazooka was born and stuck with it to this day.

A formal demonstration for Army, Navy, and NDRC officials was held at Camp Sims in June 1942, with a medium tank as the target. The bazooka entered into action during the North Africa landings that November, and was standard infantry equipment afterward.

There were two principal elements to the weapon: a tube launcher and the projectile. The tube, outfitted with shoulder stock, grip, trigger, sights, and safety switch, was 4.5 feet long, 3 inches in diameter, and weighed only 13.3 pounds. The rocket was 2.36 inches in diameter, 1.8 feet long, weighed 3.4 pounds (of which 1.57 was payload), and was fired by an electric squib igniter.

The bazooka could knock out a moving tank at 200 yards and was effective at up to 700 yards against bunkers and other stationary targets. It was one of the GI's favorite weapons; under the right circumstances, it made a foot soldier equal to a tank.

The combination of Skinner and Hickman also was effective in developing the most widely used barrage rocket of the war, the 4.5-incher that eventually was used in great numbers by every service.

The rocket started out to be an air-to-surface missile. In 1941, Skinner submitted to Hickman several sketches of rockets that would be suitable for use by airplanes. Hickman selected the 4.5-incher as the smallest that could carry a reasonable warhead and enough propellant to give it a velocity of 1,000 feet per second. Skinner then designed the rockets, improvising the first few working models from fire extinguisher cylinders. Twenty-four rockets were made at the Naval Gun Factory in Washington, and were fired satisfactorily at Indian Head.

Skinner and Uhl then designed the production model of the rocket, modifying their design to include features suggested by Hickman. In 1942, an order for 500 4.5-inchers and 500 3.25-inchers was placed with the Dresser Manufacturing Company; the smaller rocket was to serve as a test vehicle for proximity fuze development.

The rocket had been conceived as an air-to-surface weapon; but, as Skinner informed the authors:

> The reluctance of the Air Corps to try out rockets prior to their demonstrated use on aircraft by the British led the NDRC and me to approach the ground forces to see if there was any interest in that quarter. As a result of this contact the ground forces placed the first order for the production of rockets, which was shortly thereafter followed by an order from the Air Corps . . . .
>
> The original order was to have been for 780,000 but had been reduced by the time it had reached the Ordnance Department to 15,000. The ground forces order, which was for a large number (as near as I can remem-

*The bazooka became standard infantry equipment on all fronts. Below, a member of a bazooka team fires on a Japanese pillbox on Corregidor, the Philippines, in February 1945.* (U.S. ARMY AND EDWARD G. UHL)

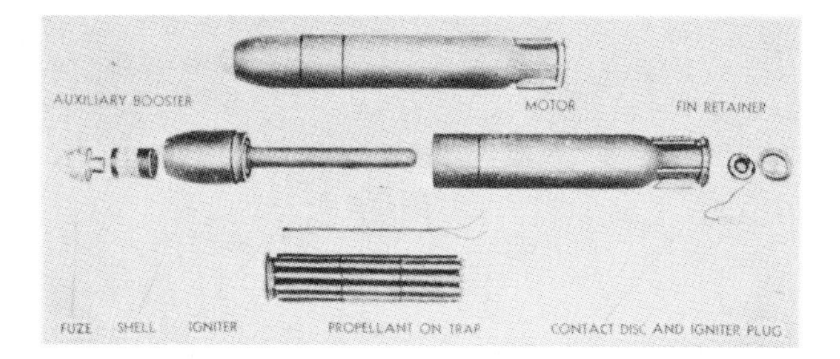


*The 4.5-incher was the most widely used barrage rocket of the war. Its component parts are shown here.* (C. N. HICKMAN)

ber, about 500,000), enabled the Propellant Section of the Industrial Division to begin planning . . . to produce solvent double-base powder strictly for rocket purposes.

The 4.5-incher eventually was made in a variety of models. There was the M-8, which was fired from truck- and tank-mounted 8-tube Xylophone and 60-tube Calliope launchers and from many different jeep-mounted launchers; for jungle fighting, the M-8 could be fired from single-shot expendable tubes. The improved M-12 had collapsible fins and was fired from plastic tripods. The M-16 was fired from 24-tube Honeycomb and 60-tube Hornet's Nest launchers, and from Navy rocket ships that softened up fortifications before Marine amphibious assaults. One model of the 4.5-incher, known as Old Faithful, was slightly shorter than the other types, and was fired from landing craft as they neared the beaches. It could carry either fragmentation or antipersonnel charges, and it helped fill the time gap between the end of long-range naval and air bombardment and the moment when the troops actually hit the beach.

The air-launched version of the M-8 was basically the same as the ground-launched rocket. It was 2.75 feet long, weighed 38.2 pounds, had a payload varying from 4.3 to 5.1 pounds, fired for .03 seconds, and achieved a speed of 600 miles per hour. This version was first used in the winter of 1943–1944 in raids on Japanese installations in Burma—the first combat employment of American-made rockets by airplanes.

A Super 4.5-incher, also launched from the air, was available by December 1944, but did not enter operations service. Designed to knock out targets that resisted the regular model, it weighed 103 pounds, including a 40-pound payload containing 8.5 pounds of high explosives. It was 6 feet long, had a range of between 3 and 4 miles, reached a speed of 900 miles per hour, and was stabilized by four large fixed fins.

Other calibers of surface-to-surface rockets were developed during the war. The 3.5-inch Spinner, developed for the Marines, had a speed of 435 miles per hour and a range of over 2 miles. It never saw combat. The Navy used a 5-inch Spinner Beach-Barrage Rocket to reach targets out of the range of the 4.5-incher. The 5-inch rocket had interchangeable warheads with high-capacity, smoke, chemical warfare, semi-armor-piercing, and pyrotechnic capabilities.

For even longer ranges—up to 5 miles—the 5-inch HVSR (high-velocity, spin-stabilized rocket) was developed. It saw action on PT boats in ship-to-ship engagements, on several types of landing craft,

*The 4.5-incher was originally designed as an air-to-surface rocket, but it was also adapted to models that could be fired from the ground or from ships. Shown here are 4.5-inch rockets in flight after being fired from a spinning launcher (left) and from projectors mounted on a 2½-ton truck (right).* (C. N. HICKMAN; U.S. ARMY)

and on the submarine *Barp,* which fired more than 70 rockets against targets on the Japanese home island of Honshu from a deck-mounted automatic launcher. This rocket was also used, and spectacularly, in ship-to-shore bombardment at Iwo Jima and Okinawa.

Some United States rocket-launching ships could fire 300 rockets a minute; the rockets carried payloads of from 1.7 to 2.8 pounds. At the end of the war, forty-eight "super" rocket ships had been developed, each of them with ten launchers that gave the capability of firing five hundred 5-inchers a minute, with loading, aiming, and firing carried on by remote control. Some of these ships were en route to combat zones when the Japanese surrendered.

NDRC Section L scientists at Cal Tech developed a short-range (.2 miles) low-velocity (120 miles per hour) 7.2-inch demolition rocket for use against bunkers and other heavily fortified positions. The Army launched them from 20-tube Whiz Bang and 24-tube Grand Slam launchers, both mounted on tanks. The Navy used Woofus 120-tube launchers, fitted to LCM(3) landing craft, to fire the 7.2-inchers.

The Navy turned to rockets to help in the fight against submarines. At the beginning of the war, it asked the NDRC-Cal Tech team to develop a weapon that could supplement the standard ship-launched depth bomb. The Navy wanted a bomb similar to the British Hedgehog, which was fired ahead of the sub-hunting vessel, but one that would not produce the Hedgehog's forbidding recoil. The scientists started work in the fall of 1941; the first rocket-propelled bombs were ready for sea firing off San Diego by 30 March 1942. The tests were successful, and in mid-April, dummy bombs were tested against an American submarine off Key West. The bombs and their launchers (called Mousetraps) were installed on many United States sub-chasers and Coast Guard vessels, starting in October 1942 in the Atlantic and April 1943 in the Pacific. Many kills of Japanese and German subs were credited to them.

One antisubmarine weapon that never saw combat was the Hydrobomb, developed at Cal Tech in response to an Air Corps request for a missile that could be launched from an airplane and would be propelled underwater to its target by a rocket motor. Two Hydrobomb prototypes were built, one by the Westinghouse Manufacturing Company, the other by the United Shoe Machinery Company. Tests of a model with a solid-propellant motor that burned for 30 seconds with 2,200-pound thrust were carried out at the Torpedo Launching Range developed by Cal Tech at Morris Dam, California, in 1943 before the project was abandoned.

*The LSM-196 sends volleys of rockets to the shores of Tokishiki Shima, 1945. This 5-inch rocket with interchangeable warheads was fired on targets out of range of the 4.5-incher.* (U.S. NAVY)

Another Air Corps idea that washed out was a 14-inch rocket-boosted, armor-piercing bomb developed in 1941–1942 by the NDRC. The idea had been that the rocket would increase the bomb's penetrating power, but by the time the weapon was ready, the Air Corps had lost interest. But the experience was put to good use, however, when the Navy asked for a rocket that could slow down a bomb; Navy patrol bombers were overshooting submarines too frequently because, sighting their targets at the last minute, their bombs would be released too late to do any harm. What the Navy needed was a rocket that would cancel the plane's momentum and permit the bomb to fall straight down.

*Beginning in October 1942, rocket launchers, called "Mousetraps," were installed on United States sub chasers and Coast Guard vessels. They proved highly effective against German and Japanese subs.* (U.S. NAVY)

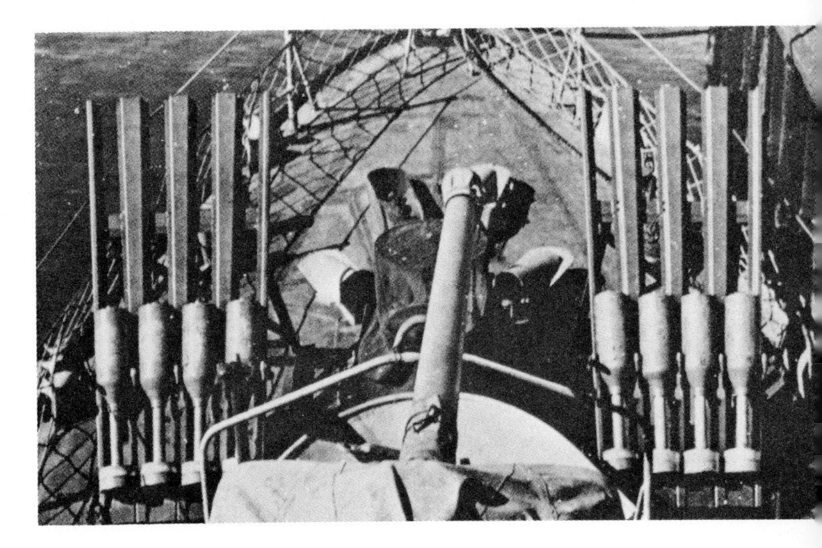

**United States Guided Surface-to-Air Missiles**

| Missile | Length (feet) | Diameter (inches) | Weight (pounds) | Altitude (miles) | Velocity (mph) | Propulsion |
|---|---|---|---|---|---|---|
| Little Joe | 11.34 | 22.7 | 1,210 | 1.5 | 400 | solid sustainer and booster |
| Lark | 18.5 | 18 | 2,000 | 4 | 600 | liquid sustainer, solid booster |

The result was a retro-firing bomb whose rocket motor permitted a nearly vertical fall. It was tested on 3 July 1942 on a lumbering PBY5A Catalina that immediately took its place in history—it was the first American airplane to fire a rocket. The weapon was 7.2 inches in diameter, carried 35 pounds of explosives, and had a maximum velocity of 200 miles per hour. Among the many kills credited to the bomb was the last German sub reported sunk in the war, on 30 April 1945 in the Bay of Biscay.

The United States was much more interested at the start of the war in firing rockets *from* planes rather than *at* them for one very basic reason: no one was bombing American cities. When the danger of Kamikaze attacks on United States ships arose, the situation changed and two surface-to-air missile programs were begun, Little Joe and Lark.

*Kamikaze attacks on United States ships intensified the need for a surface-to-air missile. The Navy began work on the rocket-powered Lark in 1944, but did not get it ready for service before the war ended.* (REACTION MOTORS)

Little Joe was gyro-stabilized and controlled by radio command through optical tracking. Its warhead was to detonate by proximity fuze as it reached its target. Several missiles were produced under the direction of the Naval Air Materiel Unit, but the long lead-time necessary before Little Joe could be used caused the program's cancellation.

Like the Little Joe program, the Lark missile effort began in 1944. By 6 February, Lark was given the go-ahead for accelerated development by the Navy Jet-Propelled Missile Board because of Little Joe's troubles. Lark was launched by two solid boosters and had an unusual two-chamber liquid-fuel rocket engine. The larger chamber, which produced 400 pounds of thrust, would be in use only when the 220-pound-thrust smaller chamber failed to keep the missile moving at a preset velocity. Lark had a command mid-course control system, with a semiactive homing device for terminal guidance; it had four wings and four fins, positioned octagonally. Lark was not ready for service when the war ended, but tests continued, first at the Naval Ordnance Test Station at Inyokern, California, and later at the Naval Air Missile Test Station at Point Mugu.

One weakness in both these weapons was guidance, a problem that American technology barely came to grips with during World War II. Air-to-surface missile development was much easier because the airplanes were shooting at targets that were either standing still or moving relatively slowly.

In addition to the 4.5-inch air-to-surface rocket, several other models saw action during the war. Many of them, like the 4.5-incher, were modified ground-to-ground weapons. The first of these was the 3.5-inch FFAR (forward-firing aircraft rocket), which was based directly on the British rocket of nearly the same diameter. The Cal Tech group, supported by the Navy, began development work in the spring of 1941. In August 1943, a rocket 4.58 feet long, weighing 54.5 pounds, and with a range of slightly less than a mile was tested. The rocket first went into action against U-boats in January, with a special head that doubled its underwater lethal range; later the first model was used against surface ships.

A 5-inch FFAR was developed by replacing the 3.5-incher's 20-pound solid head with a 5-inch, 50-pound explosive shell. The rocket, 5.4 feet long and weighing 80 pounds, was popular with the Navy for action against shore- and ship-based antiaircraft guns.

Slightly longer and considerably heavier at 134 pounds, the 5-inch HVAR (high-velocity aircraft rocket) went into action in July 1944. The Holy Moses, as it was called, was developed by an NDRC–Cal Tech–Navy Bureau of Ordnance team, but was first used by the Air Corps near St. Lô in France. About one million Holy Moses rockets had been made when the war ended.

The largest airplane-launched, forward-firing rocket developed during the war was inappropriately called the Tiny Tim. It was 10.25 feet long, 11.75 inches in diameter, and weighed 1,284 pounds. It was created primarily for use against the fortified pillboxes and bunkers that would have to be knocked out in an invasion of Japan's home islands. Tiny Tim's specifications were laid down in February 1944 by Cal Tech scientists and engineers; the first test round was fired just two months later. Its range was short—just one mile—and 30,000 pounds of thrust gave it a speed of only 550 miles per hour, but the 150 pounds of TNT in Tiny Tim's payload gave it the wallop of a 12-inch naval shell.

Tiny Tim was carried on modified bomb racks and released by standard mechanisms; it was ignited by a lanyard when it dropped several feet from the aircraft. The first test produced a catastrophic accident that destroyed the aircraft, but the ignition mechanism was redesigned and F6F squadrons on the carriers *Franklin* and *Intrepid* were outfitted with Tiny Tims by the fall of 1944, in time for the battle of Okinawa.

The United States also experimented with an assortment of guided bombs—fifteen models in all—as air-to-surface weapons. Most of them were unpowered. One of them, the GB-1, was developed starting in March 1941, entered production in May 1943, and was first used against Cologne in May 1944. GB-1 was basically a 2,000-pound bomb which had been given wings and a television-radio guidance system to increase its accuracy. About one thousand were used against targets in Germany and Austria.

The VB, or vertical bomb, series, consisted of free-fall bombs guided by bombardiers via controls in their tail fins. Azon, or VB-1, could be guided in clusters of five by one bombardier—if the weather was clear and the airplane was flying steadily. VB-3,

*The largest airplane-launched rocket of World War II was inappropriately called Tiny Tim. This 10-foot rocket, loaded with 150 pounds of TNT, was created for use against pillboxes and bunkers.* (U.S. NAVY)

or Razon, developed by NDRC and the Air Technical Service Command, had a controllable range. VB-6 homed in on the heat given off by the target. VB-10 (called Roc, as were VB's 11 and 12) had a television-radio guidance system. VB-13, last of the series, weighed 12,000 pounds, had a 54-inch-diameter lift shroud, and was more than 20 feet long. Called Tarzon, it was used against enemy battleships and heavy fortifications.

The Bat was a longer-range, radar-guided air-to-surface missile whose development won Hugh L. Dryden the Presidential Certificate of Merit. Bat was nearly 12 feet long, carried a 1,000-pound payload, and traveled at a maximum speed of 300 miles per hour. In April 1945 a Bat sank a Japanese destroyer 20 miles from the launching aircraft, which was the weapon's maximum range.

Another missile, Gargoyle, started life in November 1943 as a glide bomb, but was given a liquid-propellant rocket engine in March 1944 at the Navy's request. The McDonnell Aircraft Company delivered the first Gargoyle in December 1944; four more were delivered the next month. Production authorization was given in May 1945. When the war ended, Gargoyle became a test vehicle without ever having seen action.

One interesting weapon was an alteration of the 4.5-inch rocket into an air-to-air defense rocket.

Myths and stories suggesting different schemes for traveling to other worlds were told for nearly two thousand years, and rockets were used in war and for entertainment for at least six hundred years before men gradually began to link the two ideas. The discovery that the rocket is the key to space travel was made independently toward the end of the nineteenth century in Russia, the United States, and Germany.

The first man to really understand and develop the principles of rocketry and their application to space travel was a Russian schoolteacher, Konstantin Eduardovitch Tsiolkovsky. He stumbled on the concept of rocket flight in 1883—probably, as he later wrote, as a result of Jules Verne's influence—spent some twenty years refining his theories, and published them in 1903. On the basis of his calculations, Tsiolkovsky proposed the use of such modern rocket-propellant combinations as liquid oxygen and liquid hydrogen, and explained in detail the advantages of multistage rockets.

While Tsiolkovsky is universally recognized today as the leading pioneer of astronautical theory, his American counterpart, Robert H. Goddard, combined theory and practice in a long and extraordinary research career. Born in 1883 in Worcester, Massachusetts, Goddard developed a plan for multistage spaceships while still in high school. As a professor at Clark University, he continued to explore the theoretical implications of space flight. Although frequently hampered by lack of sufficient funds, he constructed and tested a series of progressively more sophisticated rockets at sites in Massachusetts and Roswell, New Mexico.

The third great pioneer of rocketry and astronautics, Hermann Oberth, born in 1894 in Transylvania, has been a citizen of Germany for most of his life. Primarily interested in the theory of space travel and in spaceship design, Oberth explained his ideas in two influential books. Published in 1923 and 1929, they helped inspire other rocket enthusiasts in Germany to conduct practical tests of rockets, rocket-powered planes, and even a rocket-powered car. Out of these experiments came the technical know-how that enabled Germany to produce the V-1 and V-2 missiles used during the Second World War.

The V-2, the largest and most advanced of these weapons, was developed at Peenemünde, an Army experimental test station on the Baltic coast, by a team of scientists and engineers under the direction of General Walter Dornberger and Wernher von Braun. More than four thousand of these missiles were fired during 1944 and early 1945 against targets in southern England and on the Continent.

The United States, Russia, Great Britain, and Japan also used rockets and missiles of varying sizes and effect during the war. The German V-2, however, became the direct ancestor of all major postwar military missiles and space carrier vehicles as a result of the capture of German rocket experts and many V-2 components by American and Russian armies.

Each of the military services in the United States continued to develop missiles and rockets following the war. Not until the early 1950's, however, did the United States, which had hired 127 of the German scientists, start emphasizing these programs. The Army developed the first of the larger missiles. Working at Fort Bliss, Texas, and later at Redstone Arsenal in Huntsville, Alabama, a team under Wernher von Braun developed the short-range Redstone missile, first tested in 1953. Then, by adding two special stages to an uprated Redstone, the Huntsville group built Jupiter C, America's first missile to fly an intercontinental range.

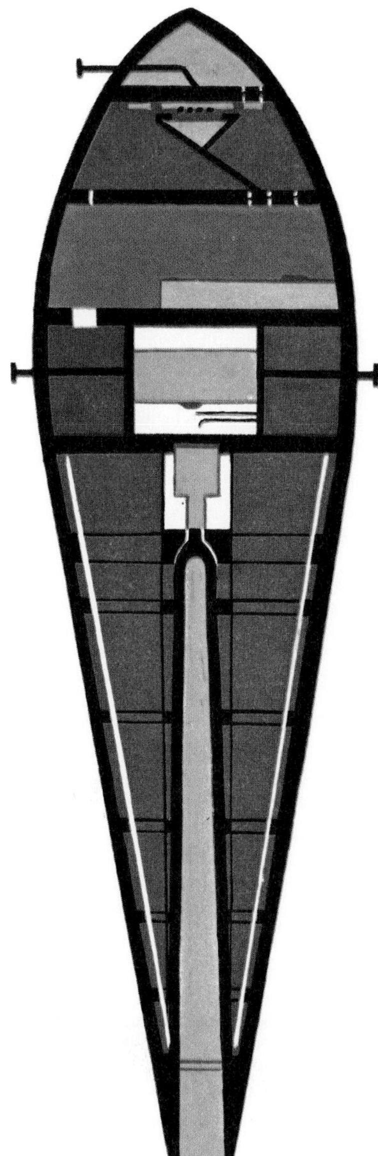

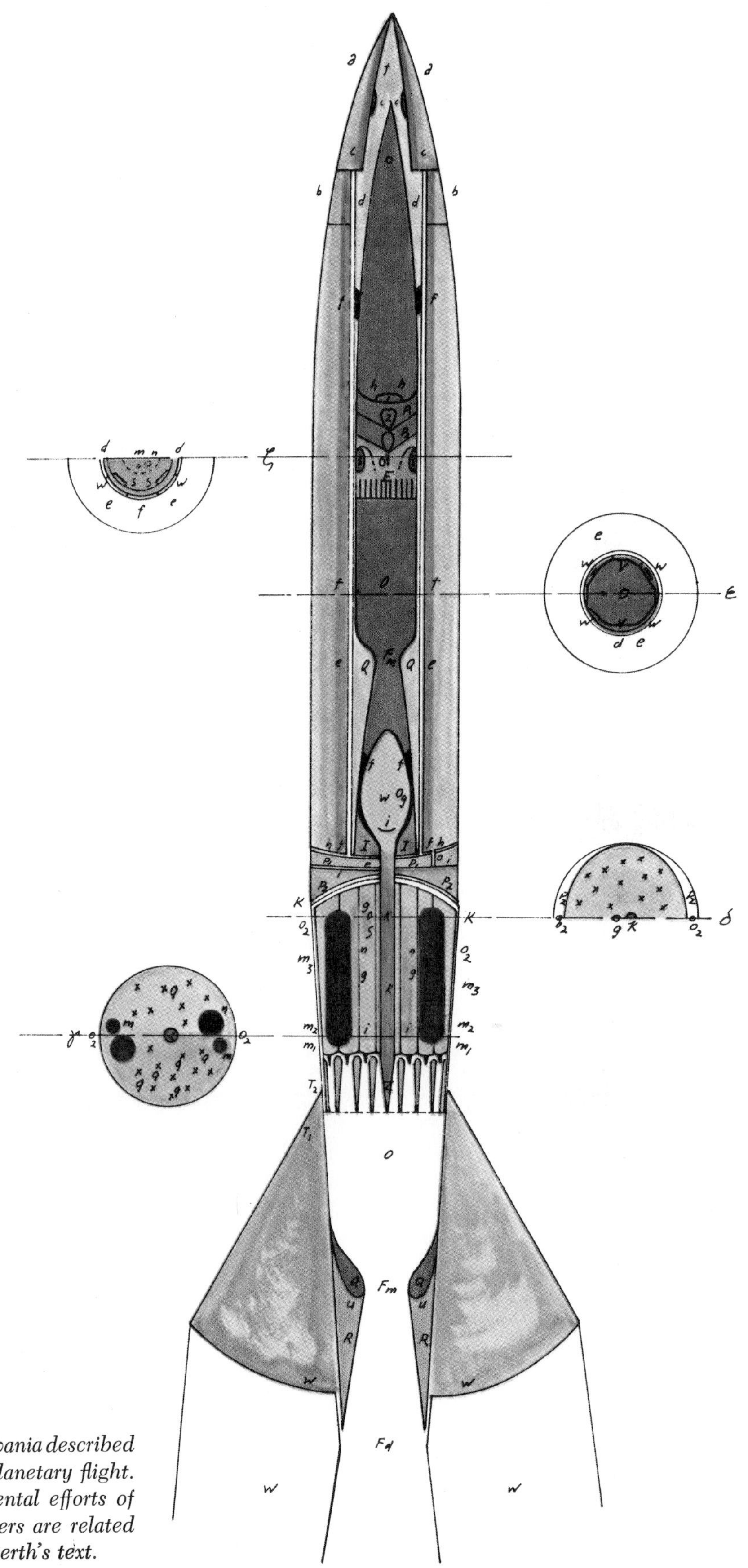

ABOVE: *The first scientific study of rocket propulsion for space vehicles was performed by the great Russian pioneer Konstantin Tsiolkovsky from 1883 to his death in 1935. This is his 1903 spaceship design.*

RIGHT: *In 1923, Hermann Oberth of Transylvania described this "Modell B" rocket in a book on interplanetary flight. Never built, it inspired the later experimental efforts of the German rocket society (VfR). The letters are related to equations and other nomenclature in Oberth's text.*

ABOVE: *Among the precursors of the German V-2 was the A-3 test rocket, shown here in its static test rig. First flight tested in 1937, it was 21 feet tall, 2 feet in diameter, and produced 3,300 pounds of thrust.*

RIGHT: *While Tsiolkovsky and Oberth developed the theory of astronautics, an American, Robert H. Goddard, constructed, tested, and fired rockets. Here, we see one of his research rockets on the test stand near Roswell, New Mexico.*

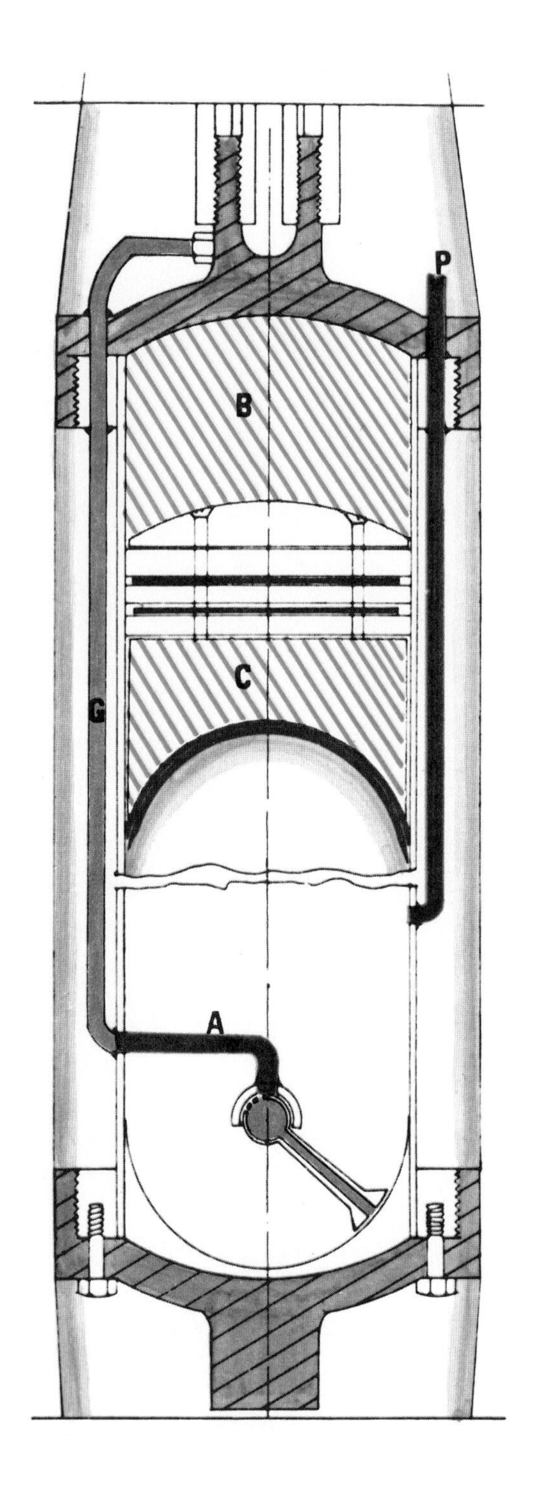


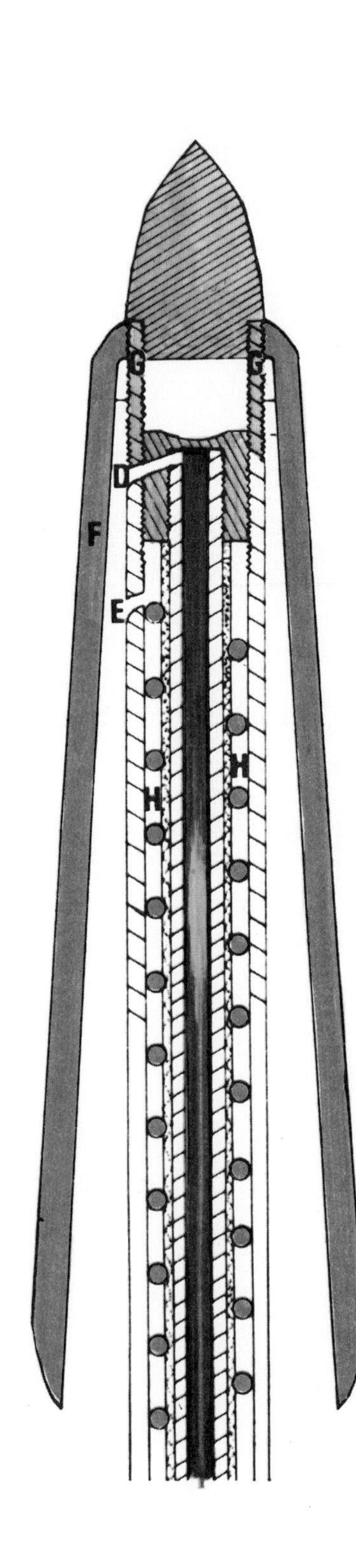


*In the mid-1930's the most advanced design of the American Rocket Society was Rocket No. 5, shown here from the outside (R) and in cutaway views. The propellants, liquid oxygen (A) and gasoline (B), were stored in the same cylindrical structure and were separated by a movable pressure piston (C). The propellants were introduced through inlets (D and E) to the motor, which consisted of a cone (F) that served as both combustion chamber and nozzle. Liquid oxygen was fed through one duct (G) and gasoline through another (H). Oxygen pressure was regulated by a safety valve (P).*

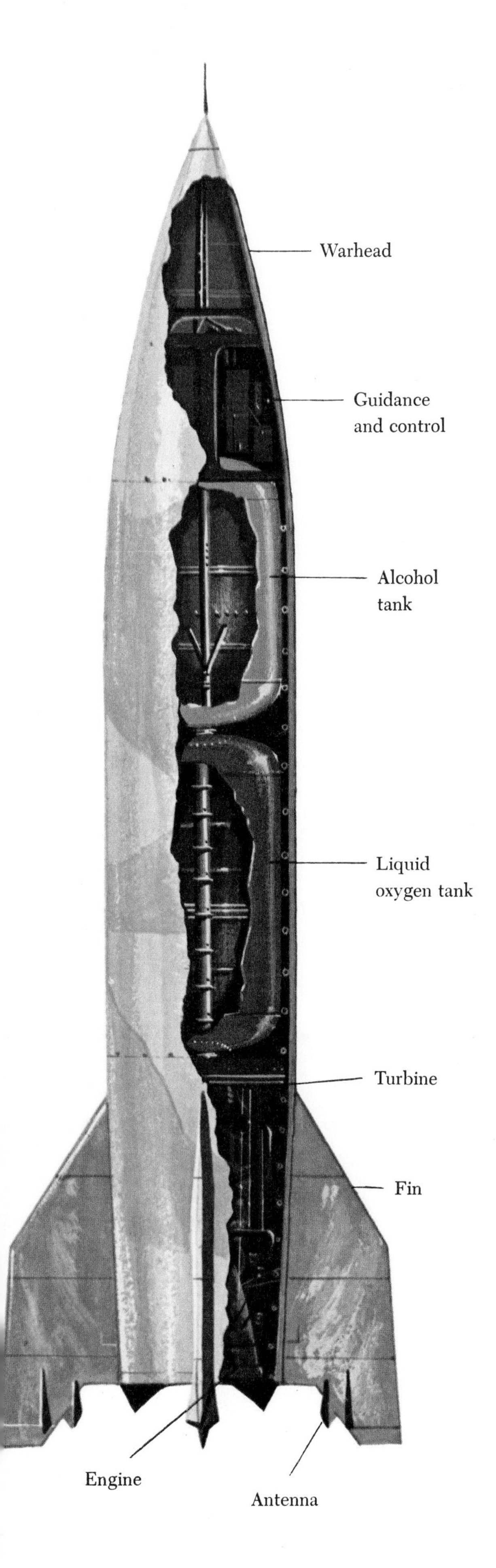


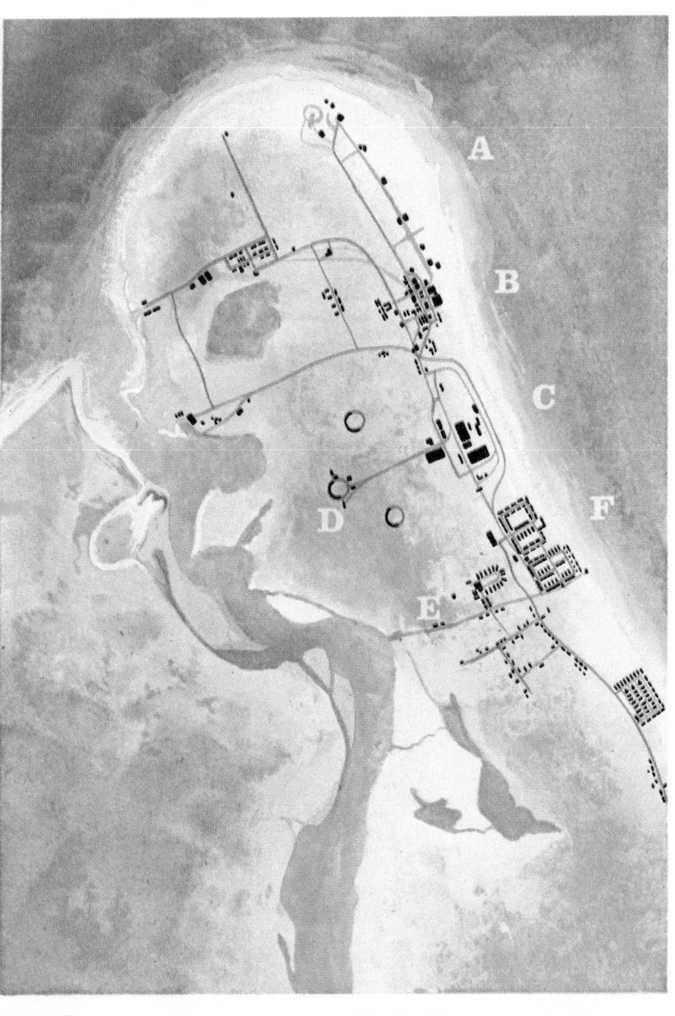


A Ordnance test area
B Ordnance engineering center
C Production plant
D Production test stands
E Military camp
F Residential communities

ABOVE: *The V-2 was designed, built, and tested at Peenemünde, the German experimental center on the Baltic coast. Launching facilities were at the northeast corner, near the sea.*

LEFT: *Developed by the Germans during World War II, the V-2 was the largest and most advanced missile of its time. Shown here in cross section, it was 46.1 feet long, 65 inches in diameter, and weighed 27,000 pounds. It became the prototype of postwar American and Russian missiles and space carrier vehicles.*

*By 1946, both the United States Navy and the United States Air Force were studying and designing carrier vehicles. Had these programs been approved, America might have been able to launch satellites in the early 1950's. The Navy proposal (left) was a single-stage carrier 86 feet long with a 16-foot diameter, designed to orbit itself. The Air Force proposed a three-stage carrier (right) for orbiting a 500-pound satellite.*

*Offspring of the V-2, the three-stage Jupiter C was designed primarily to test nosecone materials for re-entry into the Earth's atmosphere. This night exposure at Cape Kennedy, Florida, shows the upper staging and the re-entry vehicle, built to withstand frictional heat.*

NT

*Experience gained in the Jupiter C program was applied directly to the Army's Jupiter—America's first successful intermediate-range ballistic missile. An accurate weapon with a range of 1,500 miles, it was later converted to a space carrier vehicle.*

Used against enemy planes, its history has almost been forgotten—its records were burned as a matter of routine. One early attempt had been made to adapt the 4.5-incher to use by fighters, but by the time the final version, called the T-22, was developed, American control of the skies was so complete that the weapon was not needed.

In 1945, the Second Air Division of the 8th Air Force came up with a version of the 4.5-incher as a bomber defense weapon. It was to be loaded and adjusted by a bomber's waist gunner and fired by the tail gunner. Colonel John J. Driscoll, who was active in the development program, described for the authors the reticle that was used to fire the air-to-air rocket:

> The distances between the corresponding portions of the sight reticle were spaced to equal the wingspan of an average German (fighter) aircraft . . . there were normally two opportunities to fire at a tail-attacker (*e.g.*, a tail attack level might be opposed at either 850 yards or 300 yards, firing taking place when the wingspan touched the sides of the reticle or was roughly equal to the space between the reticle lines at the corresponding level of attack).
>
> Unfortunately, as with most wartime secret devices (particularly combat experiments), the reports were destroyed in accordance with U.S. Army regulations.

One of the most successful United States rocket programs of the war had nothing to do with missiles. It was the JATO program, aimed at getting heavily loaded aircraft off the ground quickly. It was notable on several accounts.

The Galcit group at Cal Tech had started work on solid-propellant JATO units before the war, developing its Galcit 27 propellant by the summer of 1941. Galcit 27, using an amide powder developed from commercial ingredients to produce 28 pounds of thrust for 12 seconds, was tested on a 753-pound Ercoupe monoplane at March Field, near Riverside, California, from 6 August to 23 August 1941. The aircraft, piloted by Captain (now General) Homer A. Boushey, Jr., took off in 7.5 seconds using 300 feet of runway, compared to its ordinary 580-foot, 13.1-second takeoff. The six JATO rocket units did not disturb the aircraft's stability or controllability.

This line of development was continued, with the Galcit group finally settling on a solid propellant whose oxidizer was a potassium perchlorate compound and whose fuel was a new type of asphalt to which was added a small amount of oil with an asphalt base. The Aerojet Engineering Corporation built motors that produced thrusts of 200, 500, and 1,000 pounds using this propellant.

At about the same time, Robert C. Truax, having completed his two years' sea duty after his graduation from the Naval Academy, went to work on JATO at the Engineering Experiment Station at Annapolis. By then, he was a lieutenant commander in the Bureau of Aeronautics Ship Installations Division under Commander C. A. Bolster. His project, known as TED ESS 3401, was staffed by ensigns R. C. Stiff, J. F. Patton, and W. Schubert, and a civilian from MIT, Robertson Youngquist.

Truax's small staff concentrated on developing a liquid-fuel rocket motor that could produce 3,000 pounds of thrust to get the underpowered PBY seaplane into the air after a relatively short run. By June 1942, a 1,500-pound-thrust engine was tested, but a lag in ignition permitted propellants to accumulate in the thrust chamber. The result was an explosion that wrecked the test stand.

Pressing ahead during the course of conducting gas-generator tests, Ensign Stiff discovered that aniline and red fuming nitric acid would ignite on contact, without needing special ignition equipment. Aniline was more difficult to come by than gasoline and it was harder to handle and store, but it eliminated so many ignition problems that it was immediately adopted as a fuel. In the spring of 1943, Truax's group tested a 1,500-pound-thrust JATO unit that weighed 325 pounds empty and 655 pounds fueled. Two units provided the thrust Truax was after.

Meanwhile, this discovery had gotten the Galcit researchers out of a hole. They too had been working on liquid-fuel rocket motors, and had also run into ignition difficulties; an October 1941 test of a 1,000-pound-thrust engine ran into ignition trouble. Galcit immediately switched to nitric acid and aniline.

The Army provided an A20-A medium bomber for flight testing, and during the winter of 1941–1942 mockups of the 1,000-pound-thrust unit were installed on the 14,000-pound aircraft. The first flight tests of the liquid JATO units were held between 7 April and 24 April 1942 at the Army Air Corps Bombing and Gunnery Range, Muroc, California, with Major P. H. Dane flying 44 test runs.

The JATO effort was further distinguished by the presence of Robert H. Goddard. Under a contract signed with the Navy in December 1941, Goddard and his crew came from New Mexico to Annapolis in July 1942, and developed a liquid oxygen–gasoline JATO. The unit, tested on a PBY-2 on 23 September 1942, was the first JATO to power the takeoff of a Navy aircraft.

The development of liquid oxygen–gasoline JATO's was continued by Reaction Motors. Thrust was raised to 3,000 pounds, and duration of firing to 60 seconds. An engine using liquid oxygen and a

*The "Bat" radar-guided missile was nearly 12 feet long and traveled at a maximum speed of 300 miles per hour. Its maximum range was 20 miles, a distance from which a Bat sank a Japanese destroyer in April 1945.* (U.S. NAVY)

*Combustion chamber developed by Robert C. Truax for the JATO (jet-assisted takeoff) program at Annapolis. Its purpose was to aid overloaded planes in taking off from short runways.*

gasoline-water mixture was successfully tested on Martin PBM3C's during 1943.

Not all American rocket work was aimed at immediate applications. Several programs of relatively pure research in rocketry were carried on during the war. They bore within them the germ of future rocket and missile programs.

The Navy's Gorgon program was conceived as early as 1937, with the aim of developing an air-to-air missile. The first designs were laid down in 1941. In May 1943, reconsidering plans for powering their missiles only with jet engines, the Naval Bureau of Ordnance decided to test a 350-pound-thrust two-chamber nitric acid–aniline rocket engine developed by Truax's group. In October 1943, two airframes were selected, one conventional and the other canard. Twenty-five of each—Gorgon 2A and 2B respectively—were ordered, although Gorgon 2B was canceled when the turbojet for which it was designed failed to become available. Other models—2C, powered by pulsejet; 3A, rocket powered, Gorgon 4, propelled by a ramjet; Gorgon 5, an enlarged model of the 4 vehicle—were used as test vehicles after the war, providing valuable information for the Navy missile program.

**Characteristics of Private A Rocket**

| | |
|---|---|
| Length (feet) | 8 |
| Span (feet) | 2.8 |
| Weight (pounds) | > 500 |
| Payload (pounds) | 60 |
| Sustainer thrust (pounds) | 1,000 |
| Booster thrust (pounds) | > 21,500 |

*In the first jet-assisted takeoff, the Ercoupe rises rapidly under rocket power.* (GEN. H. A. BOUSHEY, JR.)

The second major wartime test-vehicle program was the Private, undertaken by Ordcit (Ordnance Project—California Institute of Technology) at the Jet Propulsion Laboratory. This program, begun in 1944, was designed to develop the technology of long-range, rocket-powered missiles.

Private A was propelled by an Aerojet solid engine, with booster thrust provided by four 4.5-inch aircraft rockets attached by a steel casing. The vehicle had four guiding fins at the rear and a tapered nose. It was launched from a rectangular steel boom with four guide rails. Twenty-four test vehicles were fired between 1 December and 16 December 1944; one achieved a range of 11.3 miles.

The following spring, Private F was fired at Fort Bliss, Texas. It had a single guiding fin and two horizontal lifting surfaces aft, with two stubby wings forward; its purpose was to test the effect of lifting surfaces on guided missiles. A total of seventeen vehicles were launched between 1 April and 13 April 1945.

The United States' only wartime effort in rocket airplanes was hardly a great success. The MX-324 was originally considered as a prototype for an advanced "flying wing" fighter, the XP-79, with an Aerojet 2,000-pound-thrust nitric acid–aniline rocket engine providing the power. During 1943 it became evident that the rocket engine would not be available, and two Westinghouse turbojet engines were substituted. They didn't work out, the first airplane powered by them crashing during a test.

Eventually, three MX-324's were built. After some glide tests, the first flight with an Aerojet XCAL-200 rocket engine was held at Harper's Lake on 5 July 1944; soon afterward, the airplane was taking off and landing on skids. The MX-324 was flown for some time with these results: One plane was completely destroyed in an accident, a second was severely damaged, and the third was disposed of. For the record, the MX-324 was 14 feet long, had a wingspan of 38 feet, a range of nearly 1,000 miles,

*The MX-324, built in 1944, was America's first experimental military rocket airplane. It had a 36-foot wingspan and a "prone" cockpit in which the pilot lay flat in order to withstand higher accelerations.* (NORTHROP CORP.)

**Characteristics of Wac Corporal**

| | | |
|---|---|---|
| Length (feet) | 21 | |
| Diameter (inches) | 12 | |
| Weight (pounds) | 665 | (plus 546 for booster) |
| Payload (pounds) | 25 | |
| Velocity (mph) | 2,800 | |
| Altitude (miles) | < 45 | |

cruised at 480 miles per hour, and was designed to attain 550 miles per hour at top speed.

The first real step into pure rocket research was the Wac Corporal, which was developed by Ordcit. The program began in 1944 in response to an Ordnance Department request for a research rocket that could carry a 25-pound payload to 100,000 feet; the Signal Corps wanted the rocket.

The final design called for the vehicle to be launched by a solid booster from a tower, with a liquid-propellant rocket then taking over to sustain the flight. A one-fifth scale model was built to study the booster-sustainer combination and determine the optimum number of fins (which turned out to be three). The model, known as the Baby Wac, was flown between 3 July and 5 July at California's Goldstone Range.

The Wac Corporal managed to achieve altitudes considerably higher than had been asked, thanks to the uprated 50,000-pound-thrust Tiny Tim booster and the 1,500-pound-thrust Aerojet nitric acid–aniline sustainer; in addition, the vehicle's weight was kept below early estimates. The booster fired for .6 seconds, the sustainer for 45 seconds. Firings were conducted at the newly opened White Sands Proving Ground between 25 September and 25 October 1945; the maximum altitude achieved was 43.5 miles.

While the Wac Corporal program as such did not continue into the postwar years, it provided invaluable experience for the fledgling United States rocket industry. Even more important, however, was the know-how that was literally captured when the majority of German rocket scientists and engineers surrendered to American soldiers in 1945. For the Germans were dominant in every field of missiles and rockets during World War II. In a very real sense, they created modern military rocket technology. Virtually all postwar missile developments were based, in varying degrees, on what went on in Germany.

The major thrust of German rocket development was, of course, in the *Vergeltungswaffen*—the weapons of retaliation. The V-1, while revolutionary enough, was the more conventional of the two retaliation weapons. A winged subsonic missile developed and controlled by the technical department of the Reichsluftfahrtministerium, the V-1 used an Argus Motoren Gesellschaft pulse-jet engine that operated on gasoline and developed 1,100 pounds of thrust.

The basic concept of the engine can be traced

*The German Army rocket center was located at Peenemünde, on the Baltic Coast. Left, the engineering offices at Peenemünde-East. Wernher von Braun's office was on the second floor, above the small balcony. Right, the rocket engine research and development center as it appeared in 1942.*

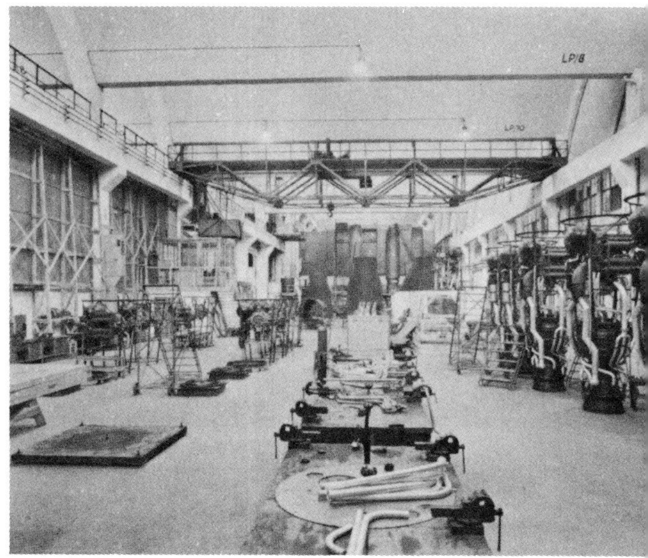

back to a man named Paul Schmidt, of Munich, whose early research was funded jointly by the research departments of the German Air Force and Army. The vehicle itself was designed by Robert Lusser, chief engineer of Fieseler Flugzeugbau, Kassel. The V-1 weighed 4,858 pounds, including its 1,988-pound payload of Amatol, a mixture of trinitrotoluol and ammonium nitrate. The missile was 27 feet long and 33 inches in diameter. It was launched from a ramp and directed to its target by a pre-set guidance system.

The first test firing of the V-1, which was then called the Fieseler Fi-103, took place in December 1941 over the Peenemunde range which had the necessary tracking equipment. By then, the idea of the V-2, a rocket that was to be launched across the Channel at Great Britain, was also well developed. There was a considerable dispute over which of the two weapons should be put in mass production and deployed operationally, since both had about the same range and payload. After a Commission for Long Range Weapons had made an exhaustive study of the two missiles in 1943, Hitler decided to accelerate the development of both weapons for an aerial offensive against southern England. From then on, both the V-1 and the V-2 were given a top priority in the German war effort.

The British intelligence service had been watching Peenemünde, with no clear idea at first of what was happening there. But the British were alarmed enough to raid the station with hundreds of Lancaster and Halifax heavy bombers on the night of 17 August 1943. The raid killed about 800 people, including Dr. Walter Thiel, who was in charge of V-2 engine development, but it did not delay either V-program seriously.

By June 1944, enough V-1's had been produced to start the attack on southern England. More than 8,000 "buzz bombs," as the British called them, were launched against London alone, and thousands of others were launched against Allied-held targets on the Continent.

While the V-1 did provide a severe nervous strain for Londoners, who were always on the alert for the sudden cut-out of the engine that meant the missile was falling, it failed to fulfill its goals. For one thing, it was too slow—with a speed of only 350 miles per hour—and could be shot down. For another, the missiles were none too reliable; only 211 of the 5,000 V-1's fired against Antwerp ever detonated on target. About one quarter of the V-1's aimed at Britain failed because of their inherent unreliability. About half were destroyed by countermeasures—barrage balloons, airplanes, and antiaircraft fire. About a quarter reached the target, and some of those did not explode.

*The British raided Peenemünde on 17 August 1943, killing 800 people including Dr. Walter Thiel, who was in charge of the V-2 engine development. But the bombs only slightly damaged the Guidance Control building, where the most vital work on the V-2 was being carried out.*

A historical curiosity that ranks with the Japanese Baka was the piloted V-1, called the V-1e. The 27-foot-long missile was outfitted with a cockpit and instruments by the Luftwaffe experimental station at Rechlin during 1944, with the aim of establishing a German Kamikaze organization, code-named Project Reichenberg. The V-1e was flown several times by a woman test pilot, Hanna Reitsch, who discovered that the 25-percent V-1 failure rate was due to engine vibrations that caused the wing skin to peel off.

Even less successful than the V-1 was a little-known rocket developed by Rheinmetall-Borsig, tested in Poland, and used operationally starting in November 1944. This rocket, called the Rheinbote, was a four-stage contraption that carried a payload of 88 pounds, only half of which was explosive. Its four stages weighed a total of 3,773 pounds, varying in thrust from the 84,000 pounds of the first stage to the more than 7,500 pounds of the fourth. The Rheinbote was unguided and used solid propellants. In one engagement sixty rockets were fired against Antwerp in January 1945 without producing significant results.

*The first successful V-2 being readied for launch, October 1942. Developed at Peenemünde, it was the largest and most advanced rocket in the world. More than 5,000 were built before the war was over.*

The Rheinbote was launched from a V-2 transport car. It had in large degree the four weaknesses that made the V-weapons something less than the terrors for which Hitler had hoped. All were introduced before they were fully developed, lacked accuracy, carried too-small payloads for their purposes, and could not be produced in the vast numbers that would make them effective.

This should not detract from the impressive technical feat produced by the Peenemünde group in making a V-2. Without a doubt, the V-2 was the largest and most advanced missile in the world. It was 46.1 feet long, 65 inches in diameter, and weighed more than 27,000 pounds. It carried its 2,200-pound payload from 180 to 210 miles, propelled by an engine using turbopump-fed liquid oxygen and alcohol, and generating an average sea-level thrust of 56,000 pounds. The V-2 had an inertial-guidance system with two free LEV-3 gyroscopes, leveling pendulums, and an integrating gyro-accelerometer.

The original design for the missile, then called the A-4, had been prepared before the war by the Dornberger–Von Braun team at the Army Experimental Station at Peenemünde. During 1938 and 1939, a smaller version of the missile, the A-5, was fired with considerable success. The launches totaled at least 25 by 1940.

While the A-5 flights went ahead, components were designed and developed for the much larger A-4. By the spring of 1939, some A-4 components were actually in production, and the rocket's design had been frozen, with Walter Riedel in charge of the design offices. Manufacturing and assembly techniques were being developed under the direction of Eberhard Rees. Rudolf Hermann had supervised the construction of a supersonic wind tunnel in 1936–1937, and vitally needed aerodynamic data had been obtained from it. Guidance and other related electronics problems were being solved under the scientific supervision of Hermann Steuding and the engineering direction of Ernst Steinhoff.

During 1940 and 1941, all these and other staff members, together with scientists and engineers at universities and in industry, worked day and night under the over-all supervision of Von Braun. By 1942, the first missiles were coming out of the Peenemünde model shops.

The first firing was on 13 June 1942, but it was a failure. Immediately after launch, the propellant feed system failed, and the huge A-4 went out of control and crashed. A second, not entirely successful, test on 16 August chalked up one notable achievement: A-4 No. 2 was the first guided missile to exceed the speed of sound.

The third launching, on 3 October 1942, was a complete success. The engine burned for nearly one minute, giving the A-4 a range of just under 120 miles and a maximum altitude of over 50 miles. Hitler, who had not seemed impressed with the potential of rockets when he viewed test-firings of two engines at Kummersdorf in 1939, suddenly became interested in what was happening at Peenemünde. A V-2 production committee was established in the Ministry of Armaments and War Production, with Gerhard Degenkolb as director. It was not entirely a blessing. As Von Braun wrote later: "The committee immediately began issuing high-handed directives and setting up a mighty production organization. Mainly composed of men of little scientific judgment, although of vast energy, this committee was a thorn in the side to Peenemünde."

Despite this interference, hundreds of V-2's were manufactured and fired over the next two years to prove out systems, train troops, and acquaint the military with the operational characteristics of the weapon.

Production of the V-2 began in a new plant built a few miles south of the Peenemünde Experimental Center. After the August 1943 air raid, which damaged this assembly plant badly, production was shifted to the underground Mittelwerk facility, a converted oil depot, near Nordhausen in the Harz Mountains. Nearly 900 V-2's a month were being produced there near the end of the war. Assembly plants under preparation near Vienna, Berlin, and Friedrichshafen were also closed and their equipment shifted to the Mittelwerk plant when it became evident that the secret was out. (The British got hold of a V-2 that had strayed from its course and landed near Kalmar, Sweden, in June 1944. As chance would have it, that particular shot was intended to test the radio-guidance system of a Wasserfall surface-to-air missile, and the British were misled into believing that the V-2 was radio-guided.)

V-2 components were produced in many parts of Germany. For example, the steam generator was made at the Heinkel factory in Jenbach, Tyrol; the guidance system at Kreiselgeräte G.m.b.H. in Berlin; propellant containers at Zeppelin Luftschiffbau, Friedrichshafen. Many universities cooperated with Peenemünde during the program. Missile training was given at a school near Koeslin in Pomerania, and military proficiency firing took place on a former Polish Army reservation near Blizna.

A program of this magnitude inevitably had political aspects. General Dornberger and his military subordinates took care of most of the contacts Peenemünde had with Nazi party officials, but Von Braun and his technical staff had to handle some of them. Until the end of 1943, the motivation for most of these contacts was curiosity on the part of the Nazi leaders, who had known little about Peenemünde's activities because of military security. But when it became obvious that the V-2 was going to be a spectacular new weapon, the SS began trying to take over the Peenemünde operation.

SS General Hans Kammler made the first attempt to take command of the Peenemünde base from Army General Leo Zanssen, but Dornberger stopped this effectively. The SS, however, kept trying. In February 1944, Von Braun was called to

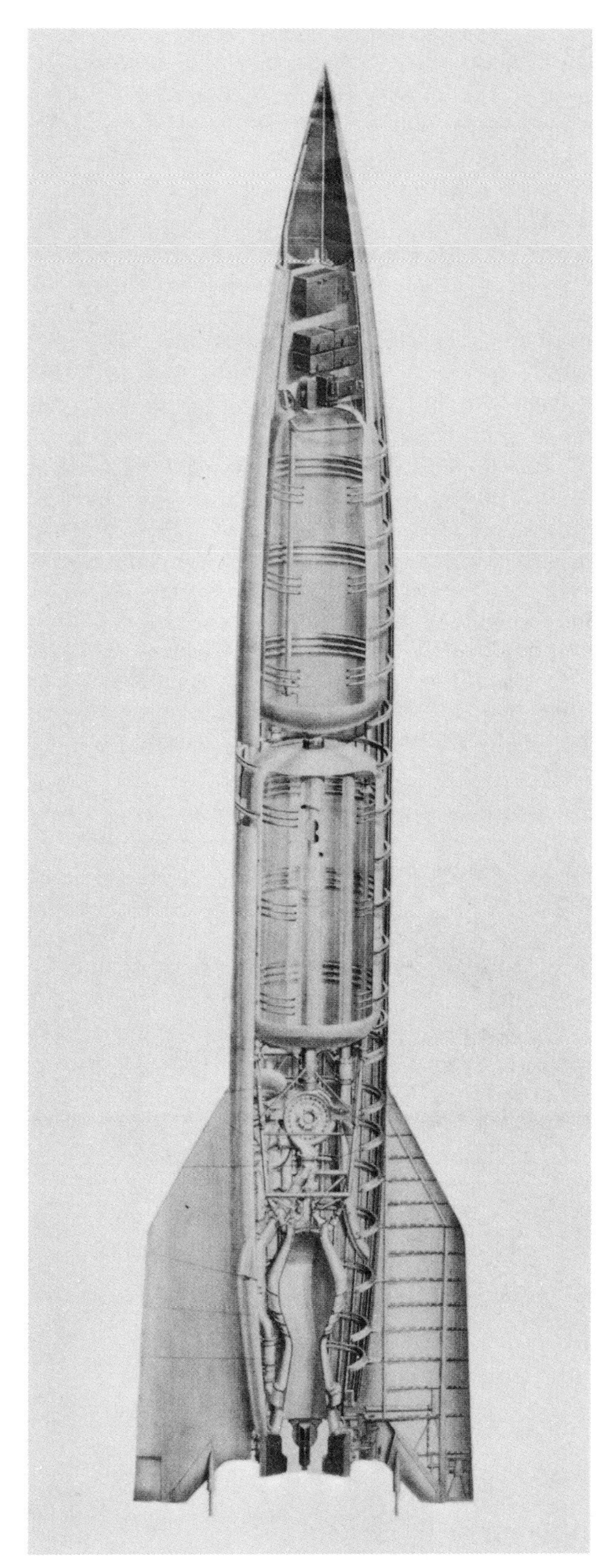

*The V-2 was 46.1 feet long, weighed 27,000 pounds, and had a range of 200 miles. It zoomed down on its target just five minutes after taking off.*

Gestapo headquarters in East Prussia, where Heinrich Himmler tried to coerce him into deserting the army and working for him. Von Braun turned down the proposal and left. A few days later, at 2 A.M., he was arrested by three Gestapo agents. After two weeks in a Stettin prison, he was charged by an SS court. The accusations: He was not really interested in war rockets, but was working on space exploration; he was opposed to the use of V-2's against England; and he was about to escape to Britain in a small plane, taking vital rocket secrets with him. Dornberger went directly to Hitler and said that without Von Braun there would be no V-2. Von Braun was released.

One problem that had to be solved was whether the V-2 should be a mobile weapon. The technical people preferred a fixed launch site where loading, repairing, servicing, and last-minute adjustments could be handled under near-laboratory conditions. But the military experts realized that any fixed site near the English Channel, no matter how well protected, would be an easy target for the thousands of Allied bombers stationed within easy reach. So the V-2 was made transportable and launchable from road or rail equipment. A typical trailer-mounted missile was supported by about thirty vehicles, including transportation trailer, launching platform trailer, propellant vehicles, and command and control trucks. A missile could be fired in four to six hours after a launch site was occupied.

*Since it was impossible to protect fixed sites from enemy attack, the V-2 was launched from rail or road equipment. Below, a launching platform mounted on a railroad car.*

The V-2 assault got off to an inauspicious start on 6 September 1944, when two missiles were fired unsuccessfully against Paris. Two days later the V-2 offensive against southern England began. The first combat missiles were launched from a site near The Hague in the Netherlands at the rate of two a day.

More than 5,000 V-2's were built before the war was over. Some 600 of them were used for training and tests, and a good part of the rest were fired against Great Britain and targets on the Continent. If everything functioned properly there was no defense against the missile. It dropped down on its target at 3,500 miles per hour just five minutes after taking off.

But a myriad of things could go wrong with the complex missile, even if the launch went as planned —and it often didn't. The guidance system could fail, causing the V-2 to miss its target. The missile could explode on its journey out of the atmosphere or break up as it returned to Earth. Even if the target was reached, the warhead could turn out to be a dud. But the V-2's took their toll. More than 1,500 V-2's were reported to have landed in southern England or just off its shores, and they were responsible for more than 2,500 deaths and great property damage.

The V-2 offensive ended on 27 March 1945, nearly seven months after it began. The Germans could no longer provide support for a weapon that clearly was going to neither influence the war's outcome nor delay its end.

Although the V-1 and V-2 were the most spectacular missiles developed by Germany during the war, they were by no means the only ones. As early as 1937, the Germans were flying a rocket aircraft powered by a 1,300-pound-thrust Walter engine. This engine was intended for a fighter, the He-176, designed by Heinrich Hertel. Easily the boldest aviation project of its time, the He-176 would probably have been able to exceed the speed of sound had it been completed. The original plan, outlined in 1936, was for Von Braun's group to develop a more powerful alcohol–liquid oxygen engine to replace Hellmuth Walter's hydrogen peroxide engine. Von Braun's engine was successfully flight tested in a He-112 fighter plane. In 1937 the flight tests were conducted with a pressure-fed system, and in early 1939 with pump feeding. However, with the invasion of

Poland in the fall of 1939, the Air Ministry lost interest in the project and it was canceled. Instead, both Walter and Von Braun's group at Peenemünde worked on JATO units, and the Walter units were placed into production and operational use. The Peenemünde unit produced 2,200 pounds of thrust for 30 seconds on liquid oxygen and water-diluted alcohol. Flight tests showed that two of these units could help an He-111 or Ju-88 to take off from a short grass strip with a heavy payload.

The He-176 project was canceled after a few highly successful flight tests with a preliminary Walter engine. But the idea of a rocket-powered interceptor was continued with a design creation of Alexander Lippisch. This was a squat, tailless monoplane with wings swept back at a 30-degree angle. The undercarriage was jettisoned on takeoff, and the aircraft landed on skids.

Lippisch had been working on rocket planes since 1932, when he designed the Delta 4 airplane, which became, in 1955, the Delta 4a. By 1940, his DFS 194, powered by a Walter 660-pound-thrust hydrogen peroxide engine, reached 340 miles per hour. The Messerschmitt Me-163A followed in the spring of 1941, when engineless tow tests were made at Augsburg. Its rocket motor was installed in the early fall, and flights were made at speeds up to 640 miles per hour from the Peenemünde airfield.

The larger Me-163B was built around a more powerful Walter hydrogen peroxide–hydrazine hydrate/methyl alcohol engine. Tests of the new aircraft were held in 1943 at Bremen, Augsburg, and Brandes, near Leipzig. Two 1,000-pound-thrust JATO units helped get the airplane off the ground. About sixty of the aircraft were built by Messerschmitt.

Junkers built three hundred C models of this aircraft, which was called first the Ju-263, then the 8-263, and finally the Me-263, when Messerschmitt continued its development. The C model was glide-tested, but never flew under rocket power. The B model was 19.5 feet long, weighed 9,040 pounds, and could fly for less than 8 minutes at 550 miles per hour; it could climb to 32,800 feet in 3 minutes. The thrust of its throttlable engine varied from 660 to 3,500 pounds. The C model was 23.1 feet long, weighed 11,280 pounds, and was planned for an engine whose thrust varied from 440 to 3,740 pounds.

The Germans had one more go at a rocket-powered airplane in the closing months of the war. In August 1944, the Luftwaffe asked the Bachem-Werke firm to undertake the development of the aircraft officially known as the BA-349 and commonly called the Natter. The concept for such a plane had been proposed by Lippisch and P. Karlson, and the German military hoped that it could destroy enemy bombers before they dropped their bomb loads on German cities. There was a large element of desperation in the proposal, since the Natter was designed for use just once.

The Natter, made of wood, was supposed to go into action immediately after an air-raid alert. It would take off with the help of two solid-propellant JATO units and its Walter sustainer engine. Ground radar control would bring it within about a mile of the enemy aircraft, and the pilot would attack from there, using either 28 electrically fired rockets or two 30-mm guns. After the attack, the plane would glide down to two miles altitude, where the pilot would bail out; the plan was to recover the sustainer rocket for use in another mission.

The Natter was less than 19 feet long, weighed 4,925 pounds, and during its few minutes of flight hit speeds of from 500 to 600 miles per hour. After a number of unmanned tests, the first and only piloted flight was undertaken in February 1945. It was a complete failure. At 330 feet, the cockpit came off, killing the pilot. The Natter continued to climb. At 1,600 feet it turned and nose-dived into the ground. That ended the history of rocket-powered German aircraft. They had been much more successful than American attempts, but their military effectiveness fell far short of expectations.

The Germans got off to a late start on surface-to-air missiles, beginning development work only when the Allies began to dominate the skies over

*The German rocket-powered airplane, Messerschmitt Me-163, is shown in rear view on ground.* (U.S. AIR FORCE)

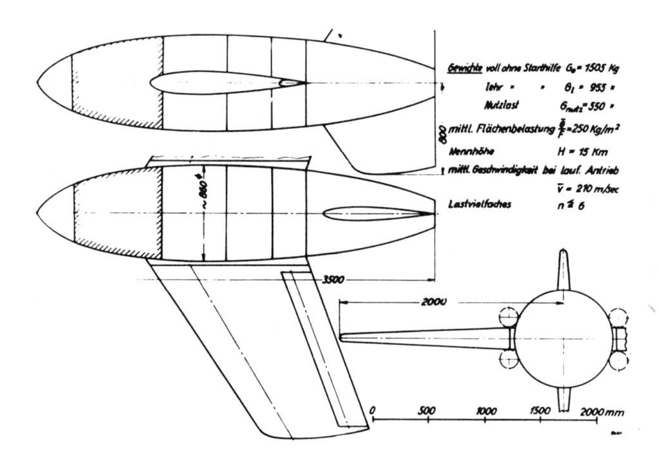

*The Germans tested a variety of surface-to-air missiles, but the war was over before they were ready for use. Some of the missiles they developed were the Enzian FR6 (top), the Feuerlilie F-55 (center), and the Schmetterling (bottom).*

Germany. A variety of models were test-fired with some success, but the war was over before they could be used against enemy bombers. The principal missiles developed were the Enzian (named for an Alpine plant), two models of the Feuerlilie (Firelily), the Hecht (a Fish), two Rheintochters (Rhinemaiden), the Schmetterling (Butterfly), the Taifun (Typhoon), and the Wasserfall (Waterfall).

Three basic series of guided air-to-surface missiles were developed by the Germans. Two were designed by the firm of Blohm and Voss, four by Henschel, and one, the SD 1400, by the Deutsche Versuchsanstalt für Luftfahrtzeug- und Fahrtzeugmotoren (German Research Center for Aeronautic and Automotive Propulsion). Though some air-to-surface missiles were used during the war, they could hardly be said to have left the testing phase.

The Germans did not have a guided air-to-air missile, but air-to-air rockets were fired on a number of occasions against Allied airplanes. Among these were an adapted ground rocket (Rz 73 or 7.3-cm Föhn) and the R4M. Air-to-air rockets were used against unescorted United States bomber formations in the summer of 1943; the first recorded use of air-to-air rockets in the war came when American bombers attacked Schweinfurt and were met by rocket-carrying German airplanes.

The Germans also had several successes with their ground-to-ground missiles in addition to the V-1 and V-2. They introduced their Nebelwerfer (Fog thrower) rocket firing device in rather large quantities on the Russian front in 1941 with considerable effect. Originally designed to produce smoke screens for infantry assaults, the Nebelwerfer's surprising accuracy led to its use as an artillery barrage rocket. A special launcher was created to fire five or six rockets. Later, an advanced Wurfgerät (Propelling Device) was developed. Both launchers fired solid-propellant rockets with diameters of 15, 21, 28, and 32 centimeters with fair accuracy. The 21-centimeter (8.3-inch) rocket, typical of the others, was 4.1 feet long, weighed 241 pounds, carried a 22.5-pound payload, and had a range of 4 miles. It was propelled by a compound of nitrocellulose and diethylene dinitrate.

The Germans also used imitations of the bazooka, developed when the effectiveness of the American weapon became apparent. The Panzerfaust (Tank Fist) fired a 2.36-inch rocket, the Panzerschreck (Tank Panic) a 3.46-inch projectile. Both could be launched by an individual against a tank, and both projectiles weighed between 7 and 9 pounds.

A more sophisticated antitank weapon, the X-7, was designed, and some components were devel-

**Principal German Surface-to-Air Missiles[a]**

| Missile | Length (feet) | Weight (pounds) | Range (miles) | Velocity (mph) | Altitude (miles) | Characteristics |
|---|---|---|---|---|---|---|
| Enzian | 12 | 4,350 | 18 | 600 | 9 | Several plastic-wood, swept-wing Enzians built by Messerschmitt in Augsburg in late 1943; heavy bombing forced move to Oberammergau. From early 1944 to early 1945, when the program was canceled, twenty-five missiles were fired, of which a third was successful. Designed by Georg Madelung and powered by a Hellmuth Walter liquid-propellant rocket engine, Enzian consisted of three compartments: (1) steel and plastic-wood warhead, (2) monocoque fuselage (including propellant tanks), and (3) tail section. Launched from high-angle, 20-foot ramp, with aid of four fuselage-mounted solid-propellant JATO's. Radio controlled. |
| Feuerlilie F-25 | 6.7 | 265 | 3 | 600 | 1.8 | The F-25 and F-55 were quite different missiles, the former subsonic and the latter supersonic. F-25 development was directed by the Luftfahrtforschunganstalt Hermann Göring E.V. at Braunschweig, fuselages being manufactured by Ardeltwerke and solid rocket engines by Rheinmetall-Borsig in Berlin. Behind warhead was main body section supporting wings and controls, and tail section containing engine and providing support for fins and stabilizer. Roll control by ailerons, stabilization by gyros and servomechanisms. First F-25 flight occurred in April 1943 at Leba, Pomerania. In 1944, after twenty missiles had been built, program was canceled, and F-55 development begun. Newer, more advanced missile was launched by a solid rocket and sustained by rocket engine operating on liquid oxygen and alcohol. Tapered, swept-back wings were positioned well to the rear. Sustainer developed by Deutsche Versuchanstalt für Kraftfahrtzeug und Fahrtzeugmotoren; booster by Rheinmetall-Borsig. First flight at Leba, May 1944; second at Peenemünde six months later. |
| Feuerlilie F-55 | 15.75 | — | 6 | 900 | 3 | |
| Hecht | 8.3 | 308 | 6 | 650 | 3–4 | Looking much like small airplane, only one was flown before project was canceled in favor of Feuerlilie series. The missile featured two swept-back wings and a high central tail fin; it was powered by a Walter hydrogen peroxide rocket engine and was ramp-launched. |
| Rheintochter 1 | 20.7 | 3,850 | 7.5 | 680 | 3.7 | Both versions of Rheintochter were test fired at Leba. Of eighty Rheintochter 1's fired, twenty-two were equipped with radio control, and of these eighteen were successful. Missile had six wooden stabilizing fins to rear, four control fins forward. Ramp-launched by solid JATO unit which separated from missile after burnout. Once aloft was controlled by radio and stabilized by gyro system. Both visual and radar tracking devices were employed. An unknown number of Rheintochter 3's were tested, some with solid-propellant sustainers, others with liquid-propellant sustainers of the Was- |
| Rheintochter 3 | 16.5 | 3,450 | 22 | 750 | 9 | |

**Principal German Surface-to-Air Missiles (continued)**

| *Missile* | *Length (feet)* | *Weight (pounds)* | *Range (miles)* | *Velocity (mph)* | *Altitude (miles)* | *Characteristics* |
|---|---|---|---|---|---|---|
| | | | | | | serfall type. Takeoff by solid JATO units mounted to fuselage. Airframe consisted of five sections: fuze, control, electric, propellant tank, and propulsion. It had four rather than six stabilizers. |
| Schmetterling | 12.5 | 981 | 10 | 540 | 6–7 | Occasionally called the V-3, and officially known as the Hs-117, about sixty missiles constructed by Henschel in Breslau. Midwing monoplane with cruciform tail. Liquid sustainer, manufactured by BMW, was mounted internally; solid boosters externally mounted, one above and one below fuselage. Fired rather successfully from rotatable platforms, it went into prototype production in 1943. Two operators were required to guide missile: one for positioning the aiming devices and the other to maneuver vehicle by control stick. |
| Taifun | 6.3 | 66 | 7.5 | 2,800 | 4–5 | Elektro-Mechanische Werke-Berlin and Peenemünde joint development, appearing as by-product of Wasserfall design study. Not guided, very small, it was a barrage weapon that was to have been manufactured and fired in vast quantities against Allied bomber fleets. Both liquid- and solid-propellant motors were in production when the war ended. |
| Wasserfall | 26 | 7,800 | 17 | 1,900 | 8 | Basically a one-third model of V-2, though of simpler construction and fitted with four small wings 11 feet from nose. Propulsion system of pressure-feed type, operating on nitric acid and vinyl isobutyl ether/aniline propellants. Designed to knock out planes at high altitudes, missile was to have incorporated radio guidance system under development by Telefunken when program terminated in February 1945. For terminal guidance, an infrared homing device was considered; the 674 pounds of explosives were to have been detonated by radio command. First firing last day of February 1944 at Peenemünde, about two years after the program began; this was followed by twenty-four others in 1944 (60 percent considered successful) and ten early in 1945. |

[a] Data derived from the Ordway-Wakeford *International Missile and Spacecraft Guide.* (McGraw-Hill, 1960)

oped, but it was never used in combat. The X-7 projectile, designed to be fired from either a spring or rail ground launcher or a Panzerfaust-type shoulder launcher, was about 2.5 feet long, 5.5 inches in diameter, and had a wingspread of 1.3 feet. It had a two-stage solid-propellant Wasag motor (fast-burning powder for quick takeoff, slow-burning powder for steady acceleration) and was guided by wire that unrolled from wingtip spools. The missile would maneuver in response to signals sent along the wires.

Almost a sidelight to the major German rocket and missile effort were the experiments conducted in the Baltic Sea with missiles launched under water. Short-range 32-centimeter Army rockets with solid propellants were used. Their nozzles were sealed and special ignition systems were installed. A simple welded steel rack, mounted on the deck of a submarine, was the launcher. Six rockets were fired with complete success in the late fall of 1943. The project was the idea of Fritz Steinhoff, a U-boat commander whose brother, Ernst, was a key

**Principal German Air-to-Surface Missiles**[a]

| Missile | Length (feet) | Weight (pounds) | Range (miles) | Velocity (mph) | Characteristics |
|---|---|---|---|---|---|
| BV-143<br>BV-246 | 19.5<br>11 | 4,000<br>1,600 | 10<br>12 | 600<br>260 | Antiship weapon. Pair V-shaped wings with elevators; two small rectangular tail fins. BV-143's flight profile called for an initial cruise at same altitude from which it was launched, then descent and level off at about 10 feet above surface. Latter maneuver was rarely successful, regardless of whether under pilot (who was guiding the missile from He-111 or He-177 control plane) or automatic altimeter control. About one hundred missiles built for Luftwaffe. An unsuccessful weapon, it was canceled and development shifted to the BV-246, a nonpowered guided ASM also used against ships. About four hundred were built, but never entered into combat service. Radar, infrared, and acoustic homing devices were tried out, as well as more standard radio guidance systems. Both missiles solid-rocket propelled. |
| Hs-293<br>Hs-294<br>Hs-295<br>Hs-296 | 12.5<br>20<br>16.2<br>17.1 | 2,300<br>4,800<br>4,590<br>6,000 | 10<br>8.5<br>5<br>4 | 470<br>580<br>500<br>500 | The most widely built and used of the series was Hs-293, a glide-bomb to which was attached a Walter hydrogen peroxide rocket motor. Control from tracking aircraft was assured by elevator and aileron action, the pilot following bomb's flight by flare or electric lamp. Launched at 1,000 to 20,000 feet altitude from several types of Henschel and Dornier planes (optimum launch speed, 210 miles per hour). First tests undertaken at the Luftwaffen-Erprobungstelle Peenemünde-West in December 1940. In November 1941 placed in production, and about a year later entered Luftwaffe inventory. Sunk several British ships in Bay of Biscay 1943. Hs-294 was an air-to-underwater torpedo missile launched by Ju-90 and He-177 airplanes and powered by twin Walter liquid rocket engines. Upon entering water the wings and engines sheared off and missile continued toward the target as torpedo. Also powered by twin Walter engines was Hs-295, designed for neutralizing lightly armored sea targets. Radio guided and armed with high-explosive, armor-piercing warhead. Missile first tested in 1943, went into production, but was canceled next year. Hs-296 was experimental, combining control system of Hs-293, rear fuselage and wings of Hs-294, and warhead of Hs-295. Liquid-rocket propelled. |
| SD-1400 | 15.4 | 5,500 | 9 | 625 | Earlier known as FX-1400X, Fritz X, and X-1, was an armor-piercing Esau bomb with four wings and a tail unit incorporating radio-controlled and solenoid-operated spoilers. Achieved world fame when it sunk the Italian battleship *Roma* following launch from a Do-217. The X-2, -3, -4, -5, and -6 models were developed with varying guidance systems. |

[a] Data derived from the Ordway-Wakeford *International Missile and Spacecraft Guide.* (McGraw-Hill, 1960)

**Principal German Air-to-Air Missiles[a]**

| Missile | Length (feet) | Weight (pounds) | Range (miles) | Velocity (mph) | Characteristics |
|---|---|---|---|---|---|
| Hs-298 | 6.7 | 265 | 4–5 | 535 | Aluminum-and-magnesium alloy missile designed by F. Nikolaus, developed by Henschel, and carried by FW-190, Ju-88, and Do-217 aircraft. Although first flight tests in December 1944 were moderately successful, and it went into pre-production, program was soon canceled. After launch by its two-stage Schmidding 109-543 rocket motor, it would accelerate rapidly past the host aircraft which, by a Fevi line-of-sight guidance system, would direct the missile to its target. Two crew members were necessary, one to follow missile and one to guide it in with joy-stick. |
| X-4 | 6.6 | 132 | 2–3 | 550 | Was guided by impulses sent along wires unspooled from missile as it flew toward prey. Program began at the Ruhrstahl firm in June 1943, with concurrent engine development at BMW, and first flight occurred in September 1944. About one hundred missiles were built and, as war ended, was considered to be in advanced development stage. Consisted of a fuze housing, warhead, main fuselage, two swept-back wings, four swept-back stabilizing fins, and control wire bobbins in housings at wing tips. Designed by Max Kramer of DVL. |

[a] Data derived from the Ordway-Wakeford *International Missile and Spacecraft Guide.* (McGraw-Hill, 1960)

department head at the Peenemünde. Von Braun participated in the project with the two Steinhoffs. Despite the successful firings, the German Navy did not become interested in the project. One reason, apparently, was the Navy's fear that the launching rack would reduce drastically the speed, maneuverability, and rough-weather capability of its submarines.

In late January 1945, when it had become apparent that the Third Reich was in the throes of its final collapse, Von Braun met secretly with his top staff members to decide whether they should remain at Peenemünde and surrender to the advancing Russian Army or to move south and make contact with American forces. The virtually unanimous decision was to head south.

The confusion rampant in Germany in those critical days made the move easier. Von Braun had received nearly a dozen uncoordinated directives from ministries in Berlin, local army and navy commanders, the SS, and a flock of Nazi party bosses. Some of the orders directed him and the entire Peenemünde staff to evacuate the site so that the "top priority research and development work could be continued at a safer location until ultimate victory was assured." Others directed the Peenemünde staff to stand firm and "defend the holy ground of Pomerania."

Von Braun and his intimates sorted out the "move" orders from the rest and disregarded the "stay put" orders. To bluff their way through the maze of military roadblocks and Gestapo check-points along the route, they equipped all the railroad cars, automobiles, and trucks to be used in the evacuation with blazing red-and-white signs reading *Vorhaben zur besonderen Verwendung*—a purely mythical Project for Special Disposition. The scheme worked. Von Braun, his close associates, and about five thousand employees and their families, along with large quantities of documents, drawings, and papers, left Peenemünde in February in ships, railroad cars, trucks, and automobiles. Heading south, dodging Allied planes and bluffing units of the Gestapo and the SS, they finally reached the town of Bleicherode in the Harz Mountains, where the Armament Ministry in Berlin had directed the rocket research work to continue.

By sheer chance, the area military commander

was the same SS General Kammler who had earlier tried to take over Peenemünde. As boss of several concentration camps, Kammler was growing increasingly jittery as the advance of Allied troops brought retribution closer and closer. He decided that he could bargain better with the approaching United States Army if he had hostages—Von Braun and several hundred of the Peenemünde scientists and engineers.

Kammler's first move in this game came on 2 April when he ordered the transfer of Von Braun and about five hundred other Peenemünde rocket experts that had just moved into the Bleicherode area to an empty army camp near Oberammergau in Bavaria, in the Alpine area that the Nazi hierarchy

*The A-4b ground-to-ground missile, shown here on test stand and in cutaway, had twice the range of the V-2. It was experimentally fired at Peenemünde in the summer of 1944.*

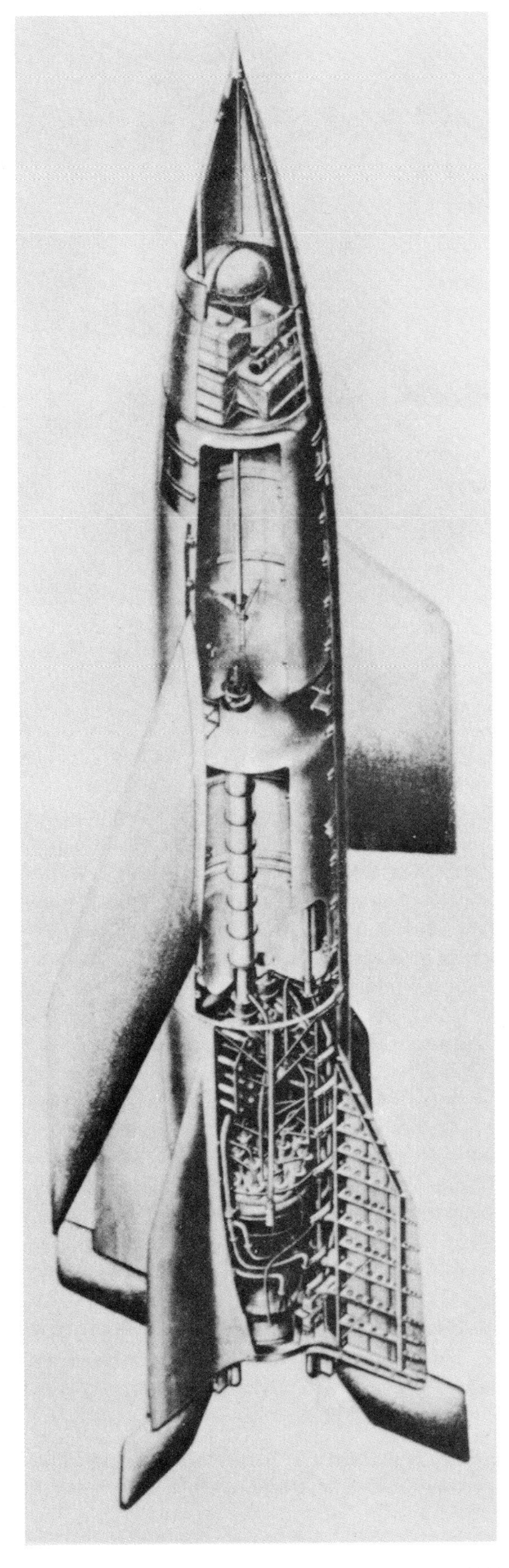

*After the war, U.S. Seventh Army troops found this Natter rocket-powered interceptor, still in the experimental stage, on an airfield near St. Leonhard, Austria.* (U.S. ARMY)

had designated its last retreat. Von Braun's countermove on arrival at the scene was to express fear about the danger of air attack on the completely unprotected, blatantly visible barracks; together with Ernst Steinhoff he managed to persuade local SS officials to get his people out from behind the camp's barbed wire and scatter them among twenty small towns in the area. Later, General Dornberger, with a small military staff, joined one element of the now widely dispersed group in the Bavarian ski resort of Oberjoch.

A few days afterward, radio news of Hitler's death on 30 April spurred final plans for surrendering to the Americans. From their mountain retreat, Dornberger and Von Braun sent off Magnus von Braun, Wernher's English-speaking brother, to attempt to make contact. Magnus von Braun surrendered to PFC Fred P. Schneiker of the 44th Infantry Division at the town of Reutte, Tyrol, informing him that he represented a group of rocket scientists anxious to turn themselves over to the Americans.

Describing the circumstances of the initial contact, Charles L. Stewart, a special agent of Army Intelligence, told the authors:

I do recall that the war was still in progress, although it was near the end and the front in our sector was relatively quiet. I was in Reutte in the Austrian Tyrol at the time, and as many of the higher German civil and political officials had sought refuge in that area, we were very busy.

Magnus von Braun came through the lines first. He was brought to our headquarters and explained that his brother and some 150 of the top German rocket personnel were lodged in an inn behind the German lines. They wished to join the Americans to continue their work in rocket development. They had selected the Americans, as they were favorably disposed to this country generally and also because this country was the one most able to provide the resources required for interplanetary travel. Furthermore, they were anxious to depart from the German side, as there was the possibility that an SS colonel in Innsbruck might eliminate them all, pursuant to last-minute Nazi orders to liquidate certain key German scientific talent to prevent them from falling into the hands of the Allies.

We made the necessary arrangements, and I went part of the way with them. When Dr. von Braun came out with General Dornberger and Colonel Axter, they lodged with us while we communicated with higher headquarters. None of us had scientific backgrounds, but the magnitude of their discoveries and their potential for the future was immediately apparent. We were dismayed when we could not arouse any interest in them at higher headquarters. Our first instructions were to the effect that they were to be thoroughly screened (a favorite solution of the military in dealing with high-ranking personnel if in doubt as to how to proceed). Our reply was to the effect that it made no difference if all were brothers of Hitler, because their unique knowledge made them extremely valuable militarily and from a national standpoint. After a few days we were able to arrange their transfer to higher headquarters.

All captured scientists were moved to German army barracks in Garmisch-Partenkirchen for continued interrogation. Several months later, Dornberger was turned over to the British. He spent two years in a prisoner-of-war camp before going to the United States.

Although the Americans were slow to react to Magnus von Braun's surrender, top United States intelligence officers—along with their British and Russian counterparts—were well aware of the importance of locating both the German rocket scientists and their rockets. Colonel Gervais W. Trichel, head of the Ordnance Corps Rocket Branch in Washington, knowing of the existence of an underground V-2 factory near Nordhausen, requested Colonel (later Major General) Holger N. Toftoy, chief of the Ordnance Technical Intelligence team in Paris, to ship V-2 assemblies out of the Mittelwerke plant as soon as it was captured. Trichel also had sent a member of his staff, Major Robert Staver, to London to work up a list of key German rocket personnel to be found and interrogated. This was the genesis of the American Intelligence project which, on 19 July

1945, was christened Overcast and which, nine months later, was given the name by which it became famous: Operation Paperclip.

V-2 production at the underground Mittelwerke plant had not been affected by the migration of the scientific team from Peenemünde into the Oberammergau area. But with the approach of Brigadier General Truman Boudinot's Combat Command B troops on 10–11 April, the remaining 4,500 workers dispersed into the neighboring villages and countryside. When Colonel John C. Welborn, Lieutenant Colonel William B. Lovelady, and Major William Castille entered the huge underground plant they found, to their surprise, that it was completely intact and that its lines of V-1 and V-2 assemblies had been left undisturbed. The Americans had come upon one of the greatest technical prizes in history. Toftoy and Staver were informed immediately of the capture.

Staver, aided by a General Electric engineering team under Richard Porter and by British Intelligence, had finished compiling the list of key personnel to be interrogated. After the capture of Nordhausen and Bleicherode, he went to Paris to see Toftoy, and the two mapped out plans to remove the invaluable materiel. Their sense of haste was heightened by the news that the section of Germany that so interested them was about to be turned over to the Soviets as a result of the Yalta agreement.

Toftoy set up "Special Mission V-2" under the command of Major James P. Hamill, who was instructed to ship from Nordhausen to Antwerp all the rocket equipment he could lay his hands on. Moving into the area, his group set up perimeter controls and, with the aid of the 144th Motor Vehicle Assembly Company, local laborers, and German railway hands, put the V-2 equipment onto flatcars and gondolas. The first trainload of materiel left Nordhausen on 22 May 1945; the last on 31 May, the day before the Russians were expected to arrive.

*Von Braun (with arm in cast) surrenders to the Americans at Reutte, in May 1945. Left to right are Charles L. Stewart, U.S. counter-intelligence agent; Lt. Col. Herbert Axter, attached to Gen. Dornberger's staff; Dieter K. Huzel; Wernher von Braun; Magnus von Braun; and Hans Lindenberg.*

In all, 341 cars were loaded, representing about a hundred V-2 ballistic missiles. Once in Antwerp, the equipment was crated and loaded into sixteen Liberty ships and transported to New Orleans—and thence to the New Mexico desert.

While Hamill removed the hardware from Mittelwerke, Toftoy ordered Staver to conduct a search for any engineering reports and other technical documents that might be in the area. Toftoy's hunch that such data were nearby was entirely correct.

Von Braun realized in early April when he was ordered to go with five hundred key rocket scientists from Bleicherode to Oberammergau that it would be impossible for the group to take with them the vast archive of documents that they had brought south from Peenemünde. This priceless collection of drawings, scientific papers, test and flight reports, supersonic wind tunnel studies, and so on, contained the essence of all the important rocket research conducted in Germany between 1932 and 1945. The materiel had been moved from Peenemünde to Bleicherode under the greatest difficulties. To leave it behind now would be to expose it to the dangers of Hitler's "scorched earth" policy. Von Braun therefore instructed two of his associates, Dieter Huzel and Bernard Tessman, to hide the archives in an abandoned mine or some similar place. Racing breathlessly against time, Huzel and Tessman secreted the documents in a tunnel near the town of Dörten at the northern edge of the Harz Mountains, just as the United States Ninth Army began occupying the area.

Staver left Paris in search of these papers on 30 April, accompanied by Edward Hull, a member of Porter's General Electric team. During the next few weeks Staver rounded up several rocket experts who had remained in the Bleicherode area after its capture, including Eberhard Rees, Karl Otto Fleisher, and Walther Riedel. With their assistance, the fourteen tons of documents were found on 21 May where they had been hidden near Dörten.

With both the missiles and the technical papers now in hand, Toftoy, who during May was made chief of the Rocket Branch in the Research and Development Division of Army Ordnance, recommended to his superiors that the German rocket scientists themselves be brought to the United States. On 23 July 1945 he was ordered to arrange such a move. Toftoy met Von Braun and his associates at Witzenhausen in early August of 1945, and offered them one-year contracts under the Paperclip program. The German scientists would be sent to the United States under Ordnance Corps custody. Their dependents would be left behind to be quartered and fed by the Army, with the costs deducted from fees paid to the scientists. All 127 scientists to whom this offer was made accepted it.

As soon as the Nordhausen and Bleicherode areas were vacated by United States troops, the Red Army moved in and rounded up some 3,500 lower echelon personnel. (Only about 1,000 of the original 5,000 Peenemünde families had managed to leave for Bavaria after the Russians moved in, not counting the original 500 sent south under General Kammler's orders.) The principal German scientist to contract voluntarily with the Russians was Helmut Gröttrup, who was soon placed in charge of the technical personnel of the Nordhausen factory and the temporary laboratory facilities set up in and around Bleicherode. A new organization was established by the Red Army, which was called Institut für Raketenbetrieb Bleicherode (RABE). Gröttrup remained there until October 1946 when, with about 200 other German rocket specialists, he was shipped without advance notice to Russia.

The British and French also obtained the services of a few German rocket experts, but were unable immediately to undertake major post-war missile development programs. In Project Backfire, the British secured German aid in firing two V-2's from Cuxhaven out over the North Sea, but this did not lead to any sustained ballistic missile effort on their part.

It was the United States which reaped by far the biggest gains from the dismemberment of the German missile establishment.

Von Braun, Dornberger, and their fellow scientists and engineers brought more than just their knowledge and talents with them when they went south to be captured by the American army; they also brought to the West a vision of the future. A vision, moreover, that had been partially worked out in their plans for the continuation of the "A" series beyond the A-5, which had served as a test model for the A-4 (the V-2). Of the later designs in the "A" series, only the A-7 had actually been built during the war.

The A-6, complete on paper, was never converted to hardware. It was to be an improved V-2, powered by an engine using a nitric-sulfuric acid mixture as oxidizer and a fuel consisting of vinyl isobutyl ether mixed with aniline. The underlying idea was to have a liquid-propellant rocket with fuels that could be stored for faster tactical response and greater ease of handling—an idea that came into its own in the postwar years.

There also was an A-7, a winged missile that was about the same size as the A-5. The first A-7's did not have propulsion systems but were used merely for air-drop tests to gather ballistic data. Later models had 3,300-pound-thrust A-5 engines for ground-launch tests. The A-7 had a 30-mile glide range when launched from an airplane flying at an altitude of 5 miles, and a 15-mile range when launched under its own power from the ground.

The A-8, like the A-6, was never built. It would have been essentially an improved A-6 adapted to the same winged configuration used for the A-7 and A-9.

The objective of the A-9 program was increased range. Instead of hurtling to earth, the winged missile would have made an extended glide of up to 400 miles toward the target. Its approach velocity would be relatively low. During the later phases of the war, this fact led many people to doubt its military usefulness, especially in view of the heavy toll taken of the V-1's by defensive forces. Nevertheless, the potential of the boost-glide technique was energetically explored.

Something resembling the A-9 was actually flown in the closing days of the war. It was the A-4b (sometimes called the "bastard A-4"), a designation that was adopted in order to get the winged vehicle the same high priorities enjoyed by the true A-4. The A-4b was a standard V-2 whose external surfaces had been modified—rather crudely—to permit the attachment of supersonic wings and enlarged aerodynamic control surfaces. The A-9 would have looked about the same, but it would have been much more polished.

The A4-b concept was the result of the loss of V-2 launch sites in northern France, Belgium, and Holland that followed the Normandy invasion. Von Braun was faced with continuing demands to keep the V-2's—by now in full mass production—flying, even though the ranges were much longer than heretofore. On 19 October 1944, it was decided to modify some V-2's to extend their range. Although five A-4b's were put under construction, only two were actually launched, on 8 January and 24 January 1945. The first was a failure, but the second proved that the designers' calculations were sound. It was launched straight up, reached a maximum altitude of 50 miles, and attained a maximum speed of 2,700 miles per hour, which made it the first winged guided missile to exceed the speed of sound. The automatic control system worked satisfactorily.

One of the more intriguing aspects of the A-9 project was the plan for a piloted version of the missile. It was proposed to fit a tricycle landing gear to the A-9, so that it could land on a conventional airstrip after its 400-mile boost-glide. Flight time from takeoff to landing was to be 17 minutes—an incredibly short time in terms of 1945 aeronautical thinking.

The final "A" program on which some design work was accomplished before the end of the war was also the most ambitious. It was the A-10, which was not a complete vehicle but the first stage of a two-stage vehicle; the A-9 was to be the second stage. The A-10 was to have produced 400,000 pounds of thrust on a combination of nitric acid and Diesel oil. The stage was to be 65 feet long, with a diameter of 162 inches. The loaded weight of the entire vehicle was 174,000 pounds, only 30,000 pounds of it contributed by the A-9 second stage. Calculations showed that the missile could carry a one-ton payload roughly 2,500 miles.

Above and beyond the A-10 were ideas aimed at exploring space. One project, which could have become the A-11, visualized a third stage *under* the A-10/9 combination. The three-stage vehicle was to place the pilot of a modified A-9 into orbit. And there was even thought of an A-12 stage, producing a minimum of 2.5 million pounds of thrust; with the A-11 second stage and a winged A-10 third stage, it could possibly have orbited a payload of up to 60,000 pounds.

One final German dream should be mentioned: the Antipodal Bomber, designed by Eugen Sänger and fellow worker (later his wife) Irene Bredt. It was described in a long-secret report translated by the United States Navy under the title of "A Rocket Drive for Long-Range Bombers."

The report envisioned a 92-foot-long, 220,000-pound craft, to be launched from a sled driven by rockets developing 1,345,000 pounds of thrust. The sled would send the bomber into the air at 1,000 miles per hour; its own 220,000-pound-thrust liquid oxygen–gasoline oil rocket engine would boost it to a speed of 13,700 miles per hour and an altitude of over 160 miles. The bomber would skip along the top of the atmosphere like a stone on a pond, reaching New York with a bomb load of 6 tons. Sänger and Bredt calculated that the whole trip, takeoff to landing, would take 80 minutes. Dubbed "skip bomber" by Allied technical experts, the concept gave impetus to projects that later evolved in the United States and Soviet Union.

The ideas of the German scientists sounded fantastic—but the reality was hardly less fantastic. As American experts sifted through the tons of information from Peenemünde, it was plain that the past was just prologue to an almost unimaginable future. Rocketry had made great strides in World War II. Man had reached the edge of space.

# 6 POSTWAR MILIT

Since the end of World War II, the combination of nuclear weapons and advanced rocketry has changed the concept of warfare beyond recognition. Thermonuclear bombs, delivered by intercontinental missiles, hold the threat of immediate destruction of the vital centers of nations. The United States and Russia now confront each other from behind deterrent forces that have established a delicate balance of power without parallel in world history.

The history of rocketry since the end of World War II is almost entirely the story of events in the United States and Russia. The vast resources, great land areas, and technological know-how needed to develop, produce, and deploy intercontinental ballistic missiles have so far ruled out of the field all but these two goliaths. Many other nations have developed smaller rockets for other military purposes, but those do not represent a great advance over the technology available in World War II. The modern military missile in its most terrifying form is a product either of the United States or of the Soviet Union.

Most details of the Soviet experience are still unknown. Even the names of leading Soviet rocket scientists are mentioned only rarely and in guarded terms. Some of the major themes of Soviet rocket development are known in outline, but for the rest, it can only be assumed that the Russian program has run generally parallel to that in the United States.

There probably is one major difference, however. Within the United States, there are really three concurrent histories of military rocket development. Because of continuing interservice rivalry, the Army, Navy, and Air Force for many years carried on what were essentially separate missile-development programs that made contact with each other only occasionally. Not until the mid-1950's did the Defense Department come to grips with the problems caused by the three services' differing outlooks and stubborn independence. The Navy had strategic ambitions in extending its striking forces to intercontinental ranges by submarine-fired missiles. The Air Force basically was more at home with the concept of long-range bombardment—although the missiles had a difficult time competing with the Air Force's own enduring love for airplanes. The Army, meanwhile, preferred to think of missiles as extensions of artillery, and argued on that basis that it should be given control of long-range missile development. Much time, money, and effort were wasted until the duplication and backbiting caused by the interservice feuds could be stopped—or at least controlled.

The Air Force was first in the field, by virtue of its concentration on long-range operations. In the early postwar days, it enjoyed a virtual monopoly on planning intercontinental strikes. The first Air Force attempt to develop long-distance missiles began in July 1944, before the end of the war. The model chosen by the then Army Air Corps did not follow the general concept of the V-2 but rather that of the V-1—an unmanned, winged, and slow-flying bomb. The result was the JB jet-propelled guided bomb series.

The first in the series, the JB-1, was basically a Northrop flying wing—which then enjoyed some transient popularity—built around two 2,000-pound general-purpose bombs and propelled by twin General Electric B1 turbojet engines. Five sled-mounted, 10,500-pound-thrust solid-propellant JATO units pushed the JB-1 into the air. The hoped-for range was 200 miles. After several unsuccessful launching attempts from the ground and the air, the program foundered.

The second try was simply a copy of the V-1. The JB-2, nicknamed the Loon, was manufactured by several auto and aviation companies, who turned out about three hundred of them before the war ended. The original intent was to use the Loon against Japan; the principal benefit of the program was to give some practical missile experience to American manufacturers.

Other members of the JB series included the 4 model and the final version, the JB-10. This, too,

# Y ROCKETRY

was based on the flying wing; it was powered by a modified V-1 pulsejet. About a dozen were built, but only a few performed at all well. The program was canceled in March 1946.

By then, the Air Corps was working on more ambitious, and ultimately more successful, programs along the same line. In retrospect, the Air Force now seems to have wasted much effort in developing relatively slow, winged missiles that cruised through the atmosphere and offered comparatively easy targets to defense forces. But the decision must be seen in the context of the times. The United States had almost no experience in building large-scale rockets and the sophisticated guidance systems they needed. There was keen awareness of the inadequacies of the V-2—despite enormous expenditures of talent and money, the missile still had a miss distance of 3 to 5 miles over its typical flight of 200 miles. And the great success of the Air Corps during World War II had conditioned it to think in terms of winged vehicles, whether manned or unmanned.

As a result, the Air Force's main efforts went into two subsonic missile programs, Snark and Matador, while development of the supersonic, longer-ranged Atlas proceeded slowly at first.

Snark was conceived in January 1946 as a cruise missile capable of traveling the full 5,000- to 7,000-mile intercontinental range. It went through a slow and often unenthusiastic development program. Built by Northrop, Snark weighed 50,000 pounds, was powered by a Pratt and Whitney turbojet engine, and was boosted to operational speed by two Aerojet 33,000-pound-thrust solid JATO's. Easily transported by land and air, Snark could carry a 5,000-pound nuclear payload. It finally entered operational service with the Strategic Air Command's 556th Strategic Missile Squadron in 1958, and was phased out as Atlas became operational.

The Matador was a smaller, shorter-range missile, designed to weigh 12,000 pounds and travel 650 miles. Preliminary specifications were established in August 1945, and the contract was let to the Glenn L. Martin Company two years later. Funding cutbacks made for slow progress, but final production design was approved in February 1951. Despite problems of production build-up and flight testing in 1954, Matador entered operational inventory during 1955. It went through several versions, one of which became the faster, heavier, more accurate, and longer-ranged Mace. Matador was phased out of inventory in the early 1960's, and Mace Models A and B were deployed in its place in Europe and the Far East.

The final Air Force cruise missile was at once the most advanced, the most useful, and, inevitably, the shortest lived. By the time the Navaho was being developed, the fate of cruise missiles was plain: they simply were not effective enough compared to intercontinental ballistic missiles. The ICBM's can travel thousands of miles along arcs that take them hundreds of miles out into space; their trajectories, once determined during the interval that the motors are in operation, are thence affected only by gravitational forces and by air resistance during their exit from and re-entry into the atmosphere. Critics of the Navaho project say it was a combination of short-sightedness, conservatism, and economics that led the Air Force to spend $690 million on this cruise missile, while Atlas and other truly effective ballistic missiles limped along on minimal budgets.

Planners were unwilling to approve the development of very large and expensive ICBM's that could carry the heavy and unwieldy nuclear bombs of the period for 5,000 to 7,000 miles; the costs, compared to those of cruise missiles, appeared far too high. (The Russians, however, believed otherwise and went ahead to develop and build the huge rockets they needed; that decision gave them a long-lasting lead in space exploration.) The Air Force continued up into the mid-1950's to concentrate on cruise missiles as an interim program while waiting for the Atomic Energy Commission to develop lighter bombs. The funds for both ICBM's and aerodynamic

*The American version of the V-1 was the JB-2, shown here after being launched from its ramp by a multiple-powder-charge launcher. Known as the "Loon," this unmanned, slow-flying bomb was intended for use against the Japanese.* (C. N. HICKMAN)

missiles were just not available. They were not available because of decisions made by key officials in the Department of Defense, in Congress, and in the executive branch.

The document that gave voice to these opinions was dated June 1947 and titled "Operational Requirements for Guided Missiles." This Pentagon study was probably the single document most responsible for stifling the nascent American ICBM program in the early postwar years. It canceled government support of a promising Consolidated Vultee MX-774 rocket program—fortunately, the company continued work, on a much reduced level, by using its own funds—and cut spending on ICBM's to a low level. There were few voices raised in protest. America felt safe with a monopoly of atomic weapons, and few people suspected that Russia was engaged in an all-out drive to develop large rockets. The eventual result was that Russia had the vehicles to launch heavy payloads into space when the time came, while the United States was caught almost completely off guard.

All this is hindsight, and the Navaho program was far from a complete waste. The vehicle itself was impressive. It weighed 300,000 pounds, was boosted into the air by three liquid-propellant rocket engines, each producing 135,000 pounds of thrust, and would have traveled to its target under ramjet power at a speed of Mach 3 (three times the speed of sound). Navaho was strangely shaped, with a huge finned booster assembly slung under a delta-wing fuselage with tip-mounted ramjets. Built by North American Aviation, Navaho was 95 feet long and guided by an all-inertial guidance system. Many of its components were checked out in the X-10, a test vehicle powered by twin turbojets. Navaho itself was flown eleven times, beginning on 6 November 1956, and ending on 18 November 1958, even though the program was officially canceled on 11 July 1957.

Navaho might have survived if its timing had been different, but it was caught in a budgetary squeeze that developed as Snark neared operational status and its ICBM successor, the Atlas, went into its test program. The technological fallout from Navaho, however, almost justified the entire program. Uprated versions of the Navaho's 135,000-pound-thrust rocket engines, built by Rocketdyne, were applied to such missiles as the Jupiter, the Thor, and the Atlas, and Navaho technology also led to the Redstone engine. Navaho left other legacies. Problems of high-speed flight and design of large supersonic vehicles were at least partially solved in the Navaho program. The experience helped in the XB-70 and X-15 high-speed airplane programs, and in developing such missiles as the air-launched Hound Dog and the Minuteman ICBM. Possibly the most important development of all was Navaho's all-inertial guidance system, whose concept was used with modifications in the Hound Dog, the Minuteman, the XB-70, and the Vigilante bomber. Nautical versions of the system were used in the nuclear submarines *Nautilus* and *Skate,* which traveled under the polar ice pack, and in the Polaris fleet ballistic missile submarine series.

While the Air Force concentrated on cruise missiles, Army Ordnance got off to a quick start in ballistic missile development after the war, picking up right where the Germans had left off at Peenemünde with the V-2.

The first Paperclip group of seven scientists, headed by Von Braun, arrived in Boston's Fort Strong on 29 September 1945. Before this group departed from Germany, Von Braun spent two weeks in Britain, where he was questioned by Sir Alwyn Crow and other Ministry of Supply officials. After processing, six of the seven scientists went to the Aberdeen Proving Ground, where they began sorting out the tons of documents that had been shipped from the Dörten mine in Germany. Von Braun, who was met by Major James P. Hamill in Boston, stopped in Washington to meet with several high Army Ordnance officers, while the other members of the vanguard group—Erich W. Neubert, Theodor A. Poppel, August Schultze, Eberhard Rees, Wilhelm Jungert, and Walter Schwidetzky—went directly to Aberdeen.

Hamill began arranging to transfer the scientists to El Paso, Texas, where they were to start establishing a United States guided-missile program at Fort Bliss. The main body of Germans began arriving at Fort Bliss in December 1945, and by February 1946 over a hundred were on hand. They were quartered in converted hospital buildings that gradually became more homelike.

The test facilities became more elaborate as the program picked up speed. The first static firing test of a V-2 power plant in the United States took place on 14 March 1946, and soon afterward a series of test flights began. The first American-adapted V-2 was flown from the White Sands Proving Ground, New Mexico, on 16 April 1946, after its static test on 14 March. Other flights of missiles equipped with instruments to test the upper atmosphere and ionosphere followed quickly. Typical firings between 1946 and the end of the program in 1952 are summarized in the table below.

To coordinate these experiments, a V-2 Upper Atmosphere Research Panel was established on 16 January 1947. As newer rockets were developed, the group became the Upper Atmosphere Rocket Research Panel in March 1948 and the Rocket and Satellite Research Panel in April 1957. The group effectively coordinated the activities of government, industry, and universities in developing payload instrumentation and gathering and distributing data from the flights.

*The last of the Air Force's cruise missiles, Navaho, flew at three times the speed of sound but was still too slow to be an effective weapon. Boosted by high-thrust rocket engines, it provided the technological basis for later ICBM's.* (NORTH AMERICAN AVIATION)

**Selected V-2 Firings from White Sands, New Mexico**

| *Date* | *Altitude (miles)* | *Range (miles)* | *Velocity (mph)* | *Engine firing time (seconds)* | *High-altitude studies* |
|---|---|---|---|---|---|
| 13 June 1946 | 73 | 40 | 2,877 | 59 | Ionosphere, Solar radiation. |
| 10 October 1946 | 102 | 12 | 3,647 | 68 | Cosmic and Solar radiation, atmospheric pressure and temperature, ionosphere. |
| 17 December 1946 | 116 | 21 | 3,683 | 70 | Cosmic radiation, micrometeorites, biological research. |
| 9 October 1947 | 97 | 28 | 3,400 | 63 | Solar radiation, atmospheric pressure and composition. |
| 5 August 1948 | 104 | 53 | 3,545 | 66 | Ionosphere; atmospheric pressure and temperature, Solar and cosmic radiation. |
| 14 June 1949 | 83 | 37 | 3,005 | 67 | Atmospheric pressure, temperature and composition; cosmic and Solar radiation, ionosphere; Earth photography. |
| 31 August 1950 | 85 | 36 | 3,136 | 85 | Ionosphere, micrometeorites, atmospheric density, sky brightness, biological research. |
| 22 August 1952 | 133 | 52.2 | 4,060 | 62.6 | Training flight, maximum altitude desired. |

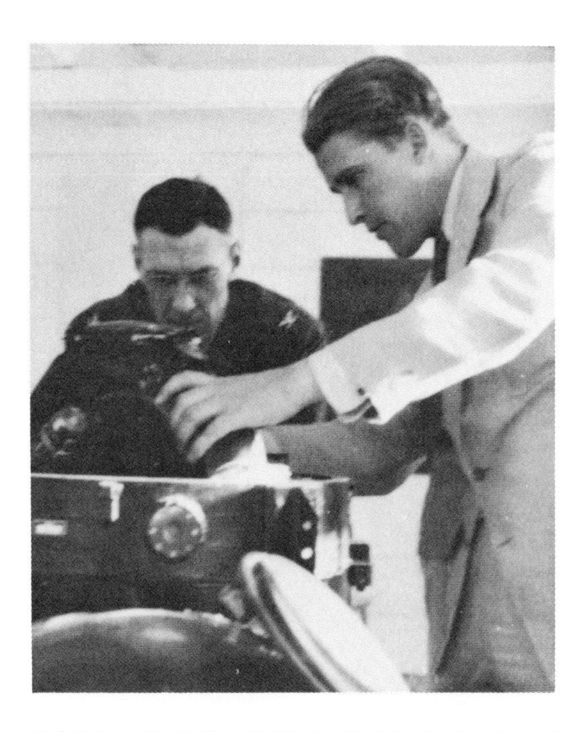

*Col Holger N. Toftoy (left), chief of the Rocket Branch of Army Ordnance, with Wernher von Braun (right) at the guided missile center at Fort Bliss, Texas, 1946. A year earlier Toftoy had offered Von Braun a one-year contract to work on the United States missile program.* (MAJ. GEN. H. N. TOFTOY)

*The first American-adapted V-2 was flown from the White Sands Proving Ground, New Mexico, on 16 April 1946. Here it is shown (above right) during engine check-out on static test stand, (below) being raised to launch position, and (right) at takeoff.* (U.S. ARMY)

To support the V-2 flights, Army Ordnance contracted for the services of the General Electric Company in what became known as the Hermes program. While the components of the missiles flown in 1946 were completely of German origin, increasing modifications were made from 1947 onward, primarily to accommodate larger and more complex payloads. By 1950 the V-2 rocket had been lengthened by 5 feet, increasing its payload capacity from 16 to 80 cubic feet.

The V-2 program, in addition to giving Americans experience in launching large vehicles, gave valuable information on every aspect of rocket flights and added considerably to information about the upper atmosphere. Most of the rockets were flown from White Sands, carrying instruments that measured atmospheric characteristics and the ionosphere. A V-2 carrying atmospheric sounding gear and a biological payload reached an altitude of 116 miles on 17 December 1946. The highest altitude attained was achieved on 22 August 1952, when vehicle TF-1, with no scientific instrumentation, flew to 133 miles above the New Mexico desert. The longest V-2 flight in the United States, 111.1 miles, took place on 5 December 1946.

For some time the Navy had been interested in the possibility of firing large missiles from ships at sea. As the V-2 was available in some quantity, it was decided to employ one of these missiles in what became known as Operation Sandy. Accordingly, on 6 September 1957 a fully fueled V-2 was launched from the deck of the aircraft carrier *Midway*. The launch was successful, but the missile exploded about 5,000 feet in the air. The program, conceived and realized by Rear Admiral Daniel V. Gallery, had obvious implications for the future—but meanwhile, the inherent hazards of shipboard launch had to be thoroughly investigated.

This led to Operation Pushover, directed by Lieutenant Commander W. P. Murphey, wherein two completely fueled V-2 rockets were exploded to determine how much damage would occur to the launch area if an operational missile accidentally blew up. The tests, begun in 1948, caused such damage that submariners were scared away from the later Army-Navy Jupiter program. "Instinctively," said Murphey, "we knew that any missile ever launched from a sub would have to be solid-fueled."

The Army began to build on V-2 technology with the program code-named Bumper, in which a small American Wac Corporal rocket was used as the second stage of a V-2 in hopes of reaching extreme altitudes. Bumper was also intended to prove out techniques needed for firing two-stage missiles—ignition and separation of the stages at high altitudes, stability of the second stage at high velocities, and the aerodynamic characteristics of the vehicle.

The Army conducted a series of the eight Bumper research firings between May 1948 and July 1950. Flight 5, on 24 February 1949, was the only complete success, reaching an altitude of 244 miles and a velocity of 5,150 miles per hour. Flights 7 and 8 were significant for being launched from what was then called the Long-Range Proving Ground at Cape Canaveral, Florida. The purpose of both flights was to determine the aerodynamic characteristics of high-speed missiles flying shallow trajectories. No. 8 hit 200 miles from the coast after breaking all speed records within the atmosphere (No. 5 had hit its top speed in space).

**Project Bumper Firing Summary**

| *Flight No.* | *Date* | *Altitude (miles)* | *Velocity (mph)* | *Engine firing time (secs)* *V-2 Stage* | *Wac Corporal Stage* | *Launch Site* | *Remarks* |
|---|---|---|---|---|---|---|---|
| 1 | 13 May 1948 | 79.1 | 2,740 | 66 | 6 | White Sands | Premature cutoff of second stage. |
| 2 | 19 August 1948 | 8.3 | 850 | 33.8 | 0 | White Sands | Premature cutoff of first stage. |
| 3 | 30 September 1948 | 93.4 | 3,160 | 64.2 | 0 | White Sands | Second stage did not ignite. |
| 4 | 1 November 1948 | 3 | 875 | 28.5 | 0 | White Sands | Explosion following cut off. |
| 5 | 24 February 1949 | 244 | 5,150 | 68 | 28 | White Sands | Successful. |
| 6 | 21 April 1949 | 31 | 1,820 | 48 | 0 | White Sands | Premature cutoff of first stage. |
| 7 | 24 July 1950 | — | — | — | — | Cape Canaveral | Low-angle firing. |
| 8 | 29 July 1950 | 10 | — | — | — | Cape Canaveral | Low-angle firing; vehicle impacted 200 miles away. |

*The Hermes A-2 rocket, at White Sands, March 1953. Under the Hermes program, the Army undertook experiments in both liquid-fuel and solid-fuel rocket technology. This solid-fuel model, with triangular tail fins, was never flown.* (U.S. ARMY)

The V-2 firings were only a part of the Hermes program. It had been evident that the supply of V-2's would soon be exhausted, and that rocket technology was advancing. Both these considerations led to plans under GE's Hermes program for new missiles based on combined German-American experience. The experimental vehicles that resulted never entered operational inventory. The program itself became somewhat controversial with many experts charging that Hermes merely repeated German technology, instead of advancing the art of rocketry. Whatever its shortcomings, the program most certainly did provide the United States with valuable experience, while it initiated American industry in the technology of large rockets.

Component development for Hermes began in 1946, and many items were flight tested in V-2's. The telemetry system for Hermes A-1, for instance, was flown in a V-2 in January 1947, and the A-1's guidance and control systems were tested in V-2's and airplanes. The new rocket motor was static tested at General Electric's Malta Test Station in Schenectady, New York, during 1948 and 1949, and five missiles were flown from White Sands between 19 May 1950 and 26 April 1951. The Hermes A-1 was very similar in configuration to the German Wasserfall surface-to-air missile. It reached a maximum altitude of 15 miles, a range of 38 miles, and a speed of 1,850 miles per hour. It was powered by a liquid-propellant oxygen-alcohol engine.

A-2, which never flew, had triangular tail fins (compared to A-1's four small midwings). It was powered by a solid motor developed by the Jet Propulsion Laboratory and the Thiokol Chemical Corporation. The motor was tested successfully in December 1951, but changing requirements caused A-2's cancellation in October 1952. However, the motor was flown in RV-A-10 test vehicles in February and March 1953.

The third member of the Hermes family appeared in two models, A-3A and A-3B, neither of which was an overwhelming success. A-3 was supposed to fly a 1,000-pound payload over a 150-mile range. The outbreak of the Korean War caused an acceleration in the A-3 program in 1951. By the time the program ended in 1954, seven A models and six B models of the A-3 had been flown. Not many experts were impressed with the results. But the A-3 did lead to development of a stable platform and radio inertial guidance system and the successful testing of rocket engines with thrusts ranging from 18,000 to 22,600 pounds, at the then high specific impulse of 242 seconds (that is, 242 pounds of thrust produced by each pound of propellant consumed per second).

Another Hermes vehicle developed by the Hamill–Von Braun group at Fort Bliss, with General Electric merely providing support, never left the study stage. Hermes II was a supersonic ramjet that was to be accelerated to three times the speed of sound by a V-2 initial booster. A full-scale model of the ramjet second stage was designed, and in November 1948 it was tested successfully in the nosecone of V-2 flight 44. Hundreds of static tests with a new "split-wing" ramjet propulsion system proved the practicality of the scheme. With the advent of the Army's Redstone missile project, however, the Hermes II program was reduced to engine development only and, in December 1952, it was canceled.

Another study program conducted by General Electric in Schenectady, was the Hermes C. This was an ambitious three-stage missile weighing 250,000 pounds and powered by six 100,000-pound-thrust rocket engines in the first stage, with one such engine in the second stage. The third stage was a glider with a proposed range of 2,000 miles. Hermes C-1 was much more modest, yet much more important in the history of American rocketry. A single-stage rocket capable of carrying a 500-pound payload for 500 miles, Hermes C-1 never got out of the study stage. But it was to lead directly to the Redstone missile, the first major American effort in the field.

In the late 1940's, it became evident that the Army's growing missile programs needed more room than was available at Fort Bliss. After a long search, the decision settled on the Redstone Arsenal in Huntsville, Alabama. The arsenal was on a large tract of government property, and its location on the Tennessee River gave it access to the power resources of the Tennessee Valley Authority. The climate was good, and the arsenal was not too far from Cape Canaveral, whose Long Range Proving Ground was growing in importance. The transfer was formally approved by the Secretary of the Army on 28 October 1949; between April and November 1950 the move was made. More than 500 military personnel, 130 members of the original Von Braun team, several hundred General Electric employees and 120 government civilian workers moved to Huntsville, where they would write a new chapter in rocket history.

The Redstone organization was designated the Ordnance Guided Missile Center. It was headed by Hamill, who had been transferred from Toftoy's Rocket Branch of the Ordnance Research and Development Division in the Pentagon. The arsenal itself was commanded by Brigadier General Thomas Vincent, who was in charge of supplying support to the missile group.

Hamill and the vanguard of the group had hardly settled in when the Korean War broke out in June 1950. Their first assignment was to conduct a feasibility study for a 500-mile ballistic surface-to-surface missile. As the Korean War grew more intense, the missile's priority increased. After being called Ursa and later Major unofficially; the project was baptized Redstone, after the arsenal where it was being developed, on 8 April 1952.

Rather than develop a new engine, the group decided to use a modification of the liquid-propellant engine developed by North American Aviation for the Navaho. As the program proceeded, the Army's requirements changed. The desired range was reduced from 500 to 200 miles, which provided a bonus: The Redstone would be able to carry a nuclear warhead and, in addition, it would be a mobile weapon capable of being launched under battlefield conditions by combat troops.

The first Redstone was fired with moderate success from Cape Canaveral on 20 August 1953. It traveled an 8,000-yard trajectory. Thirty-six more research and development models were launched through 1958, 16 of them built by Redstone Arsenal, the rest by Chrysler Corporation. On 16 May 1958, a Redstone was fired for the first time by combat-ready soldiers, members of Battery A, 40th Field Artillery Missile Group. Redstone was put into service of United States Army units stationed in Germany the next month.

The years between 1952 and 1954, during which the Redstone was designed and developed, were critical ones in the history of the entire United States missile program. The basis for every missile now in the United States armory was established during this period, in which events came with dizzying swiftness. A historian can deal with these interwoven patterns only by pointing out key developments that affected the entire program and then disentangling the threads of the ensuing growth patterns.

Perhaps the most crucial development of all was the growing realization that the Soviet Union was threatening a breakthrough in weaponry that endangered the very existence of the United States. The Soviet Union, which had its first atomic bomb in 1949, ended the United States monopoly on the

*In the Bumper series, the Army made a two-stage rocket out of the V-2 by adding a small Wac Corporal rocket as its second stage. In this flight from White Sands in April 1949, the V-2 propulsion was prematurely cut off, resulting in a maximum altitude of only 31 miles.* (U.S. ARMY)

hydrogen bomb with an explosion on 12 August 1953. Before very long, intelligence produced evidence that the Soviets were pushing hard on a ballistic missile program and might well be ahead of the American effort. All three armed services began vigorous programs to develop long-range ballistic missiles, each concentrating on its own specialty: the Army on short- and intermediate-range missiles; the Navy on missiles to be fired from seagoing vessels; and the Air Force, while continuing its long-range cruise missile programs, began work on the Atlas ICBM. The lines of action were not very clear at the start of the period, but the over-all program was well established by the end of the decade.

The Army and the Navy first began a joint program to develop a medium-range missile that both could use. On 13 September 1955, a committee under James R. Killian, Jr., President Eisenhower's special advisor on science and technology, recommended to the National Security Council that the Navy should support the development of an Army intermediate-range missile that could be launched at sea. Secretary of Defense Charles E. Wilson, on 8 November 1955, gave approval to the Joint Army-Navy Ballistic Missile Committee to develop the missile. On 17 November, the Navy established the Special Projects Office, under the direction of Rear Admiral William F. Raborn, to oversee its part of the program. Raborn took over his new position on 5 December 1955.

*America's first successful IRBM, the single-stage Jupiter, stood 58 feet tall on its launch pad at Cape Canaveral. Here it is shown at takeoff on its 18 May 1958 flight to test nose cone materials for re-entering the Earth's atmosphere.* (U.S. ARMY)

The Army organization in the program was the Army Ballistic Missile Agency (ABMA), which came into existence on 1 February 1956, taking over what was then the Guided Missile Development Division at Redstone Arsenal. It was placed under the command of an aggressive, accomplished officer from the Ordnance Corps' Industrial Division, Major General John B. Medaris.

The missile that was to emerge from this combination was the Jupiter. At the beginning, it was able to use with good effect the technology developed by Redstone. Two rockets based on Redstone came into existence to support the Jupiter program.

The first modified Redstone was called the Jupiter A. Between September 1955 and June 1958 a total of 25 Jupiter A's were fired, all within the 37-missile Redstone test flight program, to check out certain components for the Jupiter.

Jupiter C, officially known as the Jupiter Composite Re-entry Test Vehicle, was designed primarily to test nosecone materials for the forthcoming IRBM. In addition, it provided the capability of a rudimentary carrier for placing a satellite in orbit, and later was known in this capacity as the Juno 1. Since a Redstone did not have the power to generate the high speeds needed to simulate conditions of re-entry for the Jupiter nosecone, two special upper stages were added to an uprated Redstone. The Redstone first stage was lengthened to hold more propellant, and the engine was modified to operate on a new fuel called hydyne, a mixture of unsymmetrical dimethylhydrazine and diethylene triamine. The second stage consisted of a cluster of eleven solid-propellant rockets, and the third stage was three solid-fuel rockets that fit within the inner ring of the second stage. To stabilize the upper stage assembly against disturbances caused by inequalities of its many rocket engines, the whole assembly was spun rapidly during flight. The three stages fired in sequence to produce the great speeds needed for the re-entry tests.

On its first flight, on 20 September 1956, Jupiter C reached the unprecedented altitude of 682 miles, landing 3,400 miles from Cape Canaveral—a record that was not equaled by the United States for another two years. The second shot lofted a scaled-down Jupiter nosecone into space, but guidance difficulties caused the nosecone to land outside the target area. The third and last test, on 8 August 1957, resulted in recovery by parachute of the revolutionary "ablative" nosecone after a 300-mile-high, 1,200-mile-long trajectory.

As development of Jupiter IRBM continued, the Navy became increasingly reluctant to continue with the program. A huge, liquid-propellant missile—the Jupiter emerged as a 58-foot-high, 110,000-pound vehicle—could not be launched easily from any submarine. In addition, the Navy had its own particular problem: the development of a navigation system that would allow a ship at sea to determine its own position under all weather conditions with an accuracy sufficient to place a nuclear warhead on a target 1,500 miles away.

In September 1956 came another event that helped shape the future. The Atomic Energy Commission announced a breakthrough in thermonuclear technology. By 1965 at the latest, and possibly by 1963, the AEC said, small high-yield warheads would be available. This meant that a relatively small missile would be able to carry a nuclear device that could neutralize major targets. With the feasibility of small missiles in the offing, the Navy broke away from the Jupiter program and developed what came to be the Polaris missile–submarine system.

Work on the Jupiter continued, but the Army lost out in the interservice rivalry for control of the nation's IRBM-ICBM program. On 26 November 1956, Secretary of Defense Wilson issued his "roles and missions" memorandum that effectively stripped the Army of control over long-range missiles. The memorandum assigned operational control of the Jupiter to the Air Force. The Army was limited to developing surface-to-surface missiles with ranges of 200 miles or less. It could make "limited feasibility studies" of missiles with greater ranges, but that was all. Jupiter program funding was switched from the Army to the Department of Defense, and later to the Air Force. The ABMA continued to serve as the development agency, but Jupiter was destined for another service.

The first Jupiter launching occurred on 1 March 1957, with the missile flying 60 miles. The third flight, on 31 May 1957, achieved a 1,600-mile flight from Cape Canaveral, making Jupiter the first successful American IRBM. ABMA delivered its first Jupiter to the Air Force in August 1958, and more than sixty eventually were based in Italy and Turkey.

A non-military Jupiter flight, meanwhile, sounded a note for the future. On 28 May 1959, two monkeys named Able and Baker rode a Jupiter 300 miles high and 1,600 miles down range and were recovered alive. They were the forerunners of other living space payloads.

All the other Army missiles developed by the Army had shorter ranges than the Redstone, with one exception—the ABMA-developed Pershing, the Redstone's replacement, which has a range of 450 miles. For example, the Corporal, the first operational surface-to-surface missile, had a range of 75 miles. Powered by monoethylaniline and red fuming nitric acid, the Corporal was developed by the Jet Propulsion Laboratory.

The Corporal was still a rather cumbersome weapon, which took about seven hours to fire after its 250-man missile battalion had selected the launch site. It was replaced in 1957 by the solid-propellant Sergeant, another JPL product, which was considerably smaller, could be fired in far less time, and had about the same range. Like the Pershing, the Sergeant has a solid-propellant engine, but unlike the Pershing, has only one stage.

While the Army's missile role contracted, the Navy's expanded. The Navy had never pictured the Jupiter as the final answer to its highly individual needs. When it first entered the joint Jupiter develop-

**Comparison of American Short- and Medium-Range Ballistic Missiles**

| *Name* | *Length (feet)* | *Diameter (inches)* | *Weight (pounds)* | *Propulsion* | *Range (miles)* |
|---|---|---|---|---|---|
| Corporal | 46 | 30 | 12,000 | RFNA-monoethylaniline | 75 |
| Honest John | 27.25 | 24.5 | 4,700 | solid | 12 |
| Lance | 20 | 22 | 32,000 | solid | 40 |
| Little John | 14.4 | 12.5 | 780 | solid | 10 |
| Pershing | 34.5 | 40 | 10,000 | solid | 450 |
| Redstone | 69 | 70 | 62,000 | LOX-ethyl alcohol | 200 |
| Sergeant | 34.5 | 31 | 10,100 | solid | >75 |

ment program with the Army, the Navy had stated:

On a long-term basis, the Navy proposes a solid-propellant development program pointed toward surface ships and eventual submarine use. This development should be initiated immediately to alleviate the serious hazards and difficult logistic, handling, storage, and design problems associated with liquid fuels. Development of a solid-propellant missile and submarine system appears feasible, but not on the time scale of the original approach. The solid propellant is an integral part of the submarine program.

In March 1956, the Navy won permission from the Joint Army-Navy Ballistic Missile Committee to undertake both surface and subsurface missile development programs. That month, the Office of the Secretary of Defense (OSD) Ballistic Missile Committee approved a program to develop solid-propellant components to determine the feasibility of a solid-propellant missile. In mid-July, the OSD Scientific Advisory Committee went all the way, recommending that the Navy concentrate on the solid-propellant approach, cutting down on its use of Jupiter-derived hardware since "the suitability of the components was very questionable."

After the AEC announcement in September of the breakthrough in design of nuclear warheads, the next step was inevitable. On 23 October 1956 the Scientific Advisory Committee recommended that a solid-propellant missile "receive top priority, equal to that of the other [Jupiter] IRBM program . . . ." with an aim toward having an operational weapon by 1962 or 1963. The Navy officially pulled out of the Jupiter program on 8 December, with the full support of Chief of Naval Operations Arleigh A. Burke and Navy Secretary Charles S. Thomas. The joint Army-Navy committee was dissolved on 18 December, and the Navy Ballistic Missile Committee was established the next day to give broad direction to the development of the Polaris.

Although the joint Army-Navy committee had existed for only a year, cooperation between the two services during the period was responsible for great strides in the solution of the extremely difficult problem of accurately guiding a missile that had been launched from a rocking, moving platform to a target 1,300 to 1,500 miles distant. This breakthrough was accomplished by tying together the Navy's SINS —Ship Inertial Navigation System—and the ABMA-developed inertial guidance platform of the Jupiter in such a fashion that the motion and position of the ship would be reflected automatically in a bias in the setting of the missile's guidance system. As a result, the ship could move freely in any direction, either beneath or on top of the waves, while the unlaunched missile remained firmly "bracketed" on its faraway target. The guidance concepts worked out by the Army and Navy during 1956 were transferred directly to Polaris and still form the backbone of that missile's guidance system.

Raborn's Special Projects Office remained in charge of systems development after the Navy was authorized to build an IRBM of its own. In April 1956, a small study contract had been let with Lockheed Aircraft Corporation to determine the feasibility of an underwater-launched missile combined with a nuclear submarine. Lockheed recommended exactly what the Navy wanted all along—a two-stage solid-propellant vehicle that could be stored in and fired from a submarine.

The Polaris program came into existence officially on 12 January 1957. By March, the size, shape, and weight of the missile had been decided upon. Simultaneously, development of all the launch devices for the missile was undertaken. It became plain that special vessels would have to be designed, built, and tested as the Polaris was evolved. By mid-June, progress had been so great that the Chief of Naval Operations was able to approve the characteristics of the revolutionary submarine that was to fire Polaris.

As the program was accelerated, the estimates of the time needed for operational readiness were pushed forward. In October 1957, the AEC said a small-size, high-yield nuclear warhead would be ready by 1960, three to five years sooner than previously estimated. The Secretary of the Navy promptly proposed that the first-generation Polaris, with a range of 1,400 miles, could be in service by December 1959 aboard surface training ships and that the first missile-carrying submarines would be in operation by late 1960. The Secretary of Defense authorized faster progress on Polaris on 9 December 1957, and construction of the first three Polaris submarines was begun the next month. The submarines, the *George Washington, Ethan Allen,* and *Patrick Henry,* were designated SSBN—*B* for ballistic missiles, *N* for nuclear powered.

The Navy's arguments for a strong Polaris fleet were summarized in a classified memorandum dated 27 February 1958. The memorandum was prepared under the direction of J. E. Clark, director of the Guided Missiles Division of the Office of the Chief of Naval Operations. The memorandum, now declassified, listed these advantages of a fleet ballistic missile (FBM) over the conventional IRBM concept:

1. No negotiations were needed for Polaris launch sites, in the United States or abroad.

2. While liquid-propellant missiles cannot be launched immediately, a solid-propellant Polaris is

always ready for launch. "Also, each Polaris missile has its own launcher."

3. "Every unidentified submarine at sea is a potential FBM," which creates "serious antisubmarine warfare and intelligence problems for Soviets. . . ."

4. "Because FBM can be launched at Eurasia from all directions, Soviet missile defenses will have to be complex, versatile, and expensive."

5. A Polaris is less vulnerable than a land-based missile because its location cannot be pinpointed in advance, it is not easily sabotaged, and it cannot be affected by weapons aimed at other targets.

6. "Use of FBM affords greater physical safety for the U.S. and friendly nations (whereas land-based missiles) would be like magnets in drawing enemy fire on the U.S. and Allied centers of population."

As part of the Polaris program, a number of fleet test vehicles were developed to check out components. On 24 September 1958 the first AX vehicle, with Polaris's basic size and shape, was launched—unsuccessfully. Other partial or total failures followed. The first success was AX-6, which flew a 300-mile trajectory on 20 April 1959. The AX firings were followed by an A1X series. On 7 January 1960, the first inertially guided Polaris test-firing was conducted at Cape Canaveral.

The *George Washington* had been launched on 9 June 1959, and commissioned in December. Within six months of the Cape Canaveral test, on 20 July 1960, the *George Washington* made the first submerged firing of a Polaris test vehicle. In mid-November, the *George Washington* shipped out on its first

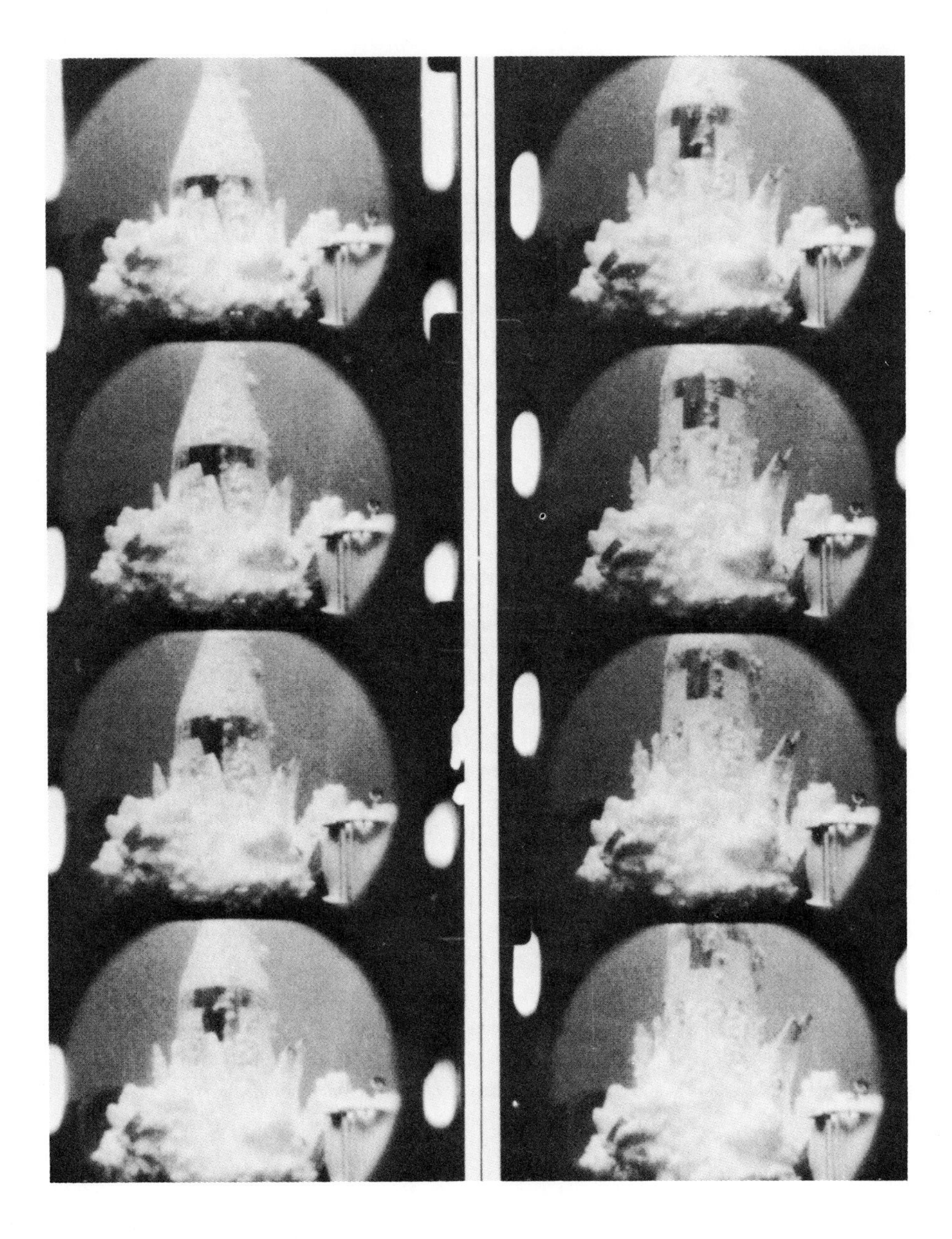

*The Navy, needing a smaller missile that could be fired from under water, began work on a solid-propellant missile. The result was the Polaris, which they fired successfully in 1960. Right, a filmstrip series showing an underwater launch from the submarine* Theodore Roosevelt *off the coast of Florida, 19 March 1963.* (U.S. NAVY)

*The A-3 Polaris missile rises from the water after its first submerged launching on 26 October 1963. In this flight a practice warhead flew more than 2,000 miles before landing on target.* (U.S. NAVY)

operational cruise with sixteen Polaris missiles aboard. The Special Projects Office–Lockheed–AEC team had beat its original deadline of 1963 by an impressive margin.

Both missiles and submarines have been upgraded since. The first Polaris has been followed by the A-2, with a range of 1,700 miles, and the 2,900-mile A-3. The last A-1 was delivered in December 1961; two months earlier, the first A-2 test vehicle had been fired from a submerged submarine. The first test flight of the A-3 took place less than a year later, on 7 August 1962, and the first submerged launch was made on 26 October 1963. By September 1964, the first A-3's were operational on the SSBN *Daniel Webster*. By 1966, the missile submarine force was fully operational, with thirteen submarines armed with A-2's and twenty-eight carrying A-3's.

The Navy has since developed the Poseidon. It has twice the warhead weight of the A-3 and has been tested with multiple independently targetable reentry vehicles (MIRV). As of 1969, four Polaris submarines had been converted to Poseidons.

The Navy has also had its subsonic missiles. Its 1946 Rigel program resulted in a series of test vehicles built by the Grumman Aircraft Engineering Corporation and tested at Point Mugu, California. The aim was a 26,000-pound ramjet-powered missile that could be launched from a surfaced submarine up to 550 miles from its target. The program was canceled in the early 1950's. The Regulus, another cruise missile, was more successful. Manufactured by Chance-Vought Aircraft, Inc., the jet-powered, sub-launched missile made its first successful flight in 1951, and later entered the fleet inventory. It was followed by the larger, faster, and longer-ranged Regulus 2, which can be fired from subs, cruisers, and aircraft carriers for shore bombardment. Both Regulus missiles are vulnerable to countermeasures. The pride of the Navy remains the Polaris, which comes as close to being the invulnerable weapon as anything ever developed.

The Navy Polaris program is a textbook example of weapons development. The Air Force program, during its early days, came close to being the opposite. The Air Force's adventures with cruise missiles have already been detailed. Its ballistic missile program, after a false start in 1947, did not really get under way until the mid-1950's.

In April 1947, the Consolidated Vultee Aircraft Corporation—now the Convair division of General Dynamics—received a one-year study contract from the Air Technical Training Command of the then Army Air Corps to develop feasibility designs for both subsonic and supersonic surface-to-surface missiles with ranges from 1,600 to 5,800 miles. The company designed a preliminary test vehicle, but Army Air Corps—and Pentagon—support vanished within months. Adding company funds to some leftover contract money, Consolidated-Vultee built three MX-774 rockets, which were static-tested in mid-1947 at Point Loma in San Diego. The rockets were launched with partial success in July, September, and December 1948 from the White Sands Proving Ground. Nativ test rockets, built by North American Aviation, Inc., were fired from Holloman Air Force Base in New Mexico to get more data on ballistic missiles.

The outbreak of the Korean War, the development of smaller nuclear weapons by the AEC, and a growing realization that the Soviet Union was active in the field changed the situation. The AEC's nuclear experiment, code-named Mike, held on 1 November 1952, made it clear that an operational hydrogen bomb missile warhead would be available within a few years. The Air Force Scientific Advisory Board established an ad hoc committee, under Professor Clark B. Millikan, the next month to review Air Force policy. Soon afterward, Air Force Secretary Harold E. Talbott named as his special assistant for research and development Trevor Gardner, a civilian who would become a leading champion of the ICBM concept.

Pressure for a policy change began to build. In April 1953 Gardner asked for a reassessment of earlier Air Force estimates that an operational ICBM might take a decade to develop. In June, General Donald M. Yates, director of research and development at Air Force Headquarters, recommended that Secretary of Defense Wilson undertake a major review of the missile program of the Air Force and the other two services. That fall Gardner set up the Strategic Missiles Evaluation Committee, which became known as the Teapot Committee, to study the Air Force program.

John von Neumann, of the Institute for Advanced Study, was chairman of the committee, which included as members Hendrik W. Bode, Louis G. Dunn, Lawrence A. Hyland, George B. Kistiakowsky, Charles C. Lauritsen, Clark B. Millikan, Allen E. Puckett, Simon Ramo, Jerome B. Wiesner and Dean E. Wooldridge. In February 1954, the committee submitted its report to Gardner, who, supported by an independent Rand Corporation study, urged General Nathan F. Twining, the Air Force Chief of Staff, to undertake a greatly increased ICBM effort. The Teapot Committee said an effective ICBM could be ready in six years or less. It urged a new organizational setup to both dramatize the urgency of the program and give the officer in charge of the program all the power he needed.

All these events led to a spurt in the Air Force missile program. The Air Force had to seek outside help, not only for development of the hardware for what became the Atlas ICBM, but for technical management as well.

First, it had to set up an office of its own with authority to call on Air Force resources and facilities whenever and wherever they were needed. Gardner recommended that the Air Research and Development Command be given that authority, with a major general at the top and a brigadier general under him to handle project direction and liaison with industry. The choice fell on Major General James McCormack, ARDC vice-commander, and Brigadier General Bernard A. Schriever, who was then with the Air Staff. On 1 June 1954, Schriever took over direction of the ICBM program, under General McCormack's command.

Since most of the industrial facilities needed for the program were on the West Coast, Schriever set up the ARDC's Western Development Division (WDD) on 15 July, and moved West himself. Procurement support was provided by a Special Projects Office within the Air Materiel Command at Wright Field, Ohio.

The Air Force still needed technical management know-how. It had several alternatives. The ARDC-WDD could try to recruit its own talent. An existing organization—the Rand Corporation or a large university—could be used. The contractors themselves could provide the management skill. Or a new organization of consultants could be established to deal solely with what promised to be the biggest development program in history.

The final alternative was chosen. The Ramo-Wooldridge Corporation was created to give technical advice first to the Strategic Missiles Evaluation Committee and later to its successor, the Atlas Scientific Evaluation Committee. By September 1954, Ramo-Wooldridge had been given responsibility for technical direction of the entire Atlas program.

The corporation, which was forbidden to obtain production contracts, was led by Simon Ramo and Dean Wooldridge, both men of proven ability in the fields of missiles and program management. The corporation grew rapidly, later assuming responsibility for the Titan and Minuteman ICBM's, the Thor IRBM, and other projects. Its 170 employees in 1954 became nearly 5,200 in 1960. In late 1957, the company became Thompson-Ramo-Wooldridge and established a subsidiary, Space Technology Laboratories, which continued advisory and technical management services for the Air Force until mid-1960. At that time the Aerospace Corporation was established to take over some of these responsibilities. WDD itself went through a name change in June 1957, becoming the Air Force Ballistic Missile Division.

The Atlas program gathered momentum starting in 1954. By mid-1955, it had the highest of national priorities (the Thor was given the same high priority that November). Decisions came rapidly, and test vehicles were fired frequently to test the new components that were being developed. The first two launches of Series A Atlases were made on 11 June and 25 September. Both were failures, but the third firing, on 17 December, was a decided success. On the early flights, only the twin booster engines that were clustered around the main, or sustainer engine, were fired, so range was measured in only hundreds of miles. The first firing of a successful Series-B Atlas, with all three engines ignited, was held on 2 August 1958 at Cape Canaveral. The range achieved was 2,500 miles.

Because the Atlas program had started late and faced early and critical deadlines, the Air Force had to scrap accepted management principles and de-

*America's first ICBM, Atlas, had to be fueled at the last minute with liquid oxygen and kerosene. Then it was raised to an upright position, where all three stages were ignited.* *This series of eight photos was taken during the 15 minutes before an Atlas launching at Vandenberg Air Force Base, California.* (U.S. AIR FORCE)

velop its own. In most previous development programs, one element of a complex system was proved out before testing of the next element began. Only when every part of the system had been tried, were all put together and the total system tested. Then decisions were made about building support equipment for the system.

There was no time for these niceties. The philosophy that emerged was expensive but effective. It was called concurrency. Major decisions were frozen early in the program, with the risks that were involved, and the higher costs resulting from setbacks being taken for granted. Concurrency meant many more tests of components and subsystems had to be made, but it also meant a faster development cycle. Work on many different components went ahead simultaneously, and all were tested at once. Concurrency might cost more, but money was secondary to the vital matter of national defense. What was most important, concurrency worked—for Titan and Thor, as well as Atlas.

One of the major problems in developing a ballistic missile is the creation of a nosecone that can take the heat and friction of re-entry into the atmosphere after a 500- to 800-mile arc into space. Many different approaches were tried at first—blunt heat-sink shapes, an Air Force concept that would slow down the nosecone and absorb heat, and ablating types, a concept taken over from the Army, that would sear down being the two major ideas. Nosecone tests began with the Atlas C series, which first used a copper heat-sink and later an ablating nosecone. A range of more than 4,000 miles was attained in flight tests that began on 23 December 1958.

Flight testing of the operational Atlas D began in April 1959, and by August the missile had been proved out. Its initial operational test was made at the Pacific Missile Range (now the Western Test Range) on 9 September by a mixed crew drawn from the Ballistic Missile Division and the Strategic Air Command. Firings that continued into 1960 saw the Atlas D stretch its range to 9,000 miles. Meanwhile, a more powerful Atlas E was launched in October 1960 and the first Atlas F was fired in August 1961.

Both these models were designed to be fired from underground "silos."

While Atlas was being developed, the Thor IRBM was also taking shape. The Air Research and Development Command, acting on a Defense Department decision to allow the Air Force to produce a missile with the same performance characteristics as the Army's Jupiter, awarded a development contract to the Douglas Aircraft Company in December 1955. It was agreed that as many Atlas components and systems as possible would be used in Thor. Thus, Thor was fitted with the Atlas D nosecone and was powered by basically the same engine as used in the Atlas booster stage. Many components of both missiles were tested on the Thor-Able two-stage rocket, a pure test vehicle.

The first four test shots, between January and August 1957, were failures. On 20 September, a Thor achieved a range of close to 1,000 miles. It was about four months behind the Jupiter, which had flown some 1,600 miles on 31 May.

With very substantial budgetary support behind it, Thor moved into the operational stage quickly. It passed its first capability tests in the summer of 1958 and was turned over to the British for training and deployment in June 1959. The first Thor squadron went into active service in Great Britain on 22 April 1960. Jupiter, meanwhile, was poised in Italy and Turkey. Both missiles were phased out of service as more advanced systems came into volume production.

One of these more advanced missiles was the Titan, whose genesis was in reports by the Von Neumann Committee and the Rand Corporation in 1954 and 1955 on weaknesses in the Atlas philosophy. One serious weakness was the fact that all three of the Atlas engines were ignited on the ground before lift-off, with the two boosters dropping off later. A more efficient procedure is to fire only the first-stage engines at lift-off, drop the entire stage when its propellants are used up, and then continue the flight with the second-stage engine. This approach had been felt to be too risky to attempt in the earlier Atlas project because little experience had been built up at the time on igniting liquid-propellant rocket engines at high altitudes. In order to keep its weight down, Atlas also had been constructed with a thin pressurized airframe, which was not ideal for advanced, higher-acceleration takeoffs.

These and other considerations led to the development of the Titan 1, which was approved early in 1957 despite arguments that the missile resembled Atlas too closely to be worth the effort of development. In fact, a major source of components for the Titan 1 was the alternate systems and subsystems developed for Atlas to prevent the failure of a primary system to hold up development of the missile. Titan 1, a two-stage missile whose upper stage was ignited at high altitude after first-stage burnout, survived its detractors. Its first flight test was held on 6 January 1959, with only the first stage being powered, and Titan 1 went into service with the Strategic Air Command at Lowry Air Force Base, Colorado, on 18 April 1962.

Titan 1 was far from the last word in missiles. Advancing rocket engine technology and changing military strategy led to the decision to develop a more advanced two-stage missile, the Titan 2. The most pressing concern was the fact that both Atlas and Titan 1 had to be fueled with tons of liquid oxygen oxidizer and RP-1 kerosene fuel minutes before they were fired. The dangers and delays this produced

**Comparison of American Intermediate-Range Ballistic Missiles**

| *Name* | *Length (feet)* | *Diameter (inches)* | *Weight (pounds)* | *Propulsion* | *Range (miles)* |
|---|---|---|---|---|---|
| Jupiter | 58 | 105 | 110,000 | LOX-kerosene | 1,600 |
| Thor | 64.8 | 96 | 110,000 | LOX-kerosene | 1,600 |
| Polaris A-1 | 28.5 | 54 | 28,000 | solid[a] | 1,400 |
| Polaris A-2 | 31 | 54 | 30,000 | solid[b] | 1,700 |
| Polaris A-3 | 31 | 54 | 35,000 | solid[c] | 2,900 |
| Poseidon | 34 | 72 | 60,000 | solid | 2,900 |

[a] Both first and second stages powered by polyurethane, ammonium perchlorate, and light metal mixture.

[b] First stage uses above propellant combination; newer second stage is powered by a double-base powder.

[c] First-stage propellant classified; second stage uses double-base powder.

made the development of storable fuels necessary. Titan 2 was built to use a mixture of unsymmetrical dimethylhydrazine and regular hydrazine fuel and nitrogen tetroxide oxidizer that could be loaded in the missile and left there for long periods of time without boiling off or evaporating. The development of storable fuels brought another bonus: Missiles could be emplaced in underground silos that could be hardened enough to resist all but an almost direct hit.

About the only characteristic Titan 2 shared with Titan 1 was its diameter—10 feet. Titan 2's first stage developed 430,000 pounds of thrust, compared to 300,000 pounds for the Titan 1. Its second stage was uprated from 60,000 to 100,000 pounds of thrust. Titan 2, at 103 feet, was 13 feet longer than Titan 1, and its weight of 330,000 pounds was exactly 50 percent greater than Titan 1's 220,000 pounds.

Helped by the management and technical lessons learned in earlier programs, Titan 2 made quick progress. By 16 March 1962 it had met all research and development objectives. Titan 2 went into operational service in nine missile squadrons at McConnell Air Force Base, Kansas; Davis-Monthan Air Force Base, Arizona; and Little Rock Air Force Base, Arkansas.

While it developed its liquid-propellant missiles,

*The gigantic Titan 2 has a weight of 330,000 pounds and stands 103 feet tall in its underground silo. Its fuels can be stored for months without boiling off or evaporating, an advantage which enables Titan 2 to be "hardened" in its site, ready for launching.* (U.S. AIR FORCE)

*The first ground test model of the Minuteman 3 ICBM is loaded onto a modified transporter-erector. The solid-fuel missile has a 1-ton payload, double that of Minuteman 1. Its initial test firing took place on 16 August 1968.* (U.S. AIR FORCE)

the Air Force kept an eye on the Navy's Polaris program. The simplicity, safety, economy, and rapid launch readiness of solid-propellant missiles impressed the Air Force planners, but their major concern was to get the liquid-propellant rockets into service. Nevertheless, early in 1956, the Western Development Division gave the Scientific Advisory Committee plans for a solid-propellant missile. Contracts were let in April, with the ARDC's Power Plant Laboratory at the Wright Air Development Center, Ohio, in charge. By March 1957, the Western Development Division was ready to start working on an intermediate-range missile, and by July the Air Force had decided that the Minuteman—the name given to the solid-propellant missile—could be upgraded to a range of 6,000 miles.

That September, a work group headed by Colonel Edward N. Hall partially defined the Weapon System Q that became the Minuteman. It would be a three-stage missile, lighter and smaller than the Titan and Atlas, and carrying a smaller warhead. But what Minuteman lacked in punch it would make up in numbers. It cost so much less than the bigger liquid ICBM's that it could be built in great quantities.

Drawing heavily on the Polaris technology, Minuteman advanced rapidly. Concurrency was used to the hilt; shortly after the missile's basic configuration had been established, plans for operational launch

*Minuteman's smaller size makes it easier to transport and install than the bulkier Atlas or Titan. Above, a Minuteman arrives at the Malmstrom Air Force Base, Montana, inside its shipping and storage container. Right, a transporter-erector places a Minuteman into its silo at Vandenberg Air Force Base, California.* (U.S. AIR FORCE; BOEING)

sites—hardened, underground silos—were drawn up.

Minuteman's first flight, on 1 February 1961, was a first in United States ICBM history. With all stages operating, the missile met all its test objectives. (Tethered Minuteman missiles with live first stages had been fired from silos at Edwards Air Force Base, California, as early as 15 September 1959.) A Minuteman was successfully flown from an operational silo in November 1961. Thirteen months later, Minuteman 1 was in service. It was followed by Minuteman 2 and Minuteman 3, the latter having a one-ton payload, double that of Minuteman 1 and 400 pounds more than that of Minuteman 2. The first test firing of Minuteman 3 took place on 16 August 1968.

Minuteman's solid-propellant system eliminated the complex procedures of loading and unloading liquid propellants. By the middle of the 1960's, Minuteman missiles were emplaced across the nation in well-protected sites. They could be fired almost instantaneously to targets that were programmed into their all-inertial guidance systems. This was as close to push-button warfare as man had ever come.

The Minuteman family was scheduled for replacement in the 1980's and 1990's with the MX Peacekeeper missile, the most powerful nuclear missile yet developed by the United States. It stands 70 feet tall and 7.6 feet wide (vs. Minuteman 3's 60 and 5.6 feet) and weighs 190,000 pounds (vs. 78,000 pounds). And where Minuteman is a three-stage missile carrying three independent warheads, the MX is a four-stage vehicle carrying up to ten warheads. Furthermore, the MX fourth stage has a restartable liquid-propellant engine and advanced inertial navigation system, so the warheads can be aimed at targets over a wider range than that of the Minuteman.

The controversy that surrounded the MX arose more over how it would be based rather than what its capabilities would be. Because of the increasing accuracy of Soviet missiles, United States leaders claimed that the Minuteman force, based in fixed underground silos, would be vulnerable to a quick knock-out strike by the Soviets. MX, they proposed, would be mobile in a large section of the western United States, so the Soviets could never be sure of targeting the right location. Up to ten hardened shelters would be built for each of the two hundred missiles. They even planned dummy missiles that outwardly would look the same, so that drivers would never know whether or not they were hauling a live missile.

But environmental objections thwarted that plan and alternative basing modes, such as placing the missiles on rails in deep, covered trenches, were proposed. Also, it was argued that making the missiles too secure would lead the Soviets to build more than enough missiles to knock out all the MX shelters, thus assuring them that all the MX's would be destroyed. An interim basing mode that finally was selected placed the missiles in the same Minuteman silos that the government said were vulnerable.

The details of how the Soviets developed their missile force are poorly documented. It paralleled the American effort only in its earliest stages; in fact, the American effort could best be viewed as a counter to the Soviet missile development program.

When the Red Army arrived at Peenemünde and Nordhausen-Bleicherode, most of the German rocket documents and scientists were already in American hands. Subsequently, the Soviets took advantage of the talents of a handful of leading rocket engineers and scientists, including Helmut Gröttrup, an authority on

**Comparison of American Intercontinental-Range Ballistic Missiles**

| *Name* | *Length (feet)* | *Diameter (inches)* | *Weight (pounds)* | *Propulsion* | *Range (miles)* |
|---|---|---|---|---|---|
| Atlas (E Model) | 82.5 | 120[a] | 270,000 | LOX-kerosene[b] | 9,000 |
| Titan 1 | 90 | 120 | 220,000 | LOX-kerosene[c] | 6,000 |
| Titan 2 | 103 | 120 | 330,000 | nitrogen tetroxide aerozine 50[d] | 7,000 |
| Minuteman 1 | 55.9 | 72 | 65,000 | solid | 6,500 |
| Minuteman 2 | 59.8 | 72 | 72,000 | solid | 7,000 |
| Minuteman 3 | 59.8 | 72 | 78,000 | solid | 7,000 |

[a] 192 inches across booster engine skirts at base.

[b] First-stage thrust 330,000 pounds for droppable booster engines, 57,000 pounds for sustainer engine.

[c] First-stage thrust 300,000 pounds, second stage 80,000 pounds.

[d] Aerozine is a blend of 50 percent hydrazine and 50 percent unsymmetrical dimethylhydrazine. First-stage thrust is 430,000 pounds, second 100,000 pounds.

*The Soviet Union is reported to have flown an IRBM in 1956 and an ICBM a year later. Shown here are three of their missiles, displayed during military parades in Moscow. They are (top to bottom) the Shyster intermediate-range ballistic missile, Frog 1, and the Skean IRBM (NATO designations).* (U.S.S.R. EMBASSY, WASHINGTON; U.S.S.R. EMBASSY, LONDON)

guidance and control, Erich Putze, a production expert, and Werner Baum, whose chief interest was propulsion. They also obtained hundreds of lower echelon personnel and tons of equipment. To this group they added their own corps of specialists, including A. G. Kostikov, inventor of the Katyusha, and Sergei P. Korolev, whose impressive achievements earned him the high honor of a state funeral when he died in January 1966.

Gröttrup's rocket organization remained in Germany for about a year, during which time V-2 production was reintroduced at the Zentralwerke in the Soviet-occupied zone of Germany. On 22 October 1946, he and hundreds of other Germans were moved without advance warning by truck and train to Russia. Near Moscow, the Germans went to work at the Scientific Research Institute 88, this time improving the V-2 rather than merely placing it back in production. A year later, on 30 October 1947, in Kazakhstan, the product of their labors underwent its first test flight, impacting some 175 miles down range. From then on, the authority of Gröttrup and the importance of the German team as a whole declined, and in the early 1950's they began to be repatriated. Gröttrup himself stayed until late November 1953 and then returned to West Germany. The story of the Germans in Russia is related in *The Rocket Team* by Ordway and Sharpe (see bibliography).

Western interrogation of the German scientists who worked on Soviet development into the early 1950's showed that they knew virtually nothing at all about the principal Soviet ballistic missile programs. The Germans were limited to working on specific elements of a given V-2 improvement program and were encouraged to submit their own ideas on more advanced rockets. But they were not given any detailed knowledge of current Soviet projects or future officially adopted plans.

Like the United States, the Soviets started by using the V-2 as a foundation. On 15 March 1947 a State Commission was formed to report to the highest Soviet authorities on the feasibility of developing long-range ballistic missiles. Its recommendations led to Gröttrup's upgraded V-2, which soon gave way to the Pobeda missile. With a range of well over 500 miles, the mobile Pobeda was a familiar sight at May Day parades in the early 1950's.

The Russians moved quickly from this limited start. One crucial stroke was to turn what appeared to be a handicap into a lasting advantage. The Soviets were far behind the United States in nuclear technology, and the Russian nuclear weapons were clumsy and bulky. United States planners decided to wait until smaller warheads were available to build ICBM's. The Soviets went ahead with the massive rockets needed to hoist their primitive bombs. The decision not only gave them a significant edge in ballistic missile technology for years, but was also a great factor in achieving early leadership in space exploration.

Soviet sources have reported that their first IRBM flew in April 1956, about a year before the United States flew Jupiter and Thor. The Soviet lead in ICBM's was even greater. Their first flight was in August 1957, 15 months before Atlas made its maiden full-range flight.

Little is known about the Soviet program except that considerable resources have been devoted to it. An estimated 200,000 to 220,000 men are attached to the Russian rocket force, and its missile inventory is in the tens of thousands. At the end of the 1960's, deployment in Siberia of the Scarp ICBM, also known as the SS-9, gave impetus to the American Safeguard (formerly Sentinel) antiballistic missile system.

Soviet military missile development continued at a rapid pace during the 1970's and into the 1980's. The Soviets have been able to improve greatly the accuracy of their ballistic missiles, giving them a reasonable chance of destroying United States missile silos with a first strike. In its 1984 edition of "Soviet Military Power," the U.S. Department of Defense listed eight ICBM's in six families that had been deployed by the Soviet Union. These ranged from the single-warhead SS-13 with a range of 5,600 miles to the ten-warhead SS-18 with a range of 6,600 miles. The SS-18, Model 4—of which 308 were deployed—was described as "specifically designed to attack and destroy [80 percent of the] ICBM silos and other hardened targets in the United States . . . [by] using two nuclear warheads against each U.S. silo." Older and less capable missiles were described as having significant capability against softer targets in Europe and China. Two more missiles were believed to be in development, the SS-X-24 and SS-X-25, which would be based in silos initially but later could be fired from mobile launchers. Large submarine-launched missiles were also developed. These carry up to seven warheads each (most carry one) as far as 5,500 miles from their launcher. In the land-based "longer-range intermediate-range" category are the SS-20 and SSC-X-4.

The only other nations to think about getting into the high-stakes game of long-range rocketry by the 1960's were Britain and France. Britain's contribution, the Blue Streak IRBM, drew heavily on American technology from its inception in 1955. Powered by twin Rolls-Royce liquid-propellant engines built under a Rocketdyne license, the Blue Streak was developed by the De Havilland organization (which later became Hawker Siddeley Dynamics, Ltd.). Blue Streak made good progress, but it was canceled in April 1960. For their part, the French developed Topaz and other test rockets as France proceeded to join the long-range ballistic missile club. Much later, while celebrating the thirty-fifth anniversary of Communist rule on 1 October 1984, the Chinese displayed new ICBM's capable of delivering nuclear warheads to distant targets. Designated CSS-4, the largest missile displayed is believed to have a 50-megaton warhead.

While ICBM's and IRBM's have dominated military planning, smaller "tactical" missiles have assumed ever-increasing importance. Their history is rooted in two families, that of the bazooka-type rockets introduced in World War II and that of the electronics revolution, which has made it possible to have these large

*The SS-20 longer-range intermediate nuclear force missile, deployed in Eastern Europe, was given as the reason for United States deployment of the Pershing 2 and Tomahawk missiles. The SS-20 is said to have a range of 3,000 miles.* (DOD)

"bullets" follow guidance commands and evolve into "smart" weapons.

The smallest of the smart weapons are wire-guided, such as the Dragon and TOW (Tube-launched, Optically tracked, Wire-command link) developed by the U.S. Army for use against tanks. Each uses a telescopic sight with crosshairs that the soldier places on the target. When the missile is launched, two strands of copper wire are reeled out of the tail to link its control fins (or rocket thrusters) to the sight. A flare in the tail of the missile is tracked by the sight and corrections are made by a small computer mounted in it. Range of the 24.4-pound Dragon is 0.6 mile; that of

*De Gaulle's decision to make France a nuclear power resulted in the scheduling of an IRBM and the launching of test vehicles such as the Topaze, shown here.* (SEREB)

TOW is 2.3 miles. In addition to being launched by foot soldiers, TOW can be fired from armored vehicles and helicopters. TOW's first battlefield use aboard a UH-1 Huey knocked out two tanks in the Vietnam War in 1967.

The missile and others like it have changed the tactics of armored warfare. Even the crude Soviet-built Sagger missiles used by the Egyptians in the 1973 Yom Kippur War devastated armored columns that lacked proper protection. More-advanced technologies now coming into play include 6-mile fiber-optic links that join a TV camera in the nose of a missile to a TV monitor, targeting computer, and operator in an armored vehicle behind the battle front. The first use of lasers in battle has not been as death rays but as target designators for missiles with special seekers tuned to the light emitted by the laser. The most accurate bombing in Vietnam was by bombs gliding onto the pinpoint of light emitted by an aircraft with a special laser pod and TV camera where a weapons officer picked targets. The Hellfire missile is designed to be fired from an AH-64 Apache helicopter that may not even see its target, which is illuminated by a forward-spotter's laser designator.

The largest armored targets, of course, are ships. Some of the most dramatic changes in naval warfare have taken place here with the development of antiship missiles. The equivalent of Billy Mitchell sinking a battleship with an airplane came in 1967 when Egyptian gunboats fired Styx missiles and sank the Israeli destroyer Eilat. Again, in the Falklands Islands conflict in 1982, the Argentine Air Force sank the British destroyer Sheffield with a single Exocet missile launched by a jet fighter almost 30 miles away. The U.S. Navy has equipped itself with the jet-powered Harpoon antiship missile that can be launched from surface ships of any size, aircraft, and submarines to targets more than 60 miles away. Unlike ballistic missiles, antiship missiles are seaskimmers, flying above the waves to avoid radar detection and then popping up to dive onto the target in the final seconds.

*A Kiev-class aircraft carrier sports SS-N-12 and SA-N-3 antiaircraft missiles, RBU-6000 and SUW-N-1 antisubmarine missiles, twin 76mm guns, and 30mm antimissile gatling guns. Other Soviet surface ships have similar arrays of missiles.* (DOD)

Cruise missiles have become the latest and most unsettling factor in the nuclear equation. Like the V-1 of World War II, these are actually unmanned aircraft programmed to fly certain courses to their targets more than 1,000 miles away. Because they are small enough to be carried by aircraft and carry conventional or nuclear weapons, it is impossible to say for sure where they are and what they carry. The United States and the Soviet Union have fielded cruise missiles that can be launched from ships, submarines, aircraft, and even truck trailers, and thus have greatly complicated each other's air defense problems.

Both the U.S. Navy and Army have been aggressive in developing air-defense missiles in the wake of the lessons of World War II. Although highly dramatic, the curtains of steel thrown up by machine guns and 5-inch rifles were found by the Navy to be inefficient strainers. A surface-to-air missile (SAM) called Bumblebee was started as a defense against Kamikazes during the war. The first flight program was Terrier, started in 1949 and first fired from a ship in 1954. The first missile-carrying cruisers, the Boston and Canberra, were converted by replacing their aft turrets with two missile mounts each, plus their tracking radar. Terrier and its cousin Tartar have evolved into the advanced Standard Missile-2 family (medium and extended range) that can strike incoming aircraft more than 30 miles away. That, of course, has spurred the longer ranges of the antiship missiles, which in turn has led to point-defense weapons like a ship-launched Sea Sparrow and an upgraded Gatling gun.

Where ships once were designed around guns, now they are designed around missile launchers and the guns have been relegated to secondary roles. The U.S. Navy's Ticonderoga-class cruisers feature only two 5-inch rifles and two twin-arm missile launchers. Even the latter will be replaced by vertical-launch tubes flush with the main deck. Dominating the ship's superstructure are the flat, octagonal faces of the AN/SPY-1 phased array radar that can scan the sky at high speeds without rotating dishes. (Special electronic lenses guide the beam.) For pure menace, though, the Soviet Navy has the most heavily armed and toughest looking ships, such as the Slava- and Kirov-class guided-missile cruisers.

It is the U.S. Army that has had the most ambitious surface-to-air missile program, with missiles ranging from shoulder-fired weapons weighing only 20 pounds and used against helicopters to giant multistage vehicles for defense against ICBM's.

The principal Army programs come under the blanket designation of Nike, developed by a Bell Telephone Laboratories–Western Electric Company team under a mandate that dates back to the closing months of World War II. The first member of the family, Nike 1 (later redesignated Nike A and then Nike-Ajax), was flown at the White Sands Proving Ground in 1951, knocking down a target drone in November of that year. Nike-Ajax, with a range of 25 miles, an effective altitude of 10 to 12 miles, and a velocity of 1,500 miles per hour, started protecting American cities in 1953.

That same year, Army Ordnance authorized development of the second of the series, Nike-Hercules. The development-test cycle was completed by 1957, and Nike-Hercules became operational in 1958. It was emplaced first around New York, Chicago, Baltimore, and Washington. Nike-Hercules, with a range of 75 miles and a speed of 2,200 miles per hour, is a considerable advance over Nike-Ajax. Both its stages used solid propellants, whereas Nike-Ajax had a solid first stage and liquid second stage. The major development problem involved guidance; Nike-Hercules relied on an elaborate ground array of three radars for acquiring and tracking enemy aircraft, and equally elaborate onboard devices. It could also be nuclear-armed.

To provide frontline troops with a surface-to-air missile that would move with them, development of Hawk (homing all the way killer) was started in the 1950's. In 1967, a Hawk missile fired by Israeli forces at an Egyptian fighter plane became the first guided missile to kill a person. A series of upgrades is expected to keep Hawk in the field through the end of the century.

Because Hawk requires several radar units to track its targets and guide the missiles, the Army in the 1960's started work on a successor, SAM-D (for development), later renamed Patriot (phased-array tracking to intercept of target). A single phased-array antenna is used to track a large number of targets and to guide the missile, which itself provides tracking information as it homes in on its target. A single command post is linked to eight launchers, each carrying four missiles sealed at the factory in shipping canisters that double as launch tubes. Patriot was intended to replace both Hawk and Nike-Hercules, but because of Patriot's rising cost—technical problems delayed its introduction from 1970 to 1983—it was ordered to be deployed in tandem with Hawk.

Because no one missile could be expected to stop all aircraft, smaller short-range SAM's were developed. The Army started a program called Mauler to provide troops with a short-range SAM on a tracked vehicle that would move with the battle front, but technical problems led to its cancellation in 1965. As an interim measure, the successful air-to-air Sidewinder missile was adapted to a tracked launcher and became Chaparral. It was to be replaced by Roland, a Franco-German system that mounted two firing arms, ten missiles, and radar and optical guidance on a tracked vehicle. But in preparing to build the weapon under license, the Army and its contractor "Americanized" the weapon and drove the price out of bounds. The handful of United States Rolands that were built were assigned to the Rapid Deployment Force. Chaparral became permanent and, like Hawk, has gone through a series of upgrades.

At shortest range there is the Stinger shoulder-fired SAM that is smaller than a bazooka yet has a sophisticated seeker that can home on the heat emitted by a jet engine. The power of such missiles was demonstrated in Vietnam when Soviet-built SA-7 missiles (equivalent to the United States Redeye which preceded Stinger) were used with success by Vietcong forces against U.S. helicopters. Helicopter pilots soon

*A Patriot surface-to-air missile flies out of its launch canister. The medium-to-long range missile was designed to replace aging Nike Hercules missiles. Sealed launch canisters like this were becoming common in the 1980's.* (U.S. ARMY)

learned that signal flares thrown out of the helicopters would divert the missiles which were, in turn, equipped with more discriminate sensors. Advanced models of Stinger are being built with ultraviolet as well as infrared sensors, because it is tougher to create a decoy in two widely separated parts of the spectrum. The war of countermeasures and counter-countermeasures is a never ending one fought between wars as well as during them.

The ultimate countermeasures in this field have been the antiballistic missiles, or ABM's. They challenged a basic tenet of military thinking, that there was no defense possible against an incoming warhead. But the increasing sophistication of SAM technologies indicated that a concentrated engineering effort might be able to make it possible to "hit a bullet with a bullet," as critics called it.

The ABM program, which evolved within the Nike framework, faced problems of such vast dimensions that for many years they appeared insoluble. Many experts argued that an ICBM could not be knocked down, either as it rose from its launch site or as it hurtled along its re-entry trajectory. The only defense, it was asserted, was a good offense—a massive ICBM force that would wreak vengeance on any attacker and thereby deter the attack.

The deterrent argument was dominant for many years, because it seemed unanswerable. The only way to prove that there was a defense against ICBM's was to inaugurate a major program, costing billions of dollars, to develop an anti-ICBM missile. One look at the basic problems involved in developing such a system would give anyone pause.

The potential enemy, the Soviet Union, was clearly capable of launching hundreds of ballistic missiles against the United States in a very short period of time. The nosecone-warhead of each missile would fly 500 to 800 miles into space, re-enter the Earth's atmosphere at more than 15,000 miles per hour, and detonate its nuclear device, each of many megatons, over a preselected target. On the way in, the warheads might release decoys whose trajectory and shape would confuse radar defense systems, which would find it difficult to distinguish them from the true warheads.

*The decreasing size of antiaircraft weapons is demonstrated by this soldier aiming the Stinger heat-seeking missile. It is the size of a World War II bazooka, but it is more accurate than conventional gun systems.* (U.S. ARMY)

Nike-Zeus did not start out in the conventional way, with a predetermined configuration into which were built propulsion, guidance, and other systems. The technology was simply not available. Instead, the engineers worked on developing individual components—structure, propulsion, guidance systems, ground radar, and so on.

During 1963 and 1964 Nike-Zeus showed that it could destroy incoming Atlas and Titan nosecones fired from Vandenberg Air Force Base over the Pacific. The Nike-Zeus anti-ICBM's were launched from Kwajalein Island and reached velocities of 8,000 miles per hour.

Despite these tests, doubts still existed about the success of the missile under actual attack conditions. One unanswered question was whether the enemy warhead should be attacked at long range or when it had re-entered the atmosphere, where radar could more easily distinguish it from decoys. To assess the two approaches, a new master program, named Nike X, was introduced. Zeus became the long-range intercept phase and Sprint, the short-range intercept concept. The

Army recommended an operational system of both Zeus and Sprint missiles mixed in the same batteries. Control of both would be exercised by radar systems that could track and control several missiles at once.

While debate swirled over the ABM, China began to emerge as a ballistic-missile power, firing a missile over a 400-mile range in October 1966 and exploding a prototype thermonuclear missile warhead in June 1967. These events, coupled with the fear that China might act irrationally and attack the United States despite the threat of massive retaliation, persuaded American planners that the United States would require at least a "thin line" ABM defense system by the early or mid-1970's. Such a program was announced on 18 September 1967 by Defense Secretary McNamara, who estimated its cost at $5 billion (by 1969 the estimate had doubled). The system, soon named Sentinel, would consist of fifteen to twenty Spartan and/or Sprint batteries, the Spartans providing long-range coverage and the short-range Sprints defending local areas.

The issue was complicated two months after McNamara's announcement by reports that Russia was developing two new weapons: an extremely long-range ICBM that might approach America's unprotected "underbelly" via the South Pole, and a satellite method, dubbed the fractional orbital bombardment system (FOBS), of delivering thermonuclear warheads. Soviet spacecraft that exploded in orbit on 17 September 1966 and 2 November 1967 are believed to have been FOBS development flights. An FOBS craft would fly at an altitude of about 100 miles, and the target nation would receive no warning until 3 or 4 minutes (compared with 30 for an ICBM) before impact, when the craft's retro-rockets would be fired to place it in its final trajectory. Accuracy of the FOBS, however, is still an open question.

On 14 March 1969, President Richard M. Nixon announced a decision to modify Sentinel, principally in terms of moving batteries away from major cities, but also in terms of scheduling. The purpose of the system, which he re-named Safeguard, now would be to preserve America's threat of massive retaliation by defending ICBM sites. Batteries would be deployed at Malmstrom Air Force Base in Montana and Grand Forks Air Force Base in North Dakota. During the year, critics continued to assert that the decision to go ahead with the system was unwise and that the system itself was unworkable. (Its radar and computer elements still required much development.) Proponents argued that Safeguard was, in the President's words, the "best preventive for war." As the controversy raged in Congress, development of Safeguard proceeded. The three-stage Spartan test vehicle was 55.2 feet long, weighed 33,400 pounds, and had a range of over 100 miles; the two-stage Sprint was 27 feet long, weighed 7,500 pounds, and had a range of 25 miles.

Safeguard was further limited by the Strategic Arms Limitation Treaty of 1972 to two sites. A later ABM accord reduced that to one site, either an ICBM base or a national command center. Additionally, the treaty restricted the ABM system to eighteen radars and a total of one hundred launchers. Despite strong opposition in the Congress, deployment of Safeguard continued and in 1976 the site at Grand Forks, North Dakota, was activated. Six months later it was shut down by direction of Congress. Only the perimeter acquisition radar, a long-range phased-array type, was left active.

The Soviet Union, meanwhile, deployed an ABM system with sixty-four Galosh launchers in hardened silos positioned around Moscow. These are reloadable, a capability that the United States considers to be a violation of the treaty but which the Soviets claim is legal because the limitation specifies launchers, not flight vehicles. This system is being upgraded to a force of one hundred launchers with additional radars. The new ABM system will use a layered defense akin to Safeguard's two tiers and using mobile launchers. Beyond that, the U.S. Department of Defense claims that the Soviets have the capability to manufacture ABM

*The Sprint component of Safeguard (formerly Sentinel) flashes away from its launcher in May 1965. While its companion Spartan is planned for longer-range defense, Sprint would attack incoming missiles once they have re-entered the atmosphere.* (U.S. ARMY)

systems that could be deployed in a few months should they decide to "break out" of the 1972 treaty. And their SA-X-12 missile can be fired as a SAM or as an anti-tactical ABM with some capability against strategic ICBM's. The Soviets are also pursuing a vigorous program in high-energy lasers and particle beams that could change the face of warfare as radically as did the atomic bomb in 1945.

In dismantling its Safeguard ABM, though, the United States did not abandon the field. The Army Ballistic Missile Defense Organization (usually called BMD) was formed from Safeguard and charged with developing the advanced technologies that would keep the United States from being caught off guard in case the Soviets did break out of the treaty or even came up with a new weapon altogether. This work was made difficult by language in BMD's funding prohibiting the development of prototype weapons. So work concentrated on development of test beds that would demonstrate the technologies required in a new system, and even on preprototype systems.

BMD also put substantial resources into develop-

*Above: A Soviet Galosh missile interceptor blasts out of its mobile launch canister. The Soviets have sixty-four such missiles around Moscow and are replacing them with silo-based missiles.* (DOD)

*Below: High-performance aircraft used by the Allies prompted the Soviets to develop the SA-X-12 air defense missile. It also is believed to be able to shoot down Pershing missiles.* (DOD)

ing infrared seeker technologies that could see the body heat of an incoming warhead or missile against the cold of space. There are two main advantages to this approach. First, infrared sensors are not limited by arms treaties. (New technologies are to be negotiated as they appear, though.) Second, the greater accuracy made possible by using shorter wavelengths than those of radio may yield greater accuracies in homing on the target—thus, the potential for killing by impact rather than by a nearby nuclear blast. And this, in turn, leaves the battle zone clear of blast effects (so additional missiles can be seen and hit) and removes the problems of nuclear-release authority.

As with the earlier ABM program, this system seemed like too much hope with too little technology until the first hit was scored on 23 May 1984. The program, called the Homing Overlay Experiment (HOE), was based on techniques proven in the Designating Optical Tracker experiments conducted aboard sounding rockets, the Homing Intercept Technology (HIT) conducted on the ground, and others. The HOE vehicle used a modified Minuteman 1 missile as its launcher (the target was launched by another modified Minuteman 1) and a modified Lance tactical missile as its upper stage. The kill vehicle itself had an infrared seeker to find the target along with a narrow-angle seeker and a laser range finder to home on the target. The "warhead" was a 15-foot-wide umbrella-like device—steel blocks on aluminum ribs—wrapped around the body of the kill vehicle and unfurled in the last seconds before impact.

The first two HOE launches in 1983 and early 1984 failed to hit their targets for various reasons, although program officials said that the HOE-2 launch validated the concept when flight data showed it was on course for a kill. Not until HOE-4 was all doubt removed, when the missile hit the target square with its body, not simply with the steel-studded umbrella. Videotapes of the intercept clearly showed the hunter and target closing in space over the Pacific Ocean between Kwajalein and Vandenberg and then disappearing in a cloud of debris from an impact with a total speed of more than 14,000 miles per hour.

But HOE was only a demonstration and, by the admission of BMD officials, much work remained to be done to turn it into a working system which would hit targets every time and which could cope with the range of countermeasures that no doubt would be developed to meet such an ABM. Other measures being developed as well were now grouped under the interservice umbrella of the Strategic Defense Initiative started by President Ronald Reagan in March 1983. Although derided by many as being a "Star Wars" defense, it was intended to examine all the technologies available for ABM systems, ranging from "point de-

*A surplus Minuteman 1 missile boosts the first Homing Overlay Experiment in a launch from Kwajalein Atoll. HOE-4 scored the first direct-impact kill of an incoming missile in mid-1984 and showed that non-nuclear ABM's are possible.* (U.S. ARMY)

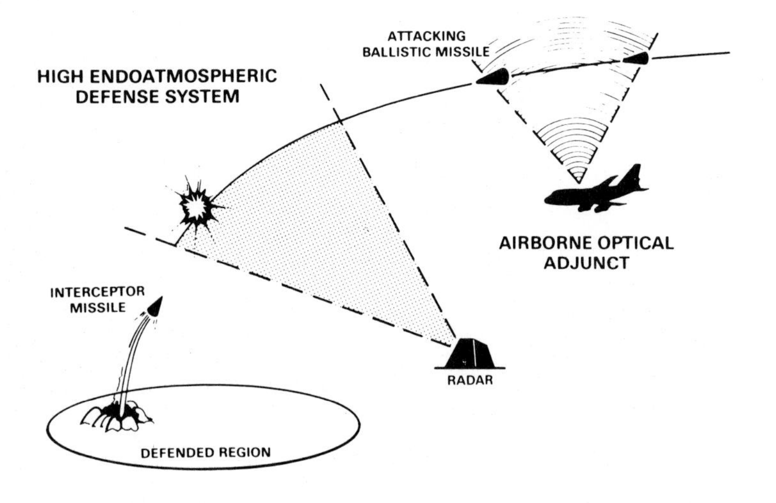


*This diagram depicts non-nuclear defense in the 1990's: an airborne infrared sensor provides early tracking data, and phased-array radar guides the interceptor to its target. This represents the outer layer of a two- or three-tiered defense.* (U.S. ARMY)

fense" at the target to midcourse intercepts high above the atmosphere to destruction during boost phase. HOE technology is a key facet of this effort.

Also being developed are HEADS (the High Exoatmospheric Defense System) and ERIS (the Endoatmospheric Re-entry-vehicle Interceptor System)—again, the layered defense. To add to the warning time that radar and satellites can provide, the Airborne Optical Adjunct (AOA) would place infrared sensors on planes flying over North America. Initially these would be manned aircraft, but later models could be piloted by drones flying at higher altitudes and for longer durations. The ADA would provide more accurate tracking data and earlier discrimination between decoys and warheads.

*Configuration for the Homing Overlay Experiment interceptor. The "homing & kill" stage used a Lance tactical missile engine and propellant tanks from an Apollo lunar module ascent stage. The sensors used heat seekers and laser trackers.* (U.S. ARMY)

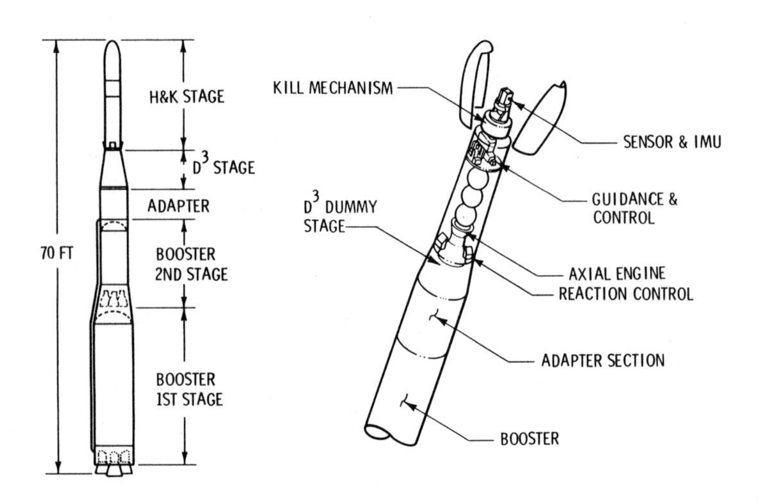


Inevitably, these activities have moved into the space arena. Where ICBM's, ABM's, and FOBS simply traversed space en route to terrestrial targets, ASAT's—antisatellite weapons—would use space as their battleground. In the early 1970's, the Soviets demonstrated a basic satellite interceptor that killed by maneuvering to within a few miles of the target and then detonating into a spray of shrapnel that would destroy any satellite that was fragile (and most are).

Spurred by those efforts and the need to be able to destroy comparable Soviet targets in case of an all-out war, the U.S. Air Force started work in the late 1970's on an ASAT that, like HOE, would kill by direct impact. Indeed, much of the technology on which the U.S. ASAT is based has roots common to HOE. Unlike the Soviet ASAT, though, the U.S. vehicle is a suborbital weapon that can be fired from an aircraft. The F-15 Eagle jet fighter was selected as the launch platform with existing rocket motors from the Short-Range Attack Missile (SRAM) and Scout third stage (Algol) used as the ASAT's second and third stages. The fourth stage is integral to the miniature kill vehicle which homes on the satellite's body heat. Because orbital velocities are much greater than ballistic missile velocities, the force of impact is more shattering.

Both the Soviet and U.S. ASAT's are limited to low altitudes, but given time and technology they are expected to be able to strike targets out to geostationary orbit. They also have raised as many political and international problems as technical ones. For example, would an ASAT attack on (or even laser-blinding of) satellites watching for Soviet missile launches necessitate a nuclear attack by the United States? At face value, a satellite is not worth a human life. But the sole function of these satellites is to give early warning of a Soviet attack on the United States, so what else could be inferred by such a move? In 1984, the Soviets were making overtures for space arms limitation talks that seemed more an attempt at making the United States appear recalcitrant to talk. And many members of the U.S. Congress were pushing for a unilateral limit on testing, fearing that a successful test flight would force the Soviets into a more intense cycle of the arms race.

Fortunately for mankind, only the smaller missiles developed in the Cold War years have been fired in anger. The warheads of the hundreds of long-range

*Engineers attach a full-scale model of the United States Air Force satellite interceptor (ASAT) to an F-15 Eagle for a flight test. While the basic technologies had been proven, firing tests were hotly debated in Congress.* (BOEING AEROSPACE)

missiles that stand poised in the United States and the Soviet Union could kill hundreds of millions of people and destroy a large percentage of the world's industry if they were ever used. The tenuousness of the balance of power that has prevented their use was shown in 1962, when the Soviets began setting up offensive ballistic missiles in Cuba. The world came close to nuclear holocaust in the confrontation between President John F. Kennedy and Premier Nikita S. Khrushchev in October 1962. There are still no guarantees that the holocaust will not be triggered by some confrontation in the future.

# 7 PROBING THE FR

The rocket enthusiasts of the 1920's and 1930's who confidently looked forward to the exploration of space did not realize the magnitude of the task ahead of them. The accelerated development of World War II and the advanced technology of the postwar years were needed to make the dream come true. But it did come true, and rockets now soar into the upper atmosphere and outer space routinely. The rocket has become a valuable scientific tool.

Rockets that are used solely for measurements within the atmosphere are called sounding rockets; those that carry scientific payloads into orbit around the Earth or to other planets are called launch, or carrier, vehicles. The distinction is somewhat arbitrary, since sounding rockets that soar 200 or 300 miles above the Earth effectively are in outer space.

Postwar sounding rockets preceded later launch vehicles by more than a decade, showing the way into space. Although the launch vehicles are more spectacular, sounding rockets are extremely useful tools for the exploration of the upper atmosphere. Their relative simplicity and low cost make it possible for most nations to use them for scientific research programs. Today, sounding rockets are flown in every part of the world—the polar regions, the equator, ships at sea, airplanes aloft, even from balloons. The nations that cannot make their own can pick the type and model they want from a long list of rockets available from many different countries.

The first rocket to be used on a sustained basis for upper-atmosphere sounding was the V-2. It did its job rather well, considering that it had not been built for the purpose. Inevitably, the supply of V-2's ran out, and new atmospheric sounding rockets were developed.

In the United States, the Navy took the first steps toward inaugurating an upper-atmosphere research program. As early as December 1945, a Rocket-Sonde Research Branch was established within the Naval Research Laboratory (NRL). The plans that were drawn up for a Navy research rocket were shelved when the Army's supply of V-2's became available, but the NRL soon began thinking of a rocket that could lift scientific instruments to altitudes of 50 or 100 miles or higher. A prime goal was low cost, so that as much money as possible could go into instrumentation and data processing.

The Navy tried two different approaches. It produced the Viking, which was derived in great part from the V-2, and it upgraded the Wac Corporal, which had flown payloads to altitudes of 40 miles toward the end of the war, into the Aerobee. Both approaches were successful. The Aerobee was not retired until 1985.

The Navy's Bureau of Ordnance awarded a contract for twenty Aerobees to the Aerojet Engineering Company in 1946, with the Applied Physics Laboratory at Johns Hopkins University supplying technical supervision. Later, the Office of Naval Research contracted for more Aerobees to be used by the NRL. As Aerobee's reliability was demonstrated, both the Army and the Air Force ordered some. Weighing slightly more than 1,000 pounds, Aerobee could carry a 100-pound payload to an altitude of 75 miles. Later models improved on that; Aerobee-Hi, for instance, carried 150 pounds between 150 and 200 miles. The first Aerobees were boosted by a 21,000-pound-thrust solid-propellant motor and had a liquid engine sustainer that used RFNA and aniline as the propellant combination.

The Viking rocket has not survived, although it gave birth to far greater programs. Plans for the rocket (which was first called Neptune) were drawn up early in 1946 by a group in the NRL under Ernst H. Krause and Milton W. Rosen. The Glenn L. Martin Company of Baltimore won the competitive bidding and was given a contract in August for building ten rockets, a number later increased to fourteen.

The subcontract for Viking's 20,000-pound-thrust liquid oxygen–alcohol controllable engine was given to Reaction Motors, Inc. At the time, the Viking was the most advanced liquid-propellant rocket under development in America. Its engine compared favorably to the 8,000-pound-thrust MX-774 and the 13,500 pounds of thrust produced by the Hermes A-1, military rockets which were being developed at the same time.

NGE OF SPACE

Reaction Motors had been founded in 1941 by members of the American Rocket Society. In October 1947, about a year after work had begun, the first engine was ready for testing. The test program was divided into three phases. First the motor was hung on a massive A-frame structure, which measured its thrust. Later, the turbo-pump, designed to feed the fuel and oxidizer to the motor, was connected. Finally, an almost complete rocket was mounted on a rotating platform for testing. After the last test, the propulsion system was shipped to White Sands in New Mexico to be static-tested in the completed rocket prior to launching.

Motor A burned itself out after a partially successful test program that yielded useful information for following engines. On 21 September 1948 a prototype engine produced 21,000 pounds of thrust for 66 seconds, a performance that satisfied the NRL. The first production engine was shipped to the Viking group at Martin soon afterward, and Viking No. 1 took shape at the company's plant near Baltimore in December. The first static test took place on 11 March. Steam from the turbine exhaust started a fire, but the rocket was not damaged. The second static test, on 28 April, was cut short prematurely when someone saw smoke, but no evidence of a fire was found. Plans went ahead for the first launching at White Sands.

It took place on 3 May 1949; at 9 A.M. Viking 1 reached an altitude of slightly over 50 miles—a highly creditable performance for the maiden flight of any rocket. American high-altitude sounding had come of age.

Since the experience of early flights was applied to the design of later models, no two Vikings were exactly alike. Flights 2 and 3, which took place in September 1949 and February 1950, reached approximately the same altitude as did the first Viking. Viking 4 was fired at sea on 11 May 1950. Viking 7, flown in August 1951, established a record height of 136 miles. More than a year went by before the next firing. The time was used to carry on design work, in process since 1950, for a larger, heavier Viking that would fire for more than 100 seconds, compared with the 50 to 80 seconds of the earlier rockets. Almost all the extra 3,500 pounds that went into the new Vikings was propellant, not structure. The new rocket could go higher than could the old Vikings or carry heavier payloads to the same altitude.

Seven of the new Vikings were built and flown. The only failure was the first one, Viking 8, which experienced one of the most unusual accidents in rocket history. On the morning of 6 June 1952, as Viking 8 was being static-tested, it broke loose from the test stand, flew 4 miles up, and crashed on the desert five miles away. The scientists and engineers on the scene could hardly believe their eyes as the rocket rose into the air. On future models, they used four tie-down points rather than two and strengthened the tail sections of the rockets.

The remaining six Vikings were fired between 15 December 1952 and 1 May 1957. Viking 11 distinguished itself in May 1954 by soaring to a record height of 158 miles with 852 pounds of instruments aboard. The last two rockets were fired not for high-altitude research but for a check on components for the forthcoming Vanguard launch vehicle. They were flown from Cape Canaveral rather than White Sands.

Viking was the largest and most ambitious of the many sounding rockets created in America. A partial roster of American sounding rockets includes—in addition to the Aerobee family—the Arcas, Arcon, Asp, Aspan, Astrobee, Dan, Deacon, Exos, Hawk, Iris, Metroc, Nike-Apache, Nike-Cajun, Terrapin, Trailblazer, Viper, and Wasp, each with different configurations and capabilities. Two unusual vehicles were the Rockaire and the Rockoon, names derived from the fact that they were flown from airplanes and balloons respectively.

The International Geophysical Year of 1957–1958, an 18-month period of sustained scientific cooperation on an international scale, was responsible for the firing of large numbers of sounding rockets by many of the major nations. The United States alone launched 210 vehicles, from Fort Churchill in Canada, the Ant-

*The Navy's Aerobee sounding rocket, an upgraded Wac Corporal, proved so successful that it was adopted by the Army and Air Force as well. Here an Aerobee 150A launches a 208-pound Australian payload of scientific instruments on 9 May 1963 from Wallops Island, Virginia.* (NASA)

*The first American high-altitude sounding rocket, Viking, reached a height of 50 miles on its maiden flight on 3 May 1949. Above, Viking 4 is seen through a hatch manhole on the USS* Norton Sound, *prior to its launch on 11 May 1950.* (U.S. NAVY)

arctic, San Nicolas Island off California, the Arctic, White Sands, Guam, and Danger Island in the Pacific. The Soviet Union flew 125 of its Meteo and geophysical rockets from locations in the Arctic, the Antarctic, and central Russia.

What the Soviets called simply a "geophysical rocket" was actually a series of different rockets that began upper-atmosphere research in 1949. One of the largest models carried a payload of 4,850 pounds to a height of 132 miles in May 1957, a feat far beyond the capability of either the V-2 or the Viking. Its payload included a capsule containing two dogs, who were parachuted safely to Earth.

Many other nations began to conduct research programs with their own sounding rockets: Britain (Skylark, Skua), France (Tactite, Véronique, Dauphin, Eridan), Australia (Aero-High, Aeolus, Long Tom), and Japan (Kappa series, Lambda series, LS-A). Japan, for example, started in the early 1960's with modest Sigma and Pi rockets, which carried 6- to 9-pound payloads. By 1965 the Japanese were launching Lambda 2's to altitudes of 375 miles. On the other side of the world, an 8-year European sounding-rocket program began in July 1965 when a British Skylark was fired from the Italian Salto di Quirra Range on Sardinia with a mixed German and Belgian payload. Other nations, such as France and Argentina, cooperated in sounding research. The United States National Aeronautics and Space Administration participated in experiments that involved some twenty nations and about 350 sounding rockets during the 1960's.

No matter how much they add to human knowledge, the sounding rockets will remain the unromantic drudges of rocketry. Much more glamorous are the launch vehicles. These are the apotheosis of rocketry—the culmination of centuries of aspirations. They have carried instruments and men from just beyond the atmosphere out to the Moon and have hurled automated probes to the planets. Several probes have already left the solar system and are cruising through the depths of interstellar space.

Expendable, e.g., nonrecoverable, launch vehicles (ELV's) are direct outgrowths of the technology of sounding rocket and military missile programs. In both the United States and the Soviet Union, missile technology in the mid-1950's had reached the point where practical planning for space missions could realistically begin.

Launch vehicles are divided roughly into four major groups. First are the so-called small ELV's derived from missiles and sounding rockets, such as Redstone and Viking, which produce less than 100,000 pounds of first-stage thrust. Second, there are those which have been developed by adding upper staging to intermediate-range ballistic missiles, such as Jupiter and Thor.

*A United States sounding rocket, Nike-Cajun, carries a Japanese payload to an altitude of 75 miles and a range of 73 miles on 26 April 1962.* (NASA)

*The French sounding rocket the Tactite, capable of reaching 120-mile heights with a 400-pound payload, is used to study infrared radiation.* (OFFICE NATIONAL D'ETUDES ET DE RECHERCHES AEROSPATIALES)

*The Soviet Union began upper-atmosphere research in 1949. Shown here, a Russian geophysical rocket at takeoff.* (U.S.S.R. ACADEMY OF SCIENCES)

*The 34-foot-long launching tube of the British Skua sounding rocket. The rocket is accelerated by a boost carriage (shown here at rear of tube) and recovered by a parachute.* (BRISTOL AEROJET)

*The Skylark upper-atmosphere sounding rocket leaves its launch tower.* (BRITISH AIRCRAFT)

Next in scale come the ICBM-based launch vehicles with first-stage thrusts ranking upward from 300,000 to 1 million pounds, exclusive of booster rockets that are often strapped on. Finally, the largest ELV's have first-stage thrusts of well over 1 million pounds. They were conceived and designed from the start as launch vehicles, and even though they may have employed military rocket technology, they are not mere modifications of missiles.

To the Soviet Union went the honor of building and firing the world's first launch vehicle to place an artificial satellite in orbit. The U.S.S.R. launched Sputnik 1 on 4 October 1957, one of history's most significant dates. Typically, the Russians did not at the time disclose very much information about this carrier vehicle. It was thought to be based on one of their military missiles.

The Russians ended much speculation in the West about the size and configuration of their carriers in May 1967 when they unveiled at the Paris Air Show the launch vehicle that had orbited the Vostok series of spacecraft. It was a five-unit, clustered propulsion system, each unit itself having a cluster of four engines. The core vehicle, thought to be a modified Sapwood SS-6 ICBM, was powered by four clustered RD 107 engines operating on liquid oxygen and kerosene. The central unit, together with four strap-on units, each with four clustered engines of the same type, provided a liftoff thrust of between 950,000 and 1,150,000 pounds. The vehicle, including the Vostok spacecraft stage, was 124 feet long. Diameter of the main stage was nearly 10 feet. Many of the structural features of the vehicle resembled those of the Scrag, an early three-stage ICBM in the Soviet arsenal. At the Paris show, the new launch vehicle was displayed on a railroad flat car erector similar to that used for the World War II German V-2 rocket.

In sharp contrast with the guesswork needed to discuss Russian carriers, many details of the American program were made public shortly after it was revealed. The record goes back to 29 July 1955, when President Eisenhower announced that plans had been approved for "the launching of small unmanned Earth-circling satellites" as part of America's contribution to the International Geophysical Year.

Almost immediately, the program went wrong.

There were two options for the selection of a carrier vehicle for the first satellites. The United States could rely on sounding rocket technology and build on the Viking and Aerobee, or it could modify existing military rockets or their test vehicles. For reasons that are still not thoroughly understood, the United States took the first choice. The Russians made the other choice. They got there first.

The American decision was not made without debate—in the Department of Defense and elsewhere. Under Homer J. Stewart, an ad hoc Advisory Group on Special Capabilities was assembled to examine several proposals for carrier rockets. The committee, whose creation had been requested by Donald Quarles, Assistant Secretary of Defense for Research and Development, studied proposals for an Atlas-based carrier, a Redstone-based vehicle, and a Viking-based carrier with an improved Aerobee second stage and a new solid-propellant third stage.

A carrier based on technology and experience gained in the Viking program was chosen. The reason for the selection of the new configuration—later called Vanguard—was that its efficiency would be better since it was being designed from the start as a carrier, and that it would make only minimum demands on the military ballistic-missile program, which was then in a desperate effort to catch up with the Russians. In hindsight, it appears logical to have bypassed the unproven Atlas. But the refusal to add a single small stage to the Redstone-based Jupiter C—which on 20 September 1956 demonstrated its capabilities by hurling a payload 3,400 miles across the Atlantic—is difficult to explain. The analogous situation would have been to refuse to use the V-2 for atmospheric sounding because it had been designed for a different role. Stewart himself favored the Jupiter C option, but was overridden by his committee—and once the decision was made, the United States had to live with it.

Scientific responsibility for the Vanguard program was given to the National Academy of Sciences, with the funds coming from the National Science Foundation. The Defense Department, which was assigned the job of launch-vehicle development, passed the management function along to the Navy.

The team put together to create the Vanguard launch vehicle was built around the NRL's Viking group, and was supported by many of the same contractor personnel.

Project Vanguard began officially on 9 September 1955, when the Department of Defense authorized the Navy to start development. Later, when defending Vanguard against criticism, program director John P. Hagen noted that he had received clear instructions from the start that he was not to interfere with the ballistic missile programs by requesting technical assistance from them—which inevitably deprived him of many flight-proven components and much readily available assistance. This kind of thinking, not weaknesses in the Vanguard team, led to a second-place finish for the United States. The NRL team actually did a very professional job that ended with the orbiting of several outstanding satellites.

The final design of the Vanguard launch vehicle emerged between September 1955 and March 1956. Its first stage, using liquid oxygen and kerosene, would develop 27,000 pounds of thrust. The second stage had a 7,500-pound-thrust engine burning white fuming nitric acid and unsymmetrical dimethyl hydrazine. The third stage, with solid propellant, would develop between 2,800 and 3,100 pounds of thrust. The overall vehicle would be 72 feet long and 45 inches in diameter and weigh 22,600 pounds—twice as long as the later Vikings, just as thick, and 8,000 pounds heavier. The first Vanguard components were flown on Vikings 13 and 14, which were designated TV-0 and TV-1. The first real Vanguard ELV was TV-2, with a powered first stage and dummy upper stages. Launched from Cape Canaveral on 23 October 1957, it sent a 4,000-pound payload on a 109-mile-high, 335-mile-long trajectory.

The next two flights failed. TV-3, with all stages powered, settled back on the launch pad, toppled over, and exploded. Two months later, on 5 February 1958, TV-3BU veered off course and broke up at an altitude of less than 4 miles.

By then, of course, it was too late for a "first"; Sputnik was already up. Vanguard TV-4 successfully orbited a small, 3¼-pound experimental satellite with two radio transmitters aboard, just "two years, six months and eight days after the initiation of the program from scratch," as John Hagen pointed out.

TV-5 failed to orbit a satellite when the third stage malfunctioned, and second-stage troubles caused the failure of the first nontest, operational Vanguard vehicles, SLV-1, -2, and -3. On 17 February 1959, the NRL placed its first full-scale Vanguard into orbit, following it with Vanguard 3 in September, after two more carrier vehicles failed. The Vanguard program, a "goat" through no fault of its own, was over.

But it did provide American engineers with valuable lessons. It bequeathed its second and third stages to later Thor- and Atlas-based ELV's, and its third stage to Scout. Later vehicles would use the swiveling techniques developed first for Viking and later improved for Vanguard's first-stage motor. But Vanguard in history was upstaged not only by the Soviets but by another American program.

This was the creation of the Medaris–Von Braun group at the Army Ballistic Missile Agency in Hunts-

*The U.S. space program reached its nadir at the very start, as the Vanguard TV-1 test vehicle—and America's initial hopes for entering space—collapsed into flame and smoke on Dec. 3, 1957.* (U.S. NAVY)

*Away from Cape Canaveral and into history, a Jupiter C carries Explorer 1, America's first satellite. The "UE" on the side is a sequence code based on Huntsville, the missile's home town.* (U.S. AIR FORCE)

ville. They evolved a launch vehicle named Juno 1—a four-stage rocket virtually indistinguishable from the three-stage Jupiter C. There was little doubt at Huntsville that Jupiter C could be quickly adapted for a satellite launching—if authority, and a rather modest amount of money, could be obtained.

After the Stewart committee made the choice for Vanguard, the Army's Office of the Chief of Ordnance turned over to Quarles's Research and Development Council a critique of the committee's decision. The council, whose members included Trevor Gardner, General Donald L. Putt, and other high service officials, turned down the ABMA again. The Redstone missilemen went ahead with their Jupiter C program, filling the fourth stage with, as Medaris put it, "sand instead of powder." During 1956 and into 1957, the Army was turned down again and again. ABMA was not arguing *against* Vanguard but *for* Jupiter C (with the fourth stage that made it Juno 1) as a back-up.

In a plan submitted in April 1957, ABMA recommended the launching of six 17-pound satellites. The first would orbit in September 1957, and the second, 2 or 3 months later. "In various languages our fingers were slapped and we were told to mind our own business, that Vanguard was going to take care of the satellite problem," Medaris recalls. "We followed in the spring and summer of 1957 with two shots with the scale-model nosecone [referring to Jupiter C firings 2 and 3], the first of which we were unable to recover—it fell too far away from the target area—but the second of which went directly into the target area, was recovered, and was the one that was shown . . . by the President [in a television presentation]."

After Sputnik 1, Secretary of the Army Wilber M. Brucker renewed his service's offer to orbit a satellite, saying that the ABMA would need "four months from a decision date to place a satellite in orbit . . . we would require a total of $12,752,000 of non-Army funds for this purpose [a six-vehicle program.]"

Suddenly, the atmosphere in Washington had changed. On 25 October 1957—just 3 weeks after the Russians orbited Sputnik 1—the Stewart committee endorsed the ABMA plan, and on 8 November the Secretary of Defense authorized the preparation of two satellites for launching in March 1958. The next week, ABMA was given $3.5 million for the mission; a few days later, the target date for the first satellite was pushed up to 30 January.

Later, Medaris described why the ABMA was so effective in the program:

> Being obviously a Government instrumentality we do not need to make contractual changes in order to make a change in our program, and therefore all that is required to meet the day-to-day exigencies of a fast-moving development program is that I make up my mind. . . .

With almost everything it needed at one installation, ABMA, assisted by the Army-controlled Jet Propulsion Laboratory, could move quickly without having to renegotiate industrial contracts and deal with subsidiary elements scattered all across the country.

Of the six Juno 1's built under an incredibly accelerated schedule, three orbited satellites. The first was Explorer 1, the first United States satellite, which went up on 31 January 1958—one day behind schedule. The United States had taken a hard look at the realities of space and vowed never to be caught lagging again.

As Juno 1 and Vanguard partially redeemed America's image, a new generation of ELV's was coming along—those based on IRBM first stages. Both the Jupiter and the Thor were brought into service, the first with Juno 1 upper staging and the second with either Vanguard staging or a specially designed Agena upper stage. These vehicles were variously called Juno 2, Thor Able, Thor Delta, Thor Epsilon, and Thor

*The Thor-Delta carrier rocket, shown here leaving the launch pad at Cape Canaveral, placed the British Ariel 1 satellite into orbit on 26 April 1962. Its first stage was based on the Thor IRBM.* (NASA)

*Juno 2 lofted a number of American satellites and probes, including Pioneer 4, which flew along a lunar bypass trajectory into orbit around the Sun in March 1959.* (NASA)

Agena. The Thor Delta, generally referred to simply as Delta, is still in operation.

All these vehicles have a place in American space history. Juno 2 launched America's first successful lunar fly-by, Pioneer 4, on 3 March 1959; Thor Able sent a much more elaborately instrumented deep-space probe, Pioneer 5, into orbit around the Sun on 11 March 1960 (Pioneer 4 also went into solar orbit, but it was tracked only to a distance of 400,000 miles). Thor Agena A orbited a 1,450-pound Air Force satellite (Discoverer 1) on its first launch on 28 February 1959; Thor Delta, on its second flight, placed the Echo 1 passive communications satellite into orbit on 12 August 1960. Meanwhile, small ELV's were developed for light payloads: the Scout, with four solid stages; a series of Argos carriers; and the Blue Scout, with either three or four stages.

The third major launch vehicle category includes those based on Atlas and Titan 2 ICBM's. The Atlas Score flight of 18 December 1958 orbited a 122-pound communications payload. Atlas Able failed three times to launch Moon probes between November 1959 and

*Neatly spaced ICBM row stretches out along Florida's east coast at Cape Kennedy. The two complexes in the foreground are Atlas Centaur LC36A (left) and B (right).* (NASA—KENNEDY SPACE CENTER)

December 1960; but, with an Agena upper stage, it orbited a 5,000-pound Midas reconnaissance satellite on its second try on 24 May 1960. Atlas has been a workhorse since, as Atlas Agena B and D, Atlas Mercury, and Atlas Centaur.

The seeds of the organization that was to manage these programs were planted in early 1958 when the President's Science Advisory Committee recommended that a central agency be created to take charge of the scientific exploration of space, both manned and unmanned. The agency was to be kept separate from the Department of Defense but was to benefit from its resources and help it to apply "space science and technology to military purposes for national defense and security." President Eisenhower sent the message to Congress with his endorsement on 2 April 1958, and the National Aeronautics and Space Act became law on 29 July. T. Keith Glennan, president of the Case Institute of Technology, became the first administrator of the National Aeronautics and Space Administration.

NASA started as a name only, but it grew by absorption. First it took over the 8,000-man National Advisory Committee for Aeronautics, an organization that had been created before World War I for aviation research. Hugh L. Dryden, NACA's director, became NASA's deputy administrator, and Robert C. Seamans became associate administrator. John Hagen's 170-man Vanguard team and the NRL's 46-man Upper Atmosphere Sounding Rocket group, under John W. Townsend, Jr., also were transferred to NASA. In December, the Army-owned Jet Propulsion Laboratory at the California Institute of Technology, with 2,800 employees under William H. Pickering, became part of NASA's resources.

NASA still had a great need for a group of scientists and engineers with long experience in the fields of missiles and carrier rockets. The only group in the nation that met the description was the Medaris–Von Braun team at the Redstone Arsenal. The transfer was made on 1 July 1960, and the facility became the George C. Marshall Space Flight Center, with Von Braun as director.

Von Braun knew that ICBM-based ELV's would just be adequate for manned missions in Earth orbit but not for manned lunar and interplanetary flights. Since the trend was toward lighter nuclear warheads, it was apparent that subsequent ICBM's would have smaller and less-powerful rockets—an appraisal later confirmed in full by the development of the Minuteman ICBM. Von Braun and his team of advance planners also recognized, however, that even the most ambitious launch-vehicle proposal would have to be built on existing technology.

Thus, the vehicle and propulsion engineers at ABMA set out to see how far they could go with the basic elements of Jupiter and Redstone.

The idea was to cluster a number of Jupiter engines around Redstone and Jupiter propellant tanks to build a large carrier vehicle that would use to the fullest the experience, hardware, and facilities of ABMA and its associated Army and industrial supporters. From this basic idea came the family of Saturns.

On 15 August 1958, the Department of Defense's Advanced Research Project Agency (ARPA) gave its approval for a research and development program whose aim was a carrier powered by eight uprated Jupiter S-3D engines. These engines, whose thrust would total 1.5 million pounds, would be mounted on a structure consisting of eight Redstone-type 70-inch-diameter tanks clustered around a single 105-inch Jupiter-type tank.

By October 1959, four Saturn configurations had evolved. Each had essentially the same first stage, with distinct variations in the upper stages. The NASA–Defense Saturn Vehicle Evaluation Committee picked one of the four configurations, the Saturn C-1, and development began in December.

The upper stages of Saturn C-1 (later simply Saturn 1) configuration were approved on 31 December 1959, and a ten-vehicle research, development, and test-flight program began. The second stage, designated S-4, would have four 20,000-pound-thrust engines; the third, S-5 stage, two such engines. After it won approval, Saturn was given the coveted DX rating, meaning it enjoyed high national priority for materials, personnel, and other resources.

Some changes were made in the upper stages during 1960. Six 15,000-pound-thrust engines, all using the high-energy combination of liquid oxygen and liquid hydrogen, replaced the four 20,000-pound-thrust second-stage engines. The 15,000-pound-thrust engines were identical to the 20,000-pound-thrust engines except for the reduction in their thrust rating. The third stage was eliminated. Testing of the first-stage engine cluster began in March, when two engines were fired. On 29 April, all eight engines were ignited for 8 seconds, producing 1.3 million pounds of thrust, an American record.

By 1961, Saturn 1's role as a test vehicle for the Apollo program, which was to take three Americans to the Moon, had been defined. Two series would be built, Block 1 and Block 2. The Block 1 Saturns would have dummy upper stages and would be fired to prove out the basic concept of the vehicle. The first stage of Block 2 vehicles, designated S-1, would carry more propellants and would be powered by upgraded engines, H-1's, which would develop 188,000 pounds of thrust each. The Block 2 Saturns would have stabilizing tail and stub fins, which the Block 1's would not; the later carriers would also have live S-4 upper stages, an improved instrument unit, and a dummy, or boilerplate, model of the Apollo capsule.

The first Saturn, SA-1, was static-tested in Huntsville in May 1961. The 162-foot-long vehicle, weighing nearly 1 million pounds, lifted majestically off the ground on 27 October in a virtually flawless maiden flight from Cape Canaveral. As it flew its short 200-mile trajectory, more than five hundred different measurements were recorded.

The rest of the Block 1 vehicles were fired smoothly during 1962 and 1963. The first Block 2 vehicle was launched on 29 January 1964; its second stage propelled a total weight of 37,700 pounds payload into orbit. Dummy Apollo capsules were orbited in May and September by Saturns SA-6 and SA-7, in flights that showed that the spacecraft and its carrier were compatible. The final three Block 2 Saturns orbited Pegasus micrometeoroid-detection satellites, as the Saturn 1 program ended with an unprecedented 100 percent successful flight-test record.

In 1962 the Air Force announced a plan to increase the payload-carrying capability of its Titan 2 ICBM by giving it an initial solid-propellant rocket boost. The new configuration became known as Titan 3C. It was developed for the Air Force to orbit payloads weighing 5,000 to 25,000 pounds.

On a typical flight, the two "zero-stage" solid boosters provided 2.4 million pounds of thrust to lift the vehicle off the ground. The first-stage engine ignited at high altitudes to give 470,000 pounds of thrust. Then the second-stage engine produced its 100,000 pounds of thrust. The twin third-stage, or "transstage," engines were rated at 16,000 pounds of thrust.

Titan 3C grew out of studies by the joint Department of Defense–NASA Large Launch Vehicle Planning Group, which reached the conclusion, in November 1961, that a large launch vehicle would be needed for military space missions in the late 1960's and the early 1970's.

Among the missions assigned to Titan 3C were orbiting military communications satellites, sending nuclear detection satellites aloft, and (with the 3M version) launching the Manned Orbital laboratory (MOL), a small space station in which two astronauts could live and work for a month. The MOL program, however, was canceled in June 1969, partly because there was a need to reduce military spending and partly because inadequate funding and other support had so lengthened its development period that the MOL promised to be obsolete before it was ready to fly.

The Titan 3C research and development (R&D)

*Dr. Von Braun (center) heads a Saturn planning meeting at Huntsville. With him are (left to right) Werner Kuers, Walter Häussermann, Willy A. Mrazek, Dieter Grau, Oswald Lange, and Erich W. Neubert.* (NASA—MSFC)

*The last Titan 3C, carrying two defense communications satellites, takes off in October 1983. It was replaced by the more powerful T34D, which uses the Inertial Upper Stage as its final stages.* (MARTIN MARIETTA)

*Saturn 1's first stage, S-1, is 21 feet in diameter and 80 feet tall. Together with its second stage, the Saturn 1B version launched experimental Apollo capsules into Earth orbit in preparation for later manned lunar flights.* (NASA—MSFC)

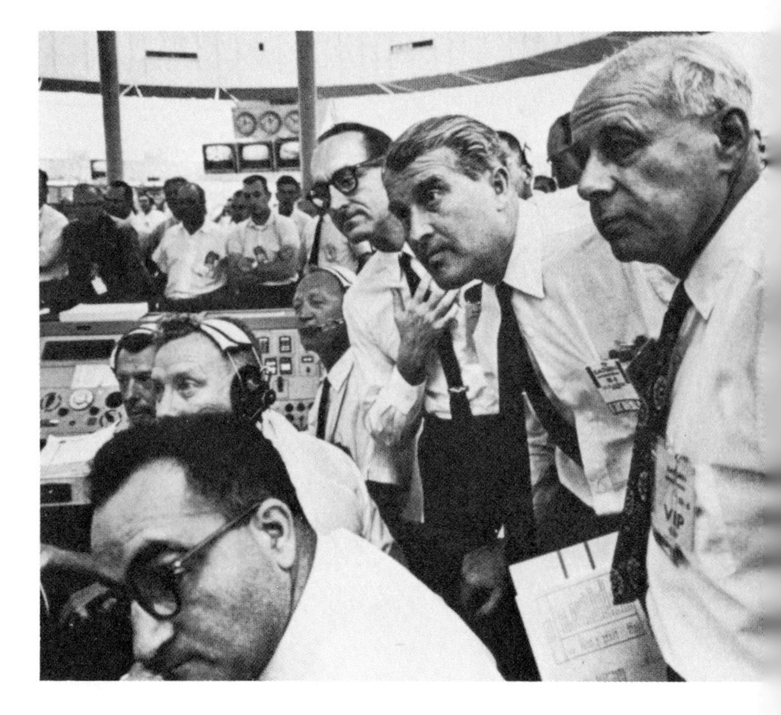

*Watching the first Saturn 1 launch from Blockhouse 34 at Cape Canaveral, October 1961. Left to right: George E. Mueller, NASA Associate Administrator for Manned Space Flight; Wernher von Braun; and Eberhard F. M. Rees, Deputy Director, MSFC.* (MITCHELL R. SHARPE)

program moved forward rapidly, often yielding benefits in addition to carrier vehicle checkout. Thus, on 16 June 1966, an R&D vehicle was launched from what was referred to at the time as Cape Kennedy (Cape Canaveral; the name change was made in November 1963 as a consequence of President Kennedy's assassination and remained such until October 1973 when it reverted to Cape Canaveral) and placed in a nearly circular orbit about 100 miles high. After about an hour, the transstage was re-ignited, lifting the apogee to nearly 21,000 miles. Just more than 6 hours later, the transstage again fired, bringing itself into a nearly circular orbit just greater than 21,000 miles high and at the same time changing the original inclined orbit to a near-equatorial orbit—the greatest plane change effected up to that time by the United States. Once in the final orbit, seven communications satellites and a single gravity gradient satellite were ejected as part of a program to deploy a global system of communications satellites for the Defense Communication Agency. Later, on 2 November 1966, five small satellites were injected into a low orbit from a single container; moreover, the carrier placed an unmanned Gemini B Capsule on a re-entry trajectory in a test of a new heat shield.

While Titan 3C was being developed, work proceeded on the larger Saturn 1B. Resembling the first Saturn, 1B had a Chrysler-built improved S-1 first stage with eight 205,000-pound-thrust engines. In the second stage, designated S-4B, there was substituted for Saturn 1's six 15,000-pound-thrust liquid engines a Rocketdyne J-2 engine which initially (it was later upgraded) developed 200,000 pounds of thrust on liquid oxygen—liquid hydrogen propellants. Saturn 1B was 224 feet high and 21.7 feet in diameter and weighed 650 tons fully loaded. The first stage was test-fired for the first time at Huntsville on 1 April 1965. Two months later, in Sacramento, California, the second stage went through acceptance testing. The first vehicle was assembled on 25 October at Launch Complex 34 at the Kennedy Space Center in Florida; less than 40 months had gone by since the development program had been authorized. Its maiden suborbital flight from Cape Kennedy, which took place on 26 February 1966 with an unmanned Apollo spacecraft as payload, was highly successful.

The second and third flights were made in July and August 1966. In the July AS-203 flight, the S-4B stage was filled with some 20,000 pounds of unneeded liquid hydrogen in a test proving that it was possible, despite the near-zero gravity environment, to seat the fuel in its tank for a second engine burn. Since the S-4B was not only Saturn 1B's second stage but Saturn 5's third, this test was highly significant for the Apollo lunar landing program. AS-202, flown after AS-203, was designed primarily to check out an unmanned Apollo 3 command module (which was put into a suborbital trajectory that caused it to re-enter the atmosphere at the maximum heating angle), the service module propulsion system (which was started in space for the first time), and guidance and control systems.

On 22 October 1968 the "101 percent" successful, long-duration Apollo 7 mission began, with the Saturn 1B performing flawlessly. Following this launch, which ended the initial fifteen-vehicle Saturn 1 series program (ten Saturn 1's and five 1B's), launch complexes 34 and 37 at Cape Kennedy were closed, awaiting the use of Saturn 1B's in the planned follow-on Apollo applications program.

Large as they were, Saturns 1 and 1B were only preludes to an even more powerful launch vehicle that was needed to fulfill the goal that had been outlined by President John F. Kennedy in a speech to Congress on 25 May 1961—landing an American on the Moon within a decade. "With the advice of the Vice-President [Lyndon B. Johnson, who was to succeed Kennedy in less than three years] . . . we have examined where we are strong and where we are not, where we may succeed and where we may not," Kennedy said. "Now is the time to take longer strides—time for a great new American enterprise—time for this nation to take a clearly leading role in space achievement, which in many ways may hold the key to our future on Earth."

On 25 January 1962, NASA approved a development program for the huge expendable launch vehicle, which was given the highest priority. Saturn 5 was to have three stages: the S-1C stage, the S-2, and the already familiar S-4B from the Saturn 1B.

The S-1C stage, developed by the staff at the Marshall Center with the support of the Boeing Company, was turned over to Boeing for production assembly at the huge NASA-owned Miehoud plant in New Orleans. The S-1C stage was about 138 feet tall and 33 feet in diameter. The 1969 model weighed nearly 300,000 pounds empty and held some 4.7 million pounds of liquid oxygen and RP-1 kerosene fuel. Saturn 5's first-stage thrust was about 7.7 million pounds, with each of its five F-1 engines capable of developing more than the 1.5 million pounds of thrust produced by the entire first stage of Saturn 1.

The S-2 stage, developed by North American Aviation, Inc., in Downey, California, was 81.5 feet long and 33 feet in diameter. It was powered by five liquid oxygen–liquid hydrogen J-2 engines producing 1,164,000 pounds of thrust. The third stage was the S-4B, 58.1 feet long and 21.7 feet in diameter, powered by a J-2 engine whose thrust was variable in flight from 184,000 to 230,000 pounds. When put together, the Saturn 5 stood 363 feet tall; fully fueled it weighed

*The eight flaming engines of its first stage launch Saturn 1B on its first flight from Cape Kennedy on 26 February 1966. It lofted a test Apollo payload to an altitude of 310 miles.* (NASA—MSFC)

*In a speech to Congress in 1961, President John F. Kennedy outlined a program for landing an American on the Moon within ten years. He is shown here touring the Marshall Space Flight Center, Huntsville, Alabama, in 1962. Flanking him are Wernher von Braun, Director, MSFC, and Maj. Gen. Francis J. McMorrow, commander of the Army Missile Command at Redstone Arsenal. Lyndon B. Johnson is on the far right.* (NASA—MSFC)

nearly 6.4 million pounds. It was able to send a spacecraft weighing about 50 tons to the Moon, or to place a 150-ton payload into orbit around the Earth.

In order to accommodate the mammoth carrier, equally mammoth facilities had to be erected at the Kennedy Space Center, located on Merritt Island next to Cape Kennedy. The 54-story, 526-foot-high vehicle assembly building enclosed 130 million cubic feet of space, making it at the time the world's largest building. It contained four huge bays, where four launch vehicles and their spacecraft payloads could be erected and assembled.

Erection of the AS-501 vehicle for the first Apollo flight began in the autumn of 1966. By the end of August of the following year, the Saturn 5, with the Apollo 4 capsule, had been loaded on a mobile launch tower, and the whole assembly was transported by a giant crawler vehicle to the launch pad. It took about 10 hours to move the rocket the 3 miles from the assembly building to the pad.

The countdown went off without a hitch and the launch took place right on schedule at 7:00 A.M. on 9 November. Each of the stages performed flawlessly, including the re-ignition of the third (S-4B) stage, which boosted the command module to 25,000 miles per hour, simulating the velocity it would reach on a return trip from the Moon. The command module splashed into the Pacific some 600 miles off Hawaii, 8 hours and 37 minutes after liftoff. The orbital weight of 278,699 pounds broke all records.

Saturn 5 was a tribute to the industrial wealth of the United States. Only the very richest of nations could afford launch vehicles on that scale; even highly advanced nations have found it advisable to pool their resources for more modest space programs.

To get into space, the European nations initially created the European Launcher Development Organization (ELDO), with headquarters in Paris. ELDO came into existence on 31 March 1962, and its first offspring was Europa 1, a three-stage ELV with a truly continental makeup. The French, meanwhile, worked on their own launch vehicles including the Diamant, Diogène, and Vulcain.

The frustrations that would develop with the Europa program and the economics of running two separate space agencies led to the formation in 1974 of a single European Space Agency from ELDO and the European Space Research Organization (ESRO). One of ESA's earliest tasks was to continue ELDO's efforts to develop a medium-class expendable launcher.

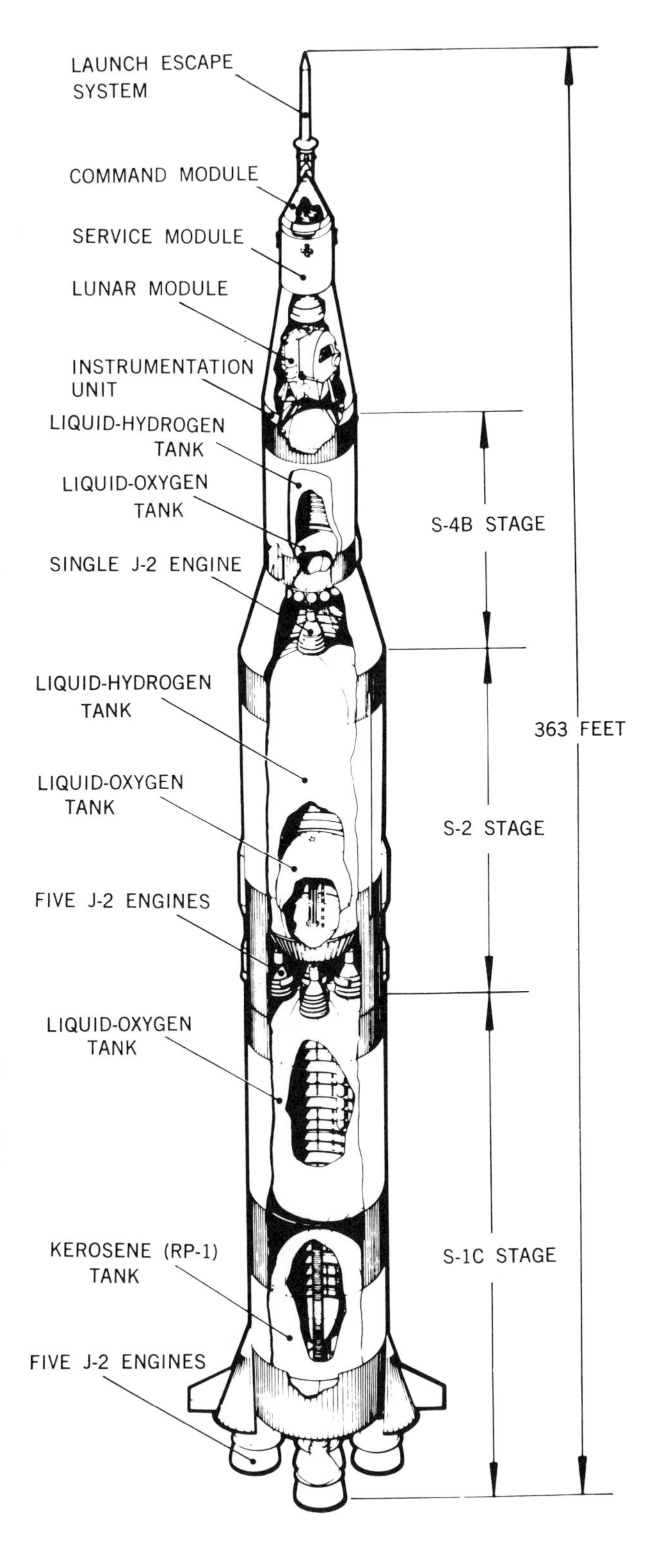


*Cutaway of the huge Saturn 5 Moon rocket, showing its three stages and modules of the Apollo payload. Fully fueled and loaded, the Saturn weighs more than 6 million pounds.* (NASA—MSFC)

*Combining the rockets of three European nations, Europa 1 has a British first stage, a French second stage, and a German third stage. It is shown in position for first-stage static testing at the Spadeadam Rocket Establishment, England.* (HAWKER SIDDELEY DYNAMICS)

*France became the third nation in space with its three-stage Diamant carrier, shown here on its maiden flight on 26 November 1965 from Hammaguir in the Sahara.* (ETABLISSEMENT CINEMATOGRAPHIQUE DES ARMEES)

One of history's most significant dates is 4 October 1957. On that day the Soviet Union put into orbit Sputnik 1, the world's first artificial satellite. It weighed 184 pounds and traveled in an elliptical orbit that took it around the Earth every one and one half hours.

The Soviets quickly followed this dramatic success with another. Sputnik 2, launched less than a month later, carried a passenger—a dog named Laika—and weighed 1,120 pounds. Sputnik 3, orbited on 15 May 1958, was even larger and heavier. It contained instruments for measuring the pressure and composition of the upper atmosphere, the incidence of micrometeoroids, and Solar and cosmic radiation.

Although the Russians had dropped a number of hints prior to 1957 that they were planning to launch a satellite, the event nevertheless took the world by surprise. Ironically, the failure of the United States to be first in space was due partly to its superiority during the early 1950's in weapons technology and strategic military position. The initial hydrogen bombs of both countries were bulky affairs, but the United States was ahead in reducing the weapon's size. United States planners, therefore, decided to defer building ICBM's until smaller warheads became available. In the meantime, their nation would be well defended by manned bombers operating from a network of bases that encircled Russia.

The Soviet Union took the opposite course. It went ahead and developed the massive rockets needed to carry its primitive bombs—the same vehicles that later gave it a significant edge in space exploration.

The first United States satellite, Explorer 1, went into orbit on 31 January 1958, four months after the first Sputnik. It was launched by Juno 1, a four-stage carrier vehicle hastily adapted from the Jupiter C by the Army's team at Redstone Arsenal. The subsequent successes of the Explorer series and of the Vanguard satellite project helped salve America's pride.

Following up their initial achievements, the United States and Russia inaugurated diversified satellite programs. Probably the most successful of the American satellites have been the Tiros meteorological series, begun in 1960. Tiros 1 eventually broadcast more than 19,000 pictures of cloud formations, literally adding a new dimension to weather forecasting. Later models in this series transmitted more than 100,000 pictures each. The United States also has been very successful with communications satellites: Echo, Telstar, Relay, Syncom, Early Bird, and Intelsat. Comparatively little is known about similar Russian projects. Nearly 300 Kosmos satellites were orbited by mid-1969, accounting for approximately three quarters of all Soviet satellite launchings.

Both nations also launched many probes toward the Moon in preparation for sending men there. In September 1959, a Russian probe crash-landed on the Moon; two months later another probe took the first pictures on the far side of the Moon; and in February 1966, after at least four unsuccessful tries, a Soviet probe achieved a soft landing on the Moon. The United States, following a series of discouraging failures in 1962, scored outstanding successes in 1964 and 1965 with Rangers 7, 8, and 9, which transmitted back to Earth thousands of pictures of craters as small as a few feet across on the Lunar surface. Ranger was followed by Surveyor and Lunar Orbiter series, which provided vital data for the six Apollos that were to land on the Moon between mid-1969 and late 1972.

During the 1960s and '70s, American and Russian spacecraft successfully probed the secrets of Venus and Mars. The latter nation pioneered the landing of capsules on both these neighboring worlds, but it was left for the stunningly successful U.S. Mariner 9 orbiter to reveal photographically the nature of the Martian surface. In late March 1974, another Mariner, No. 10, recorded an astronautical first by passing within 500 miles of innermost Mercury. Meanwhile, two other American probes, Pioneers 10 and 11, not only traversed the Asteroid Belt but photographed at close range the giant planet Jupiter. Both these amazing probes continued into the depths of the Outer Solar System.

RIGHT: *On 3 November 1957, less than a month after launching the first Sputnik, the Soviet Union marked another first by placing a dog, Laika, in orbit around the Earth in Sputnik 2. Laika's carrier was 19 feet long and 4 feet in diameter at the base.*

NOVOSTI PRESS AGENCY

BELOW: *The instruments and antennas on Sputnik 3 give a clue to its purpose—to measure and record gravity and irradiation. Launched on 15 May 1958, it too was conical in shape and measured 12 feet high by 5.5 feet in diameter.*

NOVOSTI PRESS AGENCY

NOVOSTI PRESS AGENCY

ABOVE: *While the Sputniks were being built and tested, the Soviet Union launched the first ICBM in the summer of 1957, more than a year before the United States. Some were displayed in the above parade in Red Square on 7 November 1965.*

RIGHT: *A Minuteman 2 ICBM leaves its underground silo momentarily preceded by a smoke ring. These three-stage missiles, which form a major part of the deterrent force of the United States, are placed throughout the nation, ready to be fired almost instantaneously to targets that are programmed into its guidance system.*

THE BOEING COMPANY

NASA-MSFC

*On its first flight in 1961, Saturn 1 recorded more than 500 measurements on a 200-mile trajectory. Designed as a test vehicle, it gathered basic information for the Apollo program that eventually took three Americans to the Moon. Here the first stage of a Saturn 1 carrier is being lifted onto the static test stand at the George C. Marshall Space Flight Center at Huntsville, Alabama.*

NASA-GSFC

*The tiny Explorer satellites, which discovered the Van Allen radiation belts, made measurements of cosmic rays, temperatures, and the Earth's magnetic field. Launched in 1961, Explorer 12 was only 5.5 inches high and 26 inches wide, but its complex structure included an octagonal platform and four spring-loaded Solar paddles.*

NASA

ABOVE: *The first successful probe of another planet occurred on 14 December 1962, when Mariner 2 passed within 22,000 miles of Venus. The 447-pound probe recorded Venus' surface temperature at 700°F. and sent back valuable data on previously unexplored space.*

LEFT: *By photographing the Earth from above, the Tiros meteorological satellites have added a new dimension to weather forecasting. Tiros 6, launched in December 1962, took pictures of hurricanes in the Atlantic and Pacific.*

*The Martian horizon, as viewed by Viking 2, stretches across nearly 200 degrees in this composite of three color photos taken on 4 September (center), 5 September (right), and 8 September (left) 1976. The color of the surface is predominantly rusty-red, caused by a thin coating of limonite (hydrated iron oxide).*(NASA)

## Comparison of Modern United States Carrier Vehicles (Non-Saturn Series)

| Designation | Stages | Propulsion | Thrust (pounds) | Length (feet) | Maximum Diameter (feet) | Lifting Capabilities | Typical Uses |
|---|---|---|---|---|---|---|---|
| Scout | 4 | 1. Algol 2 solid<br>2. Castor 2 solid<br>3. Antares 2 solid<br>4. Altair 2 solid | 100,950<br>60,765<br>20,925<br>6,480 | 72 | 3.5 | 300 lb. into 300-mile orbit | Explorer, Ariel, San Marco, ESRO, Secor, etc., satellites |
| Thrust-Augmented Improved Thor Delta | 3 + booster | Booster. Castor 1 solid<br>1. MB3-3 liquid<br>2. AJ10-118E liquid<br>3. FW-40 solid | 162,000<br>170,000<br>7,800<br>6,200 | 92 | 8 | 1,190 lb. into 300-mile orbit; 250 lb. to escape | Explorer, ESSA, OSO, Isis, Geos, Heos, etc., satellites; Pioneer probes |
| Thrust-Augmented Thor Delta (elongated) | 3 + booster | Booster. Castor 2 solid<br>1. MB3-3 liquid<br>2. AJ10-118E liquid<br>3. TE 364-3 solid | 156,450<br>170,000<br>7,800<br>9,980 | 105.5 | 8 | 2,000 lb. into 300-mile orbit; 575 lb. to escape | Developed as standard launch vehicle for NASA satellites, probes |
| Thor Agena D | 2 | 1. MB3-3 liquid<br>2. Agena D 8092 liquid | 170,000<br>16,000 | 76 | 8 | 3,500 lb. into 120-mile orbit | Air Force military satellites, and such NASA satellites as OGO, Nimbus, and Pageos |
| Atlas Agena D | 2 | 1a. LR-89-3 liquid<br>1b. LR-105-3 liquid<br>2. Agena D 8092 liquid | 300,000<br>88,000<br>16,000 | 104 | 16 | 6,000 lb. into 300-mile orbit; 1,450 lb. to escape | Air Force military satellites; NASA OGO, ATS, OAO satellites; Lunar Orbiter and Mariner probes |
| Atlas Centaur | 2 | 1a. LR-89-3 liquid<br>1b. LR-105-3 liquid<br>2. RL10-A3 liquid | 300,000<br>88,000<br>30,000 | 66 | 10 | 9,900 lb. into 300-mile orbit; 2,700 lb. to escape | Surveyor and Mariner probes |
| Titan 3B Agena D | 3 | 1. LR-87-AJ-9 liquid<br>2. LR-91-AJ-9 liquid<br>3. Agena D 8092 liquid | 430,000<br>100,000<br>16,000 | 124 | 10 | 8,000 lb. into 300-mile orbit | Air Force military satellites |
| Titan 3C | 3 + booster | Booster. UA 1205 solid<br>1. LR-87-AJ-9 liquid<br>2. LR-91-AJ-9 liquid<br>3. AJ10-138 liquid | 2,400,000<br>470,000<br>100,000<br>16,000 | 124 | 30 | 25,000 lb. into 100-mile orbit; 5,000 lb. to escape | Air Force military satellites; tactical communications satellites; Vela, Dodge, and other research satellites |

## Comparison of Saturn 1, Saturn 1B, and Saturn 5 Carrier Vehicles

| | *Saturn 1, Block 1* | *Saturn 1, Block 2* | *Saturn 1B (vehicle AS-205)* | *Saturn 5 (vehicle AS-505)* |
|---|---|---|---|---|
| Total length, stages + payload body (feet-inches) | 162-6 | 187-11 | 224 | 363 |
| Total weight, loaded (pounds) | 925,000 | 1,165,000 | 1,290,000 | 6,391,120 |
| **First stage** | | | | |
| Length (feet-inches) | 81-7 | 80-3 | 80-4 | 138 |
| Diameter (feet-inches) | 21-5 | 21-5 | 21-5 | 33 |
| Engine thrust (pounds) | 1,300,000 | 1,504,000 | 1,640,000 | 7,680,000[b] |
| Propellants | LOX-kerosene | LOX-kerosene | LOX-kerosene | LOX-kerosene |
| **Second stage** | (Dummy) | | | |
| Length (feet-inches) | 43-11 | 41-5 | 58-5 | 81-7 |
| Diameter (feet-inches) | 18-4 | 18-4 | 21-8 | 33 |
| Engine thrust (pounds) | – | 90,000 | 225,000[a] | 1,150,000[c] |
| Propellants | – | LOX–liquid hydrogen | LOX–liquid hydrogen | LOX–liquid hydrogen |
| **Third stage** | (Dummy) | (No third stage) | (No third stage) | |
| Length (feet-inches) | 23-4 | | | 58-7 |
| Diameter (feet-inches) | 10-0 | | | 21-8 |
| Engine thrust (pounds) | – | | | 230,000 |
| Propellants | – | | | LOX–liquid hydrogen |
| **Payload body** | | | | |
| Length (feet-inches) | 23-10 | 63-5 | 52-6 | 82 |

[a] Maximum; can be operated at lower thrust, averaging 200,000 pounds.
[b] To maximum of 9,115,000 pounds at stage burnout.
[c] Maximum; each 230,000-pound-thrust engine can be operated as low as 184,000 pounds thrust.

Spurring this effort was the refusal of the United States to guarantee that it would launch European satellites even if they competed with domestic United States firms. France worried that this might ground promising communications and other satellite projects. Soon what started as a hedge against United States refusals would become a formidable competitor for the space shuttle: Ariane. The program was approved on 28 December 1973 with 64 percent of its funding being French. The next largest share was West Germany at 20 percent, and the rest ranged from 5 to 0.5 percent.

Ariane bears a passing resemblance to the much larger United States Saturn 5. It is a three-stage launcher 154 feet tall and 12.5 feet wide weighing nearly 460,000 pounds at liftoff. The first stage is powered by four Viking 5 engines burning nitrogen tetroxide and a mixture of hydrazine and unsymmetrical dimethyl hydrazine for a total of 549,000 pounds of thrust. The second stage uses a single Viking 4 engine (basically a high-altitude version of the Viking 5). The third stage is the most advanced, an HM-7A engine burning liquid hydrogen and liquid oxygen to produce 13,500 pounds of thrust. Ariane is capable of injecting a payload into geostationary transfer orbit weighing just over 4,000 pounds.

The launch site selected for the program is on the coast of French Guiana near the town of Kourou. Like Cape Canaveral many years before, it had to be built from scratch in swampland and had a frontier atmosphere. While costly to build, it has the chief advantage of being only 5 degrees north of the equator so that its greater eastward velocity gives the launcher a faster "running start" than does the more northern latitude of Cape Canaveral.

The development program led to a successful first launch (LO1) on 24 December 1979, more than a year ahead of the United States space shuttle. But trouble lay ahead. The second flight (LO2) on 23 May 1980 failed when one of the first-stage engines developed an instability problem (caused by improper manufacture, not a design flaw) and destroyed a load of atmospheric trace chemicals and an amateur radio satellite. The LO3 and LO4 launches on 19 June and 20 December 1981 were successful, so the system was declared operational. But the LO5 launch failed when contaminants caused a malfunction in the third stage.

Four subsequent Ariane missions were successful, after which an upgraded model, Ariane 2/3, was introduced on 4 August 1984. In the same style as the United States Thor-Delta family, the upgrade for the Ariane 2 was accomplished by increasing chamber pressure in the engines of all three stages and stretching propellant tanks in the third stage. Ariane 3 is identical but has two solid-propellant boosters added to the first stage. The changes increase liftoff thrust

*Ariane, the Franco-European launcher, takes off on its sixth mission. Around the base is foam insulation, shed from the third stage. An ambitious growth program will make Ariane a competitor for the shuttle well into the 1990's.* (ESA)

*Japan entered the launch business with N-1 rockets copied (under license) from the American Delta, then went on to add their own improvements, which will lead to the more powerful H-1 series.* (NASDA)

respectively to slightly over 478,000 and 522,000 pounds, and payload in geostationary transfer orbit is raised to as much as 5,700 pounds.

An Ariane 4 version being developed for introduction in 1986 will stretch the lengths of the first and second stages and use more powerful solid- or liquid-propellant boosters (or a combination), raising the payload in transfer orbit to as much as 9,500 pounds.

In addition to the worries Europe had, Japan also wanted its own launcher program so that it could compete with the West. That nation took a different tack in both management and development. In the first case, it maintains two agencies, the National Space Development Agency (NASDA) and the Institute for Space and Astronautical Sciences (ISAS). In the latter, it purchased United States Delta launchers, then acquired the production technology under license from McDonnell Douglas Astronautics Co., the builder, and finally added its own improvements. The first N-1 flew in 1975 and was replaced in 1981 with the N-2, which is capable of placing 1,500 pounds in geostationary orbit. A more advanced H-1, with a liquid oxygen–liquid hydrogen second stage, is to be introduced in 1986. A more modest launch capability for science satellites was developed by ISAS using all-solid rockets to develop the "Mu" family of launchers.

Other nations, such as China and India, have developed launch vehicles, some of which may compete in the marketplace with the Americans and Europeans.

Finally, additional competition for the shuttle was to come from the private sector, wherein many investors believed that a straight commercial operation could launch satellites at lower costs than the shuttle charged. In 1984, NASA approved plans by General Dynamics and Transpace Carriers Inc. to take over the Atlas-Centaur and Delta rockets, respectively, and market them as a commercial service. Regulation of this new field was transferred to the U.S. Department of Transportation where a new division was established to provide the user with "one-stop shopping," as the government put it, in seeking permits from the multitude of federal agencies whose responsibilities touched this field in various ways.

Other investors saw commercial potential in developing new launchers rather than in buying old models. Two such ventures were Starstruck's Dolphin rocket, which is launched from the water and uses a solid fuel and liquid oxidizer, and Space Services Conestoga, which is based on the Delta's solid boosters. One early rocketry pioneer, Robert Truax, even started assembly of a manned suborbital rocket using surplus parts from Atlas ICBMs.

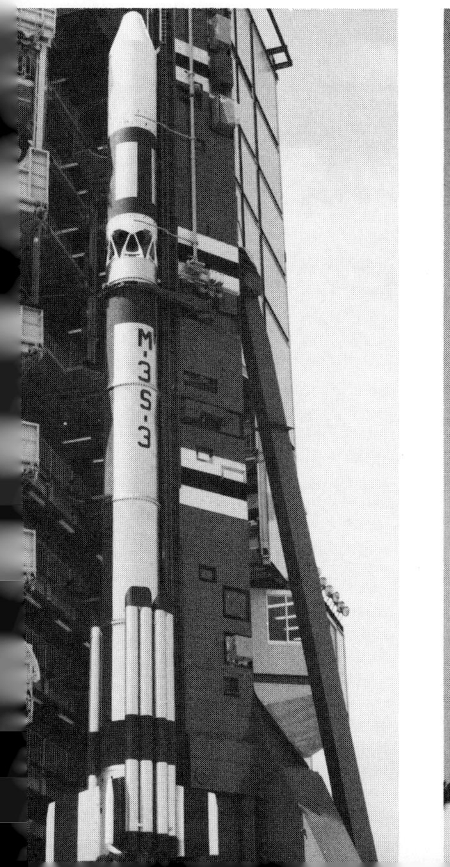


*An MU-3 scientific launcher stands on its pad at Japan's Kagoshima Space Center. Unlike the United States, Japan has two space agencies—one scientific, the other commercial.* (NEIL DAVIS)

*India entered the space age with its SLV-3 launcher series, here carrying a Rohini satellite. Based on the American Scout, it gave India valuable experience that will allow it to design its own launchers.* (ISRO)

*Several privately funded launcher projects emerged in the 1970's and 1980's, such as the Conestoga 400 orbital vehicle, based on the Delta solid-fuel rocket boosters.* (SPACE VECTOR)

# 8 THE REMOTE EXP

It is odd, in retrospect, to consider how little thought most of the pioneers of rocketry gave to the unmanned spacecraft that were to dominate the opening years of space exploration. True, Goddard did write about sending a probe to the Moon, but he was more concerned with proving the capability of the rocket than with getting information back from space. And both Tsiolkovsky and Oberth spent more time describing manned spaceships than they did on possible unmanned spacecraft.

This might seem puzzling now, when the full complexities of sending humans into space and the relative ease of getting information back from unmanned spacecraft is known. But it is understandable in terms of the technology of the time when the pioneers wrote.

Unmanned spacecraft rely completely on automatic instrumentation to gather scientific data and transmit them back to Earth. But even though the technology of rocketry and its theoretical implications had been worked out more than half a century ago, the development of techniques for sending information over long distances by remote means is a fairly recent development. The method, called radio telemetry, was not applied seriously to missile research until the initiation of the development of the V-2 (although it had been considered for sounding rockets in the early 1930's), and it did not become an indispensable part of the research that was conducted with the aid of rockets until the V-2 was used for atmospheric sounding in the United States after World War II.

Telemetry was well enough advanced by the early postwar years to spark speculation about small unmanned artificial Earth satellites. Three Englishmen, K. W. Gatland, A. M. Kunesch, and A. E. Dixon, published a paper on artificial satellites in a 1951 issue of the *Journal of the British Interplanetary Society*. And a University of Maryland physicist, S. Fred Singer, came up with the MOUSE (Minimum Orbital Unmanned Satellite of the Earth) at the Fourth International Astronautical Congress in 1953. MOUSE, details of which were published in British and American journals, was a 100-pound satellite that would operate its radio telemetry transmitter by batteries that were energized by the Sun.

While these proposals were being made, the military services in the United States were quietly studying the possibility of orbiting artificial satellites. The Navy, in 1945, apparently was the first to begin considering the idea.

The Navy's effort started with the organization of the Committee for Evaluating the Feasibility of Rocketry (CEFSR) within the Bureau of Aeronautics in October 1945. After preliminary studies CEFSR recommended that the development of an instrumented Earth satellite be started. The Guggenheim Aeronautical Laboratory at the California Institute of Technology was given a contract in December to investigate the relationship between carrier vehicle performance, the weight of the satellite, and the height of its orbit. The results showed that a satellite could be orbited, but only at a price that looked sky-high to the Navy in those postwar years of tight military budgets. The Navy went to the Army Air Forces to ask for help on 7 March 1946.

The Army Air Forces (AAF) officers noncommittally said that they would take up the project with General Curtis LeMay and other high officials. When LeMay met with Commander Harvey Hall, originator of the Navy study, later that month, it became clear that the Air Forces had no intention of cooperating with the Navy. What LeMay did not tell Hall was that the Army Air Forces also had begun a study of artificial satellites. The Navy learned about the AAF study, which was being made by Project Rand (then a part of the Douglas Aircraft Company and later the Rand Corporation) at a June conference called by the War Department's Aeronautical Board. While the Board did not discourage the continuation of either of the feasibility studies, no attempt was made to get cooperation between the services, and each went on alone.

The Navy came up with a plan for a single-stage liquid oxygen–liquid hydrogen carrier vehicle with a takeoff weight of 101,000 pounds and a thrust of 233,000 pounds. This High Altitude Test Vehicle was to boost itself into orbit, with its empty structure becoming the satellite. But the idea never got off the ground. Faced

# LORERS

with continuing reductions in appropriations, the Navy cut back its satellite studies until they petered out in 1948.

The Army Air Forces study had taken a different tack. Recognizing the difficulties of developing a single-stage orbital carrier-satellite combination, the Project Rand staff, aided by personnel from North American and Northrop, concentrated on a multistage vehicle. The results of the study were contained in a historic report—"Preliminary Design on an Experimental World-Circling Spaceship"—that was presented to the Air Materiel Command on 12 May 1946.

Rand said that a 500-pound satellite could be launched by a carrier that would utilize the technology acquired in the V-2 and other vehicle programs, and that a launch date as early as 1951 was feasible. The report recommended a three-stage carrier that would weigh 233,669 pounds and be powered by liquid oxygen and alcohol. The launching would take place at an island in the Pacific, to take advantage of the fact that the Earth's speed of rotation is greatest at the equator; this would provide a boost in putting the satellite into an eastward-heading orbit. Several uses were suggested for the satellite. Among them were meteorological research, communications, and reconnaissance—for example, observing the effects of American bombing attacks on a hypothetical enemy. The cost envisioned for the satellite carrier was $150 million.

The report included the prophetic statement that the "achievement of a satellite craft by the United States would inflame the imagination of mankind, and would probably produce repercussions in the world comparable to the explosion of the atomic bomb," a prediction whose truth was proved bitterly to the United States some 11 years later. The report also contained the following:

> It is our earnest hope that under the terms of this new study and research contract with the Army Air Forces we may be able to enlist the active cooperation of an important fraction of the scientific resources of the country to solve problems in the wholly new fields which man's imagination has opened. Of these, space travel is one of the most important and challenging.

In October 1947, the Committee on Guided Missiles of the Joint Research and Development Board (the successor of the War Department's Aeronautical Board) assumed responsibility for coordinating the work being done on artificial satellites by the various services. A technical evaluation group concluded that an Earth-satellite program could not be authorized until some definite military uses for satellites were established. Although this meant that the Rand report had to be shelved temporarily, advocates of the satellite concept continued to press for support within the now-independent (since 18 September 1947) U.S. Air Force. The best they could get, however, was a statement on 15 January 1948, from Vice-Chief of Staff Hoyt S. Vandenberg, that satellites should be developed "at the proper time."

U.S. Army Ordnance entered the picture on 15 September 1948, when the Committee on Guided Missiles recommended that the Hermes project provide a continuing analysis of the problems of developing an Earth satellite—the first official recognition of the capability of the Army Ordnance–Wernher von Braun team for contributing to satellite carrier-rocket development. And on 29 December 1948, Secretary of Defense James V. Forrestal revealed publicly for the first time that the United States was looking into the feasibility of artificial satellites.

Although receiving only meager encouragement, the military continued their studies of artificial satellites. In 1949 the Rand Corporation, which had become independent of Douglas in November 1948, concentrated on proving the military utility of satellites. There was hope that the newly created Department of Defense would approve a true development program if it was convinced that a satellite would do some useful work for the armed forces. Among the purposes listed by Rand for artificial satellites were surveillance and reconnaissance, communications, a display of American technical and scientific leadership, and psychological "cold warfare"; Rand calculated that the mere presence in the sky of an artificial satellite would have a strong psychological effect on a potential enemy.

The early postwar studies made by the services and their civilian contractors later proved to be remarkably accurate, in specific details as well as in their broad outlines. At the time, though, they were dismissed.

What little debate about artificial satellites that went on in the United States took place in a vacuum. Americans had the complacent idea that no one else was interested in placing a satellite into orbit. Little or no mention was made of the possibility that the Soviet Union was working on a satellite program. Russia was regarded as too backward technologically and too devastated by war to compete with the United States in any field. And, of course, the Russians were saying very little about their space program.

What they did say was extremely interesting.

For example, on 4 October 1951, Soviet rocket expert M. K. Tikhonravov said that Russian technology was at least on a par with that of the United States, and that the Soviet Union would be able to launch artificial satellites. And at the World Peace Council in Vienna on 27 November 1953, A. N. Nesmeyanov of the U.S.S.R. Academy of Sciences announced that "science has reached such a stage that . . . the creation of an artificial satellite of the earth is a real possibility." These were important voices that were talking, but the United States was deaf to them. It was also deaf to voices that were closer to home.

*Meeting of the Project Orbiter Committee on 17 March 1955 in Washington, D.C. Shown here are (left to right): Seated—Cdr. George W. Hoover, Office of Naval Research; Frederick C. Durant III, Arthur D. Little, Inc.; James B. Kendrick, Aerophysics Development Corporation; William A. Giardini, Alabama Tool and Die; Philippe W. Newton, Dept. of Defense; Rudolf H. Schlidt, Army Ballistic Missile Agency; Gerhard Heller, ABMA; Wernher von Braun, ABMA; Standing—Lt. Cdr. William E. Dowdell, USN; Alexander Satin, ONR; Cdr. Robert C. Truax, USN; Liston Tatum, IBM; Austin W. Stanton, Varo, Inc.; Fred L. Whipple, Harvard Observatory; George W. Petri, IBM; Lowell O. Anderson, ONR; Milton W. Rosen, NRL.* (FREDERICK C. DURANT III)

Nevertheless, the climate slowly began to change as artificial satellites began to get wider public discussion. On 4 October 1954, a special committee of the International Geophysical Year (IGY), meeting in Rome, recommended "that thought be given to the launching of small satellite vehicles, to their scientific instrumentation, and to the new problems associated with satellite experiments, such as power supply, telemetering, and orientation." The United States National Committee for the IGY approved the idea on 14 March 1955. On 28 July the Eisenhower Administration publicly approved a program "for going ahead with the launching of small, earth-circling satellites as part of the United States participation in the International Geophysical Year." The following day the National Academy of Sciences and the National Science Foundation announced the coordination of their resources in carrying out the endeavor. At about the same time, a group known as the Project Orbiter Committee met to coordinate Army and Navy talents to make the proposed satellite a reality.

Few Americans noticed that the Soviet Union was making statements of equal importance. On 15 April 1955, the Soviets announced that the Council of the U.S.S.R. Academy of Sciences had set up a Permanent Interdepartmental Commission for Interplanetary Communications, whose work included coordinating the development of meteorological satellites. On 30 July, only one day after the American announcement, the Soviets disclosed that they had their own program. At the Sixth International Astronautical Congress in Copenhagen that August, Leonid I. Sedov, chairman of the new Soviet commission, said, "In my opinion, it will be possible to launch an artificial satellite of the Earth within the next 2 years, and there is the technological possibility of creating artificial satellites of various sizes and weights." Sedov also mentioned that the satellites that could be launched within the following 2 years might be much larger than those conceived by the United States.

In June 1957, the same A. N. Nesmeyanov who in 1953 had predicted the early development of satellites said that both the launch vehicle and the instrumentation for the first Soviet satellite were ready, and that the first launching would occur within a few months. On 10 June, Lloyd V. Berkner of the American IGY Committee received a document from the Soviet Union stating bluntly that a satellite would be launched within months. The Russians announced the transmission frequencies of the first satellite on 1 October.

On 4 October, to the amazement of the world, the Soviets orbited Sputnik 1, thereby reaping the glory of opening the age of space. The world's first artificial satellite weighed 184 pounds and carried instruments to study the density and temperature of the upper atmosphere and the concentration of electrons in the ionosphere. It circled the Earth about every 1½ hours in an elliptical orbit that ranged in altitude from about 140 to 560 miles. Sputnik 1 remained in orbit, gradually losing altitude, until 4 January 1958, when it disintegrated upon re-entering the denser portion of the atmosphere. Despite the plain-spoken Soviet forewarnings, Sputnik stunned the world. Americans, who had become accustomed to laughing at Russia's technological efforts, suddenly found themselves in the uncomfortable position of second place in a two-horse race. Sputnik started a national re-evaluation whose effects in fields from politics to education were felt for years afterward.

One immediate result was an extensive series of Congressional hearings that began in November 1957 and ran well into 1958. Another was the formation by the National Advisory Committee for Aeronautics (NACA) of an advisory committee for space technology. It was headed by H. Guyford Stever of the Massachusetts Institute of Technology. The NACA soon was expanded into the National Aeronautics and Space Administration (NASA), created to handle the civilian space program, and the Advanced Research Projects Agency was set up to direct military space efforts.

But the initiative remained with the Russians. On 3 November came Sputnik 2, huge for the time at 1,120 pounds and carrying a dog, life-support equipment, instruments to measure the effects of space flight on the animal, and some unrelated instruments to measure cosmic and solar radiation. Sputnik 3, still larger and heavier, carried instruments to determine the pressure and composition of the upper atmosphere, the incidence of micrometeoroids, and solar and cosmic radiation. It was orbited on 15 May 1958.

The Soviet habit of secrecy made it impossible to determine just how much of its unmanned satellite effort had been devoted to pure science and how much had military motives. The Soviets faithfully reported the orbiting of their satellites, but they were far from generous with descriptions of their purposes. And the Russians were most reluctant to talk about their failures. Nevertheless, the broad outlines of the Soviet space program could be determined.

Students of that program believed that the period from October 1957 to October 1964 was one distinct development phase for the Soviets, and that another phase opened in November 1964. The first phase was dominated by the Kosmos series of unmanned satellites, which accounted for nearly three-quarters of all Soviet launches, manned or unmanned, during the first decade of space exploration.

After orbiting the three successful Sputniks in 1957 and 1958, the Soviets shifted during 1959 to carry out a three-shot lunar exploration program. Returning to satellites in 1960, they orbited five Korabl Sputniks during that and the succeeding year. Each was larger and heavier than the earlier Sputniks, weighing more than 10,000 pounds, and each contained a recoverable re-entry capsule. They served as test vehicles for the manned Vostok spacecraft that followed. Capsules from three of these satellites were recovered with live dogs aboard, proving that animals not only could survive the trip into space but could also return to Earth safely.

*Soviet citizens gather in Moscow to see a model of Sputnik 1, the world's first artificial satellite.* (TASS/NOVOSTI)

On 16 March 1962 the Kosmos series was inaugurated. The Soviets revealed little about these satellites, leading Westerners to think that at least some of the Kosmos satellites had a military use. The Russians disclosed neither weight nor configuration of most Kosmos satellites.

The first Molniya communications satellite was launched in April 1965; and, within only 4 years, ten others would follow. Color television programs were exchanged between Moscow and Paris with this system, which also provided multichannel telephone and telegraph communications links.

Russia also initiated its Proton series in 1965,

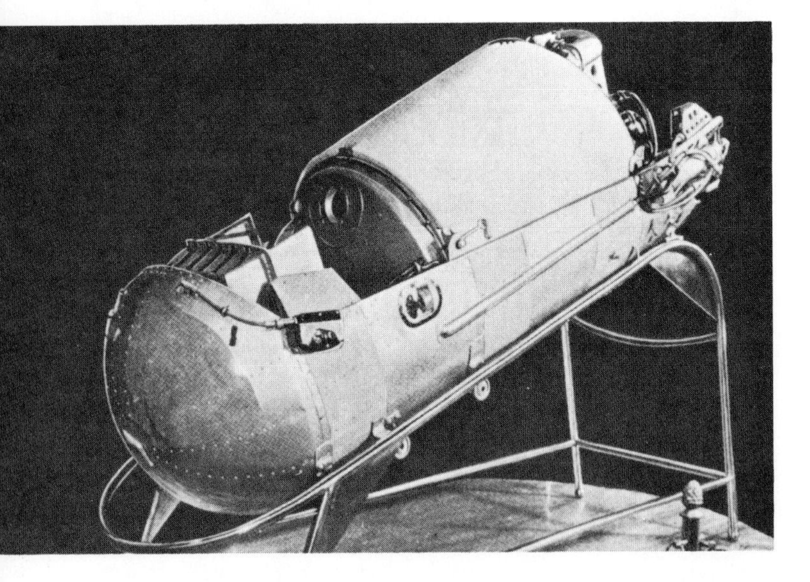

*Designed as test vehicles for later manned flights, the Russian series of five Korabl Sputniks carried live dogs into orbit. Three were successfully recovered, proving that animals could not only survive space travel but could also return safely to Earth. Shown here is the recovery capsule for Korabl 2, launched 19 August 1960.* (U.S.S.R. ACADEMY OF SCIENCES)

*Vanguard 1, the second United States satellite, was orbited on 17 March 1958.* (U.S. NAVY)

*In the National Air and Space Museum, a forlorn Uncle Sam holds the remains of the first Vanguard launch attempt. Although damaged, the device continued to transmit after the Vanguard launcher exploded.* (DAVE DOOLING)

launching huge 27,000-pound satellites on 16 July and 2 November. Proton 3 followed on 9 July 1966. Their announced purpose was to study high-intensity radiation particles, but their large size led to speculation that they were prototypes of manned space stations. This theory was supported by Proton 4, orbited in mid-November 1968. Weighing 37,500 pounds, it was put into an orbit different from those of earlier Protons and similar to that employed for manned satellites.

America's first three satellites, Explorer 1, Vanguard 1, and Explorer 2, were not in the same league with the Soviet Sputniks in size, but their miniaturized instruments gathered data of extreme value to scientists. Launched between January and March 1958, the two Explorers measured cosmic rays, micrometeorites, and temperatures, while the Vanguard beeped out signals from its transmitters, which were powered by both batteries and solar cells. Explorer 1, 80 inches long and 6 inches in diameter, was an integral part of the carrier's fourth-stage motor case. Its payload was developed by the University of Iowa under the direction of James A. Van Allen. The satellite, which weighed 31 pounds (including 18 pounds of instrumented payload), was responsible for the discovery of the radiation belts that bear Van Allen's name; the discovery was confirmed by Explorer 4. Vanguard 1 was a 6.4-inch sphere to which were attached six solar-energy converters and antennas. Explorer 1

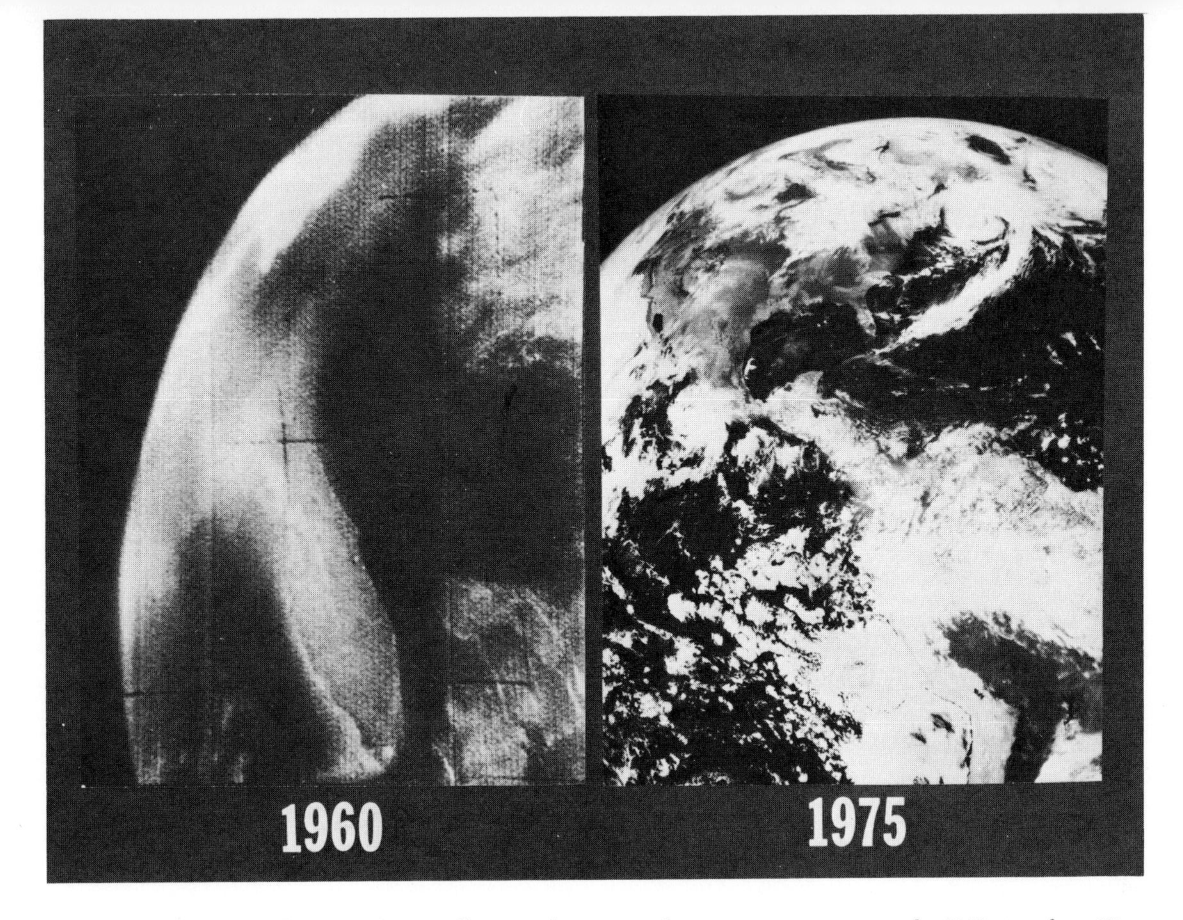


*The progress in images from weather satellites can be seen in these contrasting views—the 1960 view from Tiros 1 from low altitude and the 1975 SMS image that is clearer despite the greater altitude.* (NASA)

reentered the Earth's atmosphere on 31 March 1970, after 12 years in orbit. Vanguard 1 still circles the Earth.

More Explorers and Vanguards were orbited in 1958 and 1959. The first Discoverer went into orbit on 28 February 1959. The Discoverers, sponsored by the Air Force, were designed for space research, communications, and photographic missions. Many carried capsules that brought back to Earth film, biological specimens, and other valuable material.

One of the most successful satellite series started on 1 April 1960, when NASA orbited Tiros 1, a meteorological satellite that gave weathermen their first look at the Earth from above. Tiros 1 sent down a dazzling series of cloud pictures that literally added a new dimension to weather forecasting. Before it was through, Tiros 1 had broadcast more than 19,000 pictures to Earth.

As Tiros was progressing toward operational status, experimental equipment was being tested in a new satellite series, Nimbus. Carrying an automatic picture transmission (APT) system, an advanced vidicon camera system, and a radiometer, Nimbus 1 was launched on 28 August 1964. It provided day and night cloud pictures for about a month before the solar paddle drive failed.

Through the APT system that feeds into more than 400 ground stations, meteorologists all over the world were able to obtain Nimbus weather pictures. Nimbus satellites advanced our knowledge of the life history of storms and aided in the construction of maps of the jet stream. The pictures were also valuable to oceanographers, geologists, and geographers. The 10,000-foot-high Mount Siple, for example, was moved westward nearly 50 miles on maps of Antarctica as a result of Nimbus photographs. The boundaries of the Gulf Stream and other ocean currents were detected in Nimbus pictures, and the photographing of icebergs has led to the belief that the iceberg patrol could be handled exclusively by satellites in the future.

Eventually, a series of operational weather satellites was placed in polar orbit under the operational control of the National Oceanic and Atmospheric Administration. These satellites went by two designations, the old Tiros name and a NOAA series, and flew in sun-synchronous orbits that precessed just enough each day so they saw the ground at the same lighting angle on each orbit. The first Improved Tiros Operational Satellite (also called ITOS and NOAA-1) was launched in January 1970. This series was succeeded by the Advanced Tiros-N series (ATN or NOAA-B) starting in 1978. A military version, the Defense Meteorological Satellite Program, was launched starting in 1979 and (being built by the same firm) shared many designs.

While polar orbit provided detailed views of the

weather and even coarse-resolution images for crop studies, geostationary orbit was needed for the global view. During the 1970's an international network of weather satellites was launched by the United States (Geostationary Environmental Observation Satellite), Japan (Himawari), and Europe (Meteosat). To observe the weather from geostationary orbit, Synchronous Meteorological Satellites were introduced, followed by Geostationary Operational Environmental Satellites. When GOES-4's imaging system failed in November 1982, GOES-1's images were retransmitted by the former.

As spectacular as the weather satellites were those handling domestic and international communications. The first United States entry was Echo 1, orbited on 12 August 1960. Echo was a passive satellite—that is, it was only a balloon that reflected back to Earth the signals sent from the ground. The first active repeater communications satellite, with the ability to amplify the signals before returning them, was the Army's Courier, which went up on 4 October 1960.

The first direct television connection between continents was made possible by Telstar 1, a Bell Telephone active repeater satellite orbited by a NASA vehicle on 10 July 1962. RCA's Relay 1 went into orbit on 13 December. Relay did not work at first, but engineers on the ground were able to find the trouble and correct it. On 5 January 1963 the remarkable "remote repair job" was complete, and two test transmissions were made between receiving stations at Andover, Maine, and Goonhilly Downs, England; on 7 January, signals went from England to the United States. It began transmitting regular civilian television broadcasts between the United States and Europe on 9 January.

Telstar 2 was orbited on 7 May, and the first transatlantic color-television pictures were sent the next day. On 22 November, Relay 1 transmitted the first live television pictures across the Pacific. Japanese audiences saw NASA Administrator James E. Webb and the Japanese ambassador to Washington, Ryuji Takeuchi. They were to have seen a taped greeting from President Kennedy; instead, they were told of his assassination in Dallas a few hours earlier.

Telstar and Relay, which orbited only a few hundred miles above the Earth, had to be tracked as they appeared over the horizon, and they could transmit signals only when they were in sight of ground stations. As early as 1945, Arthur C. Clarke had suggested that if a communications satellite were placed at 22,300 miles, it would orbit the Earth at just the proper speed to appear to be hanging in space above one point on the planet's surface. The advantages of having such a synchronous orbit were plain. A synchronous satellite would always be in position to relay radio, television, and telephone signals, and no elaborate tracking devices would be needed.

*By the late 1970's, two classes of weather satellites were used. The GOES series—Geostationary Operational Environment Satellites—observed almost half the globe from 22,300 miles up. The oval port leads to the image-scanning system.* (HUGHES AIRCRAFT)

The first synchronous satellite, Syncom 1, was launched on 14 February 1963. It went into orbit, but its radio equipment failed to work. On 13 September, Syncom 2 and Relay 1 were used to link Rio de Janeiro, New Jersey, and Lagos, Nigeria, in a three-continent conversation. Because Syncom 2 was not orbiting quite in the plane of the equator, it appeared to describe a figure 8 as the Earth turned beneath it. Syncom 3 was placed into a true equatorial orbit, with no north-south swing, on 19 August 1964. Drifting over the Pacific Ocean near the International Dateline on 10 October, it telecast the opening-day ceremonies of the Olympic games in Tokyo.

By this time it was apparent that communications satellites were rapidly moving out of the experimental area into the realm of paying propositions. The Communications Satellite Corporation, Comsat for short, was set up to develop communications satellite systems, with half the stock going to the public and the other half to large communications companies. Comsat's first venture was Early Bird, a synchronous satellite providing a 240-circuit two-way voice channel link between Europe and North America; it was orbited on 6 April 1965. Comsat became the United States member, as well as general manager, of the International Telecommunications Satellite Consortium (Intelsat for short).

NASA began orbiting a series of applications technology satellites (ATS) in 1966. Cylindrical craft some 4 feet and 8 inches in diameter, they varied in length and weight (650 to 790 pounds) and incorporated thousands of solar cells that enabled them to operate for years. ATS-2 and ATS-4 never entered the desired orbits, but ATS-1, launched in December 1966, and ATS-3, launched in November 1967, were highly successful.

Since 1960, the military services orbited a variety of satellites for many different purposes. On 13 April of that year, the Navy put its first Transit navigation satellite into orbit. The Air Force inaugurated its Midas program to detect ICBM launchings (by the infrared radiation that they gave off) on 24 May, after a previous launch effort failed. And on 31 January 1961, exactly 3 years after Explorer 1 was launched, the Air Force orbited its first Samos reconnaissance satellite. With its Discoverer series, the Air Force conducted a variety of studies useful to its missions. For security reasons, little was made public about these and other military satellite programs. In 1965, the Department of Defense orbited a total of seventy artificial satellites, using forty-one expendable launch vehicles; this was nearly three times the number of satellites orbited by NASA during the same year.

*The Discoverer series, orbited by the Air Force, was designed for space research, communications, and photographic missions. Here (left to right), Col. Charles Mathieson, Gen. Thomas D. White, Maj. Gen. Osmond J. Ritland, and Maj. Gen. Bernard A. Schriever examine the Discoverer 13 capsule after its return to Earth.* (U.S. AIR FORCE)

The final spacecraft in NASA's Applications Technology Satellite series, No. 6, was placed into a stationary orbit by a Titan 3C carrier rocket on the morning of 30 May 1974. Often called an educational satellite, ATS-6's missions included broadcasting educational programs from orbit directly to receivers, first in such rural areas as Appalachia, the Rocky Mountains, and Alaska in the United States and later to some 5,000 isolated villages in India. Also carried was a health telecommunications experiment ("Telemedicine") that provided visual and voice contact between doctors and health aides, the latter working in remote villages.

Communications satellites soon became the largest element of space exploitation. In less than 20 years, the capabilities of the Intelsat satellites expanded from 240 telephone channels on Intelsat 1 (Early Bird) in 1965 to 15,000 channels on the Intelsat 5A series inaugurated in 1984. The price per channel also dropped sharply, bringing satellite communications to ever-wider populations. So common did satellites become for overseas relay of television that in the 1970's the "live-via-satellite" screen notice began to disappear. What had been unbelievable with Telstar now became as common as the long-distance phonecall. The inventor finally got to make use of his brainchild in 1983–1984, when Arthur C. Clarke used a small relay antenna at his home in Sri Lanka to work, via Intelsat, with a scriptwriter in Burbank, California, on the screen play for the movie *2010: Odyssey Two.*

Domestic communications satellites came onto the scene in the 1970's, when Canada made large commitments to the Anik (Eskimo for "little brother") series of communications satellites to link that nation's far-flung cities and towns. Indonesia and other nations with scattered populations soon followed that lead. In the United States, Western Union started domestic communications with Westar 1, relaying telex, voice, TV, and data. In the late 1970's, cable TV operators discovered the value of using satellites to relay signals to other cities; and, in time, Californians were watching Atlanta's Channel 17 and southerners were watching New York's WOR. A more interesting part of the phenomena was the discovery—with appropriate advances in electronics—that receivers could be built in backyards. At first this was done by the rich; but by 1984, some units could be bought for less than $1,000, so middle-class homes, apartment buildings, hospitals, and motels were soon sprouting satellite dishes that look like inverted mushrooms.

The value of satellite communications was not lost on the military. The Defense Satellite Communications System (DSCS) grew from the three interim satellites launched in 1967 to the advanced DSCS-3 series started in October 1983. The 2,500-pound satellite carried multibeam antennas steered electronically that have special features to defeat jamming by enemy forces.

The military also used polar orbits for a special class of communications satellites, the Space Data System, to relay commands to missile submarines below the Arctic ice pack. Early warning of enemy missile launches is provided by two geostationary satellites with the innocuous title of Defense Support Program. They have telescopes with 6-foot-wide primary mirrors focusing onto infrared detectors "tuned" to the heat of a rocket's exhaust. The satellite rotates, so that the whole Asian continent is scanned every few seconds. Although the Air Force has said little about the satellite, censored testimony before Congress has called it the key to the survival of the B-52 bomber fleet. Without its warnings, the bombers would not be able to become airborne shortly after a missile launch.

Satellites were also used in more human applications with the U.S.S.R's Kospas 1 (June 1982) and the United States NOAA-8 (March 1983). The former was dedicated to search-and-rescue (SAR) work, and the latter was a weather satellite carrying SAR equipment piggyback. Each worked the same: they listened for signals from emergency locator transmitters, carried on aircraft and meant to be activated in case of a crash, and relayed location information to SAR control centers. Within a year, this new tool was involved in saving over 200 lives. On occasion, the system was endangered by false alarms caused by haphazard care of the locators by pilots and others.

*The Kospas satellite launched by the Soviet Union has been responsible for saving the lives of several downed pilots and stranded seamen.* (SUE BUTLER-HANNIFIN)

The Defense Department has sought to ensure that the United States will be able to detect the explosion of any nuclear weapons in space, in violation of the 1963 test ban treaty. It did this with a series of Vela satellites. They were succeeded by detectors on Naustas satellites.

The Defense Department has continued to launch other scientific and engineering research satellites at frequent intervals. These include the Air Force's orbiting vehicle satellites—which typically carry cosmic ray and radiation belt detectors, satellite-to-satellite communications gear, and ionospheric research devices—as well as the environmental research satellites. The payloads of many of these, and other, satellites are often classified.

Although most satellites orbited during the past 25 years have had specific applications, the United States has by no means neglected pure science. Explorer has continued to be the mainstay of the scientific satellite field. With it, scientists have learned much about the composition of the ionosphere (Explorer 11), the micrometeoroid environment (Explorers 16 and 23), and the nature of the magnetosphere and interplanetary space (Explorer 18—which, incidentally, confirmed the existence of a "shock wave" around the Earth).

Larger and more complex experiments were mounted on orbiting geophysical observatory (OGO) satellites. OGO's were orbited annually in the 1966–1969 period. OGO-6, the last in the series, was a 1,393-pound craft with twenty-five experiments aboard. It was launched on 5 June 1969 into an orbit of from 248 to 683 miles by a Thorad Agena D carrier. At the time, four other OGO's were still in operation; they had provided more than 1,200,000 hours of scientific data involving a total of 130 experiments.

While unmanned observatories unlocked the secrets of the stars and planets, home planet Earth was not neglected. Following the success of ERTS-1 (renamed Landsat), two more identical models were launched on 22 January 1975 and 5 March 1978. Together, the three generated more than a million images of the Earth through their TV cameras and multispectral scanners (MSS). A follow-on series called Landsat D was developed incorporating MSS like the first three and a new thematic mapper. The latter is cooled so that it can detect infrared reflections deeper in the spectrum than the MSS can see. Landsat 4 was launched 16 July 1982 but soon ran into trouble when its X-band transmitter failed and the electric cable for its solar panels started decaying. (They were rated for operation in geostationary orbit, where sunrise and sunset come only a few times a year.)

The oceans were surveyed with a Geodetic Experimental Ocean Satellite (GEOS-3) launched on 9 April 1975. It used a radar altimeter and doppler-shift instruments to measure the mean heights of the oceans relative to the Earth's reference diameter. More detailed surveys were made with Seasat, launched on 27 June 1978. In addition to radar altimeters, it carried a synthetic aperture radar capable of combining echoes from a series of transmissions to give the effect of a much larger antenna. Seasat failed after 100 days (a slip ring in the rotating solar array joint broke), but not before returning striking images of the ocean surface and revealing patterns in ocean currents.

The Stratospheric Aerosol Gas Experiment (launched 18 February 1979) stared through the atmosphere at sunrise and sunset to gather data on gases, droplets, and particles in the upper atmosphere; the Heat Capacity Mapping Mission (launched 26 April 1978) used an infrared detector to map the heat patterns on the Earth as the seasons changed; and Magsat (launched 30 October 1979) mapped the variations in the Earth's magnetic field.

Work started by the OGO series was continued by two special satellite campaigns, the International Sun-Earth Explorers (ISEE) and the Dynamics Explorers (DE). ISEE-1 and -2 were launched together on 22 October 1978 into parallel orbits so that they could measure activities in the magnetosphere at different altitudes at the same time. On 12 August 1978, ISEE-3 was placed in a so-called halo orbit about a million miles from the Earth. This position enabled it to measure the solar wind before it hit the bow shock of the magnetosphere. Between 10 June 1982 and 23 December 1983, a series of maneuvers took ISEE-3 out of halo orbit and on a complex path behind the Earth and past the Moon five times. Upon emerging, it was on a trajectory bound for a 1985 encounter with the comet Giaccobini-Zinner. ISEE-3 was redesignated International Comet Explorer. Data on the solar wind have been supplied by Germany's Helios 1 (launched 10 December 1974), placed in an orbit that comes as close as 30 million miles to the Sun. Its companion, Helios 2 (launched 15 January 1976), passed 2 million miles closer but has since gone silent.

*A satellite for detecting micrometeoroids is prepared for an environmental test at Langley Research Center, Hampton, Virginia. During this test, the satellite is subjected to the widely varying temperatures it will encounter on an actual orbital flight.* (NASA—LANGLEY RESEARCH CENTER)

*The one millionth image of the Earth taken by the Landsat spacecraft was, ironically, of the Soviet Union—a farm area near Volograd.* (USGS)

The DE-1 and -2 pair (launched 3 August 1981) was similar in concept to ISEE-1 and -2. In addition to its plasma instruments, DE-1 carried an ultraviolet imager that produced the first pictures of the auroral oval seen from above the North Pole. One striking event was called the Theta Aurora because it took the shape of the Greek letter theta. The U.S. Air Force's HILAT satellite (launched 27 June 1984) also carried auroral imaging instruments that produced the first pictures of the aurora from space during the day. And the Air Force's Spacecraft Charging at High Altitudes (SCATHA) (launched 30 January 1979) monitored plasma activities in geostationary orbit.

Following the loss during the launch phase of the first two Orbiting Solar Observatories, the program recovered beginning with the highly successful orbiting of OSO-3 in 1965. Two final satellites were developed in the series: OSO-8 (launched 21 June 1975), fitted with a more powerful array of instruments; and Solwind (launched 24 February 1979), which was built

*New views of the Sun were provided by a number of spacecraft, including the Solar Maximum Mission satellite, which produced this view of the corona as a "spike" fired up from the surface.* (NASA)

for the U.S. Air Force Space Test Program using backup hardware for OSO-8 after its demise on 26 September 1978. The spacecraft continued to monitor solar phenomena; among other events, Solwind detected the apparent collision of at least two comets with the Sun.

The most advanced of the unmanned solar satellites was the Solar Maximum Mission, which went into orbit on 14 February 1980. Solar Max, as it quickly became known, incorporated seven instruments designed to study the Sun during the most violent portion of its 11-year cycle. In a sense, it was an unmanned descendant of the large array of solar telescopes carried in Skylab's Apollo telescope mount. Solar Max discoveries included vital clues to the origins of solar flares (X-ray bright spots that appear simultaneously at both feet of a coronal loop), an apparent downward drift in the solar constant, neutrons with an energy of 1 billion electron volts coming from solar flares, and gamma rays being emitted to the sides (and not vertically) by flares (indicating downward flows of energetic material). Solar Max also made a vital contribution to manned space flight by serving as the first repairable satellite. This is described in the final chapter.

A large complex unmanned satellite launched by the United States was OAO-2, part of the Orbiting Astronomical Observatory program. (OAO-1 was a victim of battery failure on its second day in orbit in April 1966). OAO-2 was launched on 7 December 1968 into a nearly circular orbit about 480 miles high. Weighing 4,446 pounds and containing eleven telescopes, it represented a major breakthrough for astronomers, who for the first time were able to make long-term observations unhampered by the disturbing effects of the Earth's atmosphere. OAO-2 was designed to study young, hot stars that emit most of their energy in the ultraviolet regions of the spectrum; to photograph some 700 stars each day as part of a stellar survey of the universe; and to observe gases in interplanetary space. During its first month of operation, it gathered a total of 65 hours of astronomical data, including twenty times more information about ultraviolet characteristics of stars than had been obtained during 15 years with sounding rockets.

Within the Orbiting Astronomical Observatory program, OAO-2 had, by 1970, discovered that an extensive envelope of atomic hydrogen surrounds comets Tago-Sato-Kosaka and Bennett, that they contain hydroxyl molecules, and that they radiate intensely in the ultraviolet. In another typical experiment, in March of the same year, the observatory was programmed to monitor the ultraviolet emission from Nova Serpentis, the first nova to be studied from a spacecraft. OAO-B, the next in the series, failed to orbit on 30 November 1970 when the protective nose covering of the observatory failed to separate at the proper time. Keen disappointment was mollified by the completely successful launch, on 21 August 1972, of OAO-3 (named Copernicus in honor of the great Polish astronomer). Among many other results, the observatory showed not only that hydrogen is the principal element in the vast clouds between the stars but that from 10 to 60 percent of it is found in the molecular form—much higher than predicted.

In the Small Astronomical Satellite category two successful orbitings occurred, SAS-A/Explorer 42 on 12 December 1970 and SAS-B/Explorer 48 on 16 November 1972. The mission of the former was to look for bright galactic and extragalactic X-ray sources as an essential step toward preparing an X-ray star catalogue. It was sent aloft from the Italian San Marco site off the coast of Kenya—marking the first time an American satellite had been launched outside the United States. Also called *Uhuru*, Swahili for freedom, Explorer 42 identified more than 200 X-ray sources by the end of 1972. Explorer 48, meanwhile, was instrumented to study galactic and extragalactic gamma radiation in the 30 to 300 million electron volt range.

The International Ultraviolet Explorer (IUE), launched on 26 January 1978, carried a 17.7-inch Cassegrain telescope and a spectrograph package. It successfully expanded on the findings made by the OAO series and was the first satellite to be operated by scientists in the manner of terrestrial observatories. Among IUE's results were the first high-resolution spectrum of a star in another galaxy; first ultraviolet observation of a supernova; discovery of apparent auroral activities in the polar regions of Uranus (as betrayed by strong ultraviolet emissions presumably caused by electrons striking the hydrogen upper atmosphere); discovery of a 180,000°F halo of carbon, sulphur, iron, and silicon around our galaxy; and proof that the star Capella has a chromosphere like the Sun's.

In recognition of the fact that the SAS series probably had seen only a small fraction of the potential X-ray universe, a larger series of High Energy Astronomy Observatories (HEAO) was started in the early 1970's. Initially, four 10-ton HEAO's to be launched by Titan 3C carriers were to have incorporated the largest X-ray, gamma-ray, and cosmic-ray instruments ever orbited. However, budget problems forced NASA to reduce the scope of the program and the size of the satellites. What evolved after an intense redesign effort was a set of three 4-ton satellites to be launched by Atlas-Centaurs. Despite this scaling down, HEAO's instrumentation still fit the category of the largest yet flown.

HEAO-1, orbited 12 August 1977, had an array of five instruments sensitive to soft- to medium-energy X-rays. The satellite was designed to rotate end over end so that the instruments could scan a thin sliver of the sky. As the Earth moved around the Sun, the sliver precessed around the celestial sphere so that the whole sky was covered after 6 months (the nominal mission). HEAO-1 trebled the number of known X-ray sources and showed what appeared to be a diffuse X-ray glow that permeated the sky. After 17 months, the satellite ran out of attitude control gas and thus could no longer serve a useful purpose. It re-entered the atmosphere in 1979.

HEAO-2, informally dubbed the Einstein Observatory, was orbited on 13 November 1978. It housed a telescope that focused X-rays to produce images of extended sources rather than scanning the sky as its predecessor had. It discovered that virtually every object in the sky seems to emit X-rays, including quasars and normal galaxies. It also showed that the glow seen by HEAO-1 was caused by a large number of small sources that could not be resolved without the high resolution available on Einstein.

The last in the series, HEAO-3, was successfully sent into orbit on 20 September 1978. Two instruments were carried to detect cosmic rays (the shards of nuclear reactions) and gamma rays (the "light" from those reactions). Like HEAO-1, the newer satellite rotated end over end to scan the sky. Because cosmic rays have electric charges, the two instruments that measured their presence could not tell where they started (their paths are altered by the solar and terrestrial magnetic fields) but revealed much about the relative abundances of elements in the universe, including a surprisingly low abundance toward the heaviest elements. The gamma-ray instrument revealed the location of the hottest objects in the universe.

The United States was not alone in exploring the vast regions of galactic space. The European Space

*The second High Energy Astronomy Observatory (HEAO-2), also known as the Einstein Observatory, carried the largest X-ray telescope yet orbited in order to study quasars, galaxies, and other objects at their most active.* (NASA)

*EXOSAT is the most advanced high-energy astronomy satellite to be launched by Europe. It carried two telescopes and other sensors.* (MBB)

Agency's Celestial Observation Satellite (COS-B), launched 9 September 1975, carried out the first all-sky gamma-ray survey of the heavens. It operated until 25 May 1982, when the satellite had to be shut down so that its frequencies could be allocated to another satellite. EXOSAT, an ESA satellite launched on 26 May 1983 aboard a United States Delta rocket, carried two X-ray telescopes similar to but smaller than the one aboard Einstein. Its mission was similar to the HEAO series in that it was designed to look for new sources and to probe those that are known.

Japan entered the field with the Tenma X-ray satellite on 20 February 1983, and Britain, with the Ariel 6 satellite on 2 June 1979. The only satellite that the Soviets identified as dedicated to astronomy was Astron, launched on 23 March 1983 and said to be comparable to the United States Copernicus.

The other end of the spectrum was pried open by the Infrared Astronomy Satellite (IRAS), a United States–British–Dutch collaboration. Launched on 25 January 1983, the craft's telescope was modest by most standards, but its design—a beryllium mirror cooled by liquid helium—allowed its detectors to sense light from the coldest objects in the solar system. Like the HEAO series, IRAS was expected to see a modest number of objects beyond those already known (largely from balloon and sounding rocket missions). In its 10 months of operation, though, the craft unveiled a universe far richer than had been suspected. An initial major discovery was Comet IRAS-Araki-Alock, the first for a satellite with non-imaging instruments. Then came the finding that the star Vega is surrounded by dark, cool material—apparently a planetary system taking shape. Since then, more comets and candidate planetary systems have been discovered. Other IRAS findings included a thin dust shell orbiting the Sun in the region of the asteroid belts, three gas shells expanding outward from the red giant star Betelgeuse, large clumps of carbon dust dubbed infrared cirrus clouds, stellar nurseries, and the first view of the galactic core.

*IRAS, the Infrared Astronomy Satellite, revealed the existence of far more objects and activities than astronomers had suspected.* (NASA)

Among the satellites launched in 1966 was the first Biosatellite, or Bios 1, which was designed to study the effects of zero gravity and radiation on fruit flies, wheat seedlings, frog eggs, and other specimens. The mission failed since the re-entry capsule was lost, but Bios 2, launched in early 1967, was successful. Bios 3, launched in June 1969, carried a 14-pound pig-tailed monkey in what was planned as a 30-day study of the effects of weightlessness on the animal's emotional, physiological, and mental processes. The flight was terminated after only 9 days, however, because the monkey was reacting sluggishly and failing to respond to signals from the ground. The animal died shortly after its capsule splashed into the Pacific on 7 July.

The first international satellite launching occurred on 26 April 1962, when an American Thor Delta orbited Ariel 1, a 132-pound British-built satellite carrying instruments to measure the ionosphere and its interactions with solar radiation. On 28 September a Thor Agena B orbited the Canadian Alouette 1 satellite, which measured electron-density distribution and variation in the ionosphere, and cosmic radiation. On 15 December 1964, Italy got into space with its San Marco 1, orbited by a United States Scout ELU from Wallops Island, Virginia. Its purpose was to measure atmospheric density.

Australia became a space nation in November 1967 by orbiting its 100-pound Wresat 1 satellite. The satellite was launched by adding two solid-propellant stages to a Redstone supplied by the United States.

During 1968, the ten-member European Space Research Organization (ESRO) began its space program by launching satellites with NASA-provided carriers. Two types bearing the ESRO name were planned, the first to study ionospheric and auroral phenomena in the polar regions, and the second to investigate solar and cosmic radiations.

Outside the United States, France, which had started orbiting its own satellites in 1965, kept up a launch rate of one or two a year. Both Japan and China joined in during 1970. Using a Lambda 4S carrier, the former nation put up its initial test satellite Osumi on 11 February to be followed the next year by Tansei and Shinsei. The first People's Republic of China's satellite was launched on 24 April 1970 from the Shuang Cheng-Tzu range near the border of Outer Mongolia. It carried a transmitter broadcasting "The East Is Red," a tune honoring chairman Mao Tse-tung. A second Chinese satellite was orbited on 3 March 1971. The

British started their own launching program when, on 28 October 1971, a Black Arrow carried Prospero into orbit from Woomera in Australia. The 190-pound satellite was instrumented to detect micrometeoroid particles in near-Earth space.

The Soviet Union continued during the 1970's to launch more spacecraft than any other nation, including the United States. It had taken the lead from the United States in this respect in 1968. For example, in 1966, the United States launched ninety-eight successful space payloads (twenty-eight for NASA and seventy for the Department of Defense) compared with forty-two for the Soviet Union. Two years later, the Soviets took the lead—seventy-four to sixty-two, and by 1972, the Russian margin was eighty-nine to thirty-six.

In October and December 1969 the Soviets orbited their first two Interkosmos satellites that were equipped with scientific instruments supplied not only by themselves but by Czechoslovakia and East Germany. Another cooperative science program began 2 days after Christmas 1971 when the Russian-French Oreol 1 went into orbit. Still another series, Prognoz, got under way in April 1972 as the first spacecraft was placed in a highly elliptical orbit reaching more than 124,000 miles from Earth. Similar to the American Interplanetary Monitoring Platform series of Explorers designed to observe the radiation environment caused by the Sun, Prognoz satellites carried fifteen instruments to measure the solar wind and the overall space environment.

While satellites were exploring space around the Earth, man began reaching out to the Moon and the planets of the solar system, in an effort to fulfill the oldest dream of astronautics—interplanetary travel.

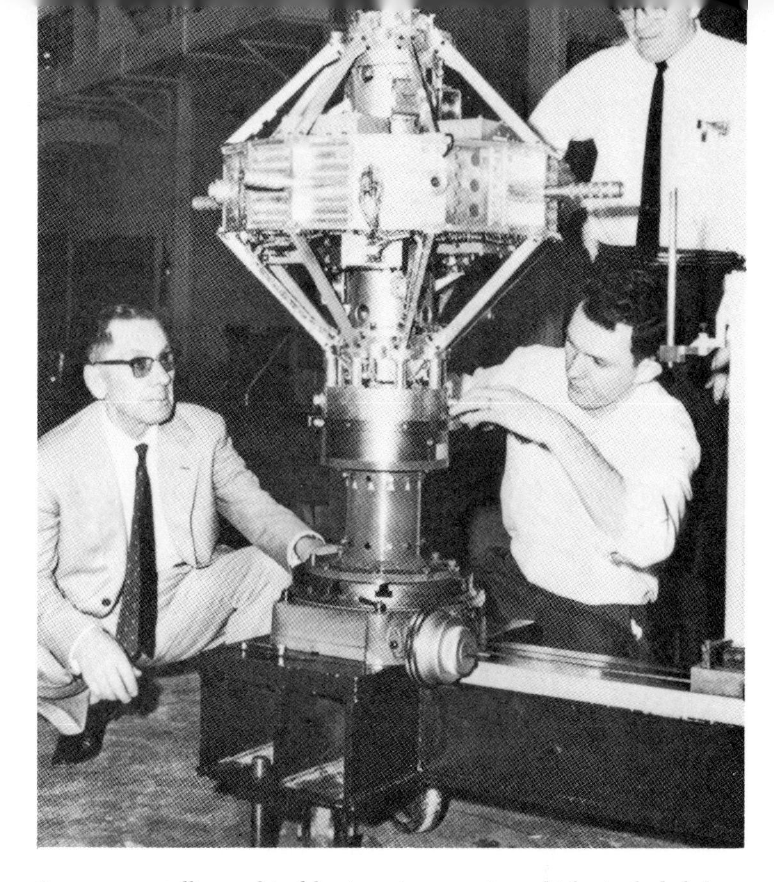

*European satellites orbited by American carrier vehicles included the Italian San Marco 1, orbited 15 December 1964.* (NASA)

Because the Moon is closest to Earth, it was the first target, and probes have been sent out to fly close by the Moon, orbit it, or even land on it. The Soviets started lunar exploration, firing their first three Luna spacecraft (inevitably called Luniks in Western nations) in 1959. Luna 1, also called Mechta, was launched on 2 January 1959. It flew within 4,660 miles of the Moon's surface and on into space, broadcasting data about the space environment from 373,125 miles from

*Prognoz 2, launched by the Soviets on 29 June 1972, was instrumented to study solar radiation. Loop is rotatable electromagnetic receiving antenna.* (ARFOR)

*The first Moon probe, Luna 1, was launched by the Soviet Union on 2 January 1959. It flew within 4,660 miles of the Moon's surface, broadcasting data back to Earth.* (U.S.S.R. ACADEMY OF SCIENCES)

the Earth. The United States did essentially the same thing 2 months and 1 day later. Pioneer 4 (numbers 1 through 3 had been failures in 1958) flew within 37,300 miles of the Moon and was tracked to a distance of more than 400,000 miles.

On 12 September 1959, the Russians launched Luna 2, a 3,000-pound probe that carried 858 pounds of instruments and transmitting equipment. Luna 2 became the first man-made object to hit the Moon, crashing between the craters Archimedes and Autolycus in the Mare Imbrium on 14 September. Luna 2 sent back signals before impacting the Moon's surface, which showed no significant magnetic field.

Luna 3, also called an Automatic Interplanetary Station, was an equally significant landmark. Launched on 4 October, it circled behind the Moon, approaching within 4,372 miles of the lunar surface, and sent back the pictures that gave man his first view of the far side. The spacecraft weighed 3,300 pounds, of which 614 pounds was the instrumented Automatic Interplanetary Station that gave Luna 3 its name.

After orbiting Rangers 1 and 2 to check out instrumentation, the United States sent Ranger 3 toward the Moon on 26 January 1962. The probe missed its target completely and sailed on into a useless solar orbit. Ranger 4 did hit the Moon on 23 April, but a failure within the spacecraft prevented it from sending back any useful information. Ranger 5, fired on 18 October, not only missed the Moon by 450 miles but also lost power when its solar cells malfunctioned. Ranger 5 was tracked for 8 hours and 44 minutes before its reserve battery went dead. After these three disappointments, the Ranger program was reorganized, and further launches were put off until 1964.

On 2 April 1963 came the next Soviet attempt at lunar exploration, an attempt that apparently was not a success. The Russian Luna 4, a 3,135-pound spacecraft, flew within 5,300 miles of the Moon.

There was speculation in the West that the Soviets had attempted a soft landing on the Moon—that is, a landing at less than 20 miles per hour. Whereas the first Rangers had carried seismometers in spherical containers designed to withstand the impact of semihard landings, NASA's growing realization of the landing difficulties resulted in a basic change in the program. All the Rangers after Number 5 were designed only to take pictures of the lunar surface and they were inevitably destroyed when they hit the Moon. Ranger 6's cameras were switched on during the last 10 minutes of the 66-hour flight, but they did not send any pictures. Later investigation indicated that the spacecraft's electric system had been burned out when the cameras were accidentally turned on earlier in the flight.

After being redesigned to eliminate this danger, Ranger 7 was launched on 28 July 1964. It was a complete success. The 4,316 pictures it sent back included some that showed craters only a few feet across; the best that Earth-based telescopes had been able to see were lunar features a half mile across. Ranger 8, flown on 17 February 1965, sent back 7,137 pictures. Ranger 9, launched on 21 March 1965, gave millions of Americans the thrill of seeing television live from the Moon, inasmuch as some of its 6,007 pictures were broadcast direct onto commercial television.

Soviet launchings came with equal frequency, but with less success than the Ranger, having picked a more ambitious goal.

The Soviets were finally successful with Luna 9, which eased onto the Moon's surface on 3 February 1966. In the next few days Luna 9 sent back a series of true close-up pictures—some showing objects only a few inches in size—that demonstrated that the Moon's pitted, barren surface was hard enough to support the weight of a manned spacecraft.

When the Russians achieved this triumph, the American program for a soft lunar landing was bogged down in difficulties. The Surveyor program, formally approved by NASA in the spring of 1960 and placed under the management of the Jet Propulsion Laboratory, started with the aim of landing a complex craft, weighing 2,500 pounds.

In addition to television cameras, the seven Surveyors were equipped with strain gauges for making soil mechanics measurements, samplers to scoop out surface materials, and instruments for analyzing subsurface materials by measuring the back-scatter from

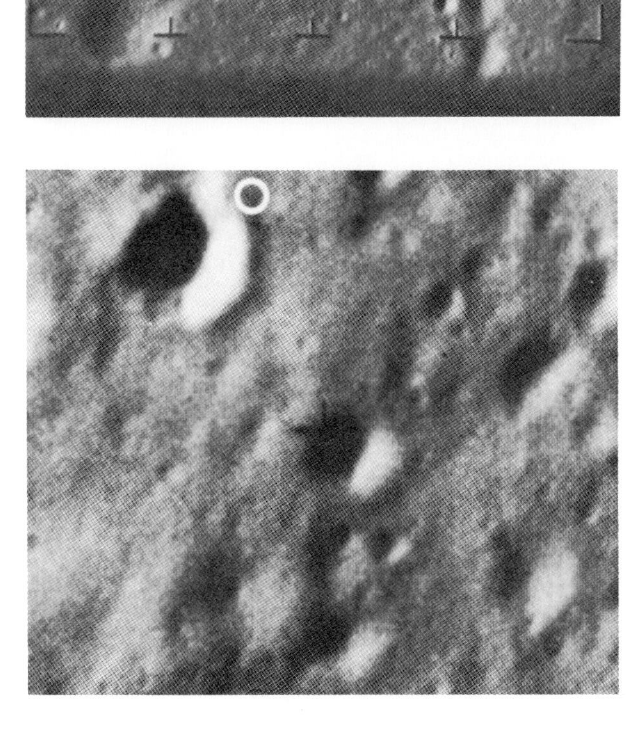

*Series of photographs of the Moon taken by Ranger 9 prior to impact in the crater Alphonsus at 6:08:20* A.M. *(PST) on 24 March 1965. The circle indicates the point of impact. Top left: 140 miles high, 1 minute 35 seconds before impact; area covered is 67.5 miles across and 62 miles from top to bottom. Top right: 50.3 miles high, 33.7 seconds before impact; area is 24 miles by 22 miles. Bottom left: 12.2 miles high, 8.09 seconds before impact; area is 5.8 miles by 5.3 miles. Bottom right: 8.3 miles high, 5.5 seconds before impact; area is 1.6 miles by 1.4 miles.* (JET PROPULSION LABORATORY, CALIFORNIA INSTITUTE OF TECHNOLOGY)

*Left, last picture taken by Ranger 9 camera at an altitude of .68 mile, .453 second before impact. Area covered in picture is 154 feet across by 125 feet high. The impact point (indicated by the circle) is on the edge of a 25-foot crater.* (JET PROPULSION LABORATORY, CALIFORNIA INSTITUTE OF TECHNOLOGY)

*On 3 February 1966 Soviet spaceship Luna 9 made the first soft landing on the Moon. A selection of remarkable close-up photos, taken by Luna 9 and transmitted back to Earth, are shown here. Notice parts of the spacecraft in the bottom frame.* (NOVOSTI PRESS AGENCY, LONDON)

radioactive sources. Some of the craft were also equipped to detect the presence of magnetic elements in the lunar soil.

The United States was also working on a probe to orbit the Moon. Lunar Orbiter, an 850-pound craft, had a dual lens camera that could detect objects 1 and 9 yards across, respectively, from an altitude of 30 miles.

Before the United States could launch either of its two lunar craft, Russia achieved another first by placing Luna 10 into orbit around the Moon on 3 April 1966. Weighing 3,500 pounds, the spacecraft was instrumented to study micrometeoritic particles, the Moon's magnetic field, solar plasma streams, and surface gamma radiations.

The pioneering voyages of Lunas 9 and 10 led off what proved to be a remarkable series of unmanned lunar missions by the United States and Russia. During the period from May 1966 to November 1968, America launched seven Surveyors, five Orbiters, and one far-ranging Explorer. Russia, meanwhile, launched four more Lunas (three orbiters and a soft-lander) and three Zonds.

The Surveyor, launched 30 May 1966, was a resounding success, landing gently on the Moon's surface on 2 June. It took more than 10,000 pictures before the onset of lunar night on 14 June and then returned to life on the following lunar day, which began on 29 June. By the time the fifth successful craft quit, in 1968, more than 87,000 photos had been transmitted to Earth.

The first Lunar Orbiter was equally successful, entering into an orbit on 14 August that ranged from about 1,150 miles to 120 miles above the Moon's surface. Picture transmission began on 18 August. Later, the craft approached to within 36 miles of the surface for close-ups. The last photo was made on 29 August; 2 months later the craft was deliberately crashed onto the far side of the Moon in order to prevent the possibility of its radio disturbing communications with Lunar Orbiter 2.

An unexpected by-product of the Orbiter program was the discovery that the Moon's gravitational field was not uniform, owing to the existence of "mascons"—mass concentrations of dense material beneath the surface. The mascons, which cause orbiting spacecraft to speed up slightly when they pass over them, are associated with maria, and they may be composed of sediments deposited in actual seas long ago when the Moon had an atmosphere.

After conducting three Luna missions during the latter half of 1966, Russia did not aim for the Moon again until 1968, when the Zond program, apparently dormant since 1965, suddenly came back to life with the launching of Numbers 4, 5, and 6. Luna 14, an orbiter, was also conducted in this period, its mission falling between those of Zonds 4 and 5. The new Zonds, unlike the first three in the series, were sent along trajectories that took them around the Moon and back to the Earth. The flights were believed at the time to have been unmanned tests of modules to be used for manned lunar flights. Zond 4 was not recovered, but the next two were. Zond 5 was picked up in the Indian Ocean; and Zond 6, after skipping off the atmosphere above the Indian Ocean, re-entered at a lower speed and descended by parachute onto Russian territory. Zond 6's method of re-entry, which reduces deceleration stress and heat loads considerably, was seen as an important step forward in a possible Soviet-manned lunar landing program.

Russia maintained an aggressive unmanned lunar effort during the time that the Moon was being visited by the scientifically more productive, more costly, and more dramatic American Apollo manned missions. For example, the circumlunar Zond 7 probe was launched only 2 weeks after Apollo 11's astronauts splashed down in the Pacific Ocean. In September 1970, Luna 16 descended from lunar orbit onto the surface and recovered 101 grams of soil samples that were subsequently brought back to Earth in the ascent stage. The spacecraft featured a hollow bit drill located in a cylinder at the extremity of an articulated boom.

*On 17 November 1970 Lunokhod 1 landed on the Moon and transmitted photographs back to Earth for a ten-month period before ceasing to operate. Pictures were taken by TV cameras located in the two portholes at the front of the vehicle. Instruments located in the hermetically sealed basin eventually froze and ended the usefulness of Lunokhod 1.* (TASS FROM SOVFOTO)

*America's Surveyor soft-landing craft showed that the Moon's surface is strewn with rocks and pocked with tiny craters. Here is a photograph of a rock 6 inches high by 12 inches long, transmitted from Surveyor 1 on 2 June 1966.*

Eight months before the United States inaugurated its lunar roving vehicle series during the course of the Apollo 15 mission, Luna 17 set down on the Moon's surface (on 17 November 1970) an unmanned rover known as Lunokhod 1. Lunas 18 and 19 followed in September 1971 and Luna 20 in February 1972.

Precisely 2.036 grams of the Luna 20's samples were turned over to NASA and thence to cooperating American scientists as part of a United States–U.S.S.R. scientific exchange program. In their preface to a special 400-page issue of the scientific journal *Geochemica et Cosmochemica Acta,* published in 1973, editors Edward Anders and Arden L. Albee of the University of Chicago and the California Institute of Technology observed that Luna 20 proved how "highly significant scientific information and conclusions can be obtained from the study of even a very small sample returned by an unmanned spacecraft. The content of this issue underscores the scientific value of such unmanned probes as a future tool for planetary and lunar exploration."

Remote chemical analyses, by the Lunokhods, were accomplished by irradiating surface materials with X-rays and then spectroscopically examining the resulting ionized radiation. Among the elements identified during a series of twenty-five tests were calcium, silicon, titanium, aluminum, iron, potassium, and magnesium. Studies also were made of solar cosmic-ray events, interstellar and galactic X-ray emissions, and natural lunar radioactivity.

The wheel base of the rover measured a little more than 7 feet, about the same as the diameter of the magnesium alloy instrument compartment. Power was supplied by batteries recharged regularly by solar cell arrays. Lunokhod 1 weighed 1,667 pounds and moved both forward and backward on four pairs of 20-inch diameter mesh wheels driven by individual electric motors coupled to a two-speed transmission. Each of the eight wheels had its own floating bearing suspension system. Distance traversed across the lunar surface was measured by an odometer.

*Luna 20 landed on the Moon near Crater Apollonius in February 1972. Drill arm (foreground) gathered samples from subsurface and returned to Earth on 25 February of the same year.* (ARFOR)

During the course of the mission, the rover traveled through craters up to 10 feet deep. Whenever it approached a slope beyond its ability to climb or descend, sensors automatically applied the brakes even if it meant occasionally overriding commands coming from the crew at mission control. The crew consisted of five men: commander, driver, navigator, engineer and radio operator.

Just before the beginning of each 2-week-long lunar night, the rover was shut down and placed in a "hibernating" state. The thermal insulating cover with solar arrays on its underside was lowered and an inert-gas heating unit, warmed by decaying isotopes, was activated to maintain internal temperatures at between 60° and 65°F. Pressure inside the instrument compartment was kept at just under 1 atmosphere.

At lunar dawn, when the Sun rose to 4 degrees above the horizon and wheel temperatures reached 0°F, the protective cover opened to permit sunlight to shine on the solar cells. Batteries were then recharged and the rover was made ready for another day of work.

In October 1971, after 11 lunar nights—the last was from 15 to 30 September—Lunokhod 1 failed to awaken. Its isotope heating fuel had given out, and the instruments and other equipment had soon frozen. All was not lost, however, because the laser reflector continued to receive signals arriving from the laser transmitters on Earth.

Twenty-six months after the launch of the first Soviet lunar rover, Lunokhod 2 was sent from Tyura Tam on its journey Moonward aboard Luna 21.

Basically, Lunokhod 2 was the same as No. 1, though it weighed nearly 200 pounds more (1,848 vs 1,667 pounds). The television system was improved, and a couple of new instruments—a magnetometer and an astrophotometer—were added. Also, the traction system on the newer rover permitted changes in direction to be made while in motion; the earlier rover first had to stop completely. Otherwise, the equipment and instruments were the same. Lunokhod 2 functioned for more than 4 months, during which time it traveled 23 miles, four times the distance traversed by Lunokhod 1.

With the completion of the Apollo 17 and the Luna 21/Lunokhod 2 programs, the astronautical exploration of the Moon ceased for the time being.

As the Russians were first into space and the first to try for the Moon, they were also the first to approach the planets. Their success in interplanetary exploration got off to a hesitant start, however. Their first two efforts to fire probes to Mars, made in October 1960, were failures.

A number of attempts to send probes toward Venus and Mars were made by the Soviets during 1962.

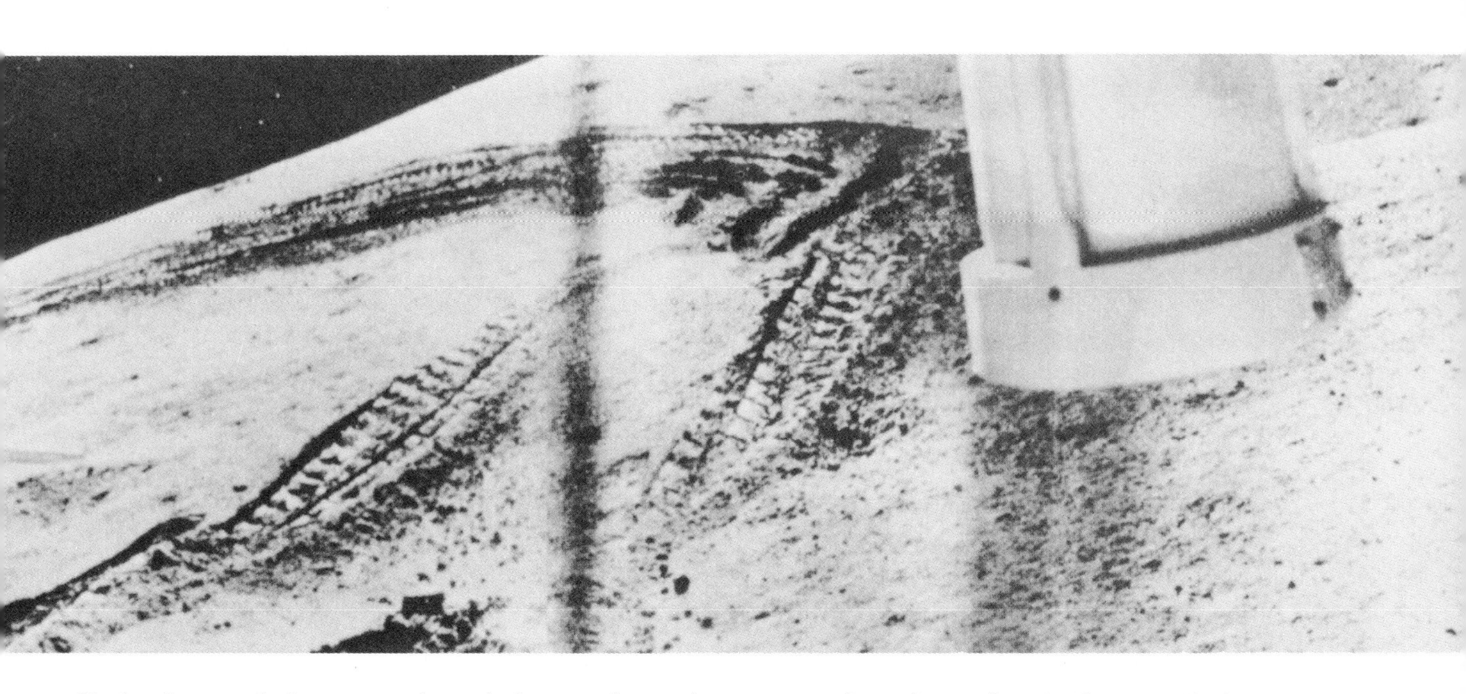

*Lunokhod 2 photographed its own tracks on the lunar surface in this composite of two photos taken 18 February 1973.* (TASS FROM SOVFOTO)

The Mars 1 spacecraft was successfully set on its trajectory, but failed 66 million miles out.

The United States interplanetary program, like its Moon effort, was initially less ambitious than that of the Russians, but enjoyed a higher percentage of successes. The interplanetary program started with Pioneer 5, which was launched toward the orbit of Venus on 11 March 1960. Pioneer's 94.8-pound payload enabled it to make valuable measurements—man's first—in deep interplanetary space. Information was sent back 17.7 million miles to Earth. Pioneer 5 was tracked to a record distance of 22.5 million miles.

The first real United States effort to probe another planet came on 22 July 1962, when Mariner 1 was launched for Venus. The probe's Atlas Agena B carrier veered off course and was destroyed. Mariner 2, a back-up spacecraft, was quickly readied and was launched on 27 August to an outstanding destiny. Mariner 2 flew within 22,000 miles of Venus on 14 December, sending back across 35 million miles of space information not only about the planet but data on vast reaches of previously unexplored space. For once, the United States had scored a major space first.

In March 1963, plans for a more ambitious Mariner voyage, this time to Mars, were defined. The Russians resumed interplanetary flights first, however. On 2 April 1964 they sent the first of a series of probes, Zond 1, off toward Venus. Again, they lost contact with the spacecraft before it reached the target. Mariner 3 was placed in a good trajectory toward Mars on 5 November 1964 by an Atlas Agena D (which, incidentally, was making its maiden flight). But the shroud that protected the spacecraft during its launch through the atmosphere unfortunately could not be jettisoned. Mariner 4 was launched on 28 November 1964. Two days later, the Soviets sent Zond 2 off toward Mars. It was initially successful, but was an ultimate failure.

Mariner 4 flew by Mars at a distance of 6,118 miles on 14–15 July 1965 after traveling a looping path of 325 million miles in a little more than 7½ months. Twenty-two pictures of the planet were taken, stored on magnetic tape, and played back to Earth at a slow rate of speed. The nineteen useful photographs—the last three were taken on the dark night side of the planet—showed a pockmarked surface that resembled the Moon more than the Earth. A preliminary report by the Jet Propulsion Laboratory–Cal Tech team that had managed the flight revealed "more than 70 clearly distinguishable craters ranging in diameter from 4 to 120 km. It seems likely that smaller craters exist; there also may be still larger craters . . . since Mariner 4 photographed, in all, only about 1 percent of the Martian surface."

Two more Soviet interplanetary probes were fired on 12 and 16 November, both toward Venus. The 2,000-pound spacecraft were at least partially successful. One, called Venera 3, impacted on Venus—on 1 March 1966, the first man-made object to touch another planet—while the other, Venera 2, bypassed the planet on 27 February 1966 at a distance of nearly 15,000 miles. Unfortunately, communications with the probes were lost when they were approaching Venus.

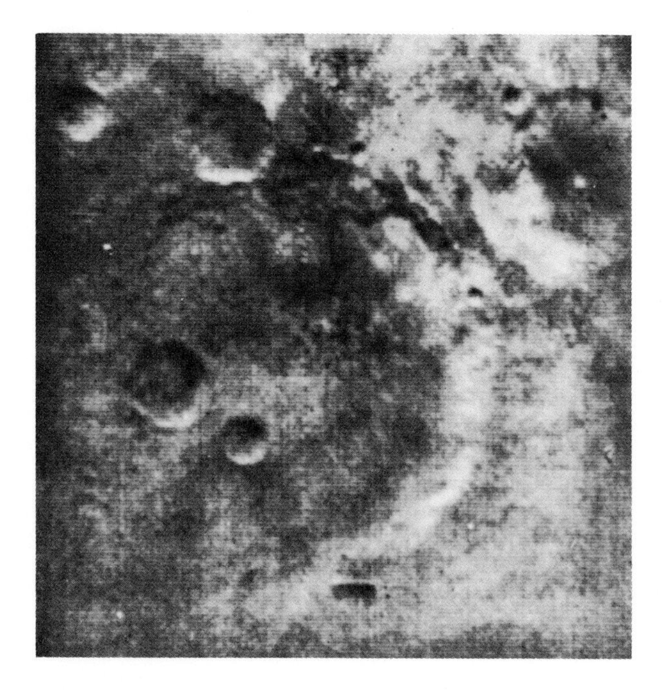

*From a height of 7,300 miles, the surface of Mars looks more like the Moon than the Earth. This photograph was taken on 14 July 1965 by Mariner 4.* (JET PROPULSION LABORATORY, CALIFORNIA INSTITUTE OF TECHNOLOGY)

In late December 1965 and mid-August 1966, Pioneers 6 and 7 were sent to study the space environment of the inner solar system. The next two heliocentric orbit probes were spaced further apart. Pioneer 8 was launched on 13 December 1967 and Pioneer 9 on 8 November 1968, their missions coinciding with the latter part of the 11-year Sun cycle that peaked in 1969.

Both nations did take advantage of an opportunity to send probes to Venus 6 months later, with Russia launching Venera 4 from Tyura Tam on 12 June. It was followed in 2 days by Mariner 5 from the Kennedy Space Center in Florida. When the Soviet probe neared Venus, it separated into two modules: the 842-pound lander parachuted slowly through the atmosphere on 18 October 1967; and the other unit acted as a relay station, continuing into orbit around the Sun. The descending module ceased communications 1 hour and 38 minutes after entering the atmosphere. The Russians have maintained that Venera 4 landed on the surface, but subsequent analysis has suggested that it may have either touched down on a 15-mile-high peak or failed while still descending. Mariner 5, meanwhile, bypassed Venus the next day at a distance of 2,480 miles.

Two more Russian probes were sent toward Venus in January 1969. Arriving just a day apart in mid-May, they descended by parachute on the planet's dark side, reporting temperatures of up to 750°F and atmospheric pressures 60 times that of the Earth. (The comparable American estimates were 900 degrees and 75 Earth atmospheres at surface.) Venera 5 and 6 instruments showed that the atmosphere contained from 93 to 97 percent carbon dioxide, and not more than 7 percent nitrogen or 0.4 percent oxygen. Both capsules were apparently crushed by high atmospheric pressures before reaching the surface.

Veneras 7 and 8 were similar to earlier Veneras 5 and 6, weighing, respectively, 2,605 and 2,596 pounds (combined spacecraft and lander portions); the latter's weight was trimmed somewhat as a result of design changes stemming from Venera 7 experience. The earlier probe's parachute deployed at about 130 miles above the surface (when pressure had reached 0.7 atmosphere), whereas Venera 8's parachute did not come out until 40 miles above the surface (when atmospheric friction had reduced the speed from 26,000 miles per hour—at the time of entry into the atmosphere—to 560 miles per hour).

Mariner 10, the only United States spacecraft sent toward the inner planets during the first half of the 1970 decade, was launched by an Atlas-Centaur carrier from the Kennedy Space Center on 3 November 1973. Its objectives were Venus *and* Mercury. Built by the Boeing Company for NASA's Jet Propulsion Laboratory, it weighed 1,108 pounds, of which 170 pounds represented instrumentation. It had two television cameras, as well as devices to study infrared and ultraviolet radiations, magnetic fields, solar and galactic charged particles, and solar plasma.

En route to Venus, Mariner 10 had been commanded to execute midcourse maneuvers on 13 November and again on 21 January. These brought the craft to within 3,600 miles at closest approach on 5 February 1974. It took excellent photographs of the structure of Venus's upper atmosphere. Other findings were that the planet's magnetic field could be no more than a twentieth of a percent that of Earth and that cloud-top temperatures varied between −12° and −64°F. The Soviet discovery that the atmosphere is made up largely of carbon dioxide were confirmed.

As Mariner 10 flew by Venus on 5 February 1974, its course was accelerated by the gravitational field so that it plunged into the inner Solar System. Two months later, on 29 March, the spacecraft made the first of three flybys of Mercury.

Photography began on 23 March at a distance of more than 3 million miles. Picture-taking continued at 42-second intervals during the arrival and departure legs of the trajectory, the closest approach being less than 450 miles. The gravitational attraction of Mercury deflected the spacecraft into an orbit with twice the planet's 88-day period around the Sun—that is, 176 days. Thus, that number of days later, Mariner 10 revisited Mercury, encountering it on 21 September at just under 30,000 miles. Finally, on 16 March 1975—after another two Mercurian years had gone by—the craft made its third sweep, approaching to within 200 miles over the northern hemisphere.

The nearly 2,000 useful, fully processed photographs revealed a lunar-like world, heavily scarred with craters. The major feature that was found by Mariner 10's TV eyes was a basin more than 800 miles in diameter. Named Caloris because it is located in the hottest region of the planet, the basin is rimmed with mountains. Numerous valleys, ridges, and cracks were also discovered over the third of the planet that Mariner 10 was able to photograph during its encounters.

The Russians continued their exploration of Venus through the 1970's and into the 1980's with ten more missions. Launched on 8 and 14 June 1975, Veneras 9 and 10 became the first Venusian orbiters following arrival at their destination on 22 and 25 October 1975. Within a day, both had released landers. Upon touchdown, the landers returned the first photos of the hellish surface.

The orbiters seemed to be standard interplanetary spacecraft (their equivalent of the American Mariner line), but the landers were unusual in that they had no legs. Instead, they rested on a ring of crushable material. The instruments were in a pressure vessel that was cooled before entry to provide extra operating time on the surface. Similar approaches were used by Veneras 11 and 12 in 1978, 13 and 14 in 1981, and 15 and 16 in 1983 (the "bus," or core, spacecraft did not become orbiters but rather continued on flyby trajectories).

TV pictures transmitted back to Earth—in color in the later models—showed a surface strewn with sharp, jagged rocks rather than rocks with rounded edges as some scientists had suggested would be produced by eons of hot, dry wind. Analytical instruments aboard the landers (including a drilling device similar to those

*Using an ultraviolet camera, Mariner 10 gave form and motion to the otherwise bland clouds of Venus. These pictures cover 24 hours; the clouds circle the planet every 4 days.* (NASA)

*Venera 14 sent back this detailed picture of the surface of Venus in 1982. The jagged edge seen at bottom of the photo is designed to prevent the probe from turning over.* (TASS/NOVOSTI)

used on the Lunas) showed that the surface chemistry is remarkably like that of the Earth's. The atmosphere, though, was worse then anyone had dreamt it would be. Temperatures ranged up to 860°F and pressures up to 88 atmospheres.

The most detailed information on the Venusian environment came from the United States Pioneer Venus mission. The orbiter element was launched on 20 May 1978 and reached its destination on 4 December 1978. Then, on 8 August 1978, the bus portion followed carrying four probes, reaching Venus on 9 December 1978. Scientific instruments aboard the orbiter included a photopolarimeter (similar to those used on Pioneers 10 and 11) to produce images of the cloud tops, and plasma physics and atmospheric instruments and a radar altimeter. With the latter, "pictures" of the surface could be built up with a resolution of a few miles. What emerged was a comparatively flat land with two major "continents": Ishtar Terra, about the size of the continental United States; and Aphrodite Terra, about the size of the northern part of Africa. The highest point, Maxwell Montes in Ishtar, rises to 6.8 miles above the planet's mean elevation. The lowest is a rift valley 1.8 mile deep on the east side of Aphrodite. At least two regions appear to have been created by volcanoes.

The Pioneer Venus Orbiter's ultraviolet scanner allowed scientists to build images of the planet's cloud system that showed its circulation as having a classic pattern once predicted in studies of Earth's weather. The four probes and their carrier bus that dropped through the clouds provided direct sampling of the atmosphere. The bus was similar to the orbiter but carried four cone-shaped re-entry vehicles—three small ones weighing 198 pounds each and a large one weighing 696 pounds. They were released 20 days before arrival at Venus and relayed their data through the probe bus. All four transmitted until they hit the surface (the bus itself quit at 71 miles altitude), with the small dayside probe transmitting for 67 minutes after touchdown.

The Venusian atmosphere was found to consist mainly of carbon dioxide but with concentrations of argon and neon several hundred times larger than that of Earth. This had been noted by earlier Veneras and was suggested as being caused by the "retarded" evolution of the planet. The concentration of deuterium on Venus was found to be 100 times greater than that on Earth, indicating that most of the lighter hydrogen (possibly in a primordial ocean) had escaped as the Sun gradually brightened and Venus had been turned into a runaway greenhouse. The clouds were found to be made up mostly of sulphuric acid. Winds as fast as 224 miles per hour were recorded in the upper levels of the atmosphere, but they dropped off and became stagnant near the surface.

More detailed analysis of the Venusian atmosphere was conducted by the two Vega spacecraft that the U.S.S.R. launched to Venus between 22 and 28 December 1984. Each released a probe during the June 1985 flyby of the planet, heading subsequently on toward a 1986 encounter with Halley's comet. The probes deployed balloons to allow their instrument packages to drift in the atmosphere for several days.

The next United States mission is scheduled to be the Venus Radar Mapper, a scaled-down version of the Venus Orbiting Imaging Radar mission that was to have carried a high-power synthetic aperture radar into orbit around Venus. The new spacecraft will incorporate backup hardware derived from the Voyager and Galileo missions. A Voyager high-gain antenna, for example, will map the surface during the low portion of the orbit and then transmit the data to Earth during the high portion.

As important as was American exploration of Venus, the nation's primary interest has been the red planet Mars. Following the somewhat disappointing Mariner 4 spacecraft, the United States dispatched twin 910-pound probes to within 2,200 miles of Mars in 1969. Mariner 6, launched on 24 February, bypassed the planet on 31 July; Mariner 7, launched on 27 March, traveled a shorter course and made its closest encounter on 5 August, less than a week after the first spacecraft.

Photographs confirmed Mariner 4's finding that the Martian surface is more Moonlike than Earthlike. Huge craters—one 300 miles across—were revealed. Many older heavily eroded craters had smaller, newer craters inside. One striking geographical anomaly was discovered: a large circular region, known as Hellas, was found to be almost completely smooth, presumably due to some process that has erased the scars of meteorite impacts in the area.

From 1970 through 1974, the Soviets succeeded in sending six spacecraft toward Mars and the Americans one. Both Mars 2 and Mars 3 took advantage of the May 1971 launch window, the former leaving Earth parking orbit on 19 May and the latter on 28 May. They reached the red planet 5 days apart, on 27 November and 2 December of the same year, and immediately released landing capsules. This done, braking engines were ignited, putting the mother spacecraft in highly elliptical orbits. The landers, meanwhile, using a combination of parachutes and descent rocket engines, headed for the surface. Possibly because of high winds blowing at the time, or thick dust layers on the surface, the Mars 2 capsule failed to communicate at all upon reaching the ground. The Mars 3 lander transmitted for only 20 seconds from a point known as Simois Strait south of the equator. Despite these disappointments, the Russian capsules were the first man-made objects to reach the Martian surface.

The orbiters produced scientific data for some 3 months, as they followed their eccentric paths around Mars. The apoapsis of Mars 2—its farthest point from the planet—was over 15,000 miles, and that of Mars 3 was 118,000 miles. Their respective periapses or closest approaches were 860 and 930 miles.

Among other things, the Russians found that "In the most 'humid' periods precipitated water may amount to 0.1 mm, i.e. 100 times as little as in the terrestrial atmosphere. But in some periods its quantity may be 1,000 times as low and the steam absorption lines disappear altogether." They discovered that atomic hydrogen forms an envelope "many thousands of kilometers" thick around Mars, and that—after carbon dioxide—the most abundant gas in the air must be argon.

The United States also took advantage of the May 1971 launch window. Its first attempt on 8 May ended in disaster: the second stage of Mariner 8's carrier vehicle malfunctioned and the spacecraft plummeted into the Atlantic Ocean. But the second in the series, Mariner 9, was widely acclaimed as the most successful planetary probe in the early history of space exploration. Even though it left Earth on 30 May, after the Soviet Mars 2, it took only 167 days to make the voyage, arriving 13 November—a full 2 weeks before Mars 2. Thus, Mariner 9 became the first spacecraft to orbit another planet. Its precision was astonishing. After a trip of 248 million miles, it reached its destination 4.4 *seconds* ahead of schedule.

As it turned out, Mariner 9 arrived at Mars when the planet was undergoing its severest dust storm since 1924. The storm, detected on 22 September, covered from view most surface detail from 25 September onward. Although momentarily fearful that the two television cameras might not hold up long enough to map the entire planet once the dust storm finally dispersed (which it did 6 weeks later), scientists soon realized that the unusual conditions would provide them with some splendid opportunities.

First, the storm itself was fascinating. By observing its movements and later its gradual abatement, they could gain a more thorough understanding of circulation patterns of the Martian atmosphere. Second, the influence of the slowly settling dust on Martian surface features could be examined. Third, the composition of surface materials suspended in the air by the high-velocity winds could be measured (silicon dioxide turned out to be the main component). And dust effects on surface temperatures could be examined. (The fact that the dust warmed the atmosphere and cooled the surface proved significant to experts wondering about what effects pollution might be having on Earth's atmosphere.) Also, the very knowledge that winds blowing at 100 or more miles per hour occur on Mars would have to be taken into account when designing future landing vehicles.

Mariner 9 proved successful beyond the most-optimistic expectations of its designers. Planned to operate for a nominal 90 days, the spacecraft continued functioning for almost a year—the last signal was re-

*As revealed by Mariner 9, Mars possesses the largest volcano known to man, Nix Olympica (Snows of Olympus). It is 310 miles across at the base and has a 40-mile-wide complex vent at the top.* (NASA)

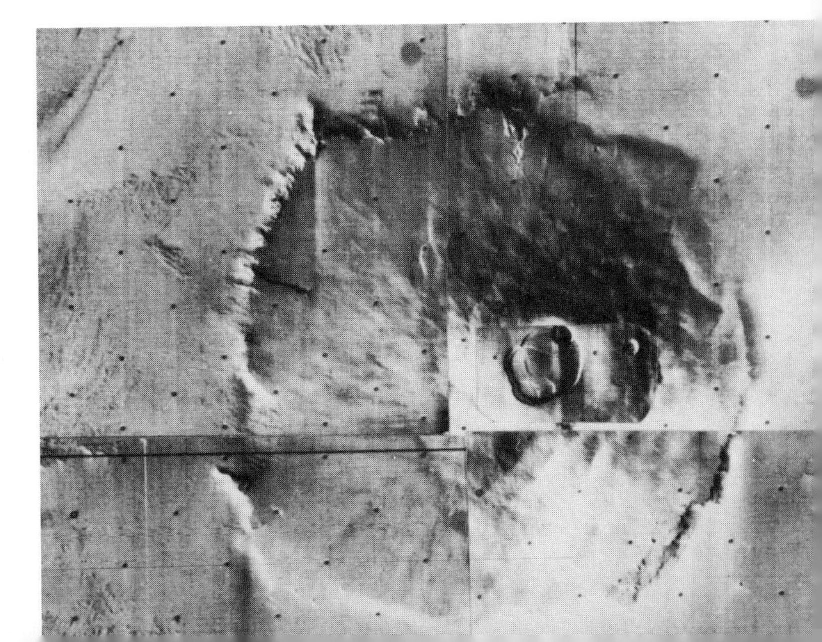

ceived on 27 October 1972 during orbit 698. On that day, engineers at NASA's Jet Propulsion Laboratory instructed the radio transmitter to switch off, the last of 45,960 commands. The reason: the propellant used to operate the spacecraft's attitude control system had become depleted.

On 2 January, as soon as the storm had cleared, mapping began. By orbit 500 in mid-July, the job was done. Every square mile of Martian surface had been photographed. Huge volcanic constructs and calderas, larger than any on Earth, were seen. One, Nix Olympica, measured about 300 miles across at its base and towered at least 10 miles above the surrounding plains. An incredible 2,500-mile-long equatorial chasm was revealed, several times deeper than Arizona's Grand Canyon. Mariner 9 also photographed what appeared to be dry meandering stream beds, indicating that water flowed across the equatorial regions of Mars during fairly recent geologic history. Large-impact craters with wind-blown dunes in their floors and the "chaotic" terrain first observed by Mariners 6 and 7 also showed up in abundance.

Scientists were at once surprised and delighted at the results. Remarked John E. Naugle, at the time NASA's associate administrator for space sciences:

> It completely changed our picture of that planet. If you think back a moment . . . Mariner 4 . . . gave us quite a picture of Mars . . . of a dead planet, Moon-like, cratered. That was followed by Mariners 6 and 7 which tended to paint that same kind of picture. But, very shortly after Mariner 9 went into orbit, it became clear that that picture was not an accurate picture of Mars . . . the old cratered dead planet has been changed to where some people say we really have [three kinds of planets]: the old cratered planet, the young volcanic active planet that you see, and then an entirely different situation in the polar regions.

As a result of the Mariner 9 experiments, scientists suspect that a considerable amount of water ice may be locked up in the polar regions. Conway Leovy of the University of Washington observed that water vapor was three times as abundant over the retreating ice cap as it was over the rest of Mars. This, he feels, "gives us a very good indication that the source of the water, which varies in the atmosphere seasonally but reappears in the atmosphere each season is, in fact, the periphery of the north polar cap either in association with $CO_2$ ice or underneath the $CO_2$ ice in that region." Ozone was also discovered in the polar regions, increasing in abundance as temperatures fall. It amounts, however, to less than 1 percent of that found in our own atmosphere.

Mariner 9's infrared interferometer spectrometer measured temperatures of about −200°F at the north pole, compared with the lowest temperature of −125°F recorded in the Antarctica on Earth. Warmer temperatures in the equatorial regions were measured by the spacecraft's infrared radiometer −81°F. The highest temperature recorded by the instrument was somewhat over 40°F, also in the equatorial regions.

Carbon dioxide was reconfirmed as the principal component of the Martian atmosphere. The radio occultation experiment showed that the lowest pressure on Mars is 2.8 millibars near the equator (compared with a sea-level average of about 1,000 on Earth). The highest readings obtained near the poles ranged from 5 to 13 millibars.

The golden era of planetary exploration, as many scientists now call it, ended with programs being either stretched out or canceled; and the prospects for new starts became dimmer. The era drew to a close following several outstanding twin missions to Mars and the outer planets: the Vikings, the Pioneers, and the Voyagers.

Each spacecraft consisted of an orbiter (based on the Mariner series) and a lander (a completely new vehicle). The mission plan called for both 7,624-pound spacecraft to go into Martian orbit and to use on-board TV cameras to scout for candidate landing sites; the resolution of the earlier Mariner 9 TV images was not high enough to guarantee safety. In addition, radar on Earth would scan the sites to gauge surface roughness. The orbiter also carried two TV cameras and infrared instruments for mapping the planet's thermal characteristics and for detecting atmospheric moisture.

Most of the science payloads went into the landers. These were squat spacecraft powered by two radioactive heat sources (selected to avoid placing vulnerable solar panels in the dusty environment) set on three deployable legs. A robot arm with a scoop for a hand was provided to pick up soil samples to be dropped into a three-in-one analysis unit where life forms would be cultivated and detected by chemical means (such as release of radioactive-labeled tracers in their food). A mini weather station was mounted on a short boom, and two light cells acted as TV cameras by means of rotating, swiveling mirrors. A high-gain antenna provided direct contact with Earth, although most pictures and data were relayed through the orbiters.

The entire spacecraft fitted inside an aeroshell for entry through the atmosphere. After supersonic braking, the aeroshell was split and a parachute slowed the lander until an altitude of 4,000 feet where three multinozzle rockets took over.

Despite having had experience with Surveyor and other spacecraft, those who designed the 1,269-pound Viking landers found it to be no easy task, because whatever went to the surface had to survive sterilization in hot nitrogen gas once sealed inside the aeroshell and the surrounding bioshell. Any chance of car-

*A panoramic view of the surface of Mars around the Viking 2 lander shows a field strewn with boulders—had the spacecraft landed a few feet to one side or the other, it would have turned over.* (NASA)

rying terrestrial life to Mars had to be eliminated for two reasons: it would always leave the "discovery" of possible life on Mars in doubt; and, it could ravage whatever indigenous life that might have developed there.

After a brief delay in which the two spacecraft had to be swapped because of a technical problem, launches were made on 20 August and 9 September 1975. Arrival at Mars came on 19 June and 7 August 1976. The landing of Viking 1 was planned for July 4—America's bicentennial—but difficulties in selecting a landing site postponed touchdown by almost 3 weeks. Viking 1 was used as the relay, though, for a radio signal that ordered a duplicate of its soil arm to extend and cut the ribbon at the 1 July opening of the National Air and Space Museum in Washington, D.C.

The original landing site was found to be too rough when surveyed by the Viking 1 orbiter and by radar, but another site was selected to the north. On 20 July the lander was deployed and settled onto the rusty surface of Chryse Planitia (Golden Plains). Within minutes, the lander's appendages had deployed and the first black-and-white TV picture was taking shape, line by line, on TV screens across the Earth. In time, such mundane objects as Midas Muffler Rock and Dutch Shoe Rock would become as familiar as Olympus Mons.

With Viking 1 safely on the ground, flight controllers decided to pick a slightly more ambitious target to the north—Utopia Planitia (Plains of Utopia) for Viking 2's descent. That both landers were successful surprised most scientists and engineers, especially when the cameras revealed close-by rocks large enough to topple the landers or break a leg.

The search for life was at once the most prominent feature of Vikings' mission and the most disappointing. No one expected to see Martians strolling across the landscape (they would have to pose, actually, because of the cameras' slow scan), but at the same time no one was prepared for the ambiguous results that came back from the miniature labs hidden within the landers. Two looked for carbon-14 in samples that were given extra carbon dioxide or a dose of sterile nutrient. The third looked for gases that would be given off by life forms as they metabolized nutrient and carbon dioxide. Some findings suggested life was present, but at no time did all three tests agree. The final consensus was that there probably was some inorganic chemistry at work in the soil that mimicked aspects of the results that the science team was seeking. Such chemistry has since been created in the laboratory back on Earth.

Overall, though, the mission was of great scientific value. The soil was found to be made up of iron-rich clay and iron oxides, long suspected to be the cause of Mars' red color. The soil was magnetic, as shown by its clinging to a small permanent magnet on the soil arm. No traces of organic molecules were found even under rocks, where they might have been protected from ultraviolet radiation.

There appears to be a large volume of water ice beneath the carbon dioxide ice caps that extend during winter at the poles. Atmospheric pressure drops 30 percent during the winter, possibly from dry ice snowing at the poles when the air supercools. Winds blew no faster than 46 miles per hour at the landers' sites.

Although designed for 6-month missions, the spacecraft operated far longer. Orbiter 2 had to be shut down when its thruster gas ran out on 24 July 1978; the same happened with Orbiter 1 on 8 August 1980. Lander 2 had a power failure in March 1980 that took it out of action. Lander 1 operated the longest, transmitting weather data and photographs until 8 March 1983. Apparently it had been given an erroneous command to aim its high-gain antenna at the ground. Ground controllers tried to circumvent that by broadcasting high-power commands in the hopes that the ground would reflect enough of the signal to the antenna, but to no avail.

Although proposals were made for a Viking 3 mission using residual hardware, and even for a mobile Viking, exploration of Mars went into a hiatus until NASA's 1985 budget started incorporating recommendations of the Solar System Exploration Committee (SSEC). The next United States mission is scheduled to be the Mars Geochemical/Climatology Observer, a modest mission based on available technology and a "production-line" spacecraft. A Mars Surface Sample Return mission, also proposed in the 1970's, was given high priority by the SSEC, but it is not expected to materialize until the late 1990's.

After the inner planets of the solar system, the next obvious target for exploration was Jupiter, the closest to Earth of the giant gaseous planets beyond the Asteroid Belt. The United States began the exploration of Jupiter by launching two flyby probes from Cape Kennedy, the first—Pioneer 10—on 2 March 1972 and the second—Pioneer 11—on 5 April 1973. Both spacecraft successfully made the 7-month journey and encountered Jupiter on 3 December 1973 and 5 December 1974, respectively.

Conceivably, as Pioneer 10 makes its lonely voyage through the unfathomable reaches of interstellar space, following its successful traverse of the solar system, it may be tracked and "captured" by representatives of an extrasolar civilization. Carl and Linda Sagan designed a 9- by 6-inch gold anodized plaque that would inform intelligent beings elsewhere in the universe concerning the origin of the spacecraft and at least the appearance of the species that created it. The plaque has a schematic diagram of the solar system, which is located in relation to fourteen pulsars within the galaxy, along with drawings of a man, a woman, the Pioneer 10 craft itself, and its trajectory out of the solar system.

Pioneer 10 weighs 570 pounds, of which 66 pounds are accounted for by eleven instruments capable of carrying out thirteen experiments. The spacecraft is 9½ feet long and consists principally of an Earth-facing 9-foot-diameter dish antenna and equipment and instrumentation compartments. Because the Sun's energy at Jupiter is only 4 percent that available at Earth's distance, Pioneer 10 was not fitted with solar cells. Rather, four SNAP–19 radioisotope thermoelectric generators producing 40 watts each at 4 volts dc were substituted as the power source. The craft was fitted with six small rocket thrusters, needed to make various velocity and attitude changes en route and to maintain spacecraft spin at 4.8 revolutions each minute. The identifying plaque was bolted to the frame. Pioneer 11 is virtually the twin of Pioneer 10.

NASA's Deep Space Network was faced with the difficult task of issuing over 10,000 commands during the course of the mission. Account always had to be taken of the long communications delay time between the sending of a command signal and its arrival at the spacecraft. For example, during the time of closest approach to Jupiter a signal took a full 46 minutes to reach Pioneer 10, meaning that the round-trip communications time lag was 92 minutes.

Pioneer 10's (and 11's) missions were divided into experiments conducted in interplanetary space, which included the Asteroid Belt, and those during the Jupiter approach, flyby, and post-encounter periods.

Pioneer 10 discovered that the solar wind and magnetic field are active and turbulent out to at least 420 million miles—farther than expected. Even though stream velocities gradually lowered with distance, temperatures were found to rise. This curious phenomenon was explained by the interaction of faster-moving streams catching up with slower-moving streams that had earlier left the Sun. Although solar wind density and magnetic-field strength declined slowly as Jupiter was approached, there was enough turbulence to block out a large portion of the incoming low-energy galactic particles. This was not foreseen, because scientists expected that decreasing shielding by the solar wind and magnetic field would lead to a higher flux of cosmic-ray particles. It is now believed that the shielding does not begin decreasing until beyond the orbit of Jupiter.

The passage through the Asteroid Belt did not pose the dangers that many scientists feared. As it turned out, the belt was far "cleaner" than was expected, with the distribution of dust particles between Earth orbit and the outer extremity of the belt depending principally on particle size. Moving outward in space, the smallest particles, those 1/1000 millimeter in diameter, are more common near the Earth than in the belt. Particles in the 1/100- to 1/10-millimeter range are about evenly distributed throughout, while those from 1/10 to 1 millimeter are much more frequent in the belt than outside. Larger particles are relatively rare everywhere.

Prior to the flight, as principal asteroid-meteoroid astronomy experiment scientist Robert K. Soberman of General Electric noted, "We might have expected anything from 1,000 events [particle impingements] per day to none during the entire passage." As it turned out, an average of one event per day occurred during passage through the peak of the Asteroid Belt.

The first major event occurred on 26 November when the bow shock was crossed at a distance of just under 4,800,000 miles from Jupiter. A day later the magnetopause—the limit of the Jovian magnetic field—was crossed, a day or so ahead of prediction. Maximum trapped particle radiation intensity was observed from periapsis to over 650,000 miles, with the highest single readings made when the spacecraft crossed the magnetic equator a little over an hour before closest approach. The radiation flux was, as predicted, far higher than that in the Earth's Van Allen radiation belts, with high-energy electrons occurring some hundred times more frequently than protons. The belts were found to be contained within the strong Jovian magnetic field, which was shown to be disk-shaped and opposite in polarity as compared with Earth's. Moreover, the magnetic poles are displaced at an angle of 15 degrees to the axis of rotation of the giant planet.

The temperature regime on Jupiter turned out to be complex. Although temperatures are about the same in the northern and southern hemispheres and fall off in general as the poles are reached, important differences were measured. For one thing, the dark belts encircling the world were found to be warmer—about 15 degrees Fahrenheit—than the brighter belts. Also, although temperatures at and near the top of the atmosphere ranged from −215 to −230 degrees Fahrenheit, somewhat lower down—but still in the upper atmosphere—readings of 260 degrees Fahrenheit were made, a complete surprise. Deeper in the atmosphere, the temperature goes up to some 800°F, suggesting that surface temperatures must be far higher. Confessed Arvydas Kliore of the Jet Propulsion Laboratory, "We do not understand these findings. These temperatures are very difficult to explain. . . . There could be some very heavy dust layers in the upper levels of the atmosphere that absorb solar radiation. It could also be absorbing heat radiation away from the surface of the planet."

Jupiter was found to be slightly more flattened at the poles than was expected, and a little heavier. Also, a relatively small red spot about 5,000 miles in diameter was discovered ("small" compared with *the* Red Spot's 15,000-mile by 4,000-mile dimensions). The densities of the four major moons were determined, Io being the densest—3.5 times the density of water, followed by Europa, 3.07, Ganymede, 1.93, and Callisto, 1.65. Io was discovered not only to have a very thin atmosphere and ionosphere, but a distinct ultraviolet glow. All the moons are cold, about −280 degrees Fahrenheit.

Pioneer 11 continued outward, each day becoming the most distant object launched and the most distant object in contact with humanity. The latter was owing, in no small part, to improvements in the sensitivities of the 210-foot-deep space network antennas. On 2 June 1983, Pioneer 11 officially left the solar system as it crossed the orbit of Pluto (then on the other side of the solar system from the Earth). Despite the enormous distance involved, the craft's instruments continued to detect the solar wind, an indication that Pioneer 11 had not yet entered true interstellar space. At the end of 1984, Pioneers 10 and 11 were, respectively, some 3.2 and 1.7 billion miles from the Sun and continuing to study the extraordinarily remote environment. The telemetry capability of NASA's Deep Space Network makes it feasible to receive data from the spacecraft well into the 1990's.

The Pioneers were followed by two Voyagers, which trace their lineage from an ambitious but never carried out mission that was too expensive for the times—the "Grand Tour." It called for a Thermoelectric Outer Planet Spacecraft (TOPS) with an advanced, reprogrammable computer and an ion engine. The Grand Tour was to have taken advantage of a rare alignment of the planets to "capture" their gravity fields, thereby accelerating the spacecraft on to the next world in an interplanetary game of billiards. A scaled-down version of a Grand Tour emerged. Initially called Mariner-Jupiter-Saturn, it was renamed Voyager in 1977 to become, in effect, a mini Grand Tour. Its objective: fly by Jupiter and Saturn and most of their moons (which are almost planetary systems in their own right) and then continue on to Uranus and Neptune.

Voyager, the latest in the Mariner series, weighed 1,797 pounds, including 254 pounds of science instruments. Because of the distance across which it had to maintain contact with Earth, its largest feature was the 12-foot high-gain antenna. Its high- and medium-resolution TV cameras and ultraviolet and infrared sensors were mounted on the scan platform deployed to one side. A magnetometer was deployed on a special mast that held it 42.6 feet away from the spacecraft; and three nuclear power generators were mounted on a boom opposite the scan platform. Plasma physics and other instruments were placed on the spacecraft body. Also carried was a special recording of scenes, music, and voices from planet Earth a celestial "roadmap" showing where the spacecraft originated.

Because Voyager 1 was programmed to travel a faster trajectory, Voyager 2 was launched first, on 20 August 1977. Its companion followed on 5 September, sending back a historic picture of the Earth and Moon in a single frame. The cruise toward Jupiter was uneventful until February 1978, when the scan platform on Voyager 1 jammed and refused to move more than a few degrees in azimuth. Insulation apparently had lodged in the gear train and had to be ground up by cycling the drive back and forth.

Closest approach to Jupiter for Voyager 1 came on 5 March 1979 and for Voyager 2 on 9 July 1979. Those dates, though, were only the midpoints of two dashes through a fascinating planetary system circling what may be regarded as a stillborn star possessing a magnetosphere that, if it were visible, would fill as much of our sky as does the Moon.

Pictures returned by the Voyagers were as great an improvement over Pioneer 11 and 12 as they had been over terrestrial telescopes. Minute details of Jupiter's swirling atmosphere became visible, showing vortices that took shape as high-speed gases swept past slower-moving structures such as the giant Red Spot. The increasing detail of the pictures showed dozens of

*Like peering into an angry wound, Voyager 2 took this photo of the Red Spot of Jupiter. With enhancement, it can be seen to be a massive hurricane, itself surrounded by other storms.* (NASA)

smaller spots that turned the Red Spot from a unique blemish on the planet's face to the largest of a complex system of storms. A special sequence of pictures taken during the inbound phase to create a time-lapse movie of the planet's weather showed intense auroral activity on the night side and lightning displays.

A thin-ring system was discovered 35,400 miles above the cloud tops. It was thinner than are the rings of Saturn (less than 20 miles) and is visible only when viewed from behind Jupiter as sunlight scatters off its particles at low angles.

The moons proved to be equally fascinating. Io was found to have a blotchy yellow face caused by billions of tons of sulphur being regurgitated by volcanoes scattered across its surface. Mars and Venus were known to have volcanic structures, but Io became the only world, other than the Earth, to have active volcanoes. In addition, its volcanism left around Jupiter a torus of sulphur and sodium gas that glowed in the ultraviolet with the power of about a trillion watts. Although too small to generate its own internal heat, Io apparently is heated by tidal stresses caused by Jupiter's intense gravitational field.

Europa, Ganymede, and Callisto turned out to have undergone extensive cratering and (resulting from that) fracture lines. Europa apparently is an ocean world with 60-mile-deep seas capped by ice and heated from within by radioactive decay. Ganymede and Callisto appeared to be blends of rock and ice. And three new moons, all asteroid size, were added to the king's retinue, for a total of sixteen.

As the two Voyagers dropped into Jupiter's gravity well, they gained speed and left faster than they arrived. Their trajectories were altered so that they were aimed at Saturn. Arrival of Voyagers 1 and 2 was on 12 November and 26 August 1980, respectively.

Saturn's atmosphere was less striking than was Jupiter's. It had similar storm structures, but the coloring was more mottled and the contrast range much narrower. The rings proved to be the source of greatest fascination, as they have been since their discovery. The images returned by Pioneer 11 did little to prepare scientists for the exquisite detail that the Voyagers revealed. Rather than facing a few major ring structures, they saw a complex array of thousands of rings ranging from less than 2 miles to more than 60 miles wide. Spokes were seen in the rings, which appeared to move around the planet almost as if they were a solid structure (a physical impossibility). Scientists theorized that the spokes might be caused by dust particles under the influence of the planet's magnetic field. Radio occultation measurements indicated that the ring particles ranged from nearly 7 to over 30 feet in size. Also discovered were two "shepherd" moons at the outer edges of the rings that appeared to be deflecting inward any massive outward leakage of ring particles. The rings themselves seemed to be on the order of 600 feet thick, and some were out of round by as much as 85 miles.

The moon system was less fascinating than that of Jupiter's had been (although six more were discovered, for a total of twenty-four). Titan was of prime interest, because for some years it has been known to have an atmosphere (of methane and nitrogen). To everyone's disappointment, visibility was zero and the surface was as shrouded as is that of Venus. Measurements indicated that the temperature and pressure are near the triple point for methane (as Earth is for water) and that there may be rivers and oceans of methane replenished by methane rains and with methane ice caps. The mystery of Iapetus's unusual coloring—brighter on one side than on the other—was solved with

*Voyager 2 provided man with his most detailed views of the outer planets and their moons. The science instruments are on the scan platform at top, and the RTG's are the three blade cans opposite the scan platform.* (GENERAL ELECTRIC)

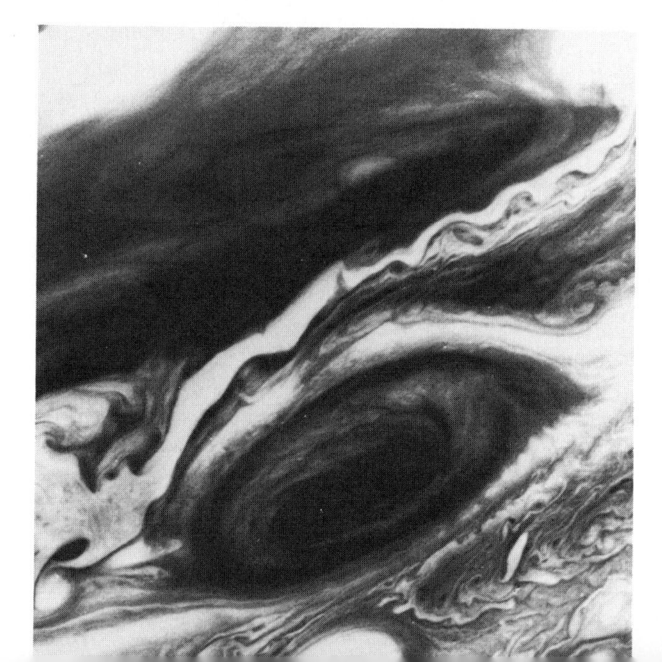

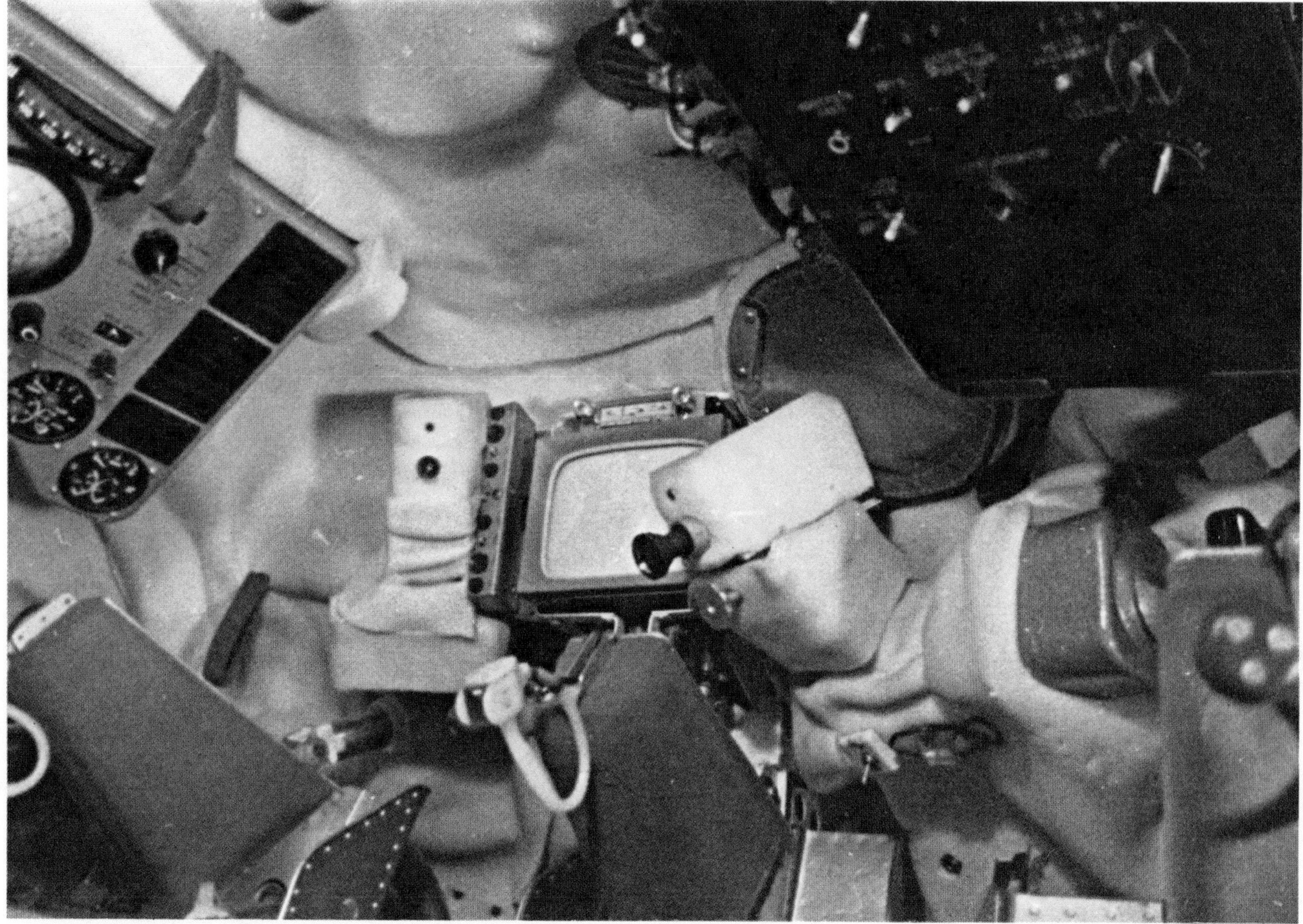

NOVOSTI PRESS AGENCY

*On 18 March 1965, Soviet cosmonaut Aleksei A. Leonov became the first man to leave his spacecraft for a ten-minute "walk" in space. This is the cabin of Voskhod 2, which he left through an airlock.*

NASA

*Floating outside Gemini 4, astronaut Edward H. White propelled himself with a compressed-air gun. After a 21-minute "walk," he rejoined pilot James A. McDivitt to complete 62 orbits between 3 and 7 June.*

NASA-MSC

*The feasibility of space-rendezvous techniques was demonstrated on 15 December 1965, when Gemini 6 and Gemini 7 came within one foot of each other. This photograph, taken by astronaut Thomas P. Stafford from Gemini 6, shows clearly the conical shape of the spaceship. Larger and heavier than the Mercury, it weighs 7,000 pounds and is 18 feet 5 inches long.*

N

*The Saturn 1B launches Apollo 7, the first manned spacecraft mission in the Apollo series, on 11 October 1968 from Complex 34 at tower-studded Cape Kennedy, Florida. During their eleven days in orbit around the Earth, astronauts Walter M. Schirra, Jr., Donn F. Eisele, and Walter Cunningham performed many demanding tasks, including rendezvous and simulated docking maneuvers with the Saturn S-4B stage.*

NASA

*The Lunar module for landing men on the Moon's surface was first tested under manned flight conditions during the ten-day mission of Apollo 9 in March 1969. While the Lunar module was docked to the command and service modules, astronaut Russell L. Schweickart went for a 38-minute space walk, during which he took this photograph of the CSM from the LM's "porch."*

*Astronaut James Irwin of Apollo 15 saluting beside the flag. Lunar landing module is in center, roving vehicle at right, and Hadley Delta in background. This mission, in the summer of 1971, was the first with the rover; the astronauts traveled more than 17 miles in it.* (NASA)

RIGHT: *Apollo 17 space vehicle, carrying astronauts Eugene A. Cernan, Ronald E. Evans, and Harrison H. Schmitt, lifting off on 7 December 1972 at the start of NASA's sixth and last manned lunar landing mission in the Apollo program.* (NASA)

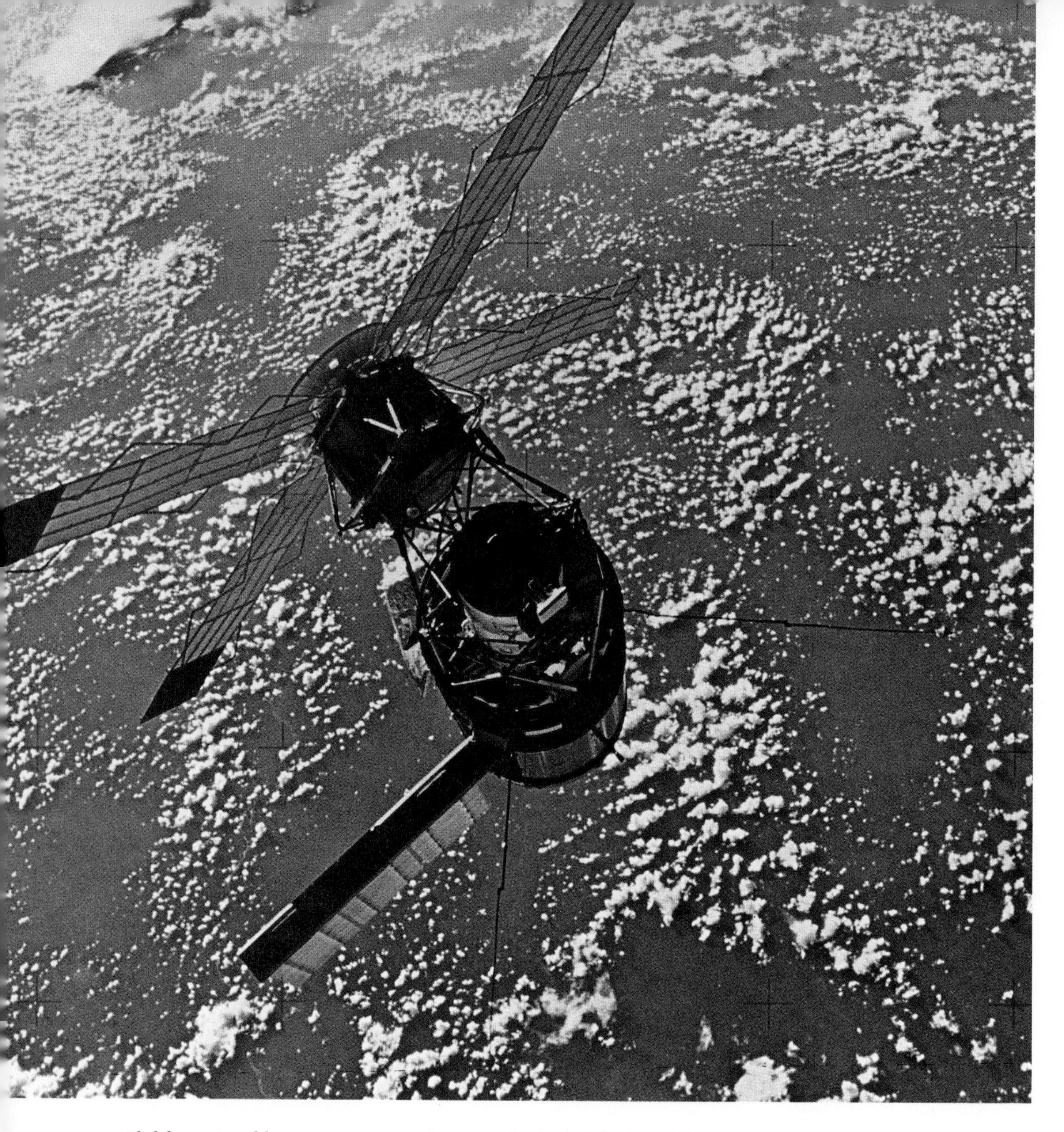

*Skylab, as viewed by astronauts approaching it on the third of the four Skylab missions. Main workshop is below and the Apollo telescope mount, with four solar panels, above. The workshop is missing a panel (it should be opposite the one at lower left) which ripped off during launch.* (NASA)

*A fantastic view of the sphere of the Earth as photographed from the Apollo 17 spacecraft extends from the Mediterranean Sea area to the Antarctica south polar ice cap. Heavy cloud cover in the southern hemisphere while almost the entire coastline of the continent of Africa is clearly delineated.* (NASA)

*Most of the well-known landmarks of New York City and its environs are visible in these 30-meter resolution Thematic Mapper images taken from the Landsat 4 spacecraft. The five boroughs are all included in the picture: Brooklyn, in the lower right; Queens, center right; Manhattan, center; the Bronx, top center; and Staten Island, bottom center. Across the Hudson (along which many piers can be seen) are Newark, Jersey City, Elizabeth, and Paterson (near the upper left corner). Manhattan's Central Park stands out in the darker area of shadow-casting tall buildings. The Brooklyn Bridge is the first bridge encountered going up the East River.* (NASA)

LEFT: *The Solar Maximum Mission spacecraft was launched from Cape Canaveral, Florida, by a Delta on 14 February 1980 and was successfully placed in Earth orbit. It was the first satellite to be repaired in orbit.* (NASA)

*The Challenger lights up the predawn sky as it lifts away from Complex 39's Pad A to begin the STS-8 mission with the first nighttime launch of the shuttle era (30 August 1983).* (NASA)

LEFT: *The space shuttle Challenger on its STS 7 mission photographed from the shuttle pallet satellite (SPAS-01) on 22 June 1983. The photograph was made by a 70mm camera.* (NASA)

*The faint light of dawn illuminates the space shuttle Columbia as it arrives at Pad 39A following a 6-hour move from the Vehicle Assembly Building at the Kennedy Space Center. This was the second rollout for the STS-9 vehicle—replacement of a nozzle on the solid rocket booster aft segment had forced NASA officials to delay the Spacelab 1 launch. Columbia with its Spacelab 1 payload flew into orbit on 28 November 1983.* (NASA)

*Astronaut Bruce McCandless, one of two 41-B mission specialists participating in an extravehicular activity (EVA), is seen a few yards away from the cabin of the Earth-orbiting space shuttle Challenger. The EVA involved the first use of a nitrogen-propelled, hand-controlled manned maneuvering unit. The device allows for much greater mobility than that afforded previous spacewalkers, who had to use restrictive tethers.* (NASA)

*This 70 mm frame shows astronaut Dale A. Gardner, wearing the manned maneuvering unit, approaching the spinning Westar 6 satellite over the Bahama banks during the space shuttle Discovery's 51-A mission. Gardner made a hard dock with the satellite, and then proceeded to recover it. Westar 6 was subsequently returned to Earth for refurbishment.* (NASA)

the discovery that its rotation matches its period and thus its leading face is scavenging debris orbiting Saturn. Enceladus, Mimas, Tethys, Dione, and Rhea revealed faces and structures similar to those seen on the Jovian moons, but none exhibited them to the same degree. Even the hint of tectonic activity on Enceladus was simply that—a hint, apparently caused by tidal interaction with Dione.

The gravity slingshot was used again at Saturn by both spacecraft, but only Voyager 2 was retargeted. It is now scheduled to arrive on 24 January 1986 at Uranus, the only other planet known to have rings and to be strange in its own right, with its poles almost parallel to its orbital plane. A slight deflection is then to be made to send Voyager 2 onward for an encounter with Neptune on 24 August 1989. The chances of such an encounter are somewhat reduced, however, because Voyager 2 suffered a scan platform malfunction while eclipsed by Saturn and much fuel was consumed in taking high-priority pictures before the problem could be resolved. Radio communication difficulties will also make the encounters more complex. Both Voyagers will continue out of the solar system, although neither will encounter Pluto, because it is in the wrong position for such a tour.

The next outer-planet mission will be the last and most complex of those conceived during the golden era. The Galileo spacecraft will be an advanced Mariner-type spacecraft designed to go into orbit around Jupiter after releasing a probe (150 days earlier) that will plunge through its atmosphere. The 5,147-pound spacecraft was to have been launched in 1981, but delays in the first space shuttle launch and problems in developing the shuttle's Inertial Upper Stage (IUS) postponed the launch until 1986 and required substitution of a wide-body version of the Centaur upper stage for the IUS. In addition, Galileo's weight grew significantly when hardened electronics had to be used because of the intense radiation belts around Jupiter.

It is planned that Galileo will operate in Jovian orbit for at least 20 months while it uses the gravity fields of the moons to "crank" the orbit ever closer to Jupiter. The spacecraft's primary instrument will again be a camera, but this time the camera will replace the vidicon tube by a charge-coupled device identical to that developed for the Hubble Space Telescope scheduled for launch in 1986. Plasma physics and other instruments based on the Voyager findings will also be carried. Galileo probably will make a short flyby of an asteroid during its cruise out to Jupiter.

Other spacecraft are being targeted in the years just ahead to study the largest body in the solar system—the Sun—and some of the smallest—the comets. The Sun is to be observed from a new angle by the Ulysses international solar polar mission scheduled for space shuttle launch in May 1986. Ulysses will head first for Jupiter, where it will make a hairpin turn through the planet's intense gravitational field, then emerge on a trajectory that will take it high above the plane of the ecliptic and more than a 100 million miles over the Sun's north pole. Plasma physics and high-energy radiation detectors should shed light on an environment previously unexplored.

*Giotto, Europe's entry in the race to study Halley's comet, was the first spacecraft to be named for an artist. The painting* The Adoration of the Magi *by Giotto (1267?–1337) depicts the Star in the East as Halley's comet.* (ESA)

Five spacecraft were built and instrumented especially to study comet Halley, the most ambitious being the European Space Agency's Giotto, which was set off toward its distant and fast-moving target in July 1985. In January of the following year, it was to close to within 320 miles of the comet, take color photographs, and make measurements with on-board plasma instruments and mass spectrometers. Meanwhile, Japan's Planet A and Tenma probes, launched in August 1984 and January 1985, were aimed within several hundred thousand and somewhat over 9 million miles, respectively, of Halley. And the Soviets sent off two Vegas in December 1984, both fitted with balloon-borne probes dropped off at Venus in June 1985. Flybys of the comet at 6,200 and 1,800 miles were scheduled between 8 and 15 March 1986, shortly after Halley's closest approach to the Sun.

In the United States, NASA tried to get budget authority for a Halley mission but was repeatedly turned down for economic reasons. It had to make do with a couple of brief observations of the comet from the Space Shuttle—one during perihelion passage and the other timed with observations made by the European, Russian, and Japanese craft.

Budgetary and other temporary restraints notwithstanding, enthusiasm for space research continues unabated. Until man can physically reach the bounds of his curiosity, he will willingly rely on the marvelous spacecraft that have served him so well in the past. But, as we shall see in the next chapter, where his instruments can go, he will attempt to follow.

# 9 MANNED SPACE FL

From the earliest days of science fiction, manned space travel has been the ultimate goal of astronautics. Ever since Lucian of Samosata made his fictional voyage to the Moon, there has never been any doubt that manned journeys into space would take place as soon as the technology made them possible.

The early writers, relying on nothing but imagination, gravely underestimated the complexities of manned space travel. Even so knowledgeable a writer as Jules Verne had his space travelers stepping blithely into their spaceship for a trip to the Moon with no previous test flights. But as scientific and engineering knowledge grew, it became apparent that man could not go to the Moon—as the nearest heavenly body, the natural goal for manned flights—without first trying out skills closer to home. Man's vision of the future was scaled down. Instead of voyaging directly to the Moon, he would initially put space stations in orbit around the Earth, such as the Brick Moon proposed by Edward Everett Hale.

As far as can be determined, Hermann Oberth was the first man to offer a scientifically thought-out idea for a space station, in his *Die Rakete zu den Planetenraümen (The Rocket into Interplanetary Space)* of 1923.

Shortly after World War II, H. E. Ross presented a paper on orbital bases to the British Interplanetary Society. His space station had a mirror to collect solar energy and was to spin on its axis to provide artificial gravity.

In 1951, Wernher von Braun proposed a 100-foot-radius doughnut-shaped space station that would be assembled in orbit from twenty cylindrical segments, each 12 feet in diameter and made of a flexible plastic. The station would have a central hub, connected to the rim by two arms, where spacecraft from Earth would arrive and depart. Later, Von Braun expanded and refined the space station proposal. He pictured the station in a 1,075-mile-high orbit that would take it around the Earth once every 2 hours.

As early as 1944, the National Advisory Committee for Aeronautics, the Army, and the Navy had begun both independent and cooperative programs to develop high-altitude and high-speed vehicles. Starting with the wartime MX-324, the services moved to the more advanced MX-653, which led to the Xs-1 (later X-1) rocket plane. The basic goal, laid down in preliminary form in March 1944, called for a one-man rocket-powered aircraft that could exceed the speed of sound in level flight.

The X-1 program was managed by the Air Force, which had the Bell Aircraft Company build the airframe and Reaction Motors design and build a 6,000-pound-thrust liquid-propellant rocket engine with a four-chamber configuration. Since throttles to control rocket-engine thrust had not yet been perfected, the four chambers gave the pilot a wider choice of power level. By cutting in chambers, he could boost the thrust by 1,500-pound increments from 1,500 pounds to 6,000 pounds.

Since the X-1's propellant supply was very limited, the rocket plane was designed for launching from another aircraft, a modified B-29. Motorless glide trials were held in 1946, with Jack Woolams at the controls, and propulsion tests took place in 1947, with Chalmers H. Goodlin and Alvin M. Johnson as pilots. On 14 October 1947 Charles E. Yeager flew the X-1 past the speed of sound, and by 1949 it had set a speed record of nearly 970 miles per hour at an altitude record of 14 miles.

The X-1B made important contributions to both aircraft and spacecraft development by allowing tests of rocket controls in the rarefied upper atmosphere, where the air is too thin for conventional aerodynamic controls. While the Air Force was developing the X-1's, the Navy was testing the D-558-2, which was also powered by a 6,000-pound-thrust rocket engine.

Plans for more ambitious flights were already underway. On 24 June 1952, the NACA's Committee on Aerodynamics recommended an increase "in research efforts on the problems of manned flight at altitudes between 12 and 50 miles and at speeds of Mach 4 through 10," that is, four to ten times the speed of sound. This NACA recommendation was the genesis of Project Mercury, the program of the United States

# GHT

for sending astronauts into orbit around Earth. But before manned satellites could be built, more airplane tests had to be made.

Following up on the NACA recommendation, Langley Aeronautical Laboratory (now NASA's Langley Research Center) identified several problem areas. Two were "aerodynamic heating and the achievement of stability and control at very high altitudes and speeds. Of the two, . . . aerodynamic heating [was considered] to be the more serious, and, until this problem was resolved, the design of practical spacecraft impractical." The Bell X-2, powered by a Curtiss-Wright rocket engine that delivered 15,000 pounds of thrust, came to grips with these problems.

Craft that flew even higher and faster were needed to fulfill the needs of research. Two approaches were taken to meet those needs. One was the manned ballistic missile approach that led directly to Project Mercury. The other was the development of the X-15 rocket airplane.

At the time, the X-15 was the closest thing to a winged spacecraft that had been built. The performance goals, laid down in the early 1950's, included an altitude of at least 50 miles and a speed of six to seven times the speed of sound, both double the records set by an earlier rocket airplane, the X-2.

The X-15 underwent demonstration tests during November and December 1960. By 1966, it had accumulated more than 4 hours of flight above Mach 3, 2½ hours above Mach 4, 40 minutes above Mach 5, and 12 seconds above Mach 6. The all-time X-15 speed record was set on 3 October 1967, when Major William J. Knight flew plane Number 2 to an altitude of 99,000 feet at a velocity of 4,520 miles per hour (Mach 6.7). The new speed topped the old record, set on 18 November 1966, by 270 miles per hour. After the flight, it was found that a hole 6 inches in diameter had been burned through the vertical tail and that a 40-pound dummy ramjet had torn loose before it was scheduled to be jettisoned.

In 1968, the basic research was completed for which the X-15 series was conceived: the last (and 199th) flight occurred on 24 October, when Pilot William H. Dana took the craft to an altitude of 225,000 feet. Having set, during its lifetime, unofficial world records of 67 miles altitude (in 1963) and 6.7 times the speed of sound (in 1967), it was the most impressive research airplane in history.

In particular, the X-15 provided unique experience in controlling a high-speed vehicle under what amounted to space-flight conditions. At X-15 altitudes, aerodynamic controls do not work because the air is too thin. Small rockets in the nose and wings of the X-15 are fired to control the aircraft's attitude and path.

Proposals to put the X-15 into orbit by using a carrier rocket were made frequently but got nowhere. The Air Force did begin an X-20 Dyna-Soar program to put a winged, reusable spacecraft into orbit, but it was killed for budgetary and military reasons. Beyond the X-20 were proposals for aerospaceplanes capable of taking off from the ground, flying into orbit, and then gliding back. By the mid-1960's, however, the X-15 approach to space flight had fallen into relative neglect in the United States.

Whereas the rocket plane remained a valuable research tool and contributed to the manned space flight program, the goal of placing a man in orbit was achieved through the rapid advances that led from ballistic missiles to sounding rockets to unmanned spacecraft launched by carrier vehicles.

Again, it was the Russians who were first. On 12 April 1961, Major Yuri A. Gagarin was orbited in a spherical 10,400-pound Vostok 1 spacecraft. While there was shock in the United States, it was not as great as that which Sputnik 1 had caused. The Russians had not only predicted a manned space flight, they had been openly practicing for it for a year.

Between 15 May 1960 and 25 March 1961 five so-called Korabl Sputniks had been launched. Four contained recovery capsules; three were retrieved. The capsules carried both instruments and animals. The second one, for example, carried two dogs, forty mice, two rats, and several hundred insects, as well as plants and biological specimens, all of which returned to Earth unharmed after seventeen orbits. Guinea pigs and frogs were added in the third and fourth flights, along with mannequins. Both were brought back to Earth after one orbit—an exact rehearsal for the flight of Vostok 1. And it was certainly no coincidence that Vostok 2, with Major Gherman S. Titov aboard, stayed up for exactly seventeen orbits, or 25 hours, on 6 August 1961.

The Russians flew four more Vostoks, in two paired missions. Major Andrian G. Nikolayev went aloft in Vostok 3 on 11 August 1962, and Lieutenant Colonel Pavel R. Popovich followed him into orbit in Vostok 4 on the same day, in a remarkable example of precision launching. The two spacecraft, often in sight of each other, circled the Earth for several days. Nikolayev came down after 94 hours and 1,645,000 miles of travel, and Popovich after 71 hours and 1,242,000 miles. Both cosmonauts left their slowly descending capsules and parachuted to Earth near Karaganda in Kazakhstan. The flight was a forerunner of later missions when spacecraft would rendezvous in orbit.

The next Soviet double-mission resulted in two close orbital approaches. Vostok 5 was flown into orbit on 14 June 1963 by Lieutenant Colonel Valery F. Bykovsky. Vostok 6, launched two days later, had the distinction of carrying the first woman, Valentina V. Tereshkova, into orbit. The two spacecraft came within 3 miles of each other at one point, although they were usually much farther apart. Bykovsky set a record of 119 hours in orbit, and 2,063,377 miles of flight, before he came down by parachute more than 300 miles northwest of Karaganda.

The United States counterpart to the six-flight Vostok series was Project Mercury. Like the Soviet program, it was preceded by unmanned flights and relied on modified ICBM launch vehicles.

*The X-15 rocketplane, seconds before touchdown, at Edwards Air Force Base, California, in a 1961 test flight. The stubby trapezoidal wings are edge-on in this view and cannot be clearly seen.* (NASA)

*Yuri Gagarin as a major in the Soviet Air Force.* (TASS/NOVOSTI)

*The X-1B, first flown in 1954, tested rocket controls in the rarefied upper atmosphere, where the air is too thin for conventional aerodynamic controls.* (BELL AEROSYSTEMS)

Even though some work on space flight had been done before Sputnik 1, the Russian success galvanized the United States into action. One important result of the research effort on rocket planes at Langley had been the publication in March 1958 of a paper by Maxime A. Faget and others entitled "Preliminary Studies of Manned Satellites, Wingless Configurations: Nonlifting." At about that time a working committee also decided that a "ballistic entry" vehicle, powered by a military missile, was the best way to get an American into orbit quickly. NASA gave its approval, and a Space Task Group under Robert Gilruth was set up at Langley to supervise the program.

A three-phase flight-testing sequence was established for Mercury. In the first phase, Redstone missiles would be used to send men on ballistic trajectories. (As far back as August 1958 the Army Ballistic Missile Agency had proposed this under the code name Adam.) In phase 2, which was later canceled, Jupiters would be used for longer trajectories. Finally, Atlas ICBM's would orbit the manned capsule in phase 3. For preliminary testing of the capsule and its parachute system, a cluster of Little Joe solid-fuel rockets would be used.

Mercury's basic mission was to put a man in orbit, test his ability to function in space, and bring him back safely. Stringent requirements were laid down for the spacecraft. It would have a reliable escape system to rescue the crew in case of an accident just before, during, or after the launch. It would be controllable by the pilot. It would be capable of landing on water. It would carry retrorockets capable of taking it out of orbit and starting it back to Earth. And it would not glide in when it landed, but would fall along a ballistic path until its parachutes deployed to control the descent.

The spacecraft that emerged was basically a capsule that would keep one man alive in orbit for more than 24 hours. It was 11 feet long, counting the package of retrorockets that slowed the Mercury for re-entry, and slightly more than 6 feet in diameter. Mercury had an emergency solid-fuel rocket escape tower that could pull it free of its carrier vehicle if trouble developed during the launch; the escape tower was jettisoned after the launch.

Mercury's test flights started on 21 August 1959 with the first of eight Little Joe capsule launches, two with rhesus monkeys aboard. Redstones were used first to send the spacecraft on ballistic trajectories and thereafter Atlases to orbit them, once with a chimpanzee on board, and twice with dummy astronauts in the capsule.

In May 1961 Commander Alan B. Shepard was launched by a Redstone on a suborbital flight, and an identical flight was made 2 months later by Major Virgil I. Grissom. Both flights enabled the astronauts to test the spacecraft's controls, evaluate their response to rocket-powered flights, and familiarize themselves with space flight.

*At an exhibit in Moscow, Soviet citizens get their first look at the Vostok spacecraft. The sphere is the re-entry vehicle. Voskhod was virtually identical, the main difference being in the interior layout.* (TASS/NOVOSTI)

The first American in orbit was Marine Colonel John H. Glenn, Jr., who orbited the Earth three times on 20 February 1962. As Glenn prepared to re-enter the atmosphere, he found that the automatic control system had malfunctioned. He handled the retrorocket firing himself and made a successful landing 166 miles east of Grand Turk Island in the Bahamas.

The second Mercury flight, under the control of Lieutenant Commander M. Scott Carpenter, took place

*On 31 January 1961 the chimpanzee Ham soared 157 miles into space in a 16-minute, 39-second flight aboard a Mercury capsule. Back on Earth, Ham reaches for an apple, his first post-flight meal.* (NASA—MSFC)

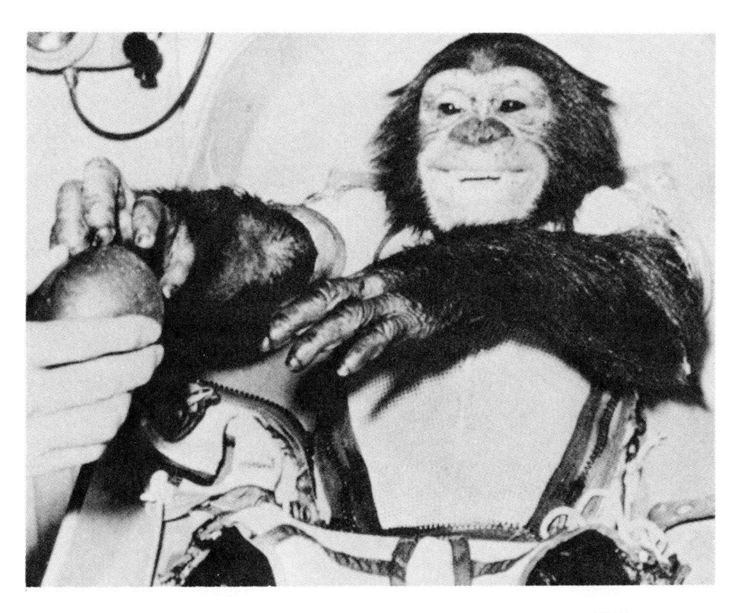

*The first seven astronauts selected for Project Mercury. Shown here in space suits, they are (front row, left to right) Walter M. Schirra, Jr., Donald K. Slayton, John H. Glenn, Jr., M. Scott Carpenter, (back row) Alan B. Shepard, Jr., Virgil I. Grissom, and L. Gordon Cooper, Jr.* (NASA)

*The small size of Mercury can be seen as the MA-5 spacecraft is mated to its Atlas booster for the last orbital monkey flight.* (NASA)

*While technicians make adjustments, John Glenn waits before boarding Friendship 7 for what turned out to be one of the scrubbed launch attempts.* (NASA)

on 24 May 1962. It was basically a repeat of the Glenn flight, although a re-entry error brought Carpenter down 250 miles off target. The third orbital Mercury flight, flown by Commander Walter M. Schirra on 3 October 1962, went for six orbits. The Mercury program was closed out by Air Force Major L. Gordon Cooper's twenty-two-orbit flight on 15 May 1963. Cooper made a pinpoint landing, coming down only 4½ miles from the recovery ship.

The next step in manned space travel was the orbiting of spacecraft with more than one occupant. Once again the Russians led the way. On 12 October 1964, they orbited the first of their Voskhod spacecraft (actually converted Vostoks), with three men aboard. Voskhod was not only heavier than Vostok, weighing 11,525 pounds, but it also represented a considerable advance in capability. Among other innovations, the craft used an ion-propulsion device for attitude control. The three cosmonauts were dressed in lightweight clothing rather than spacesuits, because of the crowding from cramming the three of them into a one-man craft. The crew on Voskhod 1 consisted of Colonel Vladimir M. Komarov of the Air Force Engineer Corps, Lieutenant Boris B. Yegorov, an Air Force doctor, and Konstantin P. Feoktistov, a civilian scientist.

Voskhod 2 provided an even more important first when it was orbited on 18 March 1965. One of its two crew members, Lieutenant Colonel Aleksei A. Leonov, left the spacecraft through an airlock and spent 10 minutes in space while the other crew member, Colonel Pavel I. Belyayev, observed him.

Voskhod 2, at 11,730 pounds, was some 200 pounds heavier than the first Voskhod. The airlock and other devices needed for Leonov's "space walk" are believed to have added the weight.

The United States' two-man Gemini program represented a similar step forward in manned orbital capability. Gemini had been conceived before the first Mercury orbital shot, and was announced publicly on 7 December 1961. The Gemini spacecraft had the same conical shape as the Mercury capsule, but it was heavier (7,000 to 8,000 pounds), bigger (18 feet 5 inches long and 10 feet in diameter), roomier by 50 percent, and it required a Titan 2 carrier, with 430,000 pounds of thrust, to put it into orbit. Gemini was also considerably more complicated.

Following two unmanned test flights, the first manned Gemini flight took place on 23 March 1965, 5 days after Voskhod 2 was launched. The flight went off perfectly. Major Grissom and Lieutenant Commander John W. Young made three orbits in the Gemini 3 spacecraft, using their onboard rockets to change the size and shape of their orbit twice and the orbital plane once. The spacecraft functioned almost without flaw.

The second manned Gemini flight, on 3 June 1965, was doubly significant. It was the first American flight whose length matched the longer Russian missions, and it allowed the United States to match the Russian space walk. Majors James A. McDivitt and Edward H. White II rode Gemini 4 for sixty-two orbits; early in the flight, White left the cabin to spend 21 minutes floating in space and maneuvering with a hand-held gas gun.

The United States captured the space endurance record with the Gemini 5 flight, which started on 21 August 1965. Gordon Cooper—who became the first astronaut to make two space journeys—and Charles Conrad, Jr., orbited for 7 days, 22 hours, and 59 minutes, a total of 120 orbits.

That record was nearly doubled in the next Gemini mission, which was distinctive for several reasons. The Gemini 6 spacecraft, which was to have been next in orbit, was held back from launch in October 1965 because the Agena with which it would practice rendezvous was destroyed after launch. Instead, Gemini 7 was sent up first, on 4 December, and the same launch pad was quickly made ready for Gemini 6. As Lieutenant Colonel Frank Borman and Commander James A. Lovell, Jr., orbited in Gemini 7, Schirra and Major Thomas P. Stafford went into orbit on Gemini 6 on 15 December.

*Astronauts James A. McDivitt and Edward H. White walk up the ramp on their way to the elevator that will take them to the Gemini 4 spacecraft.* (NASA)

*Edward H. White floating in space outside Gemini 4. The "walk" lasted 21 minutes, during which he maneuvered himself by means of the gas gun in his right hand.* (NASA)

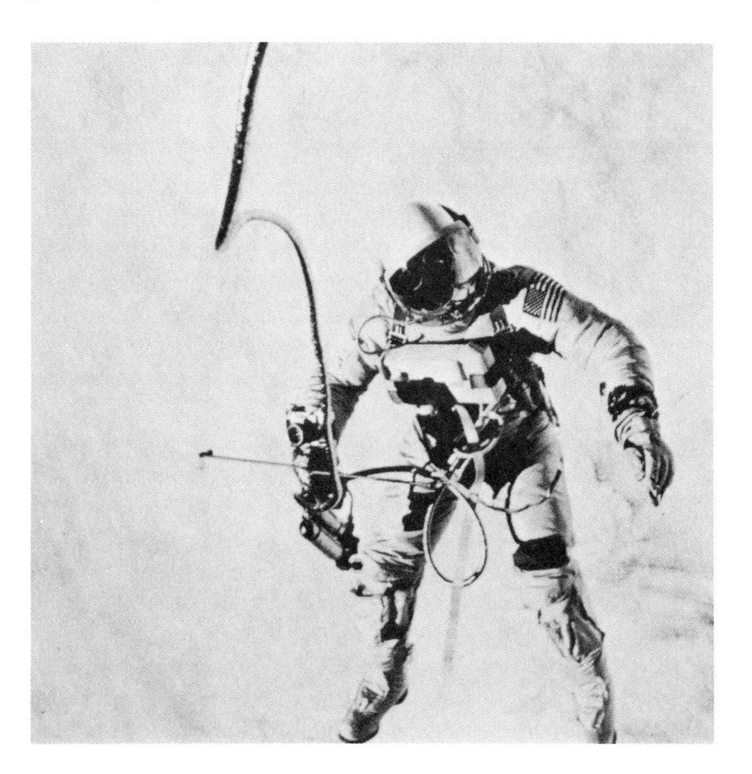

*Gemini 7 as seen from Gemini 6: the first time that two manned spacecraft rendezvoused in space. Radar used in this historic test, which helped pave the way for sending men to the moon, can be seen in the nose of Gemini 7.* (NASA)

Gemini 6 caught up with its sister craft, and in a series of impressively precise and complex maneuvers, came within 1 foot of Gemini 7, approximately 185 miles above the Earth. The two spacecraft flew in close formation for nearly 8 hours, in a dramatic demonstration of Gemini's ability to rendezvous in space. Gemini 6 splashed down in the Atlantic after 26 hours aloft, but Gemini 7 continued on for a record-breaking 206 orbits that took it more than 4 million miles in 2 weeks. The flight gave the United States nearly 1,353 man-hours in space, compared with 507 man-hours for Russia.

Gemini 8 got off to a fine start on 16 March 1966 when it went into orbit shortly after the launch of an Agena target vehicle. Late in the fifth orbit, after having accomplished the first docking operation in space, the coupled Gemini 8 and Agena commenced pitching and rolling motions, which persisted after astronauts Neil A. Armstrong and Major David R. Scott detached their craft. The mission was quickly aborted, Gemini 8 splashing down in the western Pacific Ocean instead of the Atlantic as planned.

NASA had planned to launch its Gemini 9 spacecraft on 1 June 1966, directly after an "augmented target docking adapter" was orbited by an Atlas. However, because of problems in the command data link, the launching was postponed until 3 June. Colonel Thomas P. Stafford was commander, and Lieutenant Commander Eugene A. Cernan was pilot.

In many ways, the Gemini 9 mission was the most significant to date, since it underscored the difficulties man must undergo in solving problems and working in an alien environment. Cernan's activities were so intense that he lost about 10 pounds in weight during the flight; 2 pounds of water were found in his space suit after recovery. His respiration rate went as high as 30 breaths per minute, compared with a normal 12 to 15.

Both astronauts suffered considerable fatigue during the flight because their workload was about four times greater than had been anticipated. In a rendezvous phase, for instance, they had to resort to hand calculations when the onboard computer malfunctioned. When they finally approached the target (once as close as 3 feet), they saw that the clamshell-like shrouds of the cone were open but still fastened; Stafford likened it to "an angry alligator" as it slowly tumbled in space. Ground controllers made several attempts to dislodge the shrouds by radio commands, but these were futile.

During this flight, Cernan took a space walk that lasted more than three times as long as did Edward White's extra-vehicular activity (EVA) on the Gemini 4 mission. Cernan, however, had difficulty with his maneuvering unit, and he exerted himself so greatly in his attempts to maintain position at the rear of the vehicle's adapter section that he exceeded the limits of his space suit's life-support system. This resulted in fogging of his helmet visor, which in turn so disturbed him that he was unable to use the very maneuvering unit he was trying to put into service.

Six weeks later, on 18 July 1966, both the Gemini 10 and its Agena target vehicle were orbited after flawless launches by both carriers from Cape Kennedy. The 8,248-pound Gemini 10 and its two astronauts, Commander John W. Young and Major Michael Collins, first entered into a relatively low orbit with a perigee of 99 miles. Preparations were made to raise the perigee and attempt rendezvous with the Agena 10 target. This was achieved after some 6 hours and the expenditure of an excessive quantity of propellant. Docking was successful. Using Agena 10's propulsion system, Gemini 10's perigee was raised to 184 miles and its apogee to over 476 miles. Later, rendezvous was also made with the dormant Agena target vehicle left over from the Gemini 8 mission.

During the successful 38-minute EVA, Collins used a hand-held maneuvering unit similar to the one employed in the Gemini 4 mission. He performed a number of tasks on his Gemini 10 spacecraft, and then went over to the Agena 8 target to remove some microfilm and to replace an instrument package for detecting micrometeoroids. This exchange was an important "first" that heralded a new technique in space research.

Gemini 11 was launched with great precision on 12 September 1966, less than a second later than programmed and 97 minutes after the Atlas-Agena-launched Agena target vehicle. The 8,509-pound Gemini 11 capsule carried Commander Charles Conrad, Jr., and Lieutenant Commander Richard Gordon into an orbit that ranged from 100 miles to 175 miles in altitude. Gordon's EVA went smoothly, although, as in the case of astronaut Cernan in Gemini 9, over-exertion in working outside the spacecraft resulted in an overload of the life-support system. After a completely successful rendezvous and docking maneuver, the Agena target's propulsion system was ignited to lift the two vehicles to an apogee of 851 miles, a new manned record.

The twelfth and final spacecraft in the Gemini series carried Captain James A. Lovell (it was his second Gemini flight) and Major Edwin E. Aldrin. Both Gemini 12 and its Agena target vehicle were launched from Cape Kennedy on 11 November 1966; it ended on 15 November. The flight was highlighted by a space walk by Aldrin, building up total EVA time by American astronauts to about 12 hours. He performed a number of tasks, operating specially designed space tools.

While the space walks and the docking of two vehicles were perhaps the most dramatic features of the Gemini program, each flight accomplished many different missions, including the study of prolonged weightlessness in humans, the development of rendezvous and docking techniques, the demonstration of an astronaut's ability to leave his spacecraft, and landing techniques. Gemini also had some purely scientific purposes, and other experiments with a military potential. More than twenty experiments were performed on Gemini 7 alone.

In the spring of 1967, 5 months after the end of the Gemini program and some 2 years after the Voskhod 2 mission, the Russians resumed manned space flights—disastrously, as it turned out. Their Soyuz 1 was launched from the space center at Tyura Tam on 23 April, in what was probably planned as a rendezvous flight with other cosmonauts to be orbited a day later. However, the spacecraft developed troubles in communications and stabilization. Cosmonaut Vladimir Komarov apparently resolved the stabilization problem, but the mission was terminated after 27 hours in space. Then, during the re-entry phase, the parachute recovery system of the Soyuz failed to deploy properly and the capsule crashed, killing Komarov. This was the first known death of a space flier during the course of an actual mission since manned space flight began in the spring of 1961. A monument was erected on the steppe near Orenburg where cosmonaut Komarov met his death.

The accident prompted a painstaking review of the spacecraft's guidance and recovery systems, and caused considerable delay in the Soviet manned space flight program. The Russians re-entered the field a year and a half later with re-designed spacecraft in the same Soyuz series. Unmanned Soyuz 2 was the first to be placed into orbit, on 25 October 1968, followed a day later by Soyuz 3, with cosmonaut Georgi Beregovoy aboard. He was able to rendezvous with the target satellite, but did not dock for undetermined reasons. Beregovoy also did some maneuvering in orbit and undertook extensive ground and astronomical observations before returning to Earth on 30 October. Apparently the problems that had led to the loss of Soyuz 1 in April 1967 were fully corrected, putting the Soyuz program back into full stride.

In mid-January 1969, two more manned Soyuz spacecraft were orbited. Soyuz 4, carrying Lieutenant Colonel Vladimir Shatalov of the Russian Air Force, went into orbit on January 14, followed the next day by Soyuz 5, with a three-man crew consisting of Lieutenant Colonels Boris Volynov and Yevgeni Khrunov, and civilian physicist-engineer Aleksei Yeliseyev. Both launches were given a great deal of publicity and television coverage by the Russians. After several hours of maneuvering in orbit, the two spacecraft rendezvoused and then docked. Once the vehicles were coupled, Yeliseyev and Khrunov passed first into the work compartment and then out into space through the hatch of their craft. After about an hour they entered Soyuz 4's work compartment and thence the command area. The cosmonauts then returned to their own craft, and the two vehicles were separated about 4 hours after their initial docking. The Soyuz 4 capsule came down northwest of Karaganda, while Soyuz 5 descended southwest of Kustanai in Kazakhstan. This twin flight demonstrated that the Soviets had made great progress toward their goal of perfecting space station technology.

Soyuz 6, 7, and 8, all similar to previous spacecraft in the series, were orbited successively on 11, 12, and 13 October 1971. This was the first time in the history of rocketry that three manned spacecraft were in orbit at the same time. Their crews were: Soyuz 6—Lt. Colonel Georgi S. Shonin, pilot; Master of Technical Sciences Valeri M. Kubasov, flight engineer; Soyuz 7—Lt. Colonel Anatoli V. Filipchenko, pilot; Vladislav N. Volkov, flight engineer; Lt. Colonel Viktor V. Gorbatko, research engineer; and Soyuz 8—Colonel Vladimir A. Shatalov, pilot; Master of Technical Sciences Aleksei S. Yeliseyev, flight engineer.

Soyuz 7, code-named Buran (snowstorm), and Soyuz 8, code-named Granit (Granite), carried full docking equipment, with the former slated to be the passive target and the latter the seeker. Following group flight by the three Soyuz craft, No. 8 maneuvered in toward No. 7 as if to dock. But final approach and clo-

*Soyuz 9, launched 1 June 1970, without solar panels at assembly hangar in Tyura Tam, east of Volgograd. The crew established a record-breaking 17-day, 17-hour flight as they conducted some 50 experiments on board.* (U.S.S.R. ACADEMY OF SCIENCES)

sure were not made. Perhaps something went wrong. (It is not known if the spacecraft were to have docked automatically or manually.) Soviet sources either denied that docking was intended at all or were noncommittal. Soyuz 6, Antey (Anthaeus), which did not incorporate docking equipment, made valuable compressed arc, electron beam, and consumable electrode-welding experiments. Why it was orbited during the time of the Soyuz 7 and 8 flights is not known.

Seven and a half months later (on 1 June 1970), Soyuz 9 (Sokol or Falcon) was launched from Tyura Tam with pilot Colonel Andriyan G. Nikolayev and flight engineer Vitali I. Sevastyanov aboard. During a record-breaking 17-day, 17-hour flight the cosmonauts raised their altitude twice (ending up in a 154 by 166 mile orbit) as they conducted a variety of biomedical, biological, navigational, engineering, and other tests.

A new phase of orbital operations was pioneered when Soyuz 10—called Granit like No. 8—rendezvoused and docked with a space laboratory, or embryonic space station, designated Salyut 1. Events began on 18 April 1971 when the 65.6-foot-long, 13-foot-diameter station was placed into a 125-mile perigee by 139-mile apogee orbit, inclined 51.6 degrees, and with a period of 88.5 minutes. Four days later, Soyuz 10 entered into orbit with a three-man crew consisting of Colonel Vladimir A. Shatalov (pilot), Aleksei S. Yeliseyev (flight engineer) and Nikolai N. Rukavishnikov (research engineer). Within 24 hours, ground control had shifted slightly the orbit of Salyut 1 as Soyuz 10 moved toward rendezvous. Then, after an hour and a half of maneuvering, and using new telemetry and docking systems, commander-pilot Shatalov brought his craft into a hard dock with the station. And there they remained for 5½ hours, no attempt being made by the crew to transfer into the unmanned vehicle. The next day the cosmonauts returned to Earth, landing 72 miles north of Karaganda.

Though the Soviets did not admit it, there was every reason to believe that the two-day mission was terminated prematurely. Soyuz 10 may have had systems difficulties or the vehicle's hatches and/or airlocks may have been inoperative, preventing transfer into Salyut 1.

Ground control continued to monitor and test the 40,000-pound station after the termination of the Soyuz 10 mission. The reason for doing this became clear when on 6 June 1971, Soyuz 11 was lofted into orbit. Named Yantar (Amber), its three-man crew included Lt. Colonel Georgi T. Dobrovolski, commander; Vladislav N. Volkov, flight engineer; and Viktor I. Patsayev, research engineer. The trio maneuvered their craft to within about 4 miles of the space laboratory during the course of the first day in orbit, and then allowed automatic controls to take over as closure was made. At about 300 feet, they reverted to manual control and made the dock. Once electric and hydraulic links had been established and the connection was mechanically tight, the airlocks were opened and Patsayev and Volkov crawled into Salyut 1, which thus became the first manned space laboratory in history.

The orbiting laboratory into which they entered on its orbit 794 offered about 3,500 cubic feet of space divided into several modules or compartments. Three were pressurized: the access ("transit") module through which the cosmonauts entered from their Soyuz 11 spacecraft; the 13-foot-diameter principal compartment where living and working quarters were located; and the control, communications, power supply, and life-support equipment module. Unpressurized was the final module in which reaction control engines and their control systems and propellant stores were located. The principal power source was two sets of wing-like solar panel arrays mounted outside.

For a period of 23 days, starting 7 June, the cosmonauts successfully carried out their assigned tasks, learning how to function for long periods in orbit; maneuvering, orienting, and navigating the station; making many atmospheric and Earth resources investigations, and conducting research in space physics, space biology, and space medicine.

On 29 June the cosmonauts left Salyut 1 (then in a 142- by 144-mile-high orbit) and re-entered Soyuz 11. The next morning they undocked and flew formation for a little over 4 hours. At 2235 hours (GMT), Soyuz 11's retrorocket propulsion system was fired, initiating the re-entry sequence. Firing lasted for about 5 minutes followed by a few minutes of freefall as the orbital (forward) and service (aft) modules were separated from the command (landing) module. At the moment the explosive bolts were fired, causing the separation to take place, all communications abruptly ceased. After what otherwise seemed to be a routine automatically controlled re-entry, the landing capsule touched down undamaged. But the recovery team was uneasy as it helicoptered toward the landing site, for communications had never been re-established. When team members opened the hatch a tragic sight met their eyes. All three cosmonauts had returned from their epochal 23-day flight dead. The cause: an accidentally opened exhaust valve, probably triggered by the explosive bolts during module separation. There had been no time before the capsule's air escaped for the crew to attempt to close the valve.

The Russians enjoyed momentary success when on 3 April 1973, they got Salyut 2 into a 113- by 160-mile-high orbit. The next day perigee was raised to 149 miles in apparent expectation of the arrival of a new Soyuz ferry. But no such flight was forthcoming, leading to the belief that something went wrong with what would have become Soyuz 12. On 8 April, the Soviets maneuvered the orbital laboratory to a still higher orbit (154 by 166 miles), above the traditional rendezvous altitude. Then, on 14 April it inexplicably broke up in space, with more than twenty pieces being identified by tracking stations.

A couple of months after the Salyut 2 debacle, the Soviets launched an unmanned Kosmos 573, which made a 2-day, 32-orbit flight before being recovered. That flight, it appears, was a rehearsal for a redesigned Soyuz 12, which made an identical number of orbits between 27 and 29 September 1973. Piloted by Lt. Colonel Vasili G. Lazarev with flight engineer Oleg G. Makarov, its mission was to check out new electronic and mechanical flight systems and undertake manual and automatic maneuvers. Some Earth resource photography was also accomplished, with emphasis on agricultural targets.

The Soviets refused to remained deterred by the debacle of Salyut 2, with the result that on 25 June 1974 a much redesigned Salyut 3 was orbited. Among the innovations were improved solar panels and thermal controls, a new interior layout, and better scientific instruments with which to conduct in-space experiments. On 3 July, cosmonauts Pavel R. Popovich and Yuri P. Artyukhin were sent into space aboard the Soyuz 14 craft. They subsequently docked with Salyut 3, and early on 5 July passed on board. For the next 2 weeks they successfully carried out medical, biological, and space research, and geological experiments, and engineering tests. The Soviet orbital research program appeared to have reasserted itself.

They returned on 19 July after a stay of 2 weeks. Only one more mission to Salyut 3 was attempted; the Soyuz 15 was launched on 26 August 1974 but failed to dock after several tries. These were followed by an unusual night landing. On 23 September, a re-entry vehicle was ejected (possibly carrying high-resolution film), and the station itself re-entered the atmosphere on 24 January 1975.

Until June of 1969, the Air Force had its own man-in-space program, the MOL (Manned Orbital Laboratory), which was first announced by Defense Secretary McNamara on 10 December 1963 and was placed under full development on 25 August 1965. When ready, it was to have accommodated two to three men in orbit for up to 30 days—and possibly even longer. The purpose of MOL was to shed light on the military value of man in space, to demonstrate the possibility of assembling structures in orbit, and to test the effect on man of long exposure to weightlessness. Although originally estimated to cost some $1.5 billion and to be ready for flight testing by 1970, rising costs, competing requirements for other military projects, and program re-evaluations led to funding stretchouts that would have delayed initial manned orbital testing till at least 1972. Meanwhile, spectacular progress in the Apollo program and important advances in unmanned-satellite observation techniques helped make the MOL and its Gemini B ferry spacecraft obsolete before the system could be made operational. Like the earlier Dyna-Soar, MOL never made it.

While MOL first languished and then died, Apollo moved forward toward President John F. Kennedy's goal of landing two astronauts on the Moon by the end of the 1960s. Responding to one of the oldest themes in science fiction, it began to get serious attention from scientists as the ballistic missile program expanded. As was the case with the space station idea, the proposals for lunar expeditions became less grandiose as a more refined technology for the implementation of the plan emerged from practical development.

The Apollo program was based on the development of Saturn launch vehicles. NASA first disclosed the Apollo program on 29 July 1960, at a conference it called to acquaint industry with its plans for the future. Since the Saturn designs were not firmly established at the time, the development of Saturn and its Apollo payload were interrelated almost from the beginning.

Some very basic decisions had to be made before the Apollo spacecraft could be designed. For example, how many crew members would be needed for the trip? How much equipment would they need to survive and perform scientific experiments?

And the planners had to choose from three fundamentally different flight plans.

The first, the direct-approach plan, was the simplest in concept. A launch vehicle would lift Apollo from the Earth and fly it directly to the surface of the Moon, then take off and return. The problem with this approach was that a spacecraft based on existing technology and capable of going directly to, and returning from, the Moon would have to be very heavy. To lift it initially, an extremely powerful launch vehicle would be needed—one that would require a substantially longer and costlier developmental program. Therefore, the direct-approach idea was abandoned.

An alternative flight plan was the Earth orbital rendezvous, or EOR. The Apollo spacecraft would be placed in orbit around the Earth, where it would be refueled. Only then could it fly to the Moon and return. This would require two Saturn 5 carriers, one to orbit the spacecraft attached to a partially fueled Saturn third stage and a second to launch a liquid-oxygen tanker to replenish the Saturn 5's third stage. There were advantages to this method—no huge launch vehicles would have to be developed, and the experience in orbital rendezvous and fueling could be applied to later interplanetary missions. Despite these advantages, EOR lost out to the third flight plan.

This was lunar orbital rendezvous, or LOR, an approach proposed by John C. Houbolt of the Langley Research Center early in 1962. LOR seemed complicated, but it offered one great advantage: Only one Saturn 5 would be needed to send a three-module Apollo spaceship into orbit around the Moon. One of the modules, a two-stage rocket in itself, would be detached to make the landing. After spending a short time on the lunar surface, the astronauts would return to the mother spacecraft, then discard the landing module, and return to Earth in the remaining two modules. NASA's decision to choose LOR was made in July 1962, and work on a final design for Apollo began.

The three men in the Apollo crew would spend most of the period (upward from 8 days) required for a lunar mission in the command module (CM), the only part of the Apollo spacecraft that would return to Earth. Nearly 11 feet high and 13 feet in diameter at its base, the conical module weighed approximately 12,000 pounds with the crew aboard. The CM served as the flight control center, living quarters, and re-entry vehicle at the end of the mission. It was divided into three compartments: forward, crew, and aft.

The forward compartment was built around the tunnel that connected the command and lunar modules when they were docked. It contained parachutes, recovery antennas, the beacon light, the "sling" for retrieving the capsule after it landed in the sea, reaction control engines, and a mechanism for jettisoning the forward heat shield during re-entry so that the parachutes could be deployed. The crew compartment provided the three astronauts with some 210 cubic feet of living space, vehicular controls and display panels, and other operational equipment. It was fitted with two hatches, the first employed for normal entering and exiting and the second for moving into and later back out of the lunar module once the Apollo had entered into orbit around the Moon.

Although the three astronauts were cross-trained, the flight commander (who occupied the left couch) generally operated the flight controls; occasionally the CM pilot (in the center couch) would take over. The latter's principal jobs were to guide and navigate the spacecraft along its trajectory and, when the Apollo had been placed in orbit around the Moon and the lunar module detached, to monitor the other two astronauts as they made their descent onto the surface. In the right-hand couch was the LM pilot; while in the CM, he managed all subsystems. In the event of an emergency, the CM could be handled by any of the three astronauts.

Mounted directly behind the command module was the service module (SM), a cylindrical unit more than 24 feet long and nearly 13 feet in diameter. It weighed about 55,000 pounds when loaded with propellant, and 11,500 pounds when empty. This module contained the principal propulsion system of the Apollo spacecraft, propellant, the electric system, water, and other supplies. The CM and SM flew together as a unit until the end of the lunar round trip, when the SM was jettisoned prior to atmospheric entry. The engine developed 20,500 pounds of thrust and was used to effect all major changes in velocity once the craft had left its temporary parking orbit around the Earth and had begun the voyage to the Moon.

In addition to the main rocket engine, the service module was fitted with four clusters of small reaction control engines mounted 90 degrees apart around the upper outside structure. Each engine developed about 100 pounds of thrust.

The module in which two of the three astronauts were to descend onto the Moon's surface was the lunar module, or LM. During flight out to the parking orbit, it was housed in the spacecraft-lunar module adapter (SLA), a 28-foot-long tapered cylinder be-

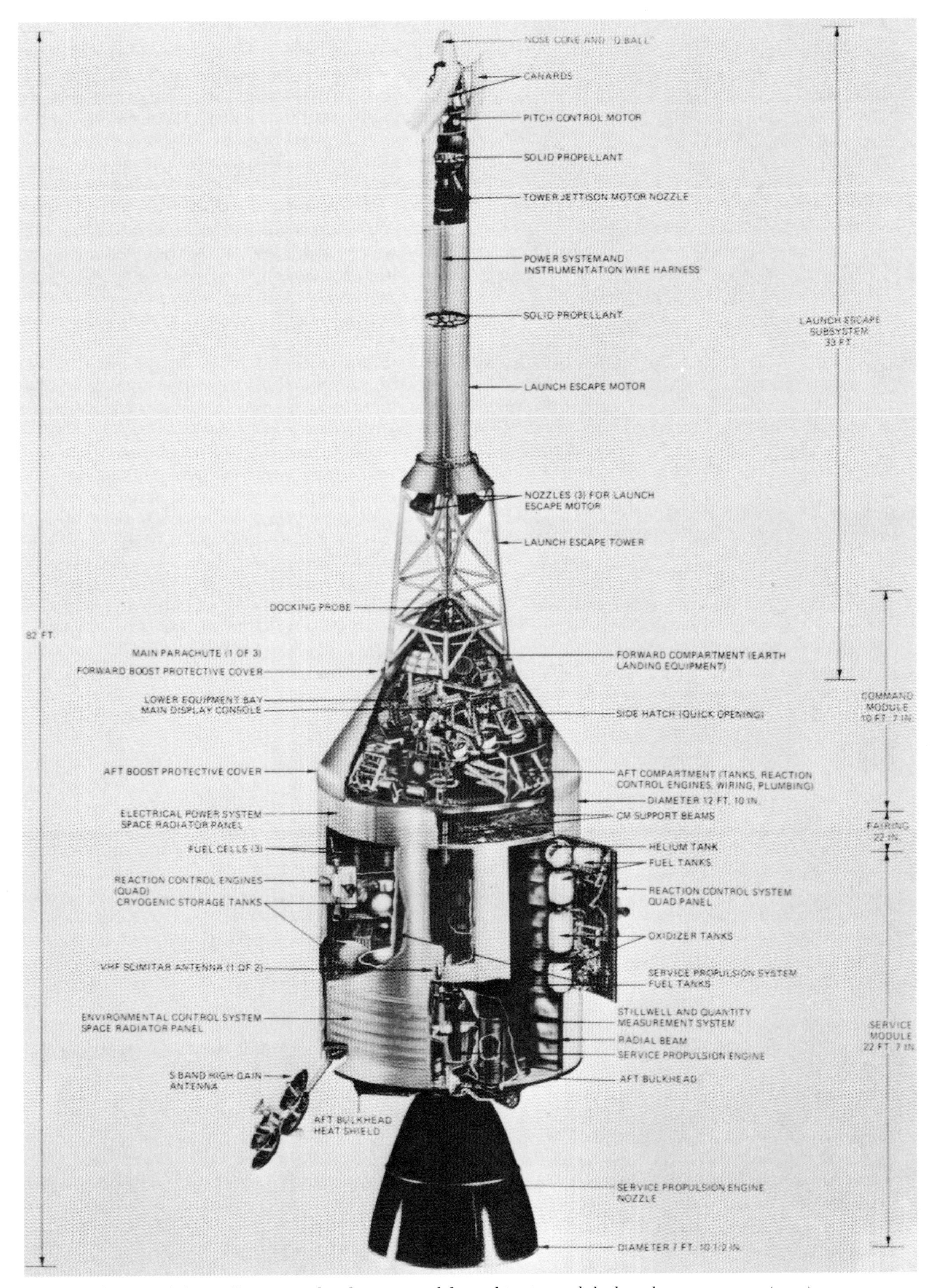


*Schematic diagram of the Apollo command and service module combination and the launch escape system.* (NASA)

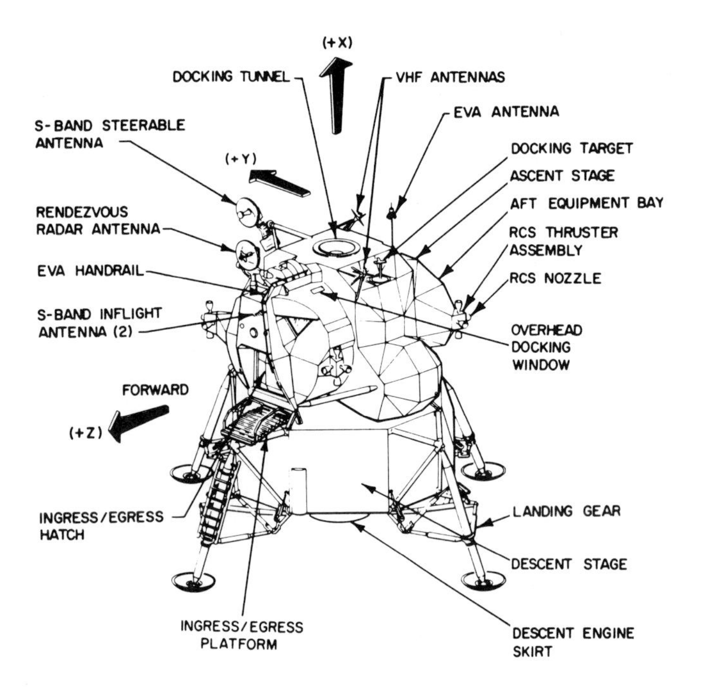


*Diagram showing the major external elements of the lunar module.* (GRUMMAN AIRCRAFT ENGINEERING; NASA)

tween the SM and the instrument unit located above the third stage of the Saturn 5 carrier. After leaving the parking orbit, the combined command and service modules separated from the combined LM, SLA, and S-4B and turned around to dock with the LM, so that the astronauts could later transfer to the landing vehicle. Once the docking was completed, the LM could be pulled out of the SLA, whose panels were then jettisoned, and the CSM and LM continued on to the Moon. The S-4B, meanwhile, was sent into a trajectory that put it into orbit around the Sun.

Built by Grumman Aircraft Engineering Corporation, the lunar module had two distinct parts: the descent stage and the ascent stage. Using the rocket engine of the descent stage to brake its landing, the LM settled softly onto the lunar surface, where it served as the astronauts' temporary home and base of exploration, communications center, and supply compartment. The descent stage became a stationary launch platform from which the ascent stage, with the two astronauts aboard, took off for lunar orbit and subsequent rendezvous and docking with the waiting Apollo command and service module combination.

The engine that powered the octagonal descent stage could be throttled from 1,050 to 9,870 pounds thrust and swiveled to permit flight control. Around the engine were located the four main propellant tanks, a variety of scientific equipment, life-support and electric system batteries, and helium, oxygen, and water tanks. The descent stage's landing gear, which was not deployed until after the LM was disconnected from the CSM, consisted of four leg-like assemblies fitted with crushable aluminum honeycomb material to help absorb the shock of the landing. The major element of the ascent or return stage was the crew compartment used for both descent and ascent, an aluminum alloy structure with titanium fittings. Of cylindrical shape, it was 7 feet 8 inches in diameter and 3 feet 6 inches wide. The total pressurized volume of the LM was 235 cubic feet, of which 160 was "habitable" space.

The LM commander, stationed on the left side of the compartment, handled engine controls, the mission timer, and lighting controls, while the LM pilot, on the right, was responsible for the abort guidance controls. Both were able to see out and below through two triangular windows canted downward and to the side. To prevent the flow of contaminants out of the hatch while it was open, an anti-bacterial filter was fitted to the cabin air relief and dump valve.

Flight testing carried out prior to the first Apollo manned mission to the Moon began in November 1963, when the launch escape system of a so-called "boilerplate" module, BP-6, was tested at White Sands, New Mexico. In 1964 and 1965, flights were made with Little Joe 2 solid-propellant rockets; and during the Saturn 1 program, engineering test models of the spacecraft were sent into orbit. By the end of February 1966, Apollo had graduated to the more powerful Saturn 1B, which carried a production model of both the command and the service modules along a suborbital ballistic trajectory. Other 1B tests in July and August were also satisfactory.

Then came tragedy. Not in the air. Nor out in the reaches of space. But on the ground. The nation, perhaps grown overconfident from continuous success, was thunderstruck. On 27 January 1967, at Cape Kennedy, astronauts Virgil I. Grissom, Edward H. White, and Roger B. Chaffee died in a flash fire that destroyed an AS-204 spacecraft undergoing routine tests at Launch Complex 34. The three Americans were to have been the first astronauts in the Apollo manned flight program. They would have been boosted into Earth orbit by a Saturn 1B carrier vehicle on 21 February to check out the Apollo in the space environment, during what could have been a 2-week trip. All three astronauts were in their couches, participating in one of a long series of pre-launch checkouts; Grissom had just stated his intention to change over from external to internal power when, at 6:31 P.M. (EST), a fire broke out in the cabin and raged for some 14 seconds.

After the fire, extensive investigations were made to learn its cause and to determine what could be done to prevent similar accidents. In April, NASA's Apollo

Review Board concluded that the "most probable initiator [of the fire] was an electrical arc," the exact location of which was unknown.

NASA moved as quickly as possible to reduce to an absolute minimum the dangers of fire aboard the command module in particular, and aboard the entire spacecraft in general. It was decided, first of all, to change the cabin atmosphere from 100 percent oxygen to a mixture of 60 percent oxygen and 40 percent nitrogen, and to employ non-combustible materials at every possible place. The matter of atmospheric composition had been considered at great length, and despite weight increases, this decision was universally approved. It was also decided to provide a new aluminum and fiber-glass access hatch that could be opened from the inside in a matter of 10 seconds. The astronauts' space suits also were re-designed to incorporate non-flammable glass fabric.

Despite temporary setbacks—the fire being by far the worst—development of the Saturn-Apollo system progressed essentially on schedule. Manned flight, the second development phase, was not destined to take place until autumn of 1968, when Apollo 7 was launched by Saturn 1B vehicle AS-205. (The AS-204 fire had temporarily slowed flight testing, although the carrier itself was later used in the Apollo 5 mission.)

The third development phase consisted of checking the flight performance of the lunar module—the one as yet untested element of Apollo. The initial unmanned orbital test of the LM was made on 22 January 1968 as part of the Apollo 5 mission.

Meanwhile, in the fourth development phase, the much larger and more powerful Saturn 5 was being prepared for its initial unmanned flight. The ability of this carrier to bring the command module to a velocity equaling that to be attained on a return trip from the Moon was a key factor in the first Saturn 5 flight test, on 9 November 1967. In this Apollo 4 mission, the huge carrier functioned flawlessly, lifting off from Launch Complex 39 at Cape Kennedy within 1 second of the planned time. It lofted CSM-017 into a 110-mile circular orbit, in what turned out to be a highly successful test of Apollo hardware. The total weight placed in orbit was 278,699 pounds, a world's record—an almost incredible increase over the 31-pound Explorer 1 satellite orbited a mere decade earlier.

Shortly after entering into its initial orbit, the S-4B stage fired a second time, inserting the CSM into an elliptical orbit with a 10,703-mile apogee. Next, the service module's propulsion system fired for 16 seconds, raising the apogee to 11,232 miles and proving that the motor could ignite and cut off in a vacuum. Later, it re-ignited and burned for a total of 271 seconds, pushing the command module to a re-entry speed of 24,911 miles per hour, to simulate lunar return reentry speeds. The CM was recovered near Midway Island, less than 5 miles from the designated target recovery point, ending a flight that was astonishingly

*The crawler-transporter carries the 363-foot-high Saturn 5 carrier vehicle (right) past the huge vertical assembly building (center) and the launch control center blockhouse (left) during "rollout" to Launch Pad B of Complex 39 at Cape Kennedy.* (NASA)

successful from almost every point of view. Not only did it represent a victory for carrier vehicle technology—the Saturn 5 was fired with all stages active on its first flight—but it proved that the vast array of Apollo ground-support equipment was capable of almost flawless operation.

The second flight in the Saturn 5 series, designated Apollo 6, was accomplished on 4 April 1968. The payload was CSM-020, which was equipped with a modified hatch, handrails, and a Block II heat shield, plus lunar module test article LTA-2R.

On 11 October 1968, 6 months after the completion of fourth-phase testing of the unmanned CSM in Saturn 5 flights, the second development phase—manned testing of the CSM with the smaller but proven Saturn 1B carrier—began with the flight of Apollo 7. This first manned Apollo flight was commanded by Navy Captain Walter M. Schirra, Jr., veteran of both the Mercury MA-8 (Sigma 7) mission in October 1962 and the Gemini 6 in mid-December 1965. His companion astronauts were Major Donn F. Eisele of the Air Force and civilian physicist Walter Cunningham. Their command and service vehicle, CSM-101, was the first of the redesigned Block II units, incorporating all the improvements made since the AS-204 capsule fire.

The routine was stiff during the 10-day and 20-hour mission, in which the crew logged 780 space man-hours. On a 16-hours-on, 8-hour-off schedule, the astronauts had to perform an impressive, and often exhausting, series of tasks that included a simulated docking with Saturn 5's third, or S-4B, stage.

Apollo 7 was as nearly perfect as one can rightfully expect a development flight to be. Of course, there were some minor problems. Schirra, and then his companions, came down with annoying colds (though their performance was unimpaired, despite their heavy workload during the 163 orbits); and they were bothered by fogging windows, an errant oxygen sensor, and tripping of a couple of circuit breakers due to current overloading. Television coverage of the activities of the astronauts aboard the craft was live—a "first" in the United States manned space program—and the audience numbered in the millions. Like the orbital flight, the re-entry went smoothly—"a nice, gentle, constant 3G trip," according to Eisele. Splashdown occurred southeast of Bermuda, less than 8 miles from the primary recovery ship, at 7:12 A.M. (EDT) on 22 October. In a relatively rough sea, Apollo 7 turned nose down so that its antennas were under water—an event that somewhat delayed recovery.

Apollo's fifth development phase called for manned operation of the three modules in orbit around the Earth, this time using the Saturn 5 as the carrier. The sixth phase called for circumnavigation of the Moon, with perhaps a pause to orbit it.

By the summer of 1968, it had become apparent that serious delays were developing in the checkout of lunar module Number 3 at Cape Kennedy, and that it would not be ready for flight until late winter or early spring of 1969. But CSM-103 could be made ready for AS-503. With these facts at hand, it was decided to reverse the fifth and sixth phases and go for the circumlunar-orbit shot first.

*Astronaut Walter M. Schirra, Jr., Apollo 7 commander, emerges from the command module in Downey, California, following testing.* (NORTH AMERICAN ROCKWELL)

Apollo 8 went out to the Moon. It went around the Moon. And it returned to the Earth from the Moon. Its virtually unqualified success made it at once a superb monument to human ingenuity, a stunning technological and scientific undertaking, and a thrilling spectacle. In a matter of days, the solar system seemed to shrink a little as mankind realized that a once impossibly distant frontier had been cracked.

The Saturn 5 performed as expected, boosting Apollo 8 onto the trans-lunar trajectory 4 days before Christmas 1968. Colonel Frank Borman, the commander; Lieutenant Colonel William A. Anders; and Captain James A. Lovell, Jr., fully understood that the lunar mission had inherent risks not present in Earth-orbital flights. The SM's propulsion system had to function to power the spacecraft out of lunar orbit and back to Earth; and when in orbit around the Moon, the three astronauts would be 3 *days* rather than an hour or so away—a grave physical and psychological factor.

The trip outward to the Moon was gratifyingly uneventful for the mission planners, for those who were monitoring it, and, of course, for the astronauts themselves. On the second orbit around the Earth, the S-4B stage's J-2 engine was fired for more than 5 minutes, bringing the craft up to escape velocity and onto the trans-lunar trajectory. And at a little more than 55½ hours after liftoff, while traveling 2,200 miles per hour, Apollo 8 passed the neutral point where the gravitational fields of the Earth and the Moon balance each other.

After the craft had swung around the rear of the Moon, the main rocket engine of the service module was fired twice. Almost immediately upon arrival in lunar orbit, the astronauts commenced photographing the spectacular scene beneath them. They became the first human beings to observe the far side of the Moon, which they reported was more rugged than the near side. Their verbal impressions were invaluable to astronomers, physicists, geologists, and geochemists back on Earth—who learned, for example, that the surface is grayish-white ("like dirty beach sand"); that many of the craters appear to be rounded (which implies that the Moon once might have had an atmosphere); and that the craters look as though they had been made "by meteorites or projectiles of some sort." Observations also were made of potential landing areas for later Apollo crews, one in particular attracting them ("it's a great spot"). Borman philosophized: "The Moon is a different thing to each of us—each one carries his own impression of what he's seen today. I know my own impression is that it's a vast, lonely, forbidding-type existence—a great expanse of nothing that looks rather like clouds and clouds of pumice stone. It certainly would not appear to be a very inviting place to live or work."

When it came time to return home, after ten revolutions in the lunar parking orbit, the service module propulsion system was fired for 3 minutes and 23 seconds. The return journey was completely routine, and re-entry into the Earth's atmosphere was so successful that the command module landed in the Pacific Ocean less than 3 miles from the predicted point of splashdown, and only 11 minutes behind schedule. Man's first trip out to the Moon—though not yet onto the Moon—was over.

Manned testing of the lunar module by simulating in Earth orbit the operations to be undertaken later in lunar orbit still had to be carried out. The step, the fifth phase in Apollo development, was a crucial, and extremely complex, one.

Apollo 9 was as successful as it was difficult. Colonel James A. McDivitt, the mission commander, and David R. Scott, pilot of the command module, experienced some illness. Russell L. Schweickart, the lunar module pilot, who fortunately did not, was faced with a particularly heavy work schedule.

During the second day, the crew fully exercised the service module's propulsion system, moving their craft out toward the optimum orbit for the series of rendezvous maneuvers that were to come. On the third day, McDivitt and Schweickart transferred from the CM to the LM, where they checked out the onboard systems and fired the descent engine over a range of thrust levels. During this time, the service module propulsion system was ignited for the fifth time to circularize the CSM-LM orbit at 155 miles.

Once in this new orbit, Schweickart took a 38-minute space walk, emerging through the LM hatch dressed in a lunar pressure suit and carrying a portable life-support system on his back. This life-support unit, tested for later lunar landings, provided communications, oxygen, and circulating water for cooling the suit.

The fifth day of the mission was at once the most difficult and the most important; upon its successful completion hinged the decision on when to go for the lunar landing itself. First, the LM and the CSM were separated to a distance of about 3 miles, and later to 15 miles. A third propulsion maneuver separated the LM by about 100 miles from the CSM. The LM then began rendezvous maneuvers that brought it close to the CSM, with which it later docked. McDivitt and Schweickart transferred back into the CM, ending more than 8 hours of exhausting separation, orbit-changing, rendezvous, and docking exercises.

Apollo 10, in broad terms, was a repeat of Apollo 8, with the lunar module acrobatics of Apollo 9 added to the excitement. It was the first flight that proved the feasibility of lunar orbit rendezvous, and the first to take man within 10 miles of the surface of the Moon. Apollo 10 also served as a test of the LM's descent propulsion system not only in the lunar landing configuration but in the lunar environment. Takeoff was on schedule, at 12:49 P.M. on 18 May 1969, and soon afterward Air Force Colonel Thomas P. Stafford (commander), Commander John W. Young (CM pilot), and Commander Eugene A. Cernan (LM pilot) were on their way toward the Moon.

As in Apollo 8, they first entered a temporary, or parking, orbit around the Earth. On the second swing around (about 2½ hours after launch), Saturn 5's third stage re-ignited and injected Apollo 10 onto its trans-lunar trajectory. A couple of hours later, the CSM separated, made the transposition maneuver, docked with the LM, and then withdrew it from the spacecraft-lunar module adapter on the third stage.

Communications most of the time were so clear, and the news coverage so intense, that the astronauts' words were soon known around the world. All three

*Apollo 11 Commander Neil A. Armstrong (left) was the first man to set foot on the Moon. Armstrong and the lunar module pilot, Edwin E. Aldrin, Jr. (right), collected samples on the lunar surface, deployed scientific instruments, and then rejoined Michael Collins (center) in the command module.* (NASA)

were awed by the experience of traveling through space. On looking back at the receding Earth, Cernan marveled, "You blink your eyes and look out there . . . and you know it's three dimensional. But it's just sitting out there in the middle of nowhere. It's unbelievable!" Almost as an afterthought, he added, "Just for the record, it looks like a pretty nice place to live!"

Instead of referring to the components of their space vehicle as command and service modules and lunar module, the crew adopted the nicknames "Charlie Brown" and "Snoopy." So, when it came time for Stafford and Cernan to transfer to the LM, they reported leaving Charlie Brown and going over to Snoopy. Once the two astronauts were there, the reaction control system (RCS) of the SM was fired briefly to separate the CSM from the LM by about 30 feet. As soon as the CM pilot, Young, reported that Snoopy was looking well, the RCS was fired to further separate the vehicles. The LM's descent propulsion system was then fired, bringing Snoopy into an orbit with a maximum altitude of about 48,000 feet at about 15 degrees from the planned landing site of the Apollo 11 mission.

The close approach was thrilling to behold. "Hello, Houston, Houston, this is Snoopy. We just saw Earth rise, and it's gotta be magnificent!" exclaimed Cernan. The Moon, he said, was beautiful . . . "There's so many things to do in such a short time. Things seem to come over the horizon at you. Okay, we're coming on Apollo Ridge . . . on my right. There's Apollo Rille right in front of my window. It appears to be just a couple of hundred feet. . . . Man, I tell you we are low. We are close, babe." Stafford joined in, observing that "there's enough boulders around here to fill up Galveston Bay. . . . It's a fantastic sight. We have different shades of brown and grays here."

Snoopy's next task was to get back to Charlie Brown by simulating the orbit that the Apollo 11 lunar module would travel when it actually ascended from the Moon's surface. By firing the descent propulsion system, Snoopy was shifted to a new orbit with a perilune of about 60,000 feet.

Just before arriving at its new perilune, Snoopy's descent stage was jettisoned; and at perilune, the ascent propulsion system took over. At the moment of separation of the descent and the ascent stages, the craft vibrated severely, causing an alarmed Cernan to yell,

"This son of a bitch! Hit the AGS! Hit the AGS!"—referring to the abort guidance system. Within seconds, Stafford had Snoopy under control, and the maneuver toward rendezvous with Charlie Brown continued. Cernan said later that he didn't "know what the hell that was, baby, but that was something. I thought we were wobbling all over the sky."

The crisis over, Snoopy's ascent stage went through the various sequences remaining for rendezvous and docking. The remaining fuel for its ascent propulsion system was burned, sending the ascent stage into orbit around the Sun.

At 6:25 A.M. on 24 May, during the thirty-first orbit, the time arrived to ignite the service module's engine to inject Charlie Brown onto the return trajectory back to Earth. "We are returning to Earth," calmly advised Stafford—following a few minutes later with "This burn was absolutely beautiful. And we've got an absolutely beautiful view of the Moon."

The return coast went smoothly, with communications being maintained until atmospheric entry. The drogue parachutes were deployed at about 23,300 feet, and splashdown occurred some 14 minutes after reentry. The entire trip lasted 8 days and 3 minutes (a minute shorter than expected), including approximately 76 hours on the outward leg, consisting of launch, Earth parking orbit, and trans-lunar coast; 61½ hours in lunar orbit; and 54 hours on the return leg. Total length of the voyage was about 700,000 miles. Landing took place within miles of the recovery ship *Princeton* (a helicopter carrier), about 400 miles east of the Pacific island of Tutuila in American Samoa.

The landing on the Moon, realized during the Apollo 11 mission, was the greatest technological triumph in the long and tortuous chronicle of mankind. The apotheosis of an age-old dream, it represented the highest motives and capabilities of the human mind. For the first time since life was spawned in the primeval "soup," creatures from Earth left their planetary cradle and strode upon an alien world. Man the infant became man the man.

President John F. Kennedy had declared in May 1961:

> I believe that this nation should commit itself to achieving the goal, before this decade is out, of landing [a man] on the Moon and returning him safely to Earth.

Apollo 11 was the fulfillment of that pledge. President Kennedy had been able to make the commitment partly because of his faith in American scientific, engineering, industrial, and management genius, and partly because he realized the time was right to translate the progress and ambitions of centuries into reality. But when Apollo 11 soared into the heavens, it rode as much on the shoulders of the giants of yesteryear as of those now living. Aristarchus, Copernicus, Newton, Einstein, Tsiolkovsky, Goddard, Oberth, Esnault-Pelterie, and countless others down the corridors of time all contributed to the epochal event.

It will be the task of future historians to assess the full impact on a struggling humanity of the first manned landing on the Moon. It seemed clear, however, that man had reached a crossroads in his quest for knowledge—that it could never again be quite the same. The limits of the new era that had dawned would be defined only by humanity's ultimate intellectual capabilities and by the energy with which it pursued its goals. As Robert H. Goddard wrote in 1922: "There can be no thought of finishing, for aiming at the stars, both literally and figuratively, is the work of generations, but no matter how much progress one makes there is always the thrill of just beginning."

At 9:32 A.M. (EDT) on 16 July 1969, hundreds of thousands of spectators in the Cape Kennedy, Florida, area and hundreds of millions of television viewers all around the world watched spellbound as the beginning passed into history. Saturn 5 performed its AS-506 mission in accordance with plan, placing the S-4B third stage, the instrumentation unit, and the Apollo 11 spacecraft, with astronauts Neil A. Armstrong, Colonel Edwin E. Aldrin, Jr., and Lieutenant Colonel Michael Collins aboard, into a temporary parking orbit approximately 115 miles above the surface of the Earth. The instrumentation unit's guidance system computed the exact moment for the S-4B stage to reignite in order to insert the spacecraft into its trans-lunar path. The course was carefully selected so that should the engine of the service module have later failed to ignite and thus not brought the Apollo 11 into lunar orbit, the craft would have swung around the Moon and returned to Earth along a so-called "free return" circum-lunar trajectory. After 5.9 minutes of operation, the S-4B stage engine cut off, leaving the spacecraft with an initial velocity of nearly 24,300 miles per hour. Mission commander Armstrong was enthusiastic: "Hey, Houston, this Saturn gave us a magnificent ride. We have no complaints with any of the three stages on that ride. It was beautiful!" Apollo 11 then began its long coast to the Moon. So accurate was the injection onto trans-lunar trajectory that three of the four planned mid-course corrections were canceled. As in earlier flights, Apollo 11's crew was inspired by the sight of the receding Earth. Said Aldrin, at 12:46 P.M. (EDT) on 17 July:

> The view is out of this world. I can see all the islands in the Mediterranean . . . Majorca, Sardinia, Corsica, a little haze over the upper Italian peninsula, a few cumulus clouds out over Greece, Sun setting on the eastern Mediterranean now.

The astronauts were kept fully informed on the progress of the Soviet Luna 15 probe, which had settled into lunar orbit while Apollo 11 was still en route. Mission control at Houston also informed them that

> President Nixon is reported to have declared a Day of Participation on Monday the [21st] for all Federal employees, to enable everybody to follow your activities on the surface. Many state and city governments and businesses throughout the country have also given their employees the day off. So, it looks like you are going to have a pretty large audience for the EVA [extra-vehicular activity.]

The trip continued uneventfully, the spacecraft rotating at a few revolutions each hour (the "barbeque" mode) to provide uniform radiation from the Sun on all parts of the hull. The first phase of lunar orbit insertion occurred on schedule, and Armstrong radioed, "It was like perfect!"

After two circuits in this orbit, the propulsion system was re-ignited to bring the craft into a roughly circular orbit of between 62 and 75 miles above the lunar surface.

Circling of the Moon continued as on-board systems were thoroughly checked out, and both commander Armstrong and the Houston controllers agreed that it was time for the LM descent maneuver to begin. The first step was to undock the lunar module, code-named "Eagle," from the CSM—which then became "Columbia." Eagle was uncoupled from Columbia—at approximately 109 hours into the mission, while the spacecraft was in orbit over the lunar far side—and the LM's reaction control system engines were ignited to effect separation from the mother ship, which remained in orbit with astronaut Collins aboard. Columbia's propulsion system then fired very briefly to place the CSM in a slightly different orbit that would put the two craft a few miles apart at one-half revolution after the start of the separation maneuver. As the LM returned to the lunar near side, Armstrong reported, "The Eagle has wings." The lunar module was ready. Man was about to descend to the Moon.

It is estimated that 500 million persons actually watched, and countless millions more either saw delayed tapes or listened to the epochal event over their radios. Only mainland China was kept in enforced ignorance of the imminent conquest of the Moon.

By firing Eagle's descent-stage rocket engine, first at 10 percent throttle and then at 40 percent, Armstrong and Aldrin changed their orbit from nearly circular to elliptical, with a perilune of about 50,000 feet above the surface. This was the descent orbit insertion maneuver. When the landing approach corridor had been identified, the engine was again fired, near the perilune point, to permit the craft to descend toward the surface. This was the first phase of the powered descent initiation maneuver.

*The Apollo Saturn 5 space vehicle carrying Apollo 11 astronauts Neil A. Armstrong, Michael Collins, and Edwin E. Aldrin, Jr., lifted off at 9:32* A.M. *(EDT) on 16 July 1969, to begin the first United States manned lunar landing mission.* (NASA)

Everything proceeded under the automatic control of the on-board computer through the point termed "high gate"—at about 7,600 feet altitude and 26,000 feet uprange from the touchdown site, south of the crater Sabine D and northwest of the crater Moltke on the Sea of Tranquillity. During this period, velocity was reduced to approximately 60 miles per hour.

From high gate, the craft was further braked and lowered to about 500 feet, or "low gate," the crew visually assessing the terrain below to pick the exact point for touchdown. At about 450 feet, the astronauts took over from the computer, having decided against an automatic landing sequence in favor of a semi-automatic mode to enable them to avoid a hazardous, rock-strewn area. The reaction control system thrusters, working under computer control, provided lateral movement over the area.

Getting even closer, Aldrin reported, "Forward, forward, good. Forty feet. Picking up some dust." And then, "Drifting to the right. . . . Contact light. Okay. Engine stop!" The contact light on the LM's instrument panel indicated that one of the 68-inch-long probes dangling from Eagle's footpads had touched the surface. Armstrong delayed 1 second after the light went on and then, looking down onto a sheet of lunar soil blowing away in all directions, turned off the descent engine. Almost as tranquil as the Sea of Tranquillity had been for aeons, Armstrong reported: "Tranquillity Base here. The Eagle has landed." The time was 4:17:41 P.M. (EDT), 20 July 1969—1 minute and 19 seconds ahead of schedule.

Man was on the Moon. More precisely (although they did not know it at the time), the astronauts were at 0 degree 41 minutes north latitude, 23 degrees 26 minutes east longitude. The exact landing spot was not determined until 12 days later.

Once Eagle was on the Moon, the world was now to get live television coverage. Ironically, the man who was nearest to the astronauts on the surface could not see what was about to happen: since Columbia was not equipped with a TV receiver, Michael Collins had to be content with listening to his fellow astronauts' words.

Originally, the mission plan had contemplated 2 hours and 4 minutes of post-landing checkout, 35 minutes for a light meal, then a 4-hour rest period and another hour for a main meal before getting ready to step down onto the surface. However, after assuring themselves that their spacecraft was in good shape, Armstrong and Aldrin requested permission to cancel the 4-hour sleep period (or at least postpone it) and go out onto the Moon as soon as they could get ready. Houston agreed: "We've thought about it. We will support it." Anyway, Armstrong and Aldrin were unlikely to get much sleep at that stage of their momentous journey.

*The second man on the Moon, Edwin E. Aldrin, Jr., left the lunar module and stepped onto the lunar surface at 11:14* P.M. *(EDT) on 20 July 1969. He had remained behind in the LM, taking still pictures and monitoring the television camera, after Neil Armstrong descended to the surface.* (NASA)

More than 3 hours were needed to get ready for the lunar EVA; donning the portable life-support system backpacks with their oxygen purge system units proved particularly time-consuming in the cramped quarters of Eagle. Then, with the cabin de-pressurized and the hatch opened, Armstrong slowly descended the nine-rung ladder. At 10:56:20 P.M. (EDT) on 20 July, he made his first step onto the surface. "That's one small step for a man, one giant leap for mankind," he said, realizing that his first words would become immortalized. While testing his footing, Armstrong began to describe what he saw:

> The surface appears to be very, very fine grain, like a powder . . . I can kick it loosely with my toes. Like powdered charcoal. I can see footprints of my boots in the small, fine particles. . . . No trouble to walk around. . . . It's quite dark here in the shadow. I can't tell if I've got good footing.

While Armstrong was down on the surface, Aldrin remained in Eagle, monitoring the television camera, taking still photos, obtaining sequence camera coverage, and observing his companion's movements outside. Hardly containing himself, he asked, "Is it okay for me to come out?" Soon he, too, was on the Moon, at 11:14 P.M. (EDT).

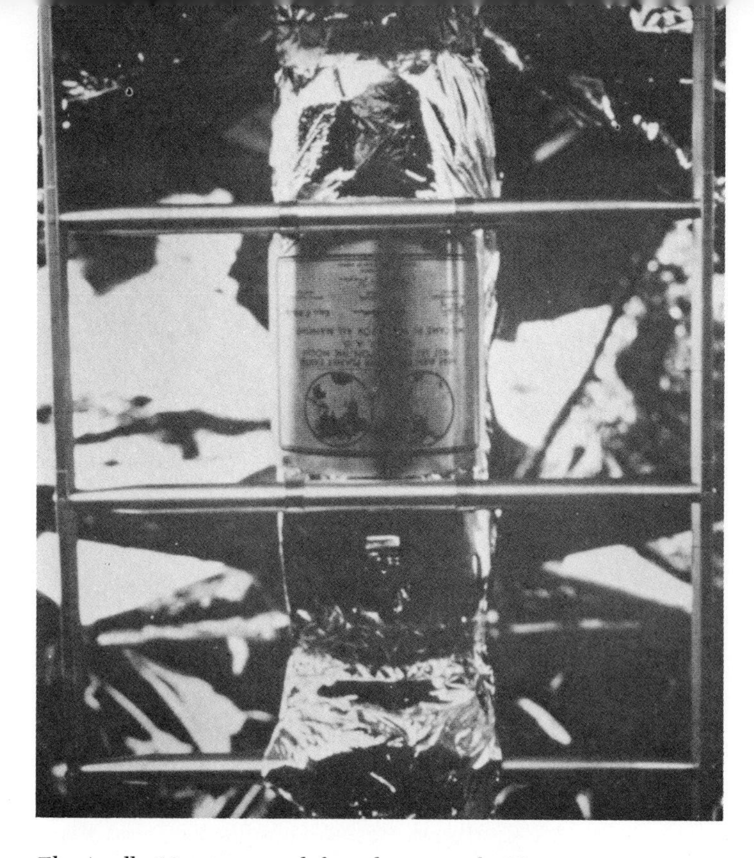

*The Apollo 11 astronauts left a plaque on the Moon commemorating their historic voyage. The plaque was attached to the descent stage of the lunar module, which remained at the landing site after the astronauts had returned to Earth.* (NASA)

Like tourists visiting a spectacular vacation area, the astronauts took dozens of photographs, and repeated "beautiful, beautiful" several times. One of their first acts was to plant an American flag on the surface. Armstrong focused the TV camera on the flag, and on the plaque attached to Eagle that read:

HERE MEN FROM THE PLANET EARTH
FIRST SET FOOT UPON THE MOON
JULY 1969, A.D.
WE CAME IN PEACE FOR ALL MANKIND

Armstrong and Aldrin had three principal tasks to perform. First, they photographed Eagle from all angles and checked it to determine if any damage had occurred during the flight and landing. Also, they gained information on the surface properties of the Moon by noting the depth (1 to 2 inches) of the depressions made by the LM footpads. Second, the astronauts had to familiarize themselves with the strange environment by simply walking and working; their evaluations of clothing and equipment were invaluable in helping to plan later explorations. Third, during the couple of hours available to them outside the Eagle, they set up three scientific experiments and collected as many operational and scientific data as possible. In order to provide panoramic coverage, they removed the television camera from the modularized equipment storage assembly (MESA) compartment in the LM descent stage and erected it on the surface. To ensure good communications, an S-band erectile antenna was deployed.

The astronauts collected rock and soil samples from the surface and hammered core tubes into the ground to retrieve sub-surface materials. The first "contingency" samples were collected by Armstrong and stowed in the LM's cabin early in the EVA. This precaution was taken to ensure that at least some material would be returned to Earth in the event the surface exploration had to be cut off for some reason prior to the collection of the main samples. The contingency samples were placed in sealed bags which were put into metal containers and transferred into the LM ascent stage by an equipment conveyor—a thin, 60-foot, continuous line of 1-inch-wide straps that looped through a support on Eagle's ascent stage and back to the ground. The samples were gathered by the astronauts with a large scoop and by tongs.

Houston was especially interested in the core samples, inquiring, "On the two core tubes which you collected, how did the driving force required to collect these tubes compare? Was there any difference?"

*The boots of the Apollo 11 astronauts left imprints about 1/8 inch deep in the layer of fine particles that covers the Moon's surface. The astronauts had no difficulty walking on this powdery substance.* (NASA)

From nearly a quarter million miles across space came the answer:

Not significantly. I could get down about the first two inches without much of a problem and then I would pound it in about as hard as I could do it. The second one took two hands on the hammer, and I was putting pretty good dents in the top of the extension rod. And it just wouldn't go much more than—I think the total depth might have been about eight or nine inches. But even there, it—for some reason—didn't seem to want to stand up straight, so that I'd keep driving it in, and it would dig some sort of a hole but it just wouldn't penetrate in a way that would support it and keep it from falling over. If that makes any sense at all. It didn't really to me.

The descriptions of the immediate area around the LM naturally were of great interest to geologists. Aldrin reported "literally thousands" of craters 1 to 2 feet in diameter, and boulders of all shapes up to 2 feet across—and some even larger. Many were above the ground, others partially or almost totally buried. Both astronauts described the ground as being similar to fine sand or silt and "slippery." During their walks around their LM, they left footprints about ⅛-inch deep. Mobility was no problem, as they quickly became accustomed to one-sixth their normal gravity and to their 188-pound (Earth weight) extra-vehicular mobility units (space suits and backpacks). While scooping out samples, Armstrong reported the ground to consist of "a very cohesive material of some sort."

The three scientific experiments deployed by Armstrong and Aldrin were the solar wind composition (SWC) detector and the early Apollo scientific experiments package (EASEP), consisting of a three-axis passive seismic experiments package (PSEP) and a laser ranging retro-reflector (LRRR). The SWC detector was a panel of very thin aluminum foil, which was unrolled so that the foil would be exposed directly to the Sun's rays. In effect, the detector was a trap for the noble-gas constituents of the solar wind, such as helium, neon, argon, krypton, and xenon. Near the end of EVA, it was folded and placed in one of the sample return containers.

The seismic experiment was deployed to monitor possible "Moonquakes" as well as meteoroid impacts, free oscillations of the Moon, and general signs of internal activity. It might also detect possible surface deformations and variations of external gravitational fields acting on the lunar mass. The seismic package was powered by a radioisotopic heater. Almost immediately, the station went into operation, first reporting tremors—made by the astronauts walking around on the ground nearby—and later what may have been quakes.

The laser reflector, the second part of EASEP, was

*Colonel Aldrin prepares to deploy a passive seismic experiments package and a laser ranging retro-reflector. The seismic package will monitor possible "Moonquakes," meteoroid impacts, free oscillations of the Moon, and general signs of internal activity. The laser reflector will "bounce" beams emitted from observatories on Earth.* (NASA)

actually an array of 100 individual reflectors upon which it was hoped to "bounce" beams emitted from observatories back on Earth. The experiment was designed to provide accurate data on the distance between Earth and the Moon, the nature of Earth's irregular rotation, the motions of the Moon, and the relative motion of Earth and the Moon. After 12 days of searching, Lick Observatory's 120-inch telescope commenced "bouncing" beams off the reflector, thus pinpointing the landing site.

After completing their work on the surface, Armstrong and Aldrin stowed the solar wind experiment, dusted off their extra-vehicular mobility units, and kicked their boots clean against a footpad. Then they ascended to the platform, swung open the hatch, disconnected the equipment conveyor, and jettisoned equipment and other items no longer needed. Among the "litter" left by Eagle were cameras, the TV unit, hand tools such as tongs and hammer, core bits, the environmental control system canister and bracket, urine bags, and the two portable life-support systems. Armstrong and Aldrin took a well-earned rest and then made ready for the return to Columbia, waiting in orbit above.

*After having deployed the seismic experiment, Colonel Aldrin walks in the direction of the laser reflector and the lunar module.* (NASA)

At 1:50 P.M. (EDT) on 21 July, Houston reported: "Eagle, you're looking good to us." Then, from Aldrin: "Roger, understand, we're Number One on the runway." A few minutes later from Tranquillity Base came the countdown: "Nine, eight, seven, . . . three, two, one . . . first stage engine on ascent. Proceed. Beautiful. Twenty-six, thirty-six feet per second up. Little pitch over, very smooth, very quiet ride." Eagle was flying. Liftoff came at 1:54 P.M. (EDT). Man's first sojourn on the Moon had lasted 21 hours and 37 minutes.

The lunar module's ascent engine burned for more than 7 minutes, providing a total velocity of 4,128 miles per hour. The first phase of the ascent involved a vertical rise to clear terrain features, followed by tipping the craft over a 52-degree angle for insertion into a lunar orbit with a perilune of about 10 miles. At apolune, the "concentric sequence initiate" occurred, targeted to place Eagle into a circular orbit nearly 17 miles below Columbia's. A terminal phase of engine burn brought Eagle along a line-of-sight path toward Columbia, to station-keeping distance. Eagle then made the terminal rendezvous and docked. Armstrong and Aldrin transferred through the LM-CM tunnel to Columbia, joined Collins, closed the hatches, and jettisoned the "Eagle" that had served them so faithfully. Columbia was Apollo 11 once again, minus its lunar module. Docking occurred at 5:35 P.M. (EDT), 3 minutes behind schedule—and 128 hours and 3 minutes into the mission (ground elapsed time).

The journey homeward was essentially a repeat of Apollos 8 and 10. The capsule splashed into the Pacific Ocean at 12:40 P.M. (EDT) on 24 July, some 900 miles southwest of Hawaii and 13 miles from the waiting carrier *Hornet*. A beacon and voice transmissions by automatically deployed antennas led the *Hornet* and its helicopters quickly to the capsule. A recovery frogman swam to the bobbing command module and passed three biological isolation garments with plastic-visored face masks in through the partially, and momentarily, opened hatch.

Donning of the isolation suits was the first step taken here on Earth in a series of rigid quarantine safety measures, developed by the Interagency Committee on Back Contamination. (During the return flight, many other precautionary measures had been taken, such as vacuuming equipment and the spacecraft interior and filtering the cabin atmosphere.) Although most scientists felt the possibility remote that any microorganisms, dangerous or benign, exist on the Moon, the quarantine was necessary to reduce to an absolute minimum the chance of epidemics occurring here on Earth, where the various forms of life might lack natural defense against alien organisms.

As soon as they emerged in their germ-proof suits, the astronauts were thoroughly washed with a decontaminant, after which they in turn washed the frogman lest he had become contaminated in the process. The astronauts were then taken by helicopter to the *Hornet* and placed immediately in a quarantine van called the mobile quarantine facility (MQF), which consisted of a lounge area, galley, and sleeping-bathing area. President Richard M. Nixon, who had conversed by phone with Armstrong and Aldrin while they were on the Moon, observed the entire procedure from the deck of the carrier, and later greeted the astronauts through the MQF window. The van was unloaded from the *Hornet* at Ford Island, Hawaii, and then flown to the Manned Spacecraft Center in Houston, where the astronauts with their precious cargo of lunar rocks and soil transferred into an 83,000-square-foot lunar receiving laboratory. The quarantine period lasted 21 days, starting from the moment that Eagle lifted off from the Moon's surface.

Within the lunar receiving laboratory, the astronauts and their attendant technicians lived in one crew reception area while physicians, medical technicians, housekeepers, and cooks occupied the other. Meanwhile, the 78 pounds of rock and soil samples gathered on the Moon and the solar wind experiment were examined in the sample operations area. Protection of these areas from outside Earth contamination, and of the world from possible lunar contamination, was afforded by "biological barrier systems"—a series of filters that treated both incoming and outgoing air. As an added precaution, air pressure inside the laboratory was somewhat lower than normal, so that flow was always inward.

While the crew were thoroughly examined and debriefed, and biological tests were continuing, the sample return containers from the Moon were brought into the vacuum laboratory for preliminary examination. Among the first group of samples tested were several lava-like rocks, indicating that volcanoes may have existed on the Moon. Another large rock sample appeared to be similar to earthly basalt. After the initial tests, the samples were repackaged and sent to the gas analysis laboratory, where the amounts and types of gases they emitted were measured. In other laboratories, the reaction of the samples to atmospheric gases and water vapor was observed, and petrological, geochemical, physical, and mineralogical analyses were conducted. After a long quarantine, the samples were sent to laboratories in the United States and to foreign countries for detailed study and analysis.

The Apollo 11 astronauts were hailed as conquering heros when they returned to Earth, but the cheers had hardly died away when doubts began to be expressed about the worth of the American space program. On 5 August 1969, less than a fortnight after Apollo 11's splashdown in the Pacific, Senator Mark O. Hatfield, a liberal Republican from Oregon, and a supporter of a continued, albeit smaller, space program, paraphrased the reactions of his constituents:

> We have unmet needs elsewhere—in education, housing, and so forth. They [his constituents] began to see the space program as the detractor, from the standpoint of meeting other needs, rather than as the complement to our total national commitment. . . . I think the euphoria of landing men on the Moon is going to fade. I admit it is one of the most magnificent things I have ever witnessed. I think you all [the NASA industry-university space team] are to be commended. But we still have to deal with . . . political realities.

Hatfield's comments apparently reflected accurately the sentiments of many voters and legislators, for NASA soon began to feel the "political realities." By 1970, the space agency's employment dropped from a high of 34,126 to a little more than 31,000. Employment of NASA contractors, meanwhile, plunged from 377,000 to 135,000. As for the agency's budget, it fell from about $5.25 billion in 1965 to $4.7 billion in 1970. And the decline continued. By the time NASA celebrated its fifteenth anniversary in October 1973, "in-house" employment had dropped to about 25,000 and contractor employment to somewhat over 100,000. The budget leveled at slightly over three billion badly inflated United States dollars.

These cutbacks forced NASA to abandon, scale down, or proceed more slowly with most of the plans that it had in development at the time of the first manned landing on the Moon. The agency had planned to launch nine more Apollo missions between 1969 and 1972; three of the missions were dropped. The agency had hoped to orbit two "workshop" space stations; only one—given the name Skylab—was built. Plans for larger, modular space stations (two in Earth orbit and one around the Moon) were canceled entirely, as was the plan to develop a nuclear rocket propulsion system that, it was hoped, would transport crews and cargo to the Moon and power spaceships to Mars by the late 1980s. A program to develop a reusable launch vehicle—the space shuttle that would reduce the cost of space operations by greatly decreasing the need for expensive one-time-use launch rockets—was maintained, but stretched out over a longer number of years than was originally envisioned. Also envisioned was a space tug, a multipurpose vehicle to maneuver between high and low orbits around the Earth.

Despite severe problems in funding, personnel, and national support, the American space program did achieve some impressive results during the first half of the 1970 decade. Between July 1969 and December 1972 twelve astronauts landed on the Moon spending some 180 man-hours working on and walking or riding across the surface (the number of men who had flown around the Moon, meanwhile, totaled twenty-seven). Through the completion of the final Skylab mission in early 1974, a total of forty-one United States astronauts became involved in thirty trips into space—starting with the short suborbital flight of Alan B. Shepard in May 1961.

To a world conditioned to blue Florida skies and warm sea breezes, the aspect presented by Cape Kennedy on the morning of 14 November 1969 was far from hospitable. But despite the low banks of clouds and persistent rain, Apollo 12—America's second manned spacecraft bound for the surface of the Moon—was committed to takeoff at 1122 hours (EST), and soon disappeared from view. Precisely 36.5 seconds later, an electric surge induced by lightning passed through the huge Saturn 5 rocket and down through the conducting exhaust plume, overloaded the circuit, and caused a temporary loss of power aboard the Apollo spacecraft. Warning lights on the spacecraft's instrument panel "lit up like a Christmas tree." As the astronauts were still busy punching in popped-out circuit breakers, lightning struck again, and the drama repeated itself. Fortunately, the rocket itself, whose own guidance system was still controlling the flight, was unaffected, and the spacecraft systems were soon back in operation. The tense, uncertain moments were over; Apollo 12 was on its way, first into Earth orbit and then along its trans-lunar trajectory.

The call signs for the spacecraft—"Yankee Clipper" for the command-service module and "Intrepid" for the lunar module—were chosen by the all-navy crew. In command was Charles Conrad, Jr., better known as Pete. The other two members of the crew were Richard F. Gordon, Jr., command module pilot, and Alan L. Bean, lunar module pilot.

Gordon remained in lunar orbit aboard Yankee Clipper while Conrad and Bean skillfully brought Intrepid down on target 1 minute 43 seconds late at 01:54:43 hours (EST) on 19 November, some 600 feet from the derelict Surveyor 3 (the unmanned soft-landing spacecraft that had sent over 6,000 closeup pictures of the lunar surface back in April 1967). During their 31½-hour sojourn on the Oceanus Procellarum plain, the cheerful astronauts made two excursions, one 4 hours long, one 4½ hours long. Shortly after 0630, as Conrad got ready to make his first step onto the surface, he exclaimed, "Whoopie, man! That may have been a small step for Neil [Apollo 11 commander Neil A. Armstrong] but that's a long one for me!" About 30 minutes later he was joined by Bean.

After studying the area in the immediate vicinity of Intrepid, they moved out to deploy the Apollo Lunar Surface Experiments Package (ALSEP). Powered by a SNAP-27 radioisotope thermonuclear generator, it was designed to send back to Earth information on the physics and structure of the surface and subsurface and the effects on the lunar environment of the corpuscular, magnetic and other phenomena emanating from or controlled by the Sun. The scientific instrumentation left on the Moon included a passive seismometer, a magnetometer, a dust detector, ionosphere and atmosphere detectors, and a solar wind spectrometer.

At the end of 4 hours they returned to the lunar module, rested, ate, and made ready for their second excursion. Some 12½ hours later they again emerged onto the surface. Enjoying the experience perhaps even more than before, Bean enthusiastically radioed that "We could work out here for 8 or 9 hours." They went back to the ALSEP array, strolled and trotted around to several small craters out to a distance of 1,300 feet, rolled a rock down one of them to test the response of the seismometer, and finally visited Surveyor 3. They discovered that the spacecraft had changed in color from tan to white, and that parts of it seemed somewhat brittle. They snipped off some tubing and cable, removed the television camera and soil scoop, and returned to Intrepid, after 4½ hours. There were a number of reasons for interest in the Surveyor. Studies of the rate of degradation of the materials of which it was composed and the outside paint—due to the solar radiations and micrometeoroid impingements—would help spacecraft designers to improve later spacecraft. Biologists and contamination experts also were anxious to know if any microorganisms that accompanied the craft from Earth had survived the craft's 2½-year exposure on the Moon's surface (they had: a dormant streptococcus colony was found and revived).

Intrepid took off and returned to Yankee Clipper without incident. After the rendezvous and docking were over, the ascent stage of Intrepid was crashed onto the surface to provide a seismic test of the Moon's internal structure. The impact occurred about 40 miles from the landing site. The return to Earth went according to plan. Just 172 hours, 27 minutes, and 16 seconds after the beginning of the journey from Cape Kennedy, the Apollo 12 command service module combination started homeward, carrying the astronauts, their samples, and the Surveyor parts. Splashdown was only a minute late and 3 miles from the

U.S.S. *Hornet* recovery ship in the Pacific on 24 November. The astronauts followed the Apollo 11 routine, ending up in the lunar receiving laboratory in Houston where they underwent the prescribed period of quarantine.

Samples brought back by Apollos 11 and 12, the only manned spacecraft to visit the flat mare plains of the Moon, were analyzed by more than 200 teams of scientists within the United States and 18 foreign countries (including the Soviet Union which, in accordance with agreements reached in January 1971, received 3 grams of Apollos 11 and 12 lunar material). As a result of the study of the samples and other data brought back from the two missions, scientists learned that the Moon is a rather quiet world—Moonquakes are infrequent—and that its interior must be quite different from that of the Earth. When the Apollo 12's Intrepid ascent stage was deliberately impacted onto the surface, the ALSEP seismometer recorded seismic signals that lasted for hours instead of minutes, as would have occurred on Earth. The seismometer studies also indicated that fewer meteoroids hit the Moon than was predicted.

The 48½ pounds of Apollo 11 and 75 pounds of Apollo 12 rock and soil samples returned were aged between 3.3 and 4.6 billion years, the latter being close to the generally accepted age of the Earth-Moon system. Based on radioactive dating of the samples, scientists believed that at least four melting periods occurred in the areas of the landing sites, resulting in near-surface intrusion of molten rock and surface extrusion of lava. The most recent such activity may have occurred a billion years ago. No evidence of water or of past or present biological activity was discovered.

Magnetic field studies suggested that the interior temperature of the Moon may be roughly 1800 degrees Fahrenheit as compared with at least 5400 degrees Fahrenheit in the center of the Earth. The surface of the Moon, which appears to be rather strong, was shown to have been eroded slowly over millions and billions of years by solar wind particles and by meteoroids and micrometeoroids. Indeed, by studying the effects of these erosion agents, much information could be gleaned of the dynamic history of the Sun and of the interplanetary environment through which the Moon moves.

Apollo 13, the third spacecraft targeted for a lunar landing, was dispatched without incident from the Kennedy Space Center on 11 April 1970 with astronauts James A. Lovell, Jr. (commander), John L. Swigert, Jr. (command-service module pilot), and Fred W. Haise, Jr. (lunar module pilot) aboard. The first couple of days the voyage was entirely routine. Then, on the third day out, a serious failure occurred in the service module's oxygen system. The command module suddenly stopped receiving oxygen and the fuel cells ceased to provide the CSM with electric power and water. A potential disaster was plainly present.

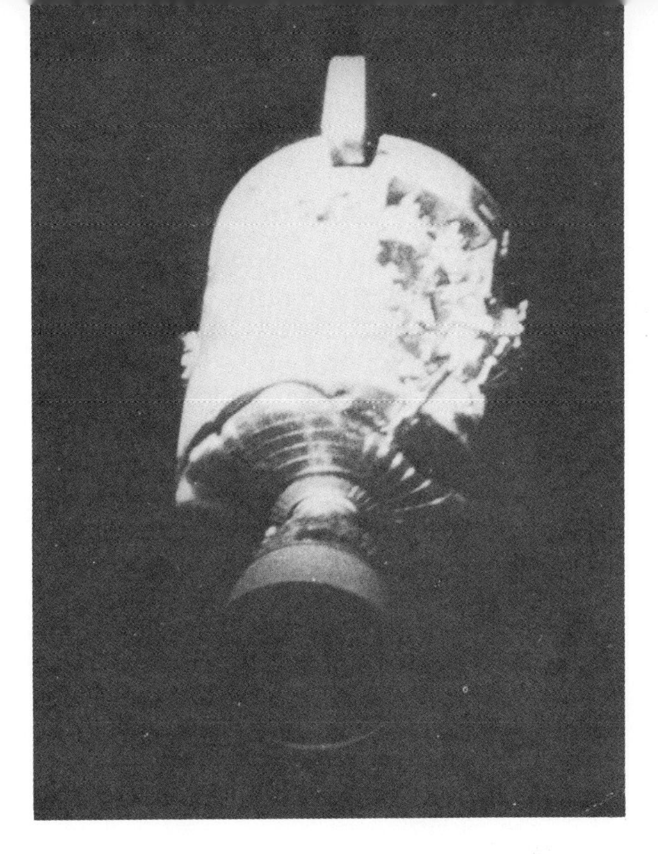

*The severely damaged Apollo 13 service module as photographed by the LM. The right-hand side of the module shows where an entire panel was blown away by the apparent explosion of an oxygen tank.* (NASA)

With the life-giving oxygen slowly being used up, the crew—with guidance from mission control in Houston—took emergency steps. The first most obvious decision was to cancel all plans for a lunar landing. Apollo 13 would change course to pass around the Moon and return as soon as possible to Earth. This decided, the crew began modifying the lunar module so that it would become a sort of lifeboat, supplying all life support until just before atmospheric re-entry when the command module's systems would be turned on.

With the continual advice and assistance of ground crews, the astronauts made the many necessary changes. Inasmuch as they had to power down equipment and heaters to conserve electricity, their stay in the LM was cold and uncomfortable. But the return voyage to Earth was completed successfully. Just before re-entry, the astronauts entered the command module, jettisoned their faithful lifeboat, and soon splashed down in the Pacific within 4 miles of the U.S.S. *Iwo Jima* recovery ship.

Apollo 13 had been targeted to land in the hilly uplands of Fra Mauro. The mission subsequently fell to Apollo 14. Launched from Kennedy the late afternoon of 31 January 1971, it was commanded by space

*Apollo 14 lunar module on the Moon's hilly upland region north of the Frau Mauro crater, February 1971.* (NASA)

veteran Alan B. Shepard, Jr., who was accompanied by command module pilot Stuart A. Roosa and lunar module pilot Edgar D. Mitchell. Apollo 14 landed on an 8-degree slope, a bare 90 feet short of the target.

The first EVA lasted 4 hours and 49 minutes, during which time Mitchell emplaced a mortar containing four high-explosive grenades to be fired by command from Earth as part of the active seismic experiment.

Another innovation brought out during the second EVA was a lightweight two-wheeled cart known as the mobile equipment transporter. It carried a portable magnetometer, various tools, and the cameras. Among many geological and other activities that occupied the 4-hour 28-minute second EVA, samples were obtained near the rim of the 1,000-foot-high Cone Crater and studies were made of a boulder-strewn field. Ascent to the waiting Roosa in CSM Kitty Hawk (from where photographic, geophysical, and astronometric investigations had been carried out) was routine, as was the return to Earth.

The seismic network confirmed that Moonquake activity peaks as the Moon passes closest to the Earth in its monthly cycle in response to tidal strains produced by the terrestrial gravitational field.

With Apollos 11, 12, and 14 successfully completed, NASA prepared for the final three landing missions, Nos. 15, 16, and 17. Redesigned spacecraft systems, improved space suits and life-support equipment, and the introduction of lunar roving vehicles allowed for a greatly augmented scientific return from each flight. Whereas the Apollo 14 astronauts could traverse 2½ miles of surface, the new lunar rovers permitted them to travel over 20 miles with 250 pounds of additional equipment and samples. Meanwhile, modified CSM's orbiting above were fitted with more sophisticated equipment, permitting more detailed studies of the physical and chemical nature of wide regions of the Moon. Even so-called subsatellites—small artificial satellites ejected from Apollo 15 and 16 CSM's—were introduced. Weighing slightly over 80 pounds each, they had the job of investigating the mass of the Moon, gravitational anomalies (including the location of mascons), particle composition in lunar space, and the interaction of the lunar and terrestrial magnetic fields. Apollo 15's subsatellite retrieved the first magnetic data from the far side of the Moon.

Justifying the final three missions, NASA pointed out that

the relative gain of science at a given site is greater than its proportion of the total number of sites. To place our accomplishments in proper perspective: the moon is about the area of North and South America. The first crew investigated an area the size of a suburban lot, the second a city block, and the third ventured about a mile away from the landing site.

The three advanced, J-class extended lunar exploration missions took place between July 1971 and December 1972. Apollo 15 was launched on 26 July 1971 with David R. Scott (commander), Alfred M. Worden (CSM pilot), and James B. Irwin (LM pilot) on board. Not until 16 April 1972 did Apollo 16 head for the Moon, with John W. Young, Thomas K. Mattingly II, and Charles M. Duke, Jr., filling the respective crew positions. The final flight in the series, Apollo 17, got off on 7 December, and turned out to be the longest of all the missions, lasting 12 days and 14 hours. The commander was Eugene A. Cernan, CSM pilot, Ronald E. Evans, and Harrison H. Schmitt, LM pilot. The CSM's and LM's for the three missions were called Endeavour and Falcon (No. 15), Casper and Orion (No. 16), and America and Challenger (No. 17).

As the initial extended lunar exploration flight, Apollo 15 was bound to chalk up some impressive firsts. Three different types of terrain were analyzed during the course of exploration of the Hadley-Apennine site—a mountain front, a mare basin, and a rille—and 168.5 pounds of samples were brought back to Earth. Among them was the Genesis Rock, a piece of anorthosite believed to have come from the early lunar crust and to be about 4.1 billion years old. In addition, an 8-foot deep drill core was returned that, among other uses, provided scientists with a record of solar activity far into the past (solar particles having been "frozen" in to the subsurface selenological record).

The voyage of Apollo 16 to the hilly, grooved Descartes terrain in the lunar highlands resulted in a successful landing 650 feet to the northwest of the target on the evening of 20 April 1972. The lunar roving vehicle was taken from Orion onto the ground and soon afterward the ALSEP array was emplaced. During the course of three EVA's totaling 20¼ hours, 213 pounds of samples were gathered, many geological investigations made, and hundreds of photos taken (including one of a rock as large as a house). The rover proved capable of traveling at a speed as high as 11 miles per hour and of climbing 15- to 20-degree slopes at nearly 5 miles per hour. Total traverse distance on the Moon's surface was 16.8 miles, mostly accomplished by the rover (though Young and Duke did spend slightly more than 9 hours on foot).

Apollo 17—the last in the series—set off for the Moon (at 0033 hours [EST]) on 7 December 1972. Landing took place at 1455 hours (EST) on 11 December at the Taurus-Littrow site in a deep, narrow valley in the mountainous highlands at the southeastern rim of the Serenitatis basin. The lunar rover was removed from the Challenger LM, and soon afterward still another ALSEP was set up on the Moon's surface. Three successful EVA's were conducted, totaling a record-breaking 22 hours, 5 minutes, and 4 seconds. Some 250 pounds of samples were retrieved (including 97 major rock samples and 75 soil samples), two double cores, and a deep drill core. More than 2,300 photographs were taken. The lunar rover carried Cernan and astronaut-geologist Schmitt for a distance of nearly 22 miles. Up in orbit, Evans operated CSM America's scientific equipment, establishing a record for longest time in lunar orbit: 75 revolutions during 6 days, 3 hours, and 48 minutes.

The premature ending of American expeditions to the Moon with Apollo 17 was a bitter disappointment to a large sector of the space community. The dismantling of the Saturn 5 and Apollo spacecraft production lines just as flight to the lunar surface was becoming somewhat routine seemed an incredible decision by the world's most powerful spacefaring nation. Yet once the decision had been made, it was irrevocable. Once again, the Moon would remain beyond the reach of man.

With the termination of Apollo, America resumed activities in Earth orbit with its embryonic space station Skylab. The origin of Skylab can be traced back to November 1962, when the Douglas Aircraft Company (later McDonnell Douglas Corporation) published a briefing manual entitled "1965 Manned Space Laboratory." In it, engineers called for modifying the Saturn 1B rocket's second (S-4B) stage into an orbital laboratory. After the stage had completed its propulsion role, an astronaut work crew (orbited by another Saturn 1B) was to go on board and convert the liquid hydrogen tank into a habitable laboratory.

From the beginning, the idea behind Skylab was to derive maximum utilization of hardware and techniques emerging from the Apollo experience. An early Apollo Extension System Program was established to synthesize and coordinate all Apollo-related concepts and proposals beyond the initial series of lunar landing missions. This was followed in August 1965 by the Apollo Applications Program set up within NASA's Office of Manned Space Flight. On 1 December of the same year, NASA headquarters gave the formal go-ahead for Skylab (then simply called "Orbital Workshop") as a major element of AAP.

Budget cuts and program stretch-outs between early 1967 and mid-1968 led to the decision by NASA that the OWS would follow the Apollo lunar landing program rather than be undertaken concurrently with it. Along with this decision, NASA announced that the huge three-stage Saturn 5 would substitute for the smaller two-stage Saturn 1B as the launch vehicle. This meant that the OWS could be sent into orbit "dry" rather than "wet." The S-4B stage, from which OWS

*Apollo 17 commander Eugene A. Cernan prepares to board the lunar rover in the Taurus-Littrow mountainous region of the Moon.* (NASA)

*Program completed: The Apollo manned lunar landing program ended successfully with the splashdown of Apollo 17 in the Pacific on 19 December 1972. Accuracy of the landing was so fine that a helicopter was in position to photograph the spacecraft as it popped up to the surface of the water after initial splashdown.* (NASA)

*The Apollo 17 command and service module as photographed from the lunar module Challenger just before it docked with the CSM, bringing the astronauts back from the Moon's surface.* (NASA)

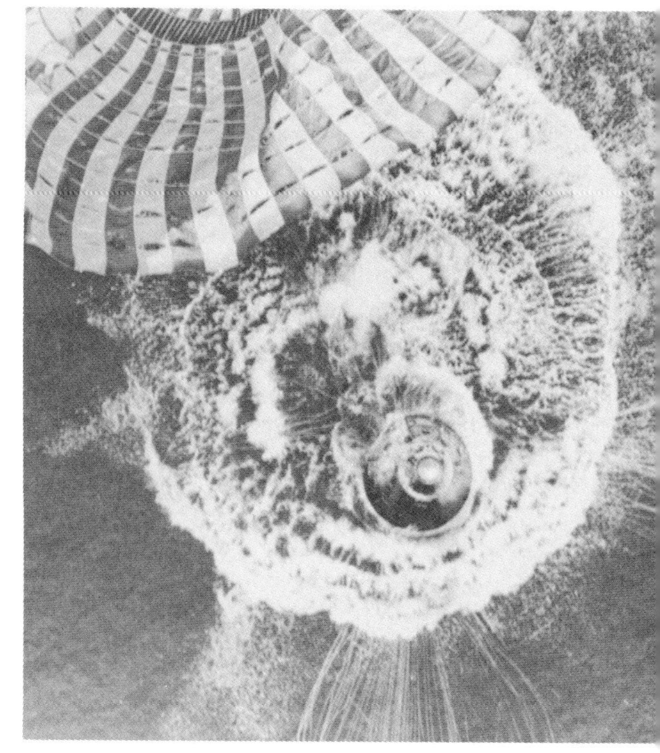

was to be fashioned, would not be employed for propulsion purposes as was the case when it served as the upper stage of Saturn 1B; rather, Saturn 5's first and second stages would orbit an S-4B already completely outfitted for its orbital laboratory mission. Since the OWS was to be prepared on the ground rather than refurbished in orbit, the installation of more elaborate equipment and crew provisions became feasible. Also, the greater lifting capability of the Saturn 5 as compared with Saturn 1B meant that a larger overall payload could be accommodated, including the Apollo telescope mount.

The name Skylab, made official on 20 February 1970, was understood to include a cluster of four elements: the main OWS, the airlock module (AM), the multiple docking adapter (MDA), and the Apollo telescope mount (ATM). Inside the orbital laboratory, which was nearly 120 feet long and weighed under 200,000 pounds, it was planned to carry out 270 scientific and technical investigations. These investigations, which required ninety different pieces of experimental hardware, covered the fields of space physics, stellar and galactic astronomy, solar physics, biological sciences, space medicine, Earth resources and meteorology, materials science and technology, in-space manufacturing, and general studies of the functioning of human beings in an orbital laboratory environment. Nearly 250 participants were involved, including some one hundred principal investigators and eighty-two foreign scientists representing forty-five institutions in eighteen countries. To undertake all the work implied by the amazing array of experiments, the crew was provided with a total living and working area of 12,398 cubic feet (in the OWS, the AM, the MDA, and the Apollo CSM—which, of course, would remain docked during each manned mission).

Prospects for a rich data collection seemed good indeed, as the unmanned Skylab cluster headed toward orbit from the Kennedy Space Center shortly after noon on 14 May 1973. At first all seemed to go well as SL-1 streaked skyward. Events ticked off on schedule, and injection into orbit was accomplished. The cover that protected the spacecraft from contamination during the launch phase was jettisoned. And the ATM was deployed without incident. But just as Skylab arched eastward following the curvature of the Earth and temporarily went out of ground communications contact, it was noticed that the micrometeoroid shield deployment signal had not been received. After 15 minutes, when communications resumed, there was still no indication that the shield had locked into position. Even more ominously, the "on" light for OWS solar array beam deployment remained dark. It seemed that the micrometeoroid shield (which also provided thermal protection) was inhibited from deploying by an interlock. (Since the solar array beam lay over the top of the undeployed shield, the latter would not be able to lock into position until the beam had been extended.)

Saturn 5 launch vehicle personnel reported "a very strange lateral acceleration at approximately 65 seconds after liftoff on the S-2 second stage." This occurred when SL-1 was traveling at a velocity just over Mach 1 and precisely 13 seconds before it experienced "max Q" (its severest aerodynamic strain). Almost immediately after this, the thermal monitoring crew noticed that OWS temperatures were starting to rise out of control. Considering that the micrometeoroid shield protected the OWS not only from particle damage but from the direct rays of the Sun, engineers began to put two and two together. Obviously the shield must have been ripped off.

As far as the micrometeoroid hazard was concerned, this was not really serious. Without the shield there were still only five chances in a thousand that a particle puncture would cause a loss in OWS pressure during the mission. The lost thermal protection of the shield was another matter: the OWS's pressure vessel now was deprived of its sunshade and hence exposed to merciless solar radiation. To put Skylab into a slow roll—"barbeque it," as space engineers would say—was not compatible with its solar research and Earth resources missions.

The lost shield clearly spelled trouble, as did the undeployed OWS solar array panels, which were to furnish half of Skylab's electric power. Mission controllers soon became aware that one of the two panels was completely gone, probably having been torn off at the same time as the micrometeoroid shield. A trickle of charge going into some of the batteries suggested that the other panel was still there and had tried to deploy but was somehow blocked in a partially opened position.

An investigating team immediately was established under the leadership of James E. Kingsbury, deputy director of Marshall's Astronautics Laboratory, and William Horton, deputy director of the Astrionics Laboratory. The team's first task was to orient the Skylab in such a position with respect to the Sun that the dangerously climbing interior temperatures could be controlled; otherwise, invaluable food, film, and other supplies would be damaged irretrievably. Essentially, the problem was to maneuver Skylab into a position where interior temperatures would remain stable without allowing the partially deployed solar panel and the main ATM panels to point away from the Sun, since then the vehicle would not receive any power.

It took about 2 days for engineers to orient Skylab to be in a position in which internal temperatures seemed to stabilize at 105 to 110 degrees Fahrenheit.

*Dangling cables atop Skylab 2 are all that remain of the solar panel, torn off at the same time as the micrometeoroid shield during launch. Despite this mishap, Skylab proved immensely successful, serving as a base for three consecutive three-man missions totaling 171 days.* (NASA)

A kind of game had to be played, pitching and rolling the vehicle until, as Kingsbury explained:

> the attitude control people were so confused that they really didn't know where they were. They had not had an opportunity to go back to a zero point and set everything, so when someone would say go back to 42 degrees, for example, they would say, "I don't know where 42 degrees is." Then that person would say, "Go back to wherever you thought it was yesterday or two hours ago. Wherever that was, go back to that." That is the kind of thing that was done.

No matter how clever ground controllers might be at staving off a disastrous temperature rise, they could not maneuver the orbital laboratory indefinitely. As Kingsbury put it, they had to put something up there that would "knock the Sun, the hot Sun, off the vehicle." One possibility that emerged from round-the-clock work sessions in Huntsville, Houston, and elsewhere was to send the first crew up to Skylab to insert, through the small scientific airlock in the OWS, a canister that contained a parasol with a center post and four legs. Another was for the astronauts to extend a couple of poles over the exposed area where the micrometeoroid shield had been and run out a curtain "much as mother used to hang clothes." These and other schemes were checked out by engineers and astronauts alike in Marshall's Neutral Buoyancy Simulator, a huge tank of water in which zero-gravity operations could be simulated.

As it turned out, both the parasol-type Sun shield and the twin-pole Sun shield were used, the first being deployed by the SL-2 crew and the second by SL-3. When SL-2's astronauts, Charles Conrad, Jr., Paul J. Weitz, and Joseph P. Kerwin, reached Skylab (which was in a nearly circular 270-mile orbit, inclined 50 degrees to the equator) after a flawless launch on 25 May 1973, they discovered that the micrometeoroid shield and one solar panel had indeed been ripped away. The remaining OWS solar beam assembly had been jammed by a piece of aluminum strip remaining from the lost shield. After completing a fly-around in their Apollo CSM they docked and the next day entered Skylab. The first order of business was to deploy the parasol Sun shield through the scientific airlock. This done, Conrad radioed to Earth: "She's out, and temperatures seem to be coming down."

There was still the matter of the undeployed solar panel whose 3,000-watt electric power generating capacity was needed vitally. During a 3-hour EVA on 7 June, Conrad and Kerwin assembled a 25-foot aluminum pole on the aft end of the OWS to one end of which they attached a cable cutter. They maneuvered the pole to clamp cutter jaws on the strip and then fastened the other end of the pole to the ATM truss structure. This "trail" was used by Conrad to work his way hand over hand to the solar panel beam. Once at the beam, he lay down across it while Kerwin did the actual cutting, closing the tool's jaws with the aid of a lanyard. He calmly reported, "We got the wing out and locked."

The $2.5 billion Skylab space station program was now truly operational. Conrad, Weitz, and Kerwin completed the SL-2 mission and were followed by SL-3's Alan L. Bean, Jack R. Lousma, and Owen K. Garriott, and finally by SL-4's Gerald P. Carr, William R. Pogue, and Edward G. Gibson. Those nine astronauts stifled conclusively any lingering doubts about the value of man in space during their respective 28-, 59-, and 84-day missions, for they not only preserved a costly and highly instrumented spacecraft but also made significant contributions to man's understanding of Earth and the universe. The chief of Marshall's ATM Experiments Branch, William C. Keathley, proudly reported that the eight solar instruments "worked extremely well—met all of our expectations and, in some instances, exceeded those expectations." Solar astronomer Richard Tousey of the Naval Research Laboratory was, if anything, more enthusiastic when he said after only the *first* manned mission that "I think it's perfectly obvious that they [the instruments] have recorded by far the best results we've ever seen, both in quality and in quantity. And I hate to say so but I sometimes think that with these experiments we have recorded more solar information and better solar information than everything we did before, rolled together."

Both in terms of man-hours in space and in terms of EVA time, the astronauts exceeded the *combined* totals of all other flights undertaken by the United States and the Soviet Union. Perhaps most important of all was the fact that the Skylab experience, in the words of program manager Leland Belew, "is providing a solid base of factual data and scientific information for use in planning space programs, including the shuttle and advanced space stations. . . . Man's performance on Skylab, in providing his capability for long-duration missions and his ability to save a mission; the successful demonstration of advanced designs in hardware and systems; and the accumulation of very high quality data with major scientific significance, all more than fulfill the program expectations."

The resounding success of Skylab would have been compromised if the ability of the astronauts to perform for long periods of time under zero gravity conditions had been impaired. It was not. The astronauts returned to Earth hale and hearty, even after the long 84-day SL-4 mission. Some temporary changes in physical condition were registered. Small drops were observed in all crew members' red blood cell mass and in the calcium content of their bones. Also, mild cardiac arrhythmia, or irregular heartbeats, were observed. But in all cases, tests indicated "rapid and stable adaptation to the zero-g condition." A few weeks after return to Earth, all abnormal readings had returned to regular preflight levels. One interesting observation was that the crew members grew from 3/4 to 1 3/4 inches taller as the effect of gravity compressing the disks in their spines was removed. They shrank to their original size shortly after returning to their domestic environment.

Skylab solar observations were conducted with the aid of seven solar telescopes comprising the Apollo Telescope Mount (ATM). Six of these recorded their observations on film and the seventh relied on the telemetric transmission of photoelectric data. During the course of the Skylab mission, the six photographic instruments returned a total of 182,800 exposures of the Sun or the solar corona while the transmitted data represented the equivalent of an additional 12,000 frames, mostly taken in the extreme ultraviolet portion of the spectrum. The astronauts demonstrated conclusively that they could do everything that was expected of them—and more.

One of the most interesting discoveries made by ATM was that the corona behaves quite differently than was previously believed. Huge, rapidly expanding loops of material ejected by the Sun, known as "coronal transients," and previously believed to be rare phenomena, were observed forty times during the three missions. These loops were found to be ejected at speeds exceeding a million miles per hour.

"Coronal holes" became a subject of great interest as the flight progressed. These holes indicate to solar physicists the local absence of 2 million-degree coronal gases radiating in the X-ray and extreme ultraviolet, with the much cooler solar surface beneath being exposed as black splotches. Coronal holes are now interpreted to be regions in which the magnetic field lines leave the Sun radially rather than looping back to another spot on the surface. The resulting weak outbound radial fields appear to serve as conduits for relatively low temperature gases puffed up by the Sun's surface, which then stream outward ultimately to form the solar wind. It may be that all the solar wind blows out of the holes rather than being a general expansion of the corona. The coronal holes, which seem to be rooted deep inside the Sun, may reveal the internal magnetic structure and provide us with a deeper understanding of the nature of, and variations in, solar energy production.

Skylab's astronomy experiments concentrated on two major objectives: (1) large-scale sky surveys in the ultraviolet portion of the spectrum and (2) comet Kohoutek, a highly publicized visitor from outer space. Other activities included studies of the composition and distribution of interplanetary dust, the incidence of cosmic rays, and the behavior of heavy ions in the Earth's magnetosphere.

Skylab's Earth Resources Experiment Package, EREP, consisted of an array of cameras, multispectral scanners, and a microwave radar scatterometer altimeter. During the course of the three missions, EREP provided many images and data that were subsequently distributed to 116 principal investigators and other interested parties in the United States and abroad. In addition, the three crews made numerous visual observations of special meteorological, flooding, and volcanic events, and took large numbers of photographs with hand-held cameras. The photos are used in a host of activities including crop and forest surveys, detection of locust and other insect damage, search for ore deposits, oceanography (including barrier reef observations), water and air pollution assessment, snowfall and related storage dam water management, and volcano and earthquake studies.

Fourteen experiments and nine science demonstrations were conducted in Skylab to study the effect of zero gravity on metallurgical processes, crystal growth, and the alloying of mixes of normally uncooperative materials. (In space, the immiscible components of materials no longer tend to separate because of differing densities, and the gravitational settling of heavy particles ceases.) Expectations were more than borne out. For instance, an indium-antimodine crystal of unprecedented smoothness and perfection was created in the furnace. By heating above vaporization temperature and subsequent cooling, germanium-selenium and germanium-tellurium crystals were made that were ten times as large and much more homogeneous than those grown in laboratories here on Earth.

No less successful were the attempts to alloy normally immiscible material compositions. When gold-germanium and lead-zinc-antimony mixtures were melted and resolidified under zero gravity, metallurgical regions of such fine dispersions were produced that elated experimenters labeled them "space mayonnaise." Such alloys cannot be made in our terrestrial laboratories.

Besides the principal scientific and technological experiments conducted on board Skylab, nineteen student experiments—selected by the National Science Teachers Association on the basis of a nationwide competition—were undertaken. The most publicized of these involved two spiders, Anita and Arabella, who learned to spin their webs in Skylab's zero gravity environment. Peter Witt, a researcher at Dorothea Dix Hospital in Raleigh, North Carolina, who had studied behavior patterns of spiders for years, was greatly surprised. He commented: "I think that now we can expect that many nonintellectual functions in animals, including human beings, are far more adaptable and able to be reprogrammed and reorganized than we ever thought. If you think about the astronauts, how they adapted, they did it all by planning, by reading books, by making calculations, by having done it in previous flights. None of this is true for the spiders."

As Skylab's workshop operations were brought to a close, the Soviets continued to work hard to stay in the space station business. Salyut 4 was launched on 26 December 1974, the day before it was announced that operations with Salyut 3 had ended. Among the science facilities carried on board was the OST-1 optical solar telescope mounted in the same location as was the Orion telescope on Salyut 1. Photographs and spectrographs were made in the ultraviolet spectrum and the 10-inch primary mirror was resilvered at least once during the manned periods. Other equipment included the Stroka (line) teletype for sending messages to the crew without the crew's having to take them by dictation, and the Delta autopilot.

Soyuz 17, carrying Alexei A. Gubarev and Georgiy M. Grechko, was launched on 11 January 1975. The two men stayed aboard for 29 days, returning to Earth on 9 February. A replacement mission that started off on 5 April did not reach orbit because of an upper-stage failure, and the two-man crew of Lieutenant Colonels Vasiliy G. Lazarev and Oleg G. Makarov were recovered safely in western Siberia. Many observers termed this Soyuz 18A; the Russians, however, preferred to think of it as simply an anomaly.

Soyuz 18B had better luck, placing Lieutenant Colonels Pyotr I. Klimuk and Vitaliy I. Sevastyonov in orbit on 24 May. They returned to Earth on 26 July after 62 days in space. No more manned missions were flown to Salyut 4, but in a curious departure from the norm, Soyuz 20 was launched on 17 November with no crew but only a cargo of biological specimens. It docked with Salyut 4 and remained for 89 days before returning to Earth. The intent was apparently to qualify the Soyuz spacecraft for on-orbit storage periods of 3 months in anticipation of long manned missions. On 15 February 1976 the spacecraft propulsion system was used to maneuver the docked craft into a better position for re-entry, which followed the next day.

In the brief interim between Salyuts 3 and 4, a single manned mission was flown in preparation for the upcoming joint mission with the United States. Soyuz 16, flown during the 2 to 8 December 1974 period by cosmonauts Colonels Anatoliy V. Filipchenko and Nikolai N. Rukavishnikov, tested the docking and pressurization systems that would be used in the Apollo-Soyuz Test Project (ASTP).

This mission became the ironic note with which the Apollo era was closed: cooperation with the very nation that had spurred the program into being. As much a creature of political convenience—détente was in vogue—as Apollo had been of political necessity, ASTP was officially hailed as opening an era of United States–U.S.S.R. cooperation in space.

ASTP's origin lay in talks that started back in 1969. The project was formally set into motion following the 24 May 1972 signing of an "Agreement on Cooperation in the Exploration and Use of Outer Space for Peaceful Purposes" by United States President Richard M. Nixon and Soviet Premier Alexei N. Kosygin. The engineering work required to bring this about was no easier than were the negotiations that led to it: the two spacecraft to be used were incompatible in virtually every respect other than their having been designed to support human life in space.

The plan that evolved placed the burden of work upon NASA, because Apollo was a more capable spacecraft. It could remain longer in orbit and had more maneuvering capability. In addition, the Saturn IB carrier had enough lift capability to haul a docking module; the Soviet launch vehicle did not. The module served two purposes: it held the docking adapter, which could not be fitted within the launch shield over Apollo; and it was a decompression chamber to allow transfers between spacecraft with different atmospheres. (The Soviets still had to reduce their cabin pressure in order to reduce transfer times and the complexity of the module.) Common procedures had to be worked out, including having the crews speak the language of each other's nation. Although slower than using one's native tongue, the practice forced the astronauts and cosmonauts to think about each phrase and be more precise in what they said. Despite what at times seemed to be insurmountable problems, not the least of which was gaining access to Soviet facilities, working groups from both nations hammered out compromises and kept the mid-1975 target date continually in sight.

Three Americans and two Russians were designated to man the spacecraft on the unique mission. Thomas P. Stafford was named as commander of what was sometimes informally called "Apollo 18." The command module pilot was Vance D. Brand, and the docking module pilot was Donald K. ("Deke") Slayton, the last of the original seven astronauts to make a flight into space. (The heart arrythmia that had kept him from Apollo flights had not recurred for several years and NASA felt confident enough to let him make the mission.) The Soviets named two veterans—commander Aleksei A. Leonov, who had made the first space walk, and flight engineer Valeriy N. Kubasov.

Launch was scheduled for 15 July 1975. Soyuz 19 took off first to make sure that the Soviet partner, with its limited capabilities, was aloft and operating. Launch vehicle and spacecraft trim maneuvers resulted in an orbit 140 miles high, well within the limits allowed for rendezvous. Just 7½ days later, the Apollo crew took off and entered into a 92.5- by 103.8-mile orbit. They efficiently extracted the docking module from within the spacecraft's lunar module adapter. By late morning on 17 July, the gap between the two spacecraft had shrunk to the point that at 12:10 EST, Leonov called out, "We have capture. . . . OK, Soyuz and Apollo are shaking hands now." After the two were pulled together and the seals made solid, Stafford asked the Soviets to "Tell Professor Bushuyev it was a good soft docking," a reference to Konstantin Bushuyev, the Soviet project manager, and the many hours he and others had spent in working out the androgynous docking system that allowed the two craft to become one.

The first crew transfer came 3 hours later over Metz, France, with Stafford and Leonov shaking hands where the docking module joined Soyuz. Four crew transfers were made before the Soviet closed up their spacecraft and disengaged from Apollo. The Soyuz subsequently redocked as the active partner. Before the Soviet craft returned to Earth, a unique science experiment was carried out. First, Apollo positioned itself so that Soyuz was between it and the Sun. Then with the Soviet spacecraft serving as a giant occulting disc, a partially successful attempt was made to view the solar corona from Apollo.

*Soyuz 19, as viewed from the Apollo spacecraft during the Apollo-Soyuz Test Project in 1975.* (NASA)

Soyuz 19 returned to Earth on 21 July, followed by Apollo 3 days later. It was during the recovery of the latter that the only blot on the mission occurred. Toxic nitrogen tetroxide (a thruster oxidizer) was sucked in by the spacecraft air system during parachute descent. Astronaut Brand passed out from the fumes and Stafford had difficulty in breathing. During the 2 weeks of hospitalization that followed, Slayton's X-rays showed a spot on his lung that was unrelated to the fumes but

*The Apollo spacecraft, with the docking module attached, as seen from Soyuz 19. The box-like extensions hold air tanks.* (NASA)

*A gathering that once seemed impossible: astronauts Stafford and Slayton with cosmonaut Leonov (center) in the Soyuz orbital module.* (NASA)

which doctors suspected might be cancerous. The spot was later removed and proved to be benign.

In addition to the engineering data returned on the mission, ASTP also carried out a number of science experiments. An extreme-ultraviolet survey camera, a helium glow detector, and a soft X-ray telescope were mounted in an equipment bay of Apollo's service module (their data were recorded on tape in the command module). Measurements of atomic oxygen in the near-space environment were undertaken by a lamp tuned to two frequencies corresponding to atomic oxygen and nitrogen resonance lines and a spectrometer mounted on the docking module along with a set of retroreflectors mounted on Soyuz. Apollo tracked the jettisoned docking module with a doppler-shift radio while the ATS-6 unmanned satellite tracked Apollo. The purpose was to determine that satellite-to-satellite tracking was possible and that it could be used in geodetic mapping. Several experiments on the effects of space conditions on living organisms were carried out.

The most dramatic science results came from the materials processing field, where eleven experiments added 125 hours to the processing time accumulated on board Skylab. Among the results achieved were segregation of kidney cells that produced significantly more urokinase (an anti-clotting factor) than the best terrestrial separations, and production of magnets with properties better than those manufactured on Earth. These results produced great optimism, some of it premature, about the future of materials processing in space.

The Apollo-Soyuz Test Project marked the end of manned space flight in the United States based on expendable launch vehicles. The value of the science accomplished was dubious. As "make work" for flight controllers, however, ASTP was invaluable in that it gave them training and purpose in the lean years between the end of Skylab and the advent of the space shuttle. Also, as a foray into Soviet territory, it was unforgettable, teaching the West much about Russian operational capabilities and management styles.

Inasmuch as ASTP was a product of politics, any benefits from it would become a victim of politics. Cooperation in space would soon be chilled as détente died and agreements were allowed to lapse in the wake of the Soviet invasion of Afghanistan and pressure exerted on the internal affairs of Poland. Exchanges of data began to take place between scientists on an informal basis, as the climate for major new joint missions became uncertain pending improvement in international affairs. Whether ASTP was worth the expense as a joint mission will have to be judged by future historians; the seeds it planted have yet to germinate.

Meanwhile, museums began to collect the remnants of the Apollo era. Three unused Saturn 5 launchers went on permanent display at the Alabama

Space and Rocket Center in Huntsville, Alabama, at the Kennedy Space Center, Cape Canaveral, Florida, and at the Johnson Space Center in Houston, Texas. The backup Skylab space station was cut into three sections so it could be squeezed through the doors of the National Air and Space Museum in Washington, D.C., and then reassembled. Remaining Apollo command, service, and lunar modules ended up in various museums and tour centers in the United States and abroad.

The closing chapter on the Saturn-Apollo era occurred 4 years later, when hopes were raised that Skylab might be reactivated and used by space shuttle crews. Increased solar activity so heated and raised the upper layers of the atmosphere that Skylab's decay from orbit was unfortunately accelerated at a rate faster than NASA had anticipated. Because of tight budgets at the time, Skylab was built without inexpensive on-board rockets that could have raised its orbit. In 1978, NASA began to work toward preventing a premature re-entry. Engineers tried to orient the station in an optimum attitude to reduce drag while leaving enough sunlight on the solar arrays to power the space station. To their pleasant surprise, the on-board systems responded as well as they had when launched. The only major problems were lower power from the solar arrays after several years in sunlight and a control moment gyro bearing that was acting up.

In time, NASA managers acknowledged to themselves and the world that Skylab was coming down faster than the first space shuttle could go up. A small unmanned space tug was rushed into production for use on the third shuttle mission. This tug would fly from the shuttle to Skylab, dock, and then use its own thrusters to boost the space station to higher orbit. But Skylab was falling ever faster and the mission plan was moved up to the second shuttle flight. Finally, in late 1978, NASA announced new shuttle delays that clearly put its first launch well beyond the most optimistic lifetime predicted for Skylab.

"Skylab is falling" became a fad saying during 1979 that held little joy for many space engineers and scientists. And the falling was progressing faster than before, with NASA now orienting the station in a maximum drag attitude to accelerate its re-entry into the atmosphere. This would allow space agency engineers to avoid populated areas by reducing drag and shifting Skylab's impact point down range. During its final hours, engineers predicted that the space station would fall on New England. So changing to low drag, the revised predicted impact point was shifted farther east. As it turned out, the station exhibited more lift than was expected and held together until it had descended to only 18.6 miles above the Earth's surface, far closer than anyone had expected. There it broke up, with pieces raining down on western Australia's Indian Ocean coast eastward to midcontinent.

The Soviet program continued at its own pace during the intervening years between ASTP and the decay of Skylab. While the ASTP astronauts and cosmonauts were shaking hands, the Soyuz 18B crew was busy aboard Salyut 4. That set a new record (by one) of seven for the number of persons in space at one time. Almost a year elapsed before Salyut 4 was commanded to re-enter on 3 February 1977.

Salyut 5, launched on 22 June 1976, seemed to follow the military pattern of Salyut 3: the orbit was low—only 129 by 144 miles high. Few details or photographs were ever released; and at the end, a re-entry vehicle was ejected to return unspecified material to Earth. Only two crews used it: Soyuz 21, flown 6 July to 24 August, by Colonel Boris V. Volynov and Lieutenant Colonel Vitaly M. Zholobov; and Soyuz 24, flown 7 to 25 February 1977, by Colonel Viktor V. Gorbatko and Lieutenant Colonel Yuri M. Glazkov. Soyuz 23, which flew from 14 to 16 October, was intended to dock with the Salyut but was thwarted by a failure in the automatic docking system. Lieutenant Colonels Vyacheslav D. Zudov and Valeriy I. Rozh-

*A Soviet A-2e launcher takes off with a Soyuz spacecraft at top. This reliable launcher has been to the Soviets what Atlas and Titan 3 have been to the United States.* (TASS/NOVOSTI)

*The Soviet Union's "mission control" is laid out in a similar manner to NASA's. Note the United States flag on the table at center foreground—this photograph was taken during the Apollo-Soyuz flight.* (TASS/NOVOSTI)

destvensky made the only water landing of the Soyuz program (showing that contingency training pays) when they were blown off course by a snowstorm and splashed down, in sub-zero temperatures, in Lake Tengiz. Boats and helicopters rescued them in short order.

The Soyuz 21 crew was expected to stay aloft longer than actually occurred. The cosmonauts' hurried departure—they skipped the standard change of diet and exercise plus 2 days of spacecraft preparations—revealed to the world that something was amiss. Reports later reaching the West indicated that there was some sort of failure in the life-support system (suggested by reported acrid odor). The crew retreated from Salyut 5 into the safety of Soyuz and soon returned to Earth.

The Soyuz 24 crew enjoyed a quiet stay in orbit while they ran the various on-board experiments. The next day, the station's unmanned re-entry capsule landed. Little more was heard of Salyut 5 until it was commanded to re-enter the atmosphere on 8 August 1977.

Soyuz 22 was flown separate from the Salyut program between 15 to 23 September by Colonels Valeriy F. Bykovskiy and Vladimir V. Aksyenov. A major element of the mission was the MKF-6 multispectral camera system built by Carl Zeiss in East Germany. It actually consisted of six cameras, four operating in visible light and two in infrared. The system was mounted at the forward end of the orbital work module in place of the docking port. Earth resources surveys were announced as part of the mission objectives, but it has also been speculated that the flight involved surveillance of NATO military maneuvers.

The Soviet space station program entered a highly successful—and visible—era with the launch of Salyut 6 on 29 September 1977. In it (and its enhanced successor, Salyut 7) the Soviets seem to have incorporated all the lessons learned from the earlier five space stations. Salyut 6 was visited by eleven crews, including the first international crew, and saw introduction of two new variants on Soyuz—the Progress unmanned resupply craft and the Soyuz T manned transport craft.

Although it was not a radical departure from its predecessors, Salyut 6 had a number of layout and subsystems changes. The aft module was widened from merely being a Soyuz service module (done in the past as an economy measure) to becoming a structure sufficient for an aft docking port. The attitude thrusters were modified to accept use of the same storable propellant supply as did the maneuvering engines, eliminating the bulky hydrogen peroxide tanks and pumps. The docking ring at this end also had interfaces for resupplying the station's propellant, air, and water tanks. The Delta guidance computer tested on previous Salyuts became operational on Salyut 6. No change was made to the internal volume; but there were a number of improvements to the subsystems, including a water regeneration plant and a collapsible shower. The science complement consisted of an MKF-6 multispectral camera like that tested on Soyuz 22, the BST-1M millimeter-wave radio telescope, and equipment for materials and life sciences experiments.

Operations aboard Salyut 6 got off to a rocky start on 9 October 1977 with the Soyuz 25 mission, carrying Vladimir V. Kovalenok and Valeriy V. Ryumin. The docking mechanism refused to operate when the two craft came together, and because Soyuz still operated on batteries, the crew had to return the next day. Another attempt was made on 10 December with Soyuz 26, carrying Yuri V. Romanenko and Georgi M. Grechko. As a precaution, the spacecraft docked with the rear port in case the problem turned out to be with the forward port. This was successful and the crew soon had Salyut 6 operating. On 20 December they performed the first Soviet extra-vehicular activity (EVA) in 9 years. Wearing new space suits designed in standard sizes and having an integral life-support pack, the cosmonauts emerged through the forward docking port and found it to be in perfect condition. Apparently, the failure lay in Soyuz 25's docking gear, which of course had been destroyed during re-entry into the atmosphere. The crew then settled in for a 96-day stay, breaking the United States 84-day record set by the final Skylab mission.

On 11 January 1978, the Soyuz 27 crew—Vladimir A. Dzhanibekov and Oleg G. Makarov—joined the Salyut 6 hosts for 5 days, returning to Earth in Soyuz 26 and leaving behind a fresh spacecraft. This procedure would be repeated many times in the future. While docked, the Soyuz 26–Salyut 6–Soyuz 27 combination was the largest Soviet assembly of spacecraft elements placed in space at the same time. It weighed 71,060 pounds.

The Progress series was introduced on 20 January 1978, making an automated docking with Salyut 6's aft port. Progress retained the basic outline of Soyuz, with the orbital work module being used to store dry goods and propellants. Other fluids were stored in tanks in an unpressurized compartment where the re-entry vehicle would otherwise be. In what became the normal mode of operation, the crew unloaded the dry goods—food, film, mail, clothing—and then stuffed the compartment with waste they had generated. About 2½ tons of supplies could be brought up. That done, the hatches between the two were sealed and the fluid transfer was made under remote control. Progress then undocked, backed away to 10 miles, and repeated the automatic docking sequence before descending for re-entry over the Pacific.

In a flight taking place from 2 to 10 March 1978, Soyuz 28 inaugurated the Intercosmos program of flying foreign guest cosmonauts. Although some experimental work was accomplished, for the most part these flights were undertaken for motivational purposes with the satellite nations. The first such flight was made by Captain Vladimir Remek of Czechoslovakia. During the course of the Salyut 6 program, he was followed by cosmonauts from Poland, East Germany, Bulgaria, Hungary, Vietnam, Cuba, Mongolia, and Romania.

*Salyut, with a Soyuz spacecraft docked at the rear (this is a replacement for one that just vacated the opposite port and took this photo). The third solar array protrudes toward the viewer in this photograph.* (TASS/NOVOSTI)

Three more Soviet crews visiting Salyut 6 set endurance records: Vladimir V. Kovalenok and Aleksandr S. Ivanchenkov, launched 15 June 1978 aboard Soyuz 29, stayed aloft for 139 days; Lieutenant Colonels Vladimir A. Lyakhov and Valeriy V. Ryumin, launched 25 February 1979 aboard Soyuz 32 stayed aloft for 175 days; and Lieutenant Colonels Leonid I. Popov and Ryumin launched 9 April aboard Soyuz 35 stayed aloft for 184 days—half a year.

Soyuz T was introduced during the Salyut 6 program. It incorporated solar panels that had been used and then discarded by earlier Soyuz spacecraft. The T series also possessed advanced electronics and improved maneuvering and navigation capabilities, and could accommodate three instead of two crew members in space suits (shirtsleeve launch and descent having been ruled out after the Soyuz 11 tragedy). The inaugural T-series flight was the unmanned Soyuz T-1 spacecraft. It docked with Salyut 6 under automatic control and returned to Earth 100 days later. The flight verified the craft's readiness to stand by for long crew stays. No doubt having a good supply of older-model Soyuz spacecraft on hand, the Soviets used both old and new models until Soyuz 40 was flown in 1981. That was apparently the last of the old series.

Also introduced was the Kosmos 929 class of supply ships. Not only did these bring up fresh supplies, but they had a re-entry vehicle attached that allowed them to return to Earth with payloads larger than Soyuz could handle. Kosmos 929 also had its own life-support system, guidance, and propulsion, raising speculation that it was intended to support experiments that had to be separate from the Salyut station. It was tested on 17 July 1977 as a free flier. Kosmos 1267 docked to Salyut 6 in June 1981 and stayed attached until the two were destroyed on re-entry into the Earth's atmosphere on 29 July 1982.

The Soyuz 32 crew also took part in the first space-based very-long-baseline interferometry experiment. On 18 July 1982, the cosmonauts deployed KRT-10, a 10-meter (33-foot) dish antenna through the aft docking port. It was electronically joined to a radio telescope at the Crimean Astrophysical Observatory to produce the net effect of a radio telescope slightly wider than the globe. KRT-10 was also used in Earth observations, making soil moisture measurements. After completing its operations, KRT-10 was to be jettisoned; but it became tangled instead in the station's structure. On 15 August the Soyuz 32 crew went out to cut it loose.

The Soviets launched the Salyut 7 space station on

19 April 1982. Outwardly it appeared to be the same as the Salyut 6 and indeed continued using the general layout of Salyut 1. However, it incorporated a number of improvements based on the experience gained during the Salyut 6 mission. The thermal control system included isolation valves that permitted repairs to be made without the need to chase globs of fluid. Also, handholds and cleats were added to make it easier for the cosmonauts to repair the station. Salyut 7's windows were given removable external covers to protect them from the pitting and surface crazing caused by micrometeoroid dust and the chemical cloud that built up around the station by its own thrusters and waste dumps. Inside, crew accommodations were made more comfortable by the substitution of zero gravity restraints for the airplane-type chair used previously and by the introduction of a buffet-style dining system that allowed each cosmonaut to choose his meals rather than draw on packages prepared and scheduled before launch. Various subsystems—including life support—were improved and made more compact, thereby freeing the space station's interior working volume. Finally, the positioning of Salyut's lighting was improved.

An enhanced version of the Delta system permitted start-and-stop communications with ground stations without crew operation as well as the use of portable personal radios that let crew members speak from wherever they were inside the station rather than having to go to the main console. Delta also improved station pointing, important in carrying out many scientific experiments.

On-board scientific equipment was also changed. The millimeter-wave telescope carried on Salyut 6 was replaced with an X-ray telescope in the same location. A gamma-ray detector was added, along with Czech and French upper-atmosphere instruments. Czech, East German, Bulgarian, Hungarian, Polish, and even Cuban and Vietnamese materials-processing facilities were added to those devised by Soviet scientists.

The first crew members to occupy Salyut 7, cosmonauts Lieutenant Colonel Anatoliy N. Berezovoy and flight engineer Valentin Lebedev, were launched on 13 May by Soyuz T-5. They docked and boarded the next day, staying in orbit for an astonishing 211 days. Not only did the crew establish a world record, but it took maximum advantage of the breaking-in period of the new workhorse of the Soviet manned space program.

During those 7 months, the cosmonauts welcomed six visitors flying two separate missions. The first of these was also the first to carry a free-world passenger on a Soviet mission. Jean-Louis Chrétien, a French Air Force pilot, was one of two trained for the mission under the Intercosmos program. This Soyuz T-6 mission, with Vladimir A. Dzhanibekov and flight engi-

*A full-scale mockup of Salyut 7 on display at the Paris Air Show, with Soyuz docked at left and Progress at right. The upright solar array was deleted because of the ceiling.* (SUE BUTLER-HANNIFIN)

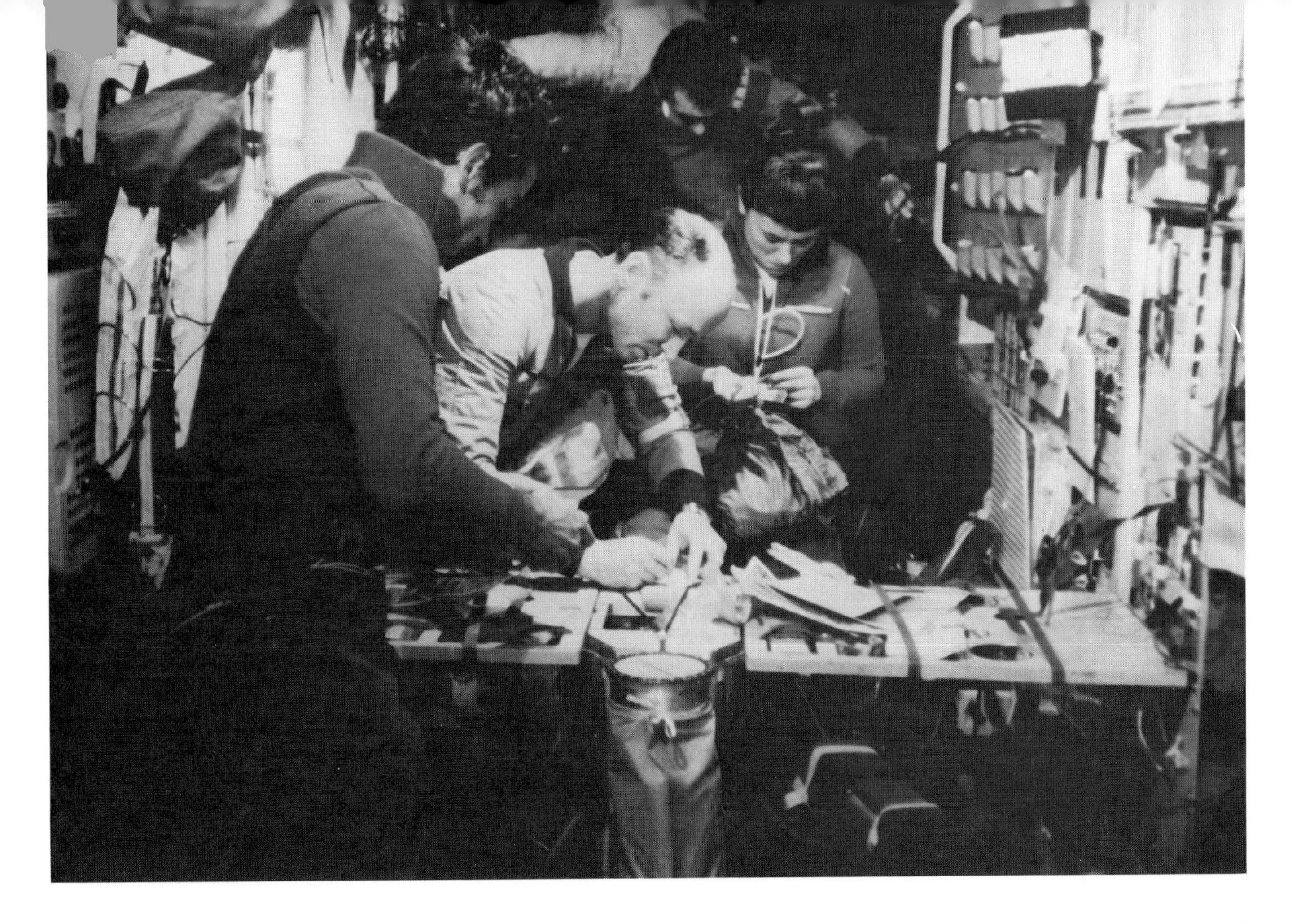

*Above: Life aboard Salyut 7 can be crowded, as can be seen in this view taken in 1982. The T-shaped platform is the crew's workbench and mess table.* (TASS/NOVOSTI)

*Below: As on American spacecraft, clutter was inevitable aboard Salyut 7, with gear being tethered to anything convenient. Here Oleg Atkov prepares to run on the treadmill during his 237-day stay.* (TASS/NOVOSTI)

neer Aleksandr S. Ivanchenkov aboard, was launched on 24 June. The three cosmonauts docked early on 25 June and stayed aboard Salyut 7 for 7 days. Experiments were conducted, and the Echograph, a French-built echocardiograph that uses sound to produce an image of the heart's action, was utilized. This was the first use of such an instrument for the study of human adaptation to space.

The second visit to Salyut 7 carried another guest cosmonaut—the second woman in space. Svetlana Y. Savitskaya, a noted aerobatics pilot and holder of several records, followed Valentina V. Tereshkova's flight by almost two decades, and preceded by almost a year (to no one's surprise) the flight of the first United States woman in space, Sally Ride. Also aboard the T-7 mission were Leonid Popov and Anatoly Serebrov. Upon her arrival on Salyut 7, Savitskaya was presented with an apron and asked to fix dinner (this seemed typical of the Soviet attitude; official comments were that women would not be equal crew members for some time and that even then it would be nice to add the feminine touch to balance long missions). Savitskaya declined, reminding Berezovoi and Lebedev that it was their duty as hosts. The visiting crew stayed aboard only 8 days.

Berezovoi and Lebedev continued on alone to set the 211-day world record for space endurance. They finally returned to Earth on 10 December 1982. Although their adaptation to gravity was understandably slow, the two cosmonauts apparently suffered no ill effects from their months-long orbital sojourn.

*Svetlana Savitskaya before she became the second woman in space. Unlike Valentina Tereshkova, Savitskaya was recognized as a skilled pilot before she was selected to fly in space.* (TASS/NOVOSTI)

In a series of missions indirectly related to Salyut, the Soviets relied on late-model versions of their Vostok spacecraft carrying animals to measure the effects of space flight on life. Kosmos 1129, for example, lofted into orbit five rats on 25 September 1979. The experiment, conducted in cooperation with the United States, lasted 18½ days. Kosmos 1514 was flown from 14 to 19 December 1983, with two rhesus monkeys and several rats on board.

Salyut 7 remained relatively inactive until 2 March 1983, when Kosmos 1443 was launched. On 10 March, it docked to the still-unmanned station and delivered 4.4 tons of cargo and a re-entry vehicle for returning experiment samples from the space station.

Kosmos 1443 was pronounced to be a combination of space tug and freighter. Launched by a Proton rocket (the same carrier as that used for Salyut), it was 13.2 feet wide by 43 feet long with solar arrays covering almost 4,800 square feet. It incorporated its own life-support system and had almost 13,000 cubic feet of pressurized cabin space. Trolleys were repeatedly used to help in transferring cargo. Missions announced for Kosmos 1443 involved an exoatmospheric observatory, a biological greenhouse, and a materials processing facility.

An attempt to reman Salyut 7 was aborted on 19 April 1983. Though the Soyuz T-8 launch, with cosmonauts Vladimir Titov, Gennediy M. Strekalov, and Aleksandr A. Serebrov aboard, was uneventful, problems began to crop up upon arrival in orbit. For one thing, it appeared that the on-board rendezvous radar system failed to function properly. Also, engine problems were reported. After a couple of days, the crew decided to return to Earth along a steep ballistic re-entry rather than take a more leisurely and smoother glide home.

On 27 June 1983, Soyuz T-9 carried Vladimir A. Lyakhov and Aleksandr Aleksandrov aloft to reactivate the station to which Kosmos 1443 was still coupled and to transfer cargo. In an attempt to exchange Soyuz spacecraft at Salyut, as was done with T-7 and T-5, the Soviets scored another unwanted space first when, on 27 September, the crew of an undesignated Soyuz (mistakenly called T-10 in some reports) survived an on-the-pad launch abort. A propellant leak developed in the launcher's tanks as the engines were starting. The escape tower was fired, pulling the spacecraft away seconds before the launch vehicle exploded. The Soyuz parachuted to Earth 2 miles away. Cosmonauts Strekalov and Titov were reported to be in good shape physically but emotionally shaken by the experience.

The loss created speculation about the fate of Lyakhov and Aleksandrov. The normally staid BBC

*Jean-Louis Chrétien, a French Air Force pilot (center), was the first Westerner to fly on a Soviet spacecraft.* (TASS/NOVOSTI)

even went so far as to say that they were "lost in space," because their aging T-9 spacecraft would soon exceed its nominal 90-day orbital life. In addition, Salyut 7 had experienced a thruster leak that required shutting down half the system and reducing control (without endangering the crew, though).

While the investigation of the failure continued, the unmanned Progress 18 carried out—in October—a resupply mission to Salyut 7, including the delivery of two fresh solar arrays. On 1 and 3 November, during EVA's that each lasted less than 3 hours, the cosmonauts replaced two solar arrays, thus restoring Salyut 7 to the power level it had at launch and demonstrating a degree of orbital maintenance not previously attributed to the Soviet space program. Finally, on 22 November, the Soviets reported that the crew had returned safely to Earth.

The orbiting space station was next inhabited by the crew of Soyuz T-10, which docked on 9 February 1984. Spacecraft commander Colonel Leonid D. Kizim, flight engineer Vladimir A. Soloviov, and cardiologist Dr. Oleg Atkov received visits in mid-April by the crew of Soyuz T-11: Yuri V. Malyshev, Gennediy M. Strekalov, and Rakesh Sharma, an Indian Air Force pilot and the first space flight participant from a non-Communist third-world nation. During the mission, which lasted from 3 April to 12 April, Sharma measured the effectiveness of yoga in helping the body to adapt to space. The crew returned to Earth in Soyuz T-10, leaving the newer T-11 spacecraft for the resident crew's eventual use.

After the three visitors departed, Kizim and Soloviov carried out a series of four EVA's that seemed to restore Salyut 7 to full operational status. On 23 April, they set up equipment for repair work in the aft engine compartment. Three days later, they prepared the engine for repairs and replaced the leaky valve that had caused earlier problems. On 29 April the cosmonauts replaced the manifold that was also involved in the leak. And on 4 May, they completed all repairs. Their combined EVA time in space was 14 hours, 45 minutes. Although lacking the drama of the American Solar Max repair accomplished a few days earlier by a shuttle crew (see next chapter), the Russians' actions demonstrated their ability to handle fluid systems in zero gravity and a vacuum.

The resident crew aboard Salyut 7 was joined on 18 July by Soyuz T-12 cosmonauts: mission commander Vladimir A. Dzhanibekov, flight engineer Svetlana Y. Savitskaya, and research specialist Igor Volk. The three returned to Earth on 29 July. Then, on 2 October 1984, after setting a new space endurance record of 237 days, cosmonauts Kizim, Soloviov, and Atkov returned to Earth, landing about 100 miles east of the town of Dzhezkazgan. For its part, their Soyuz T-11 spacecraft had spent about half a year in space, indicating its ability to withstand the space environment for at least that period of time.

Continuity of the Salyut program seemed assured with the success the Soviets have enjoyed over the past few years. Salyut did not appear to be an end in and of itself but rather a stepping stone to a larger station and a grander plan for using the space environment. This will be discussed at the end of the next chapter.

# 10 THE ERA OF THE

The idea of human beings being able to fly to and from space in the same spacecraft is not new and can be found throughout science fiction literature as well as in pre-space age conceptual studies. To re-enter the Earth's atmosphere from orbit or following excursions to the Moon and planets, the returnable spacecraft must be fitted with wings.

As we recall from Chapter 5, the earliest carefully developed proposals for such vehicles date from wartime Germany. One involved a winged rocket bomber that was investigated by Eugen Sänger and Irene Bredt beginning in 1937. The two scientists were considering targets from 600 to 12,000 miles from the point of takeoff. To achieve the latter around the world range, they proposed a rocket-boosted configuration that would soar to an altitude of 30 to 90 miles. "At the end of the climb," they wrote in the top-secret report *Über einen Raketenantriebe für Fernbomber* (A rocket drive for long-range bombers), "the rocket motor is turned off, and the aircraft, because of its kinetic and potential energy, continues on its path in a sort of oscillating gliding flight with steadily decreasing amplitude of oscillation. This type of motion is similar to the path of a long-range projectile which from similar heights follows a descending glide-path."

They explained that because of its wings, the bomber "descending its ballistic curve bounces on the lower layers of the atmosphere and is again kicked upwards, like a flat stone ricocheting on a water surface." As it continued its skipping flight toward the target, its kinetic energy would gradually be consumed. Sänger and Bredt concluded that "with a group of 100 rocket bombers, surfaces the size of a large city at arbitrary places on the Earth's surface can be completely destroyed in a few days." Needless to say, the victorious Allies carefully read the report when it became available to them at the end of World War II.

While Sänger and Bredt were completing their Fernbomber studies, General Walter Dornberger, Wernher von Braun, and their teammates at Peenemünde were pursuing long-range bombardment investigations of their own. They came up with a two-stage missile configuration designated A-9/A-10 that could reach targets 2,500 miles distant or even farther. The smaller component, the winged A-9, would be boosted by the much larger A-10 launcher along a trajectory that would reach into space. Upon re-entry into the atmosphere, the A-9 would undertake a glide descent toward its target.

The development of the A-10 was discontinued early in 1944 because, in von Braun's words, "it was foreseen that this work would take two more years." With the outcome of World War II no longer in doubt, it made little sense to pursue what was clearly a long-term development. Sporadic work did continue, however, on the much smaller A-9 (redesignated the A-4b).

These pioneering studies gave rise, during the post-war years, to a variety of rocket-power aircraft developments in the United States and elsewhere. High-speed flights were soon being made ever closer to the border of space. Most famous, as we saw in Chapter 9, was the X-15.

Meanwhile, ballistic missile technology was advancing rapidly; it was only a matter of time before it was adapted to the task of lofting man into space. The U.S. Air Force started off with its Man-in-Space-Soonest project, which became incorporated into Project Mercury when the fledgling National Aeronautics and Space Administration began searching for missions.

Although they bore little physical resemblance, the early manned space capsule configurations were similar in their technology and operation to nuclear warheads. Any chance to depart from this line of development into recoverable winged space vehicles evaporated when President Kennedy committed the nation—in the spring of 1961—to landing a man on the Moon within the decade. Expendable-launch-vehicle-design philosophy would prevail.

Here we go back to the post-World War II period, when the London-based British Interplanetary Society (BIS) outlined proposals for manned space exploration. A plan was put forward to achieve suborbital flight using a modified V-2 launcher. From this, larger vehicles with wings would evolve so that the booster and upper stages could return to Earth to be

# SPACE SHUTTLE

used again. The next step, the BIS suggested, would be to build a space station from which expeditions would depart for the Moon.

In 1952, in a series of articles in *Collier's* magazine that were expanded into a book, *Across the Space Frontier,* Wernher von Braun proposed an enormous three-stage rocket launcher that stood 265 feet tall, or about as high as a 24-story office building. "And the overall weight of this monster rocket," wrote von Braun, "is 14,000,000 pounds, or 7,000 tons—about the same as a light cruiser." The third or upper stage would incorporate wings that would be used during the descent to Earth. There, the third stage could be refurbished and used for a later mission. (The unmanned booster stages, incidentally, would be recoverable by a combination of parachute and braking rockets.)

Assorted proposals for reusable spacecraft continued to be made during the 1950's and 1960's, but they soon were overshadowed by Apollo. However, one configuration did make some progress: the U.S. Air Force's X-20 Dyna-Soar. It was a two-passenger winged craft that would prove that the technology was capable of ferrying crews into orbit and back. Lack of missions and funding killed it in 1962.

Designs were offered and models tested through the 1960's, but there was little hope for a reusable craft until after man had landed on the Moon and most of the associated development funding was past. NASA had plans for expanded space programs in Earth orbit, on the Moon, and out to the planets in the 1960's. One theme that continued to recur, though, was the cost of getting those payloads off the ground; the first 100 miles are the most expensive.

What was needed, NASA reasoned, was a truck that would haul cargo up to orbit and then return for another trip. In what was known as the "integrated space program plan," a NASA task group chaired by then United States Vice President Spiro Agnew recommended in 1969 a wide variety of goals for the nation's future in space.

Central to them would be four primary vehicles: a shuttlecraft for launches from Earth, a space station for operations in orbit, a space tug for switching orbits around Earth, and a nuclear propulsion module for missions to the Moon and planets. Variations and combinations of these four would serve most purposes foreseen by the NASA planners.

Studies by NASA and aerospace companies offered assorted designs around 1968–1969. This included the so-called Phase A studies that developed NASA's basic desire for a space launcher with two fully reusable stages—booster and orbiter.

Competition started in earnest in February 1970 with issuance of a request for proposals on a Phase B study to be awarded to two contractor teams. It was to be the start of the space shuttle, in effect the selection of semi-finalists who would offer their best concepts. One would be selected for Phases C and D—detailed design and construction.

But that was not to be. Congress and President Nixon were unwilling to provide the money required to build such a craft. Cost was projected at about $12 to $14 billion.

One thing that increased the size of the shuttle was its payload. Although NASA had looked at several sizes, it was persuaded by the Department of Defense and its own studies to settle on a payload bay measuring 15 feet wide and 60 feet long and holding up to 65,000 pounds of cargo. So whatever shape the shuttle assumed, it would have to encompass that size payload.

NASA started scaling down the size of the shuttle and consequently its price. But in order to do that, they also had to cut away at the system's reusability. And that, in turn, drove the cost per flight up, something that NASA did not like.

The first step was to put the second-stage fuel in drop tanks over the wings, reducing the size of the ororbtiter and therefore its price. But that was not enough, so the oxidizer was also put in a drop tank.

This resulted in several designs using the same orbiter but different boosters—a recoverable Saturn 5 first stage, clusters of 10- and 13-foot-wide solid rockets, and a "big dumb booster" using simple engines.

When NASA finished its analysis of the proposed designs, the winner was an odd-looking beast with the orbiter and two solid rocket boosters strapped to the side of a large propellant tank.

The design plan was to reduce the technical risk as much as possible. Putting the orbiter's main engine propellants in a drop tank reduced the size of the orbiter as much as possible. The tank would have as few valves and electronic systems as possible so that they would not be wasted. Solid rockets, although never used to launch a manned mission, posed a lower technical risk than did a liquid booster.

Roles in the project were assigned as might be expected, but not before some jockeying for position. The Johnson Space Center in Houston was named overall project manager under NASA's "lead center" concept. It was also responsible for the orbiter.

The Marshall Space Flight Center in Huntsville, Alabama, was placed in charge of the engines, tank, and boosters, plus major vibration testing. Kennedy Space Center at Cape Canaveral, Florida, was named as the launch site, with Vandenberg Air Force Base, California, as the second such site to be developed.

Nonetheless, the design was not really final. Countless changes would be made—some as refinements, others as fixes that were added to space shuttle (SS) Columbia even after it went to the launch pad.

The configuration that finally emerged stands 184 feet tall and weighs about 4.5 million pounds at liftoff. The orbiter weighs about 185,000 pounds empty. It is 122 feet long and 78 feet across the wing tips and stands 56.7 feet tall on the runway with wheels down. (These figures, and those that follow, refer to basic design parameters; actual numbers may vary somewhat.) The key feature is the payload bay in the middle, 15 feet wide and 60 feet long. Ahead of it is the two-level crew cabin with the flight deck on top and living quarters at middeck. A lower deck contains the electronics. Three main engines, 7.5 feet wide and 14 feet tall, and their associated plumbing are mounted in the aft section of the orbiter. The engines burn liquid hydrogen and liquid oxygen to produce 375,000 pounds of thrust each (at sea level, 100 percent power). Each pound of propellant produces 455 pounds of thrust. The engines can be throttled from 60 to 109 percent of "rated" power, depending on mission needs, and can be reused without overhauls. (Initially only 10 reuses were possible; the goal was 55.) The engines work on a preburner cycle which powers turbines that run the main fuel and oxidizer pumps which rotate at up to 33,000 revolutions per minute and up to 6,174 pounds per square inch of pressure. The exhaust from the turbines is a hydrogen-rich steam that is ducted into the main combustion chamber where the remainder of the oxygen is injected. In this manner, virtually all the energy available is captured and used as thrust. Overall, the demands placed on the shuttle main engine are ten times greater than are those on the Saturn engines, which had to work only once. To control the engines, each has its own two-in-one computer so that the orbiter need only issue thrust commands; the engine controller translates those into valve positions that change pump speeds.

Auxiliary propulsion is provided by the orbital maneuvering engines which burn hydrazine and nitrogen tetroxide, the same storable propellants used by the Titan launcher and Apollo spacecraft. These two engines produce 6,000 pounds of thrust and are used to make the final injection into orbit, to slow the shuttle for re-entry, and for any major orbit changes in between. Also mounted in the same pods as the maneuvering engines, and in a third pod in the nose, are the primary reaction control thrusters with 870 pounds of thrust, and the vernier reaction control thrusters with 25 pounds of thrust.

Electric power for the shuttle's systems is provided by fuel cells that combine hydrogen and oxygen, just as in the Apollo spacecraft. The extra power needed to steer the shuttle engines and to move the aerodynamic control surfaces is provided by three auxiliary power units (standard in the aircraft industry) that drive hydraulic pumps with the hot gas from hydrazine decomposition.

Protection during re-entry is provided by a complex thermal protection system, more commonly known as the heatshield tiles. The most severe heating, up to 3,000 degrees Fahrenheit, is encountered on the nose cap and leading edges. These are coated with a reinforced carbon-carbon material, a special nylon cloth impregnated with pyrolized carbon that is too heavy for use on the entire orbiter. Elsewhere, protection is provided by the now famous heatshield tiles, special silica tiles that absorb the heat of re-entry and subsequently radiate it away even as more heat is generated. This works so well that it is possible to hold a piece of the tile seconds after pulling it from a furnace. Columbia initially was outfitted with almost 32,000 tiles of two types—black-coated tiles for extra protection on the belly (the black material has special heat emission properties), where temperatures reach 2,300 degrees; and white on the upper surfaces, where heating is less severe, up to 1,200 degrees Fahrenheit. A flexible heatshield was made of Nomex felt for temperatures below 700 degrees. Holding the tiles in place requires a complex arrangement of strain isolation pads and room-temperature vulcanized glue—fancy rubber cement.

The computers that make it possible to fly the shuttle through the demanding steps planned for it are a modified version of the AP-1 used in the F-15 jet fighter. Only one computer is needed to operate the shuttle, but the demands of manned flight safety require more: four are used in the primary flight system, and all four vote on each other's health and can ignore a machine if it is out of line. Additionally, a fifth computer serves as the backup flight system. It is identical but has a program written by a different team, so an undetected error in the primary system would not appear in the backup.

Most of the expense for the shuttle went into the orbiter. The tank and boosters were designed for low cost and risk. The tank—measuring 27.5 feet wide and 154 feet long and weighing 73,861 pounds—is really two tanks together, the 139,623-gallon egg-shaped oxygen tank at top and the 378,378-gallon barrel-like hydrogen tank at bottom. Because liquid hydrogen is one-seventh as dense as is liquid oxygen, the masses carried on the shuttle are 224,000 and 1,332,000 pounds. Joining the two tanks is an intertank section with a large box beam spanning its width; its ends are the attach points for the shuttle boosters. The outside of the tank is covered with a spray-on foam insulation to keep the supercold liquids inside from boiling away while awaiting launch or being superheated by air friction during launch. Attach struts are on one side for holding the orbiter in place while on the launch platform. During flight, the orbiter pushes against the struts with the force of its main engines. (Contrary to the illusion, the tank is carried by the orbiter and boosters; the orbiter is not dragged along by the tank.) At the tip of the liquid oxygen tank there is a valve that is popped after separation to make the tank tumble during re-entry and thus ensure its destruction.

The boosters are the most powerful and heaviest part of the shuttle. Each one is 149 feet long and 12 feet wide and weighs 1,293,004 pounds. The solid rocket motor makes up the bulk of this, with electronics and parachutes being mounted in the nose cone and steering gear for the nozzle being mounted in the aft skirt. Each uses a total of 1,110,190 pounds of propellant chemically identical to the mix used in the Castor 2 booster for the Delta rocket. Because of the size of the motor, it is built in four segments for transport by rail. These are joined on the launch platform and held in place by inch-wide metal pins.

Solid rockets cannot be throttled in flight, so any changes in thrust must be planned before the motor is cast. Since a motor burns along its surface, the right geometry can accomplish this. In the shuttle booster, the inside of the motor is a tube from nozzle to the top half of the fourth segment. Here the grain is cast in the shape of an eleven-pointed star. The increased surface area provides extra thrust at liftoff and is burned out in less than a minute, just before maximum dynamic pressures are encountered by the shuttle. Thrust then increases slowly as the size of the burning area increases in the remainder of the motor. Average thrust is 2,900,000 pounds for the 2 minutes that the boosters burn. The nozzle is steered by two actuators in the aft skirt operating at right angles to each other. Each actuator is driven by an auxiliary power unit identical to those used in the orbiter.

Unlike the tank, the boosters are to be recovered. Eight small rocket motors, four each at nose and aft skirt, push it away from the tank shortly after burnout. The boosters fall sideways through the atmosphere. At 15,400 feet a pilot chute is deployed, followed by a

*Dwarfing workers at Kennedy Space Center is a shuttle external tank, which is being prepared for hoisting in the VAB. After the second mission, the tank's foam insulation was left unpainted to save money and weight.* (NASA)

*Openings left by uninstalled tiles reveal the complexity of the heat-shield system that protects the space shuttle orbiter during re-entry.* (ROCKWELL INTERNATIONAL)

drogue and then the three 115-foot-wide mains (enlarged to 130 feet in 1984). The latter slow the booster's impact to 60 miles per hour, nozzle down. The boosters float upright, partially filled with water. Recovery teams plug the nozzle, pump out the water, and tow the boosters back to land for reuse.

Although a study was done to analyze the subject, to no one's surprise, NASA's Kennedy Space Center in Florida was picked as the primary launch site for the shuttle. Appropriate changes would be made in the vehicle assembly building and the launch control center to handle the new craft. Payloads would be handled in the operations and checkout building a few miles away. New buildings would be the vertical processing facility for payloads with rocket stages that would be installed in the shuttle on the launch pad, and the orbiting processing facility, a twin hangar for recycling the shuttle for each mission. To the northeast a 3-mile-long, 100-yard-wide concrete runway would be built to receive the shuttle at the end of its mission.

And the old Saturn 5 launch pads would also be reused. The mobile launch platforms were stripped of their launch umbilical towers, which were then reassembled on the concrete pad as the fixed service structures. A rotating service structure hinged on the fixed one was added so that the entire payload bay of the shuttle could be enclosed when the doors were opened for payload installation.

The shuttle could reach only orbits inclined no more than 57 degrees to the equator, so a third launch pad was planned for Vandenberg Air Force Base, California, on the site of the incomplete Titan-Manned Orbiting Laboratory's Space Launch Complex 6 (SLC-6). From there, orbits crossing the poles can be reached.

Johnson Space Center's role as mission control was also affirmed. The first of the contractors to be selected for the shuttle program—in 1971—was Rocketdyne Division of Rockwell International. This early date anticipated the expected lead time needed to develop the complex engines. Much of that lead time was lost, however, when Pratt & Whitney Aircraft filed a protest over alleged contractor selection irregularities, and the formal award was withheld until March 1972.

Rockwell International won the biggest plum—integration contractor—in July 1973. Rockwell was to assemble the forward and aft fuselage sections and the cargo hold doors. General Dynamics Convair, Grumman Aerospace, and Republic Aircraft were contracted to build the central fuselage, wings, and vertical stabilizer. Martin Marietta was selected in August 1973 to build the large external tank, a task that is a major project in itself. Up to 500 tanks would have to be built, each with equal precision and at rates approaching two a month in later years.

There was no prime contractor for the solid rocket booster. NASA delayed setting its requirements so that it could make up for any major changes in the orbiter or tank. The Marshall Space Flight Center issued separate contracts to Thiokol for the motor (physically the biggest part), McDonnell Douglas for the forward and aft structural skirts, and Martin Marietta for the recovery system. Last to come aboard was the integration and assembly contractor United Space Boosters, a company established simply for that purpose by United Technologies.

But this configuration was only part of the space shuttle that NASA wanted. Still lacking was a third stage to carry payloads from the shuttle's low-orbit regime—normally lower than 200 miles—to geostationary orbit, and to place them on trajectories to the planets. NASA had wanted a reusable space tug (now called orbital transfer vehicle) for this task, but the money was nowhere to be found, nor was the White House willing to back increases in NASA's funding to accommodate what could be a $1 billion craft.

A series of half measures was then studied, centering mostly on existing upper stages adapted for the shuttle. Among the more promising was a wide-body version of the Centaur stage first used in the early 1960's. A chief problem with it (and others) was that it required carrying large volumes of liquid rocket propellant in the payload bay. Managers at Johnson balked at that idea. At the same time, money was still not available for any kind of upper stage. Without one, the shuttle would lose many of the customers it was expected to capture from expendable single-use launchers.

The United States Air Force, expected to be the major user, was asked to take over development of the upper stage. The same studies were held again and in 1975 the Air Force settled on an upgraded version of the Burner II rocket stage built by Boeing Aerospace. This was to be the Interim Upper Stage (IUS), a two-stage configuration designed to place up to 5,500 pounds in geostationary orbit. Development cost at first was estimated to be $100 million, but it soon grew beyond $500 million as the Air Force placed ever more stringent standards on its reliability and guidance. Weight also grew to the point at which only one IUS could be carried by the shuttle rather than two. Eventually, as skeptics predicted earlier, the IUS was no longer "interim." It became the *Inertial* Upper Stage, a subtle change implying that it would stay for a long time.

But commercial customers found that the IUS was too expensive and powerful for their needs, so that development of a smaller upper stage was begun by NASA and later taken over by McDonnell Douglas Astronautics as a purely commercial development. This was called the Spinning Solid Upper Stage, because it was intended to have a simple spin-stabilized guidance system. In the design that emerged, called PAM-D (for Delta-class Payload Assist Module), the 4-foot-wide rocket motor and its satellite sit upright in a special cradle in the payload bay. In this way, they take less length than would be the case if they lay horizontal; length as well as weight is the rule by which shuttle launch charges are tallied.

Upper stages were only part of the payloads expected for the shuttle. The rest were attached, or "sortie," modules that would be returned to Earth for reuse. Among other things, it was expected that this would lead to lower cost experiments, because instruments could be developed with less assurance of success and a lower degree of autonomy—a man would be there to repair and operate it—and could be improved with each successive mission. Two key facilities at opposite ends of the complexity spectrum emerged for accommodating payloads on the shuttle, the Getaway Special (GAS), and Spacelab.

The Getaway Special was announced in 1976 and soon became a best seller. Officially called the Small Self-Contained Payload, it is an aluminum can 19 inches wide and 20 inches deep, about the size of a sounding rocket payload, holding up to 200 pounds of equipment. It was conceived as being able to hold experiments that would be completely autonomous and get nothing more than turn-on and -off commands. The fees were $3,000, $5,000, and $10,000, for 2- to 5-cubic-foot canisters. NASA had not counted on the ingenuity and persistence of American high school and university students, who became the most frequent user by way of civic and professional groups donating payloads, or on the interest of other government agencies. In time the GAS cans, as they were called, would have opening lids, radio aerials, and even deployment mechanisms.

Spacelab is the most complex payload accommodation, carrying both on-board crew and computers. It was developed by the European Space Agency because it wanted a major role in the United States shuttle program. The role it had requested was to develop the space tug, but this request was denied by the Department of Defense. The concept that evolved for Spacelab was to build two basic segments—a pressure module and an open U-shaped pallet—that would be assembled in various combinations to meet mission needs. The value of such a system is that it would allow sortie payloads to be assembled without tying up the orbiter for months at a time; the completed Spacelab could be inserted a few weeks before launch.

Both the module and pallet were about 10 feet long and spanned 14 feet. The module could be flown either as one (short) segment or as two (long) segments bolted together. Racks 19 and 40 inches wide on each side of the module would hold experiment gear and, at the forward end of the lab module, the control computers. Also, a high-quality viewport was provided in the short module's roof, and a 40-inch-wide science airlock was provided in the extension module's roof. A transfer tunnel would link the module to the orbiter middeck. With this lablike environment and the easier ride that the shuttle offered, NASA offered to fly scientists as "payload specialists," under the assumption that they would know best how to conduct specialized experiments. NASA's career scientist astronauts were now designated as mission specialists.

Canada, also wanting a role in the shuttle program, offered to develop a robot arm that would be used to deploy and pick up certain payloads. The arm—officially called the Remote Manipulator System (RMS)—weighs 1,000 pounds yet can gently link the 190,000-pound orbiter and a 65,000-pound payload, somewhat like joining two bricks with a soda straw. It has shoulder, elbow, and wrist joints driven by a complex precision system of gears and motors. An orbiter computer acts as a midbrain responding to commands made by a crewman using a hand controller at the aft flight deck.

*Investing in the future of his nation: Gilbert Moore (left) of Ogden, Utah, bought the first Getaway Special and gave it to area students to fill. GAS-1 stands in the background as James Elwell works on the control system.* (MORTON-THIOKOL)

Meanwhile, the shuttle was taking shape as assembly started on the first orbiter, Enterprise. Perhaps because the task seemed so mundane, and because NASA was more than reassuring in selling it to Congress, both the public and the press anticipated no major problems. After all, the shuttle was only going into orbit and back, a task that had been accomplished for years by expendable launch vehicles.

The difference was in the way that the shuttle would accomplish its missions and in the demands that NASA was placing both on itself and on the vehicle.

The money would be slow in coming. After promising that the project would cost only $5.15 billion, NASA saw the costs creep upward until they hit around $9 billion when development was complete.

But the agency had said, in effect, that President Nixon and the Office of Management and Budget (OMB) reneged on their promise of funding levels. OMB deferred some funding and in so doing pushed the shuttle's initial launch date back a year.

Technical problems added another 2 years, problems that pushed the overall cost up but which NASA said could have been solved had the money been available earlier. A fleet of five shuttle orbiters was planned by NASA, although by the end of 1984 the last in the series (known only as OV-105) had yet to materialize. The development program authorized two orbiters, with the other three to be procured in the course of follow-on production. Actually, three airframes were to be built in the development program: OV-101 and -102, the first two orbiters; and STA-099, a structural test article.

Originally, the OV-101 was to have been named Constitution, but the Ford administration, bowing to pressure from science fiction fans, ordered it named Enterprise after the ship of the same name in the "Star Trek" TV series. Enterprise was to be used first for landing tests and then in launch vibration tests. After that, it was to be outfitted as a space-worthy craft and become the second operating orbiter, joining OV-102.

In the approach and landing tests, two pairs of astronauts—Fred W. Haise and Charles Gordon Fullerton, and Joe H. Engle and Richard H. Truly—were to ride Enterprise while mounted atop the 747 carrier jet, then glide to a landing at Edwards Air Force Base, California, the site where all the early shuttle missions were to end. These tests were needed because wind-tunnel models could not predict everything about the shuttle's behavior at low speeds. At high speeds and altitude, there would always be time to correct one's path. Nearer the ground, there would be less time.

For these tests, Enterprise was fitted with enough gear, including fuel cells, power units, and electronics, to make it behave as an operational shuttle would. A covering of styrofoam blocks served as an ersatz heatshield, the real thing not being needed or available at this time. The crews would wear standard flight suits and ride in ejection seats borrowed from the Air Force's SR-71 Black Bird reconnaissance plane.

A series of "captive" flights were made with Enterprise atop the 747 so that the pilots of the larger carrier jet could learn how to fly "the world's largest biplane." Enterprise was at first inert and then active and manned. After minor problems delayed the whole test program, Enterprise was launched on its first free flight on 12 August 1977, with Haise and Fullerton at the controls. The 747 carried its heavy load to an altitude of 22,800 feet and then went into a shallow glide with engines throttled down so that the plane was actually hanging from the Enterprise. This ensured that at release the two would pop away in opposite directions and reduce the chances of collision. As a final safety measure, the flight was made with a streamlining tailcone covering the three dummy main engines. After release, Haise put Enterprise into a dive to build up as much speed as the craft would have on returning from space. A preflare maneuver at high altitude simulated how he would pull the nose up in the final seconds before touchdown. After two right turns, Haise lined Enterprise up on the dry lakebed at Edwards and brought it in for a smooth touchdown only 5 minutes and 23 seconds after release.

Two more "tailcone-on" flights were made on 13 and 23 September, with the crews alternating. Then on 12 October, with Engle and Truly at the controls, Enterprise was carried aloft with the tailcone off. This configuration matched the craft's shape as it would return to Earth—engines exposed and blunt tail creating more drag. Because of that drag, the 747 could carry Enterprise to only 17,000 feet and the flight lasted 2 minutes, 34 seconds. But it went as smoothly as the first three. A final "tailcone-off" flight was made on 26 October.

After that, Enterprise's systems were stripped out, because they would not be needed in the next round of work. Again atop its 747, it was shipped to the Marshall Space Flight Center for a series of mated vertical ground vibration tests. In these, Enterprise was mated to the external tank and two boosters for a series of tests in which key points on the shuttle would be pushed by actuators (rather like loudspeakers with arms sticking out) to see how the vibrations would dampen or build through the whole craft. Mounted in the test stand where similar experiments were run on the Saturn 5 more than a decade earlier, tests were carried out with boosters full of inert propellant (simulating liftoff) and empty (burnout), and with the boosters removed and the orbiter-tank combination hanging by the booster attach points on the tank. Water simulated various levels of liquid oxygen in the external tank.

Among the findings was the need to "delay" the launch by 3 seconds because the whole shuttle was more flexible than was predicted. As the main engines came to life, they would bend the shuttle, turning it into a giant leaf spring. Booster ignition was pushed back by 3 seconds until the shuttle was again pointed vertical, thus reducing the chances of a mishap.

In April 1979 Enterprise was again shipped by air, this time to the Kennedy Space Center, where it served as an "iron bird" pathfinder to let engineers go through the motions of preparing a shuttle for launch. On 1 May, Enterprise, mated to boosters and tank, was rolled out to the launch pad. This was as close as it ever got to space flight. As the test program proceeded, design of the orbiter was refined. No plans had been drawn for outfitting Enterprise with space equipment, because they were not needed at the time. But subtle changes in the structure for the orbiter meant that the designs for OV-102 were not compatible. NASA was faced with the prospect of paying a high bill to have a custom set of drawings made only for Enterprise. A cheaper way out was to outfit STA-099, the structural test article, which was identical to OV-102. This required putting it under less stress than was planned (40 percent instead of 25 percent beyond normal flight limits). But it was feasible, and STA-099 became OV-099, SS challenger. Enterprise would continue to serve the space program, though, by becoming a traveling museum model of the real thing and making traffic-stopping tours of Europe in 1983 and being a prime exhibit at the New Orleans World's Fair in 1984.

One other shuttle test model was built that bore little resemblance to the real thing. This was the main propulsion test article (MPTA) erected on the Saturn 5 test stand at the National Space Technology Laboratories in Bay St. Louis, Mississippi. The MPTA consisted of a shuttle tank mated to a high-fidelity model (designated MPTA-098) of the orbiter's aft section and a trestle-like representation of the rest of the fuselage. Mounted in the aft section were three main engines and all the associated plumbing and electronics needed to carry the liquid hydrogen and oxygen from the tank to the engines. It was not meant to test the engines per se (this was done singly on two stands nearby) but to test the shuttle's main propulsion system as a whole. Even the propellant loading procedures and countdown were copies of those that would be used at the Kennedy Space Center during a launch. Firings ranged from a 1.5-second ignition test (initially thwarted, after 10 hours of trying, by a valve sensor problem) to a 12-minute simulation of a return-to-launch-site abort, where only two engines were available and the shuttle had to make a wide U turn over the Atlantic Ocean. There were many failures in single-engine testing, some of them destroying engines; but, while frustrating, they were fewer (in proportion to the test time involved) than had occurred during the Saturn 5 program.

Assembly of the heatshield turned out to be the most difficult task of all. As the shuttle slipped ever farther behind schedule, NASA and Rockwell decided to move Columbia to Kennedy, where final assembly would be completed in the new orbiter processing facility (a special hangar) so that other tests could be held simultaneously rather than afterward. With gaps in its heatshield that gave it a ragtag appearance, Columbia was ferried across the nation on the 747 airplane and arrived on 25 March 1979. Styrofoam blocks were used as gap fillers in some sections and were ripped off by the wind or by special tape that refused to stay in place. From this arose the rumors of heatshield tiles falling off Columbia. They would be fueled in the coming months because technicians had to remove tiles and rebond them in position. The problem was that they were doing their jobs, but aerodynamic engineers repeatedly came up with new predictions of tougher flight loads, and many tiles had to be removed for stronger rebonding.

The crew for the first mission was named in 1979: John W. Young, who had flown twice each on Gemini and Apollo; and Robert L. Crippen, who had been selected for the Air Force's Manned Orbiting Laboratory program. (The pilots of the first six missions were MOL transfers.) Launch dates were set and then reset, sliding from March 1978 to March 1981.

But the last year or so of work went surprisingly well. The last chance for a major delay came in July 1980 when NASA officials pondered whether to postpone the flight for another 6 months so that more heatshield tiles could be strengthened. Administrator Robert Frosch and others decided that the time had come to fly. The launch would slip no more.

Columbia was rolled from its steel hangar on 23 November 1980. As it emerged into the glare of floodlights that split the night, the multitude of tile workers—many of whom would soon lose their jobs—cheered. For them it was as good as the launch.

Within a few days Columbia was mated to the waiting boosters and tank which had been "stacked" earlier in the vehicle assembly building. Shuttle interface tests were accomplished with only a few problems. On 26 December 1980, the first true space shuttle was rolled to the launch pad. Even more tests were conducted there—"plugs out" to make sure that umbilicals could be dropped without disrupting systems, power-out to make sure that losing power from Florida utilities would not damage the shuttle, the first loading of supercold liquid oxygen and hydrogen, and others.

In the first tank loading, engineers encountered a problem that would delay the launch yet another month. As the tank was cooled to as much as $-423$ degrees Fahrenheit, portions of the exterior insulation loosened, or debonded. One official likened it to a vinyl floor that had come loose in the middle. Analysis would later show that some sections of the cork-epoxy insulation had been bonded with old glue. Engineers were able to do on-the-pad repairs, but not until a few weeks before launch did they know that the $2.6 million repair job had worked.

The two most important tests were the countdown demonstrations—one wet (i.e., propellants loaded into tanks) and unmanned, the other dry and manned—following the procedure set during Apollo. For the shuttle there was a major difference: the wet countdown test was to end with a 20-second firing of the three main engines. Although they had been fired extensively in two acceptance tests and all of Columbia's systems had undergone extensive tests, the shuttle as a launcher had never been fired. The name described the importance of this trial by fire: flight readiness firing. If all of Columbia's systems could not be exercised in this manner, then a launch surely would fail.

The firing that was set for 16 February slipped to 20 February 1981. With an hour's delay because of minor ground problems, the three engines started bellowing at 8 A.M., then shut down as planned. Data analysis would take a few weeks, but only confirmed the obvious—the shuttle was ready for flight. Actually there was a minor problem that would have prevented liftoff. The ground launch computer became overloaded, leading to premature shutdown of three of four booster auxiliary power units. But such a finding was a goal of the flight readiness firing, so it was a complete success. The dry countdown test with Young and Crippen aboard went equally well, and repairs to the tank were validated.

For its first mission Columbia carried no cargo other than development flight instrumentation designed to record how the craft reacted to the stresses and strains of liftoff, orbit, and re-entry. The sensors were spread through the spacecraft but fed into three electronics boxes mounted on a T-shaped frame in the payload bay and an electronics rack in the middeck. This array of instruments would remain there through the first five missions and then be replaced, on Columbia and other orbiters, with operational instrumentation that would serve in less complex experiments as the shuttle's capabilities were expanded.

Launch was targeted for 10 April. More than 2,000 reporters, photographers, and technicians descended on Kennedy Space Center to convey the launch to the public, and more than 300,000 persons gathered outside the spaceport to watch. The countdown went smoothly until the final 10-minute hold at $T-9$ minutes. The primary and backup flight systems were not talking to each other. Twice the launch teams tried to recycle the count and get the computers to converse, and twice they failed. Kennedy director Richard Smith finally scrubbed the launch attempt because both the astronauts and the launch team were becoming tired and no solution was in sight.

But Johnson Space Center soon offered one. A timing "skew up" had the primaries talking to the backup 4/100ths of a second too soon. The problem had occurred in simulations but was handled by restarting. No one thought it would occur on the launch pad. The solution was simple: immediately after starting the primary and backup systems, their timing would be checked. If it did not match, then they would be restarted.

The countdown was rescheduled for 12 April—exactly 20 years after Yuri Gagarin had been hurled into space in his spherical Vostok 1.

This time there were virtually no problems in the letter-perfect countdown. At 3.5 seconds before 7 A.M. the main engines started their ignition sequence at 12/100th-second intervals. At 7 A.M. they hit 90 percent power; 3 seconds later the twin boosters were ignited and eight explosive bolts were severed. As if casting off a decade of doubts, Columbia rose steadily toward the sky. A striking contrast was made by the blinding-white exhaust of the solids and the ghostly blue flame of the main engines

*The space shuttle Columbia lifts off from Pad 39A (the pad from which man also departed on his first journeys to the Moon) on its first orbital flight test.* (ROD WHITED)

The 4,458,000-pound spaceship climbed upward, cleared its fixed gantry, and then executed a 110-degree pirouette with the grace of a jet fighter. At 10,000 feet it pitched sharply to start its race downrange. That hid part of the flight as Columbia appeared to dive behind its own column of smoke.

About a minute after launch the main engines throttled back and booster thrust tapered to reduce aerodynamic loads—max Q—on Columbia. Then thrust climbed and was tailored to limit acceleration to 3 *g*.

Two minutes after liftoff the twin solid rocket boosters burned out. Columbia carried the dead weight for 12 seconds while the last propellant spurted inside the motors; then the empty casings were jettisoned to parachute into the ocean. They could be seen, through binoculars, as tiny silver dots falling to either side of Columbia, now a pale blue dot receding in the distance. Still under the power of its three main engines, Columbia continued to race skyward. At $T$ + 520 seconds the main engines shut down. Half a minute later the tank was jettisoned and Columbia was at last in its element.

A series of four orbital maneuvering engine burns (two on later flights) were made, raising Columbia's initially elliptical orbit to a nearly circular one about 170 miles above the Earth at an inclination of 40.3 degrees to the equator. But when the payload bay doors were opened, Crippen immediately spotted—and televised to Johnson—heatshield tiles either missing or damaged on the OMS pods on each side of Columbia's vertical stabilizer. The immediate worry was whether there were tiles missing on the belly, where the heating would be the worst during re-entry. Because of the symmetry of the damage, engineers suspected that it was caused by an unexpected shockwave during launch. The pod tiles were thin and had not undergone the extensive proof-testing that the belly tiles had seen. Heating during launch was greater than was expected during re-entry.

*Astronaut John Young, seen here commanding STS-1, gives mute testimony to the growing ease of space flight: glasses were not permitted on astronauts in earlier years.* (NASA)

*A booster from the STS-5 launch splashes into the Atlantic Ocean. Its three 115-foot parachutes can be seen starting to collapse.* (U.S. NAVY)

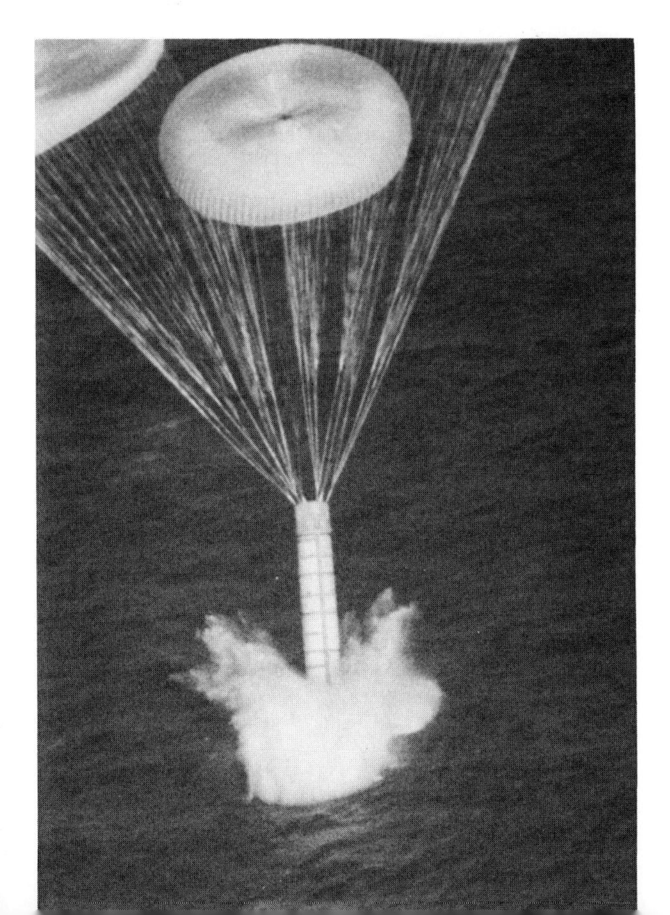

Optimism and joy rather than undue concern seemed to reign as flight director Neil Hutchinson and deputy flight operations director Eugene F. Kranz talked with the press. "We've got a super vehicle up there," Hutchinson said. "We can't say enough about the launch phase and how nominal everything is. . . . It performed absolutely admirably."

Thus did the maiden voyage of the shuttle settle down to routine chores. Young and Crippen put the craft through its paces, testing attitude control thrusters, computers, radiators, and fuel cells. Each function, even though done before by similar equipment on earlier craft, was new, because this was Columbia's first flight. The main thrusters sounded "kind of like a muffled howitzer roar and at night it put out a 40-foot flame," Crippen said after the mission.

The crew members found life on orbit much easier than expected and films showed them moving around as if zero gravity were the environment in which they had been born. The crew members also found the

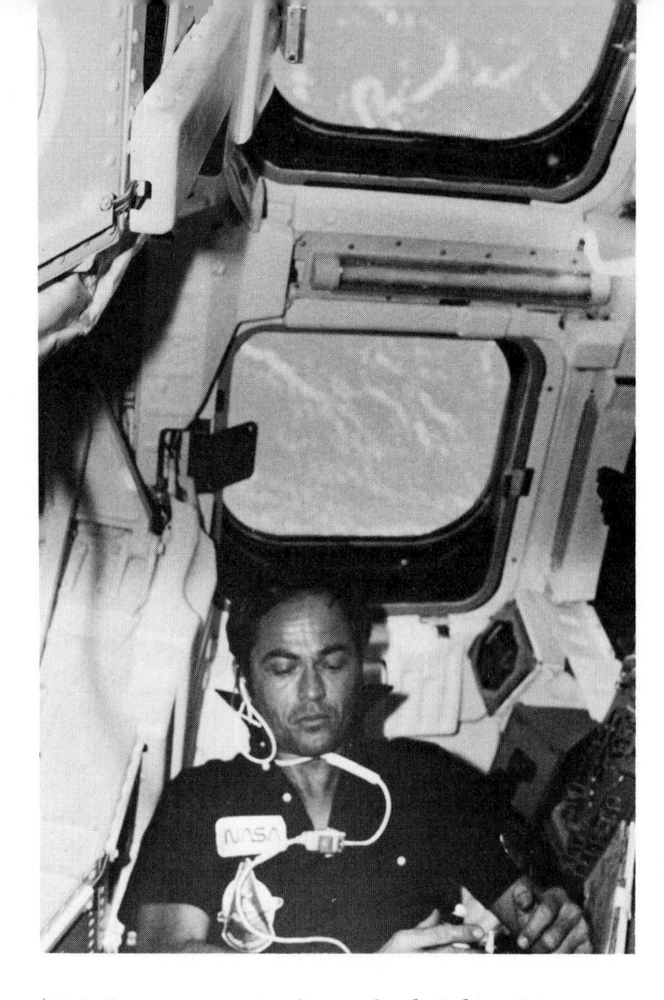


*With the oceans passing by overhead, Robert Crippen prepares a meal during STS-1. The bulky ejection seats can be seen at left.* (NASA)

view from space enthralling and they commented that film could not capture the subtleties or boldness of the scenes that appeared to have been made by a master painter.

There were few failures on board. A data recorder refused to shut down after launch (a washer later was found jammed in the tape drive), a heater on an auxiliary power unit was lost (meaning a slow start), a chemical canister for cleaning air became stuck in place, a thermostat had to be locked open to keep the crew warm, and so on. Perhaps the worst failure of all was the zero gravity toilet, a problem that would recur on almost every mission for 3 years. This forced the crew to resort to older, less comfortable procedures.

But the important systems worked as advertised; even the computers showed not the slightest problem.

"Nothing failed," Crippen said afterward. "Statistically, I didn't think that was possible. We've been working 3 years to learn how to handle catastrophies and all we did the whole time was sit back and enjoy it."

On the second day of the mission, the crew rehearsed re-entry, including closing the payload bay doors, donning space suits, and strapping into their ejection seats (work done for them by ground crews before launch). Ground controllers maintained their confidence that the heatshield tiles would be no problem. "We have reviewed all of the data available to us and we see no reason to alter our plans for re-entry this morning," Hutchinson said. Nevertheless, not since Apollo 13 limped home had a spacecraft's re-entry been awaited with such anticipation. Even a lack of visible damage would have left doubts in many minds simply because of years of difficulty in getting the tiles bonded perfectly.

The next morning the crew awoke and started re-entry preparations. At 11:21 A.M., belly up and tail pointed high and into the line of flight, Columbia's two maneuvering engines fired, acting as retrorockets. Then the craft swung over and was pointed nose high into the line of flight. The last word from mission control as Columbia sailed out of tracking range and headed for radio blackout was, "Everything looks nice going over the hill. Nice and easy does it, John. We're all riding with you."

Columbia was falling steadily and the atmosphere was becoming thicker. At 400,000 feet the air was thick enough to be ionized by the force of Columbia's impact, forming a plasma sheath around the craft and blocking all radio contact for 17 minutes. Temperatures outside soared to 2,700 degrees Fahrenheit, but the crew later reported they could only see a faint pink glow that disappeared as soon as the sun rose.

Columbia was turning from a spacecraft into an aircraft, becoming a hypersonic glider at twenty-five times the speed of sound, making rolling S curves to kill speed. At 12:08 P.M., Edwards Air Force Base made two radar contacts.

Seconds later, "Hello, Houston, Columbia here." The laconic call meant everything was as it should be.

"You're coming right down the chute," was the reply from capcom Joe Allen. "You're coming right down the track."

Without jet engines Columbia was making a controlled fall to Earth along a path that had to be right the first time. As the craft zipped over Edwards, it finally dropped below the sound barrier. A wide 225-degree turn was made around the heading alignment cylinder (an imaginary cylinder in the sky) to line up on Edwards' runway 22. Followed by a retinue of T-38 chase planes, Columbia was on a 20-degree glideslope that looked more like a strafing run.

Young flared to make the approach shallower, a planned maneuver, then dropped the landing gear as he flew over the dotted line down the middle of the runway. Columbia overshot the touchdown point by some 3,000 feet—it had more lift than expected—as a chase pilot counted down the altitude in feet: 5-4-3-2-1. Just as the main wheels were about to kiss the runway, capcom told the world, "Prepare for exhilaration." Touchdown at 215 miles per hour came at 1:21 P.M. EST, 14 April 1981. Then the nose gear touched and Columbia, now weighing 196,500 pounds, rolled to a stop in 8,993 feet. Mission duration was 54 hours, 21 minutes, 57 seconds. Distance traveled was 1,074,567 miles in 36½ orbits.

Ground crews rushed to Columbia to attach cooling and power equipment. In Houston, Johnson director Christopher Kraft announced, "Suddenly we are infinitely smarter."

Within 45 minutes Young emerged, grinning and punching the air with his hands. This sharp contrast to his laid-back personality said more about the mission's success than any press release could tell. His post-flight walkaround looked more like a jig as he found that NASA engineers were right—not a single belly tile had fallen off. A few minutes later Crippen followed and the two were whisked away for a quick physical before the flight back to Houston. As he was leaving, Crippen spoke what many Americans were feeling that afternoon: "We're back in the space business to stay."

In the weeks that followed Columbia's flight, analysis of data continued to show that all had gone well. When the two boosters—jettisoned 2 minutes after liftoff—were towed back into Port Canaveral, Florida, the Monday evening after launch they hardly looked as though they had been to the edge of space and back. Most blackening was soot from separation motors and burning insulation inside the casings. Structural rings inside the aft skirts were ripped away, though, apparently by a "water hammer" effect. The damage was repairable. Two parachutes were lost, but overall the boosters appeared to be in good condition.

A camera in Columbia's belly showed the tank to be in better shape than was expected when it separated. Insulation at its base was blackened by recirculating exhaust gases, and there was some charring in the upper areas from shockwave heating. But most of the insulation looked pristine and the repaired areas were intact. A tumble valve failed to open, but that was of minor concern.

Data from the main engines indicated that each had performed as if making a perfect run on a test stand. Only a single pressure sensor was lost during launch. And inspections of Columbia showed it to be in near-perfect condition. Only two major tile damages were found—an 8-inch crease by the nose-wheel door and a large gouge on the body flap at the tail. These apparently were caused by ice or debris falling during launch and had no effect on re-entry. One hundred tiles were replaced—far less than the 1,000 predicted—and another one hundred or so needed repairs akin to filling a crack with putty.

Columbia was returned to Kennedy and the first "turnaround" began. With STS-2, Columbia would become the first reusable spacecraft and a shuttle in fact as well as name.

Columbia's second journey into space did not hit the level of near perfection that made the first one a dazzling success. But it proved that a large, complex spacecraft can be reused even though the task remains difficult and lengthy.

Selected to fly the mission were the men who had piloted Enterprise on two of its landing tests, Joe Engle and Richard Truly. Preparation for the flight appeared shaky for more than a month as the launch teams at Kennedy worked to get Columbia ready for the return to space. The mission had been set for late September, but various items of work put the launch crew behind schedule and pushed the target date back to 9 October.

A more serious problem appeared—the spillage of a few gallons of toxic oxidizer across the right forward side of Columbia's fuselage that ate away at the glue holding the fragile heatshield tiles in place. A total of 370 had to be replaced without taking Columbia from the launch pad, a move that would have delayed the launch by 2 months or more. Instead, on-the-pad repair procedures held the delay down to a month, and the second launch attempt was set for 4 November.

But again it was delayed. A flawless countdown was stopped cold at $T - 31$ seconds by a minor software problem that could have been overridden had a human finger hit the right button only a second earlier.

While engineers recycled the countdown and prepared for another attempt that morning, flight directors at Johnson started worrying about high pressure in the gearboxes of the three crucial auxiliary power units.

They decided to be prudent, flight director Neil Hutchinson later said, and wait for another day rather than run the risk that gunk might clog the oil system and leave Columbia unable to steer during launch or landing. With the oil drained and filters replaced, it turned out that the gunk was not as thick as feared. The countdown was quickly recycled for 12 November and what appeared to be another flawless countdown began.

But the day before a multiplexer-demultiplexer unit, which mixes data for easier transmission, had a failure in its backup section. A replacement had to be borrowed from SS Challenger and flown in from California in a literal race against the clock. A planned hold was extended so that the new unit could be installed and tested. Shortly before midnight, the count was resumed, aiming at a 10 A.M. launch.

All went smoothly; the only delay was when launch director George Page added 10 minutes to the last 10-minute hold only to give his launch team a chance to relax and assure them that he would scrub the launch if they had any doubts about their systems.

There were none.

The count resumed at 10 A.M., and 9 minutes later Columbia's three main engines and two solid rocket boosters roared into life, literally pushing the shuttle off the launch pad and into space. But shortly after reaching orbit, a problem developed that would force mission control to cut the flight short by 3 days.

One of the three fuel cells that produce electric power developed a leak and let the electrolyte potassium hydroxide trickle away with the water that is a by-product of the cell's operation. This would have ultimately left the cell depleted and the crew with a bit of electrolyte in their drinking water. More important, it would have posed an explosion hazard in the cell with uncombined oxygen and hydrogen. So mission control told the astronauts to shut off the valve leading to the cell and to let it "burn" up the remainder of the gas already inside. That left Columbia with only two fuel cells, enough to fly the mission. It forced a one-orbit delay in the series of thruster burns that raised Columbia's orbit to an altitude of 152 miles.

But that was not enough to give mission control and NASA managers the confidence to continue the mission. If a second fuel cell failed, there would only be one to power Columbia, and that would eliminate the recording of re-entry data. On the morning of 13 November, NASA announced that the mission would be cut short. Despite the problem, the mission was not considered by NASA to be a failure. "We've got 90 percent of what we flew for," said project manager Glynn Lunney.

NASA always has backup plans drawn up before any spacecraft leaves the ground. The minimum mission plan was written with an eye toward what happens if there is a failure that does not endanger crew safety right away but puts them at some greater risk. The answer was to try to meet the most important test objectives and then come home.

First among the objectives was a desire to launch and land safely and test several on-board systems in space. Additionally, on this flight the shuttle started a gradual transition from tests to operations. If the shuttle was to be advertised as a science platform and this was to be proven, NASA had reasoned earlier in the program, then it must carry some instruments that would provide a database verifying those expectations. It was decided to take two engineering models of the Spacelab pallet and rate them for one flight each and outfit them with science instruments, one Earth-oriented and the other space-oriented, to match the test program. The first of these was the OSTA-1 payload, so-named for its sponsor, NASA's Office of Space and Terrestrial Applications. Its major feature was the backup synthetic aperture radar from the unmanned Seasat ocean-mapping satellite, modified to scan the land rather than the water. Other instruments were the feature and identification landmark experiment to test a theory about how to tell instruments not to take pictures of cloud-covered terrain, the measurement of air pollution from space experiment to test a method of sampling carbon monoxide concentrations around the globe, and the shuttle multispectral infrared radiometer to try a new means of prospecting from space for minerals on Earth.

Included in the OSTA-1 grouping but stored in the cabin were a special movie camera for filming lightning from above the cloud tops, a plant-growth unit for cultivating sunflower seedlings in zero gravity, and a chemical reactor for growing microspheres of latex. The latter, the first materials processing experiment carried by the shuttle, is developing into the first made-in-space product to reach the marketplace. The spheres are used in calibration of cell pores and of electron microscopes, and may even find use as drug-carrying "smart bombs" that might be entrapped in tumors.

Despite the shortened length of the mission, most of the observations planned for the OSTA-1 instruments were crammed into the available time ranging from 337 minutes of observations versus 120 minutes planned for the radar to 32½ of the 65 hours for the feature experiment. Later, NASA would announce the discovery of new mineral deposits in Mexico (with the infrared scanner) and ancient river beds in Egypt (with the radar).

Also added on this flight was an induced environmental contamination monitor designed to measure how much pollution the shuttle generated around itself. The monitor was mounted atop the flight instrument pallet and was made up of several instruments, including mass spectrometers, stereo cameras, and gas-capture bottles. The monitor would be carried on the remaining two test flights as well.

Of greatest interest to the public, though, was the robot arm, or Canadarm as it was dubbed before launch. Nine hours of tests in 3 days were planned for Canadarm, but these were cut to 3 hours of the most crucial. The only failure was when the crew tried to move the arm in yaw in the backup mode, a minor problem traced to a loosened wire harness. Television from space showed the arm in various positions over Columbia's yawning cargo hold, and in turn used its wrist cameras to send pictures of the cargo hold.

Ironically, the end effector—or hand—worked well when it grappled an imaginary fixture. It was to have grappled the contamination monitor, but a test model failed and the test was dropped as a precaution. A number of tests of the arm's finer motions and activities were dropped, but enough was accomplished to verify the basic design and operation.

After gathering all the data that the shortened flight would allow, the crew on the evening of 13 November started stowing equipment for the flight home and just before turning in for the night, replaced the data display screen in front of Engle.

14 November was spent with final preparations for the return to Earth. The Earth-observation gear on the pallet was turned off in the morning, and the rest of the day involved turning off or on everything that needed to be that way during re-entry. In particular, engineers wanted the development flight instrumentation operating so that they could collect data that was missed during the first re-entry on 14 April. The 2-minute 47½-second de-orbit burn started at 2:23 P.M. while out of radio contact. A few minutes later the crew was picked up by the Yarragody, Australia, tracking station which stated, "the burn looks good."

Contact resumed at 4:08 P.M., when a C-band radar at Buckhorn picked up Columbia off the California coast, moving at seventeen times the speed of sound. The numbers rapidly shrank as Columbia made its long shallow dive for Edwards. Voice contact was made at 4:10. There was no hurry to talk: The data showed that Columbia was flying much as planned, but 25 miles south of the desired ground track, a difference the crew quickly made up. During entry the crew made its "roll reversals," rolling the wings up and down to check the handling for later missions.

The California coastline was passed at 4:13 P.M. As a small white dot barely visible in the blue sky, Columbia appeared on TV screens at 4:15. Again, Columbia was a little off track—3,000 feet low, capcom said. As the picture from a T-38 chase plane grew larger, a white trail periodically streamed from Columbia's tail, as yaw thrusters helped change Columbia's course.

Then there was the command, "Go for autoland," using the sophisticated automatic pilot. Earlier it had been decided not to try for the crosswind landing, because there was too much wind on runway 15.

The landing gear doors popped open almost at 4:23, as Columbia glided over the end of runway 23 at 227 miles per hour. At 10 seconds past 4:23, the main wheels touched the runway, followed 15 seconds later by the nose gear. Engle steered Columbia straight down the centerline, slowing as the recovery convoy closed behind. Then there was a flurry of calls between spacecraft and mission control as the crew started turning systems off and the convoy crew checked for hazardous gases.

Almost as if to make up for STS-2 being shortened, STS-3 demonstrated that the shuttle had the flexibility to stay up longer if necessary.

The mission was commanded by Jack Lousma, a Skylab veteran. The pilot was Gordon Fullerton, who had flown Enterprise on landing tests. Their cargo was the OSS-1 pallet (so called for NASA's Office of Space Sciences).

OSS-1 included a heat pipe canister (the prototype of a new thermal control concept eliminating the mechanical pumps common to spacecraft); an X-ray flare polarimeter to measure polarization in X-rays emitted by solar flares (none was detected); an induced atmosphere photometer to measure contaminants in the shuttle environment; and a cosmic dust catcher. The most dynamic experiments, and most demanding of crew time, were those with the plasma diagnostics package (PDP) and the vehicle charging and potential (VCAP) experiment. These were first proposed for Spacelab 2 but given a preliminary run on an early shuttle flight to help in understanding how the shuttle affects the environment immediately around itself. The PDP was shaped like a large can and had several plasma (electrified gas) detectors; it was designed for release from the shuttle, but on STS-3 was used at the end of the robot arm to map plasma patterns. The VCAP included an electron gun to probe the plasma around the shuttle.

Launch was at 11 A.M. on 22 March 1983. Despite the opportunity to advance the schedule by a few days, top NASA officials declined to do so, deciding instead to let the launch crews work at a deliberate rather than frantic pace with days off, and to assure commercial customers that a launch date could be set and met. It was missed by only an hour, because a temperature sensor on a fuel-loading system in the launch platform gave a false reading of a high temperature.

Other than that problem, the countdown was remarkable in that it was unremarkable, free of the computer-software holds that had frustrated the first and second launches minutes before $T - 0$.

Total mission duration was to be 7 days 3½ hours, with a landing at Northrup Strip at the White Sands Missile Range in New Mexico. Heavy rains had flooded the desert at Edwards, giving NASA the choice of postponing the flight for days or weeks or switching to Northrup, an alternative landing field since the earliest days of the program. And the shuttle was more colorful, with its external tank a sporty reddish-brown because it was not painted to save money and weight.

The launch phase was fairly normal, except that shortly after liftoff, flight controllers at NASA's Johnson Space Center saw high-temperature readings on the No. 3 auxiliary power unit, a problem later traced to cooling difficulties. It was shut down after 2 minutes, requiring the No. 3 main engine to be locked at 72 percent power and then shut down pneumatically rather than hydraulically. But the problem caused no great concern at mission control.

"The first part of this ride is a real barn-burner," Lousma called out shortly after the twin solid rocket boosters burned out and separated.

After being inserted in an orbit almost 150 miles high and inclined 38 degrees to the equator, Lousma and Fullerton started activating payload systems aboard Columbia.

The heatshield again became an issue when Fullerton, peering over the coaming on the forward instrument panel on the first day in orbit, saw several areas with missing tiles on the nose shortly after going into orbit. Ground inspection teams then reported finding both white and black tiles on the launch pad and the beach. A total of thirty-seven missing or damaged tiles was counted, twenty-five white and twelve black. For a while it seemed that the nightmare of the shuttle shedding its tiles at launch had come true. But again, NASA-Johnson officials said that there was no need for concern. The affected areas had seen their worst heating during launch and had partial protection from the glue pads that held them in place. Maximum heating during re-entry was to be 600 to 800 degrees on the nose and 400 degrees on the body flap.

Launch pad cameras showed black tiles, designed for higher temperatures than were white tiles, popping loose from the upper side of the body flap facing the three main engines as the latter were powered up. No belly tiles, the most crucial, were seen coming loose, and the tiles found on the ground accounted for most of the open positions on the nose and flap. The nose tiles were believed to be lost to ice falling off the external tank or the higher aerodynamic pressures of launch. It soon became apparent that the tiles presented a greater hazard to the turnaround schedule than to re-entry.

Loss of TV cameras on the robot arm and in one corner of the payload bay meant that the contamination monitor could not be picked up by the robot arm for tests with a full load at the end. The wrist camera is essential to the grappling phase, and the aft starboard camera was needed to check positioning while reberthing the monitor. Ironically, the flight plan had called for the monitor to be grappled for thruster plume mapping. Without the wrist camera, Fullerton had to use the naked eye and binoculars to grapple the plasma diagnostics package (part of the OSS-1 payload) for tests with it on the arm. To everyone's surprise, he did this handily and was able to reberth the package in only 4 minutes.

About midway through the flight, mission control became worried about the weather at Northrup Strip. It was windy too often, and there was a distinct possibility that the landing could be delayed a day. Here, the engineering conservatism that went into Columbia's design paid off. The craft did not cool as much as was expected when in the tail-to-Sun and nose-to-Sun attitudes, thus requiring less use of the heaters. Use of hydrogen and oxygen powering the electricity-producing fuel cells was lower than was anticipated, offering the chance to stay in space longer. Mission control told the crew to cut back on smaller systems and save more consumables. In this manner they would be able to extend the mission by 3 days and not touch the mandatory 24-hour reserve. It paid off on 29 March when the weather at Northrup became unpredictable. Astronaut John Young, flying the shuttle training aircraft at the strip, gave the crew updates until 30 minutes before retrofire, when he said, "Let's knock this thing off." The winds were gusting to 48 knots and gypsum dust in the air reduced visibility to a few hundred yards. Even the crew could see it from orbit as they passed over.

With Northrup in question, mission control weighed the possibility of a landing at NASA-Kennedy. The strip there had not been planned for use until the fifth mission, but the center is equipped with the same serving gear as the Edwards' crew had shipped via train to Northrup. The weather prediction was good for NASA-Kennedy on 30 March but bad for the next day. The decision was made to try for re-entry on revolution 129 to Northrup. If the weather was bad, the crew would reload the computer and try for Kennedy on revolution 130.

The weather held at Northrup, and the 2½-minute retrofiring of the maneuvering engines was made at 10:13 A.M. on 30 March, followed by a near-perfect re-entry. Columbia exited the blackout 2 minutes early and the crew resumed normal conversations with mission control, a sure sign that all was in order. The approach to Northrup was less complex than were those at Edwards—straight in, turn right 93 degrees, then down onto runway 17. The rate of descent was a little faster than normal, bringing the landing gear down closer to touchdown than was preferred. But they locked in time and the main wheels hit the gypsum dust at 11:04 A.M., only 8 days and 4 minutes after launch.

Post-landing inspection showed the craft to be in excellent shape except for the missing tiles. About 1,500 tiles had to be removed and strengthened on the gluing face before the next mission.

In addition to further tests of the shuttle as a spaceship, the STS-3 mission demonstrated that the shuttle is a good platform for performing space science. Not only did the OSS-1 pallet show that Columbia is clean enough for space physics experiments and observations of Sun and stars, but the crew showed that people, too, can be valuable tools in performing experiments.

A disturbing discovery made on the mission was the so-called shuttle glow effect. This was noticed in time-exposures shot with 35-millimeter film cameras and showed a dim but distinct glow on the windward surfaces of the shuttle. There were immediate implications for astronomical observations, and plans were made to carry film cameras with spectral gratings on later flights and to make measurements with atmospheric instruments planned for Spacelab 1. A possible cause was atomic oxygen, formed by sunlight splitting oxygen molecules, which also ate away at the mylar covering on the TV cameras and other mylar surfaces in the payload bay. This, too, has led to a new line of investigation because of the implications for long-term space missions.

In sharp contrast to the open daily briefings given reporters on the success of the STS-3 payload, NASA attempted complete secrecy about the makeup of the STS-4 payload. This was a Department of Defense (DOD) cargo known officially as DOD 82-1 (a reference to the fiscal year in which it flew). It had been scheduled for a later mission, but the Air Force was eager to have it flown on time despite delays in the shuttle program, so it prevailed upon NASA to postpone flight of a barbell-like payload that would be used to test the robot arm.

The mission placed NASA in an uncomfortable position. The agency was, by law, open, yet this major customer was insisting on complete secrecy about the mission. In time a compromise was worked out: there would be no television pictures of the payload nor any open references by the crew to the payload. The rationale that DOD gave was that only by concealing all the defense-related payloads could they hope to protect those that were truly sensitive. And for NASA the policy was an extension of that which they followed for commercial customers: press kits contained only what the customer provided. In the case of DOD, nothing would be provided.

Because DOD waited until a few months before launch to announce this policy, the media was able to deduce details of the DOD 82-1 payload from technical papers and congressional testimony that had been released in the previous 2 years.

The principal payload element was CIRRIS (Cryogenic InfRared Radiation Instrument for Shuttle), a telescope cooled by liquid helium and designed to measure heat emissions and patterns from the upper atmosphere (as seen when looking at the horizon) and the environment around the shuttle. It was mounted on a special gimbal that allowed it to scan in a fore-to-aft direction relative to the shuttle. With CIRRIS, defense scientists hoped to extend their knowledge of the infrared sky so that data could be used in improving sensors for missile detection and warning.

At the other end of the visible spectrum was a horizon ultraviolet scanner with a similar role. It was the forerunner of a more complex array of sensors to be tested on later shuttle missions. Other elements of the payload were two sets of plasma physics instruments—Shuttle Effects on Plasmas in Space, and Sheath, Wake and Charging—akin to those on OSS-1. And there was a space sextant, the prototype of a navigation device that would allow satellites to determine their position with great accuracy and without relying on ground control, thus giving them a degree of autonomy and safety.

A degree of secrecy also surrounded a special cargo in the middeck, the Continuous Flow Electrophoresis System (CFES) developed by McDonnell Douglas Astronautics and Ortho Pharmaceuticals, under a joint-endeavor agreement with NASA. Electrophoresis is a common process by which electric fields separate biological fluids somewhat like a prism separating light into colors. It is limited, though, because the electric fields also generate heat which leads to convection that blurs the separation between materials. In space, McDonnell Douglas and Ortho calculated, zero gravity would prevent this blurring and allow separation of materials that could not be obtained otherwise. This raised the potential for curing the incurable by separating hormones and cells that could not be obtained on Earth.

In order to assist companies to start space processing operations, the joint-endeavor plan was devised whereby NASA would fly research equipment for free use of the machine or for royalties later when the product reached the market. Basically the government became a risk partner with industry, helping to cultivate new customers for the shuttle.

The CFES was the first piece of hardware to be developed in this manner. It had been scheduled for Spacelab 3, but when faced with extensive shuttle delays, NASA agreed to accommodate it instead in the shuttle middeck, thus keeping it on schedule. Because of the competitive nature of the drug market, though, the identity of the candidate material Ortho wanted refined was not released even as late as mid-1985.

Also being flown was the first Getaway Special, a package of experiments developed by students at Utah State University. In developing this package, the dozen or so students received practical engineering experience that would have taken years of employment to obtain. Most of the experiments failed for various reasons, but it was the total experience—including some degree of failure—that was valuable.

Other aspects of the mission plan were conventional and would complete most of the test phase for the shuttle program. Thomas K. Mattingly II, Apollo 16 veteran, was named as commander, and Henry W. Hartsfield, a rookie, as pilot. Launch was set for 27 June, so that the landing, to be attended by President Reagan, would come on the 4th of July.

The countdown went so well that press briefings lasted less than 10 minutes, an unusual measure of overall confidence in the shuttle. One NASA official jokingly offered to hold a glitch, simply to give the media something to do. The afternoon before launch the worst thunderstorms in 20 years smashed through central Florida, pelting the launch pad with half-inch hailstones. There was brief uncertainty about whether the launch would be delayed. Inspection on the pad showed that most of the damage was superficial, easily fixed by applying a slurry of silica solution and fibers. Repairs were completed with an hour to spare before loading the external tank with supercold liquid oxygen and hydrogen began.

Liftoff came about 1/7th second before 11 A.M. The performance was good, with the only two problems occurring in the twin solid rocket boosters. For the first minute of ascent they delivered less impulse than was expected, leaving Columbia a few miles low and slow, and then they picked up. The variation was within allowable limits, though.

The only other major failure of the flight test program occurred when the parachutes separated from the boosters before they even deployed, and the empty rockets plummeted straight to the ocean and quickly sank. Although the boosters were not recovered—their crumpled remains were photographed by an unmanned submersible—the fault was traced to a switch that was intended to release the parachutes after impact with the water. The inner mechanism of the switch was filled with hydraulic fluid that in the June heat was less viscous than when the previous three launches took place, and thus was triggered when the drogue parachutes snapped open. It just as easily could have happened on the previous missions.

After tank separation, Columbia's two orbital maneuvering engines fired twice to nudge the craft into a circular orbit 175 miles high. Initially Columbia was placed in the gravity-gradient mode, nose to Earth, so that the military cargo could scan the horizon. But Columbia developed a tendency, still unexplained, to drift out of position. So the craft was stabilized for the night in the bottom-to-Sun attitude to bake out any water that might have been absorbed by tiles with waterproof coatings cracked by the hail. Ground tests showed that up to 100 pounds of water might have been absorbed by the entire heatshield. During entry it could boil out and crack the surface of the tiles, but it was not believed to be more of a threat to repair costs after the mission than to the crew during entry. Although NASA officials did not say so at the time, data from the mass spectrometer on DOD 82-1 showed that the water was indeed leaving the tiles.

The next day four thrusters in the nose were turned off because one of them was leaking hydrazine fuel; the other three were on the same manifold. And the Getaway Special refused to turn on, giving an indicator light that no signal was being received.

The crew conducted the first run on the CFES and the latex reactor was turned off after completing its run.

The following day, the contamination monitor was picked up by the robot arm for two sets of tests, one to map possible pollutant patterns outside the orbiter, and the second to map thruster exhaust pressures. The first was completed after some minor difficulty in getting started, but the second was lost to electric noise. The monitor, designed with a simple control, does each successive task based on how often a toggle switch has been hit. When the switch was thrown to start the plume survey, there was also noise from other electric systems, and the signal was masked. Another 24 hours of "bake out" was also conducted for the heatshield.

At the end of the bottom-to-Sun soak on the third day of the mission, the crew tried to close the cargo doors but the latches were misaligned because the uneven heating caused a slight "banana effect." This was seen on STS-3, but some engineers thought it was foreign material jamming the latches. This time, TV broadcasts (with the cameras carefully pointed away from the DOD cargo) showed the doors overlapping the cargo hold edge. The problem was cured by heating the shuttle in top-to-Sun attitude.

The second CFES run was made, and the Getaway Special was started when the crew hot-wired the hand controller. In order to preserve the biological experiments, it was later decided to leave the Special on until 40 hours after landing rather than turn it off before entry.

A couple of days later, with Columbia still coasting tail-to-Sun, Mattingly entered the airlock and donned the new space suit that was developed for shuttle missions. Mattingly later said that he was surprised the donning posed no problems. The only disappointment was that he could not go outside. The next morning, the crew started stowing equipment and winding down the mission in preparation for landing. Only a few tests were conducted. On the seventh and final mission day, all gear was stowed and the cargo doors were closed by 9 A.M., so that the crew donned their ejection suits and strapped in.

At 11:30 A.M. the maneuvering engines were fired to shave about 209 miles per hour off Columbia's speed. Touchdown came at 9:09 A.M. Pacific time. When the crew emerged less than an hour later, they were greeted by President and Mrs. Reagan. At the same time, in Washington, the White House released unclassified details of a new national space policy. It reaffirmed the basically civilian nature of NASA and the shuttle, and the exploration and exploitation of space as national goals. Disappointing many, there was no commitment to a manned space station in the policy. But the wording reversed decisions by President Jimmy Carter that had limited NASA to modest goals for the future.

Columbia was responsible for much of the policy. In four missions, the ship and its ground and flight crews completed most of the objectives that had been planned for six flight tests. "We have demonstrated a system maturity that is far in excess of what you would expect after four flights," Hartsfield said.

With its fifth mission space shuttle Columbia officially became operational. In truth, this was simply another milestone in a gradual transition that started when a science payload was carried on STS-2 and which will continue through a number of missions in the future. While Columbia still carried the development flight instrumentation that measured its in-flight performance, it carried its first two satellites for deployment. And it accommodated two additional crew members, thus requiring that the rocket charges be removed from the ejection seats and that the commander and pilot no longer wear their SR-71 suits. Instead, all crew members now wear a protective helmet that provides them with fresh air in case of a propellant leak after a hard landing.

The crew was made up of commander Vance D. Brand, who had been command module pilot on Apollo-Soyuz; pilot Robert F. Overmyer; and mission specialists William B. Lenoir and Joseph P. Allen IV. In addition to deploying two satellites, Lenoir and Allen were scheduled to make the shuttle program's first space walk. The first paying passengers were SBS-3 and Anik C, communications satellites owned, respectively, by Satellite Business Systems and the Canadian telecommunications agency. Each rode in a PAM-D module, described earlier.

Liftoff came on 11 November 1982 at 7:19 A.M. as planned, an important demonstration for the operational shuttle. Ascent was held back slightly by headwinds at high altitude (they approached the limits of the shuttle's capabilities the night before launch) and by lower-than-expected thrust from the boosters. The main engines were able to make up the difference, though, and Columbia was inserted in a 184-mile-high orbit after its twin maneuvering engines made their burns. The boosters parachuted into the ocean as planned and were recovered later after some difficulties with high seas.

About 2:32 P.M. on the day of launch the SBS-3 satellite was spun up, and at 3:17 P.M. it was released to coast away from Columbia for 45 minutes when its solid motor fired to place it on a transfer trajectory to geostationary orbit. Allen later said the PAM-D had "a large and a loud bang when it let loose"; and Lenoir said, "It's quite a sight when you see it going away from you."

Mechanically, the only problem was failure of a cathode-ray display tube on the forward flight deck. Its display electronics was replaced later in the flight with one from the aft flight deck.

The second day went as smoothly as did the first with respect to the satellite deployment. Anik C was released at 2:25 P.M.

When Columbia released SBS-3, it was within 5,000 feet of the target point; for Anik C that was cut to 496 feet, both a fraction of the promised "error box." Attitude was misaligned by only 0.4 degrees versus a tolerable error of 2.0 degrees.

"Two for two," Allen called from space. "We deliver." That would become the shuttle program's unofficial motto.

Following deployment, Columbia went into the side-to-Sun attitude to see if a "sideways banana" effect would result from long exposure in that attitude. None did.

*SS Columbia made its first commercial deliveries on STS-5. Here the Anik C satellite, with rocket motor at its base, is released. Its launch cradle is hidden by an identical cradle used the day before to launch SBS-3.* (NASA)

The crew was suffering various degrees of space sickness as their bodies adjusted to zero gravity so flight surgeons and directors decided to postpone Lenoir's planned space walk for 24 hours.

The flight plan was flexible enough so that they could do this, and the added day would give Lenoir's vestibular system more time to adjust. Experience showed that most space passengers adjusted within 3 or 4 days. The problem had nagged earlier shuttle crews, so that NASA was moved to add medical doctors to the crews of STS-7 and -8 to study the problem firsthand.

The space walk was attempted on the 14th, but was quickly thwarted because of problems in the extra-vehicular mobility units, or EMU's. The EMU's were a new design that came in dozens of different-sized pieces. Rather than building a custom-fitted suit for each astronaut, a suit would be assembled from the appropriate size segments: hard upper torso (with a life-support pack mounted on a fiberglass shell), pants, gloves, helmet, and liquid-cooled undergarment. Sizing inserts were laced into position to change the fit of the sleeves and pants, and the upper torso came in three sizes. The result cost more than a single suit but was expected to save more in the long run by eliminating the multitudes of suits built in the past.

Allen and Lenoir were to spend 3½ hours in the EMU's breathing pure oxygen to purge nitrogen from their blood so that they could drop pressure to 4.3 pounds per square inch without risking the bends. However, Allen's life-support pack developed a noisy fan and Lenoir's held pressure at 3.8 instead of 4.3 pounds per square inch. Although either suit could have been used if an EVA was necessary, mission managers decided to be cautious and not take unnecessary risks. (All things are comparative; the EMU risk would be minimal against not coming back because the cargo doors jammed open.)

Entry day, 16 November, was as uneventful as such a day can be, with Columbia gliding back to a landing only 4 minutes after sunrise over the Mojave Desert. (Its orbit had been tweaked just a little to let it land then rather than 1 minute before sunrise.)

The de-orbit burn was made at 8:36 A.M. over the Indian Ocean. "On time, good burn," was the word from Brand as Columbia passed over the Guam tracking station.

Re-entry was smooth, and about 9:10 A.M. Columbia dropped through a thick overcast that shrouded Edwards, made its turn to runway 22, and touched down at 9:33 A.M. It was so smooth that Brand called out, "Hey, Roy, are we on the ground now?"

The last problem came just before Columbia stopped. After easing off on the max-braking test, one brake locked up and may have been the cause of a tire blowing while Columbia was being towed. Engineers knew that would destroy the brakes but said that they needed the data. The brakes would continue to be a nagging problem for many more missions.

At a post-landing briefing, Lieutenant General James A. Abrahamson, NASA associate administrator for space flight, admitted that the flight was almost boring because Columbia worked so well. But he was excited.

After five missions Columbia was briefly retired for modifications to remove gear needed only for tests and to add items that would make it fully operational. Taking its place was SS Challenger, the revamped STA-099, now OV-099. It had a number of improvements, including a "heads-up display" to allow a pilot to look through a special mirror that gives the illusion of having flight data projected over the scene ahead and not have to take his eyes off the runway to see crucial instruments. Challenger also had improvements that made it lighter and the heatshield tougher.

Like Columbia before it, Challenger was given a flight readiness firing (FRF) before its first liftoff to make sure that all the elements of the complex craft worked together as planned. This precaution paid off when data from the first FRF showed excessive amounts of hydrogen in the aft section of the ship. Tests were unable to trace the cause, so a second firing was ordered. This also was inconclusive. Finally, a ground technician working inside the tail isolated the leak on one engine. A tiny crack was growing in the manifold and could conceivably have grown to the point of catastrophe in flight. The engine was replaced and launch was scheduled for April.

Making up the crew were commander Paul J. Weitz, pilot Karol J. Bobko, and mission specialists Franklin Story Musgrave and Donald H. Peterson. Only Weitz, who spent 28 days aboard Skylab in 1973, was a veteran. The principal cargo was the first tracking and data relay satellite (TDRS-1), a 5,000-pound switchboard in the sky designed to pick up radio signals from satellites, including the shuttle, in lower orbits and relay them to a central ground station at White Sands, New Mexico. With two TDRS's (and a third in orbit as a spare), NASA would replace the necklace of ground tracking stations built up over the past quarter-century. (Exceptions were made for special satellites and the Deep Space Network.) TDRS-1 sat atop the first Inertial Upper Stage (IUS) built for the shuttle. (A slightly different version had a successful debut atop a T-34C Titan the previous October.) An additional goal was for Musgrave and Peterson to attempt the EVA's which Allen and Lenoir were unable to undertake on the previous flight.

In contrast to the many problems that had preceded it, the countdown was almost uneventful. Liftoff came at 1:30 P.M. on 4 April 1983. The solid rocket boosters, the first using lighter casings to increase cargo by 400 pounds, burned 0.3 percent low early in the flight but provided extra thrust later.

The three main engines, the first to be rated for 104 percent of normal power, also worked to perfection, laying to rest concerns that arose as leaks were detected earlier. The only problem was failure of a temperature sensor. The external tank, first to be lightened so that the shuttle could carry another 8,000 pounds of cargo, also worked as planned and splashed down in the Indian Ocean a few miles short of the aim point but within allowable bounds.

After tank separation, two firings of the orbital maneuvering engines placed Challenger in a 177-mile-high orbit and the crew started preparations to deploy TDRS-1. It was elevated in its launch cradle to an angle of 57 degrees, then pushed away gently by four springs at 11:32 P.M. Launch almost was delayed by failure of two of five gyros in the redundant inertial measurement unit; they were brought back into service at 9:15 P.M. but failed again after deployment. First-stage ignition came at 12:27 A.M. and placed the TDRS-1 in an elliptical orbit 177 × 22,335 miles high. The burn was a few seconds short and the IUS reaction control system made up the difference. After confirmation of burn, Challenger's crew bedded down at 2:45 A.M. but not until Musgrave, wound up by the day's activities, did some preliminary work on the space suits.

Second-stage ignition came at 5:41 A.M. and went well until telemetry was lost after 70 seconds. An Air Force ground optical tracking station in New Mexico saw the plume from the second stage swing off axis at that moment. Both IUS and TDRS-1 appeared to be lost until ground controllers at the Air Force's Satellite Control Facility in Sunnyvale, California (which controls the IUS), transmitting blind through the Goldstone, California, station, commanded separation of the IUS and TDRS-1 just as the IUS's batteries were running out. Within 3 hours, TDRS-1 had canceled a 30 revolutions-per-minute spin accidentally given it by the IUS and deployed its solar arrays and antennas.

Instead of being in a 22,335-mile-high orbit directly over the equator, TDRS-1 was left in one 13,450 × 21,950 miles and inclined 2.37 degrees to the equator. NASA officials soon announced that they still had hopes of saving the satellite by using 1.1-pound thrust rocket engines and much of their 1,300 pounds of fuel to gradually raise the orbit.

There was nothing that Challenger's crew could do about TDRS-1, so they started 2 days of modest experiment activities. Musgrave conducted the first three of six runs on the continuous-flow electrophoresis system that had first flown on STS-4 in 1982. The monodisperse latex reactor, flown on two previous missions, was turned on for a 20-hour run. And the night/day lightning camera was set up so that the crew could film thunderstorms below when they had the chance.

Also started was a series of rendezvous phasing maneuvers, small thruster burns that would have Challenger rendezvous with a phantom target to test that capability for later missions with real targets. The next day Musgrave and Peterson checked and donned their space suits in preparation for the EVA. As a precaution against the double suit failures on STS-5, a third suit was brought along. That proved to be wise, because Musgrave's suit developed a minor electric problem in a prelaunch check and had to be swapped with the backup.

Preparations for EVA began early on 7 April and briefly raised the possibility of starting EVA an hour ahead of the flight plan. But the crew soon slipped back and at 4:30 P.M., Musgrave opened the airlock hatch and started America's first EVA in 9 years.

As per plan, the two moved over to the cargo bay storage box and removed 25-foot tethers that would keep them attached to Challenger. Musgrave then moved quickly down the starboard side and across the aft bulkhead. Pausing in front of a TV camera there, he offered a two-fingered wave to the viewers. He was soon joined by Peterson, who had moved down the port side, and the two practiced procedures for winching

*Astronauts Storey Musgrave and Donald Peterson took the first shuttle spacewalk during STS-6, as they tested the new spacesuits, or EMU's, which are assembled from parts, thus avoiding having custom-built suits for each astronaut.* (NASA)

the IUS cradle into position should it refuse to do so on future missions. At one point the pulleys locked up and Musgrave had to yank the line free so that the gear could be stored; it was the kind of problem the EVA was designed to find. He had also went atop the cradle and drifted into its 9-foot-wide O-shaped opening.

The pair then moved along the port side to the midcargo bay where Musgrave evaluated the ability of the tether to reel him back in. Next came a test of how they would winch the cargo hold doors shut in case they should refuse to do so. (A space suit had been carried on each mission since STS-1 for this contingency.) They then went to the cargo box to evaluate some work tasks and suit mobility.

The suits worked so well that mission control authorized an extra hour of EVA. But Weitz, being cautious on what he regarded as the suits' first test flight, vetoed the idea. At 7 P.M., Musgrave and Peterson entered the airlock. There was a brief moment of concern when the airlock refused to repressurize, but that was because a valve had not been resealed properly. After doffing their suits, the pair recharged them in case they would be needed again.

As the mission drew to a close, the crew enjoyed a more leisurely schedule of cleaning up for the return. For a phone call from Vice President George Bush, the entire crew—which had dubbed themselves the Geritol Bunch because of their average age of 48 years—donned glasses and held up a sign proclaiming "111 years of aviation experience."

In keeping with the rest of the mission, re-entry on 9 April was "nominal" almost every step of the way. Unbeknownst to anyone on the ground—and to the world at large until revealed at a post-flight briefing—Musgrave unfastened his seat belt as re-entry started and remained standing through landing. It was not that formidable a task for a person in excellent condition—the force is only 1.5 *g*—but it had never been done before.

Work for TDRS system operators was just beginning as Challenger landed. Officials at NASA and Spacecom (the owner that contracts TDRS system services to NASA) decided to take their time analyzing some minor problems aboard the spacecraft before committing themselves to raising the orbit. Two thrusters were damaged—and showed different kinds of leaks—apparently from a collision with the tumbling IUS second stage at separation.

A plan was soon devised for raising the orbit in a series of thirteen 3-hour burns, but this had to be revised because of minor difficulties in keeping one thruster firing. In what has been hailed as one of the engineering accomplishments of the year, a series of thirty-nine burns succeeded in placing TDRS-1 in geostationary orbit by 29 June, only 59 days after launch. Following checkouts, it was shifted to its permanent location at 41 degrees west longitude, where it subsequently relayed images from Landsat 4 and data from Spacelab 1.

Solving the IUS problem proved to be a more formidable chore, however. The limited amount of telemetry indicated that a leak had occurred in the inflated seal on which the steerable nozzle sat, somewhat like a weight riding on a tire. When it went flat, the nozzle locked over to one side and could no longer be steered. Extensive testing and materials changes had to be made, thus delaying TDRS-2's launch.

The first deployment and retrieval of a satellite was accomplished on the STS-7 mission in mid-1983. It was also the first flight to carry a woman and the first to attempt a landing at Kennedy Space Center.

This was the second flight for SS Challenger and for Robert L. Crippen, this time the commander. The rest of the crew was made up of pilot Frederick H. Hauck and mission specialists Sally K. Ride, John M. Fabian, and Norman Thagard, M.D.

The first half of the mission was a repeat of the STS-5 flight, starting with the 7:33 A.M. liftoff on 18 June and continuing through the deployment of Anik and Palapa communications satellites almost identical to those on STS-5.

Launch was nominal, but Ride had a different view, noting that, "If you've ever been to Disneyland, then you'll know that definitely was an 'E' ticket." A "D"

*Sally Ride became the first American woman in space, and the first woman to fly twice. Although trained as an astrophysicist, she became skilled at operating the shuttle's robot arm.* (NASA)

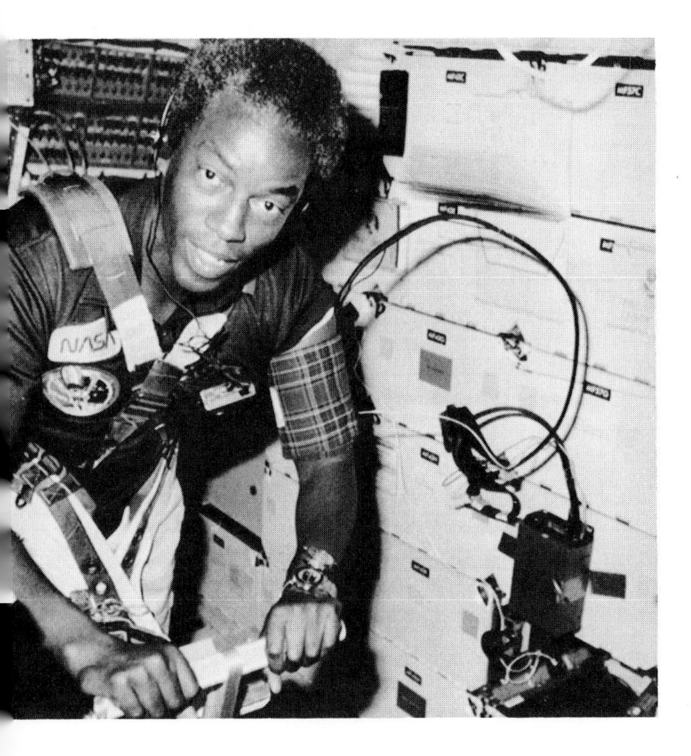

*Guion Bluford, a member of the STS-8 mission, became the first American black in space. Here he works out on the treadmill (while wearing a blood pressure cuff), a key part of preventing deconditioning in space.* (NASA)

ticket, with the most rides, used to be the most exciting at Disneyland.

Although the crew was a record—five—for a mission, Crippen said that, "We didn't have problems operating with five people at all. . . . That was a 'plus' to have so many hands available."

Two small payloads were the OSTA-2 materials experiment assembly and the SPAS-01 subsatellite.

The SPAS-01 satellite was a commercial venture by Messeschmitt-Boelkow-Blohm and based on Wernher von Braun's dictum that "Better is the enemy of good." Engineers designing SPAS used components that were "good enough" rather than seeking the best possible, such as sailboat masts for the carbon-epoxy struts that made up the structure, modified diving tanks for the attitude control gas bottles, and off-the-shelf electronics and lights. The result was a successful satellite carrying complex, meaningful experiments.

OSTA-2 was developed from designs proven on sounding rockets in the lull between Apollo-Soyuz and the shuttle. It carried two general-purpose rocket furnaces and a single-axis acoustic levitator (for containerless processing). The two furnaces processed their samples as programmed, but the levitator quit after two samples were melted and solidified and one was moved but not processed; the other three samples were not touched. The three West German MAUS cans worked, too. One was a reflight of the first MAUS that failed on the STS-5 flight because of a battery problem. It and a carbon copy on STS-7 made X-ray photographs of the inner actions of a metal when melted and solidified. One ran 78 hours and the other quit after 20 hours. A third MAUS, designed to study a phenomena called Marangoni convection, worked as planned.

CFES made two sets of three runs each for McDonnell Douglas Astronautics and NASA. The monodisperse latex reactor also made a 20-hour run, manufacturing the largest latex microspheres yet. Some have been sold to the National Bureau of Standards, and another series of six flights is being considered for commercial production.

All went well on the fourth day except that Challenger used only half as much propellant as expected, and the space vehicle's radar worked well without the aid of an enhancer placed aboard SPAS. Exhaust from Challenger's thrusters did have a stronger effect on SPAS than was anticipated, though, making it rotate about once per minute when hit with a glancing blow from a few dozen feet. In the first test sequence, Challenger released SPAS from the robot arm, dropped below, and then pulled ahead to make an approach from 1,000 feet out along the velocity vector, the so-called V-bar approach. Color TV cameras aboard SPAS transmitted stunning pictures of Challenger with the Earth and later deep space as backdrops. The second test used the more demanding inertial approach, which requires closing along a certain angle with relation to the target, as planned, for the Solar Max repair mission. SPAS was removed from the shuttle on the fifth day for test of the robot arm with a moderate mass at the end.

Thagard operated a number of head-and-eye motion and other experiments intended to expand what is known about how the human body adapts to space and what causes, or resists, space sickness. The data are being evaluated, he said, and will take several flights to become meaningful, but "It doesn't show any large difference between up there and preflight and so far postflight."

The landing on 24 June was to have been the first at Kennedy Space Center, but it was thwarted by ground fog in the Cape Canaveral area. Retrofire was delayed for an orbit while John Young flew the shuttle training aircraft and looked for signs that the fog would burn off in the morning sun. But it simply hung at the north end of the runway and Challenger finally was ordered to wait another orbit and then land at Edwards. "The good news is that the beer is good and cold," mission control radioed. "The bad news is that it's 3,000 miles away."

The STS-8 mission saw the first night launch and landing for the shuttle and the first flight of an American black, Guion S. Bluford. (The Soviets had scored that first with a Cuban.) The mission was commanded by Richard H. Truly, the only veteran on the mission. The pilot was Daniel C. Brandenstein, and the mission specialists were Dale A. Gardner, Bluford, and William E. Thornton, M.D.

The only paying customer was India's Insat 1B. Because the second tracking and data relay satellite had to be bumped by delays in fixing the Inertial Upper Stage, some secondary cargoes were added to fill the void—a payload flight test article (PFTA) for testing the robot arm, boxes, and Getaway Special cans carrying 250,000 U.S. Postal Service commemorative covers, and an atomic oxygen interaction with materials test. Already scheduled were the third flight of the CFES, medical experiments to examine the space adaptation syndrome, a shuttle glow photography experiment, and four Getaway Specials.

The launch and landing will always be remembered by those who saw it. As Apollo 17 did almost 11 years earlier, Challenger turned Florida night into a false dawn. Liftoff was delayed 17 minutes—half of the hold period—because of light rain around NASA's Kennedy Space Center. The launch was on the verge of being scrubbed when astronaut Robert Crippen, flying the shuttle training aircraft, noticed a slight clearing trend in the area and finally gave the go-ahead.

Ascent was nominal except for the darkness split by the 500-foot column of flame from the twin boosters. "You should have seen it from here," Truly called out. "It was daylight almost all the way up."

Flight day 1 was devoted to miscellaneous on-orbit activities, including the priority runs on the CFES. Flight day 2 saw launch of Insat 1B. "The deployment was on time . . . and the satellite looks good," Bluford said as Insat spun away. He reported that the crew felt a slight "clunk" as it left. Although it appeared nominal, engineers determined that it may have been hit with some debris within a few seconds of springing from its payload assist module. Deployment of the folded solar array was hampered after Insat arrived in geostationary orbit. The manner in which it appeared and was fixed led to the suspicion that something in the area of the shuttle, perhaps from the PAM, struck Insat.

Robot arm exercises with the PFTA (a 7,460-pound aluminum-and-lead "dumbbell" designed to test the operation of the Canadian robot arm) were scheduled for flight days 3 and 4; additional exercises were added on flight day 5 (an "extra day" was inserted between flight days 5 and 6), an option the Mission Control held from the start. Tests with the SPAS-01 satellite on STS-7 were the heaviest to date, about half the PFTA's mass. Most of the lead ballast was concentrated in a can at one end so that measurements could be made of how the robot arm moved when holding the PFTA by the heavy and light ends (like holding a hammer by the head or the handle). Screens at the front and rear blocked the view, to give the same effect as a satellite 14½ feet wide and 20 feet long. "It works like a champ," Gardner commented on the first day of tests. Mission commentator Terry White added that, "Challenger has just demonstrated that it can bench-press . . . without the aid of steroids," a reference to that summer's scandal over athletes and drugs. The second day of arm-PFTA tests went so well that there was no time left for "shopping-list" activities. Some of those were added to flight day 5.

The six rats were carried in a small locker container so that the animal enclosure module could be tested for later uses; but no experiments were run on them. Thornton checked on their health periodically and commented that at the start they "were all asking for their money back on the ticket." But after 3 days "they seem to have settled in rather nicely. All of them are in excellent condition."

Testing with the TDRS system ranged from excellent to poor. TDRS-1 and the shuttle's Ku-band antenna performed as planned, but there were a number of "high-tech learning problems" at the ground station in White Sands, New Mexico. The first results were frustrating: "Since it's a brand new system," flight director Harold Draughon said, "we expect it to be clumsy." But gradually the system and its operators improved. They moved data of higher quality than was expected via the S-band link and made several unplanned TV broadcasts via the new Ku-band link. On flight day 3, flight director Brock R. ("Randy") Stone said that testing was being suspended temporarily because of "normal high-tech growing pains." Two hours later, high-quality TV was being relayed via TDRS-1. On flight day 5, the crew held a 20-minute press conference with six reporters in Houston via TDRS-1—the first since Apollo 17 was returning from the Moon.

The landing on 5 September had a brilliance that in its own way matched the launch. Challenger has no running lights, and no search lights probed the sky to find it. Instead, xenon floods—the same as were used on the launch pad—illuminated the runway. A special ball-bar light, similar to that used aboard aircraft carriers, and a new precision approach path indicator gave Truly and Brandenstein cues to whether or not they were on the glide path as they sailed in for the landing.

Challenger was picked up on a special TV camera equipped with a military infrared scanner that not only outlined the spacecraft but showed its nose cap still glowing from the residual heat of re-entry.

Challenger was not visible until it dropped into the glare of the lights at 80-foot altitude, then settled onto the concrete runway at 12:40 A.M. Pacific time, as one crew member called out, "OK, 3," a carrier pilot's reference to correctly snagging the third wire on the flight deck.

One problem emerged days after the mission. Insulation on one of the booster nozzles had nearly burned through, which forced a check of all nozzles made from the same batch of material.

On the next flight space shuttle Columbia returned to duty with the most complex manned payload launched by NASA since Skylab. Indeed, the Spacelab 1 mission was, in many ways, like a quarter-scale abbreviated Skylab mission. This flight did not have science as its primary goal, however. Rather, it was intended to demonstrate Spacelab's capabilities, just as the OSTA-1 and OSS-1 pallets had done on the STS-2 and -3 missions. In designing this so-called verification mission, NASA issued an "Announcement of Opportunity" to the entire space science community rather than to only a few disciplines. Spacelab 1 was consciously designed for as wide a spectrum of users as was possible, so that its utility could be demonstrated to all. Additionally, half the payload mass went to the European Space Agency and half to NASA. Europe also was asked to select one of the two payload specialists for the mission. Experiments would be accommodated in a long lab module and on two pallets (later reduced to one). NASA's Marshall Space Flight Center was placed in charge of the mission.

More than 200 replies to the announcement were received; more than 70 were selected for flight (there were only 38 experiment facilities, when one considers several samples using the same furnace, etc.; the experiments are discussed later). A working group made up of the principal investigators who developed the experiments considered candidates for the two payload specialist slots. For NASA, the candidates came from among the investigators, a procedure repeated many times on subsequent payloads. On its side, ESA issued a call to European scientists as a whole because of the desire to start building a small cadre of what they would consider to be mission specialists.

Five men were selected: Dr. Byron K. Lichtenberg, a biomedical engineer at the Massachusetts Institute of Technology; Dr. Michael Lampton, an astrophysicist at the University of California at Berkeley; Dr. Ulf Merbold of West Germany; Dr. Wubbo Ockels of The Netherlands; and Dr. Claude Nicollier of Switzerland (later detached from the group). In the final year of training, Lichtenberg and Merbold were selected as the primary payload specialists who would make the flight, and Lampton and Ockels were picked as their alternatives or backups who would work in the payload operations control center, called Marshall Ops, at Mission Control. Two mission specialists were named to the crew, Drs. Owen Garriott and Robert Parker, veterans of Skylab in flight and on the ground, respectively. Finally, John Young and Brewster Shaw were named as commander and pilot, respectively. For Young it would be a record sixth flight into space.

Planning for the mission started in 1977 and the launch slipped in pace with slips in the shuttle program. The most serious delays, though, came in 1983, when TDRS-1 was late in going on station and then when the near burn through on STS-8 was discovered. These resulted in the postponement of Spacelab 1 from 28 September to 28 November 1983. The problem posed was that many of the atmospheric and astronomical instruments were now placed in less-than-ideal lighting conditions, including constant daylight for the latter part of the mission. The alternative was to fly in February when the lighting would be better but at the expense of keeping the teams assembled for several more months.

Launch, aboard an upgraded SS Columbia, finally came off at 11 A.M., 28 November, at the start of a narrow 9-minute window. Although the boosters burned a little bit hot, the launch was so good that flight director Jay H. Greene called it "meaningless to compare it to others." Spacelab 1 was placed in an orbit 155 miles high and inclined 57 degrees to the equator—just grazing the auroral oval in Canada—to satisfy the needs of the atmospheric instruments.

*Owen Garriott and Ulf Merbold pause during their work aboard Spacelab 1. This view shows the space available for working and the complexity of the equipment that lines the walls.* (NASA)

After launch, Parker started turning Spacelab on from a control station on the aft flight deck. Entry into Spacelab was delayed for 24 minutes because of a balky hatch. Already there were signs of experiments yet to come, with both Merbold and Lichtenberg wearing headbands that held small accelerometer packages on the backs of their heads for space-sickness studies. The crew finally floated into the Spacelab module at 2:42 P.M., with Garriott and Lichtenberg together, back to back, and then shaking hands as they emerged through the transfer tunnel. Merbold was close behind, and Parker joined them a few minutes later.

They immediately went to work turning on lights and starting to awaken Spacelab's computer and experiments. By 4 P.M. one could readily see that the lab was in use: a file drawer of experiment handbooks was halfway open and a manual flapped gently in the air as it bounced on a short tether. The payload operations control center came on the air, and at 5:40 the word passed that Spacelab was operating. "Marshall Ops, this is Spacelab," Garriott called out, "and we think payload activation is complete. Outstanding. No glitches, no nothing."

A glitch did, however, soon emerge—one that would hamper experiments on the pallet throughout the mission. A remote acquisition unit, RAU 21, refused to pass data from four instruments connected to it to the main experiment computer. The problem appeared only when RAU 21 was hot from operating and from exposure to sunlight, so that the suspicion emerged that there was a loose connection that opened when heated. It did not affect data going to the experiments, and fortunately not all the science data coming back had to pass through the RAU. The problem later was traced to heat-sensitive resistors.

The crew's heaviest workload was in the first day, when a multitude of space-adaptation experiments had to be performed before the crew adjusted to space. The experiment regimen involved extensive testing of reactions in the weeks before launch, immediately after going into orbit and just before returning, and in the weeks after landing.

Meanwhile, the mission was going so well that mission control was considering adding a day to the mission, making it 10 days long, because Columbia wasn't using its fuel cell supplies as rapidly as had been expected. The low rate continued and the extension was later granted. The extra day was filled with experiments that were not completed during the "nominal" mission and with repeats of those found to be interesting on their first runs. There was already a twice-daily replanning operation under way in recognition of the fact that the mission could not be expected to run precisely as was written.

Indeed, a number of minor problems did occur. None in Spacelab proper hampered the mission, and Parker turned out to be quite capable of fixing several of them. A high-rate data recorder was written off and would have forced juggling of the timetable until Parker opened it up and freed a stuck drive mechanism. And later, when the second film magazine on the metric mapping camera jammed, he turned a bunk into a darkroom and (in the blind) rethreaded the film.

Most of the mission was devoted to science work, so that the payload crew seemed to be in their element. Many experiments did have difficulties that can occur in laboratories on the ground or aboard unmanned spacecraft anywhere, and a few achieved 100 percent of their planned activities.

The data haul, simply put, was tremendous. It included more than 2 trillion bits of data, 20 million frames of TV, thousands of photographic frames (including 900 on the metric mapping camera Parker repaired), plus the many materials processing and biological samples that were grown in orbit.

In astronomy, three instruments were carried. The most attention went to the far ultraviolet space telescope on which all the film was so fogged that only a few bright known stars could be seen rather than the faintest, oldest galaxies. This fogging was mistakenly attributed at first to the shuttle-glow phenomena, but it was later shown to be caused by two bands of ionized oxygen that circle the globe. They were photographed from the Moon in 1972 by an ultraviolet camera carried aboard Apollo 16.

A very-wide-field camera photographed all its primary targets, including a thin bridge of stars and gas between the two Magellanic clouds. An alignment problem, though, thwarted spectroscopy experiments. An X-ray spectrometer, identical to one used on Exosat, revealed iron lines in the Perseus cluster of gal axies and Cygnus X-3 and measured X-ray emissions from the Cassiopeia A supernova remnant.

Measurements of the solar constant were taken with three different instruments supplied by United States and European scientists, but shuttle contamination obscured some wavelengths.

The most advanced space plasma experiments yet were performed with the space-experiments-with-particle-accelerators (SEPAC) apparatus developed by the University of Tokyo. Early loss of the high-power electron-beam gun (caused by a loose nut that should have been removed after ground tests) and intermittent problems with RAU 21 reduced the total science return to about 80 percent of that anticipated before launch. An attempt to create an artificial aurora failed. But what collected was far more than that derived from all sounding rocket flights over the previous 10 years. Measurements included the beam plasma discharge phenomena that had long been suspected but could not be proven in chambers on Earth. Additionally, electrons emitted by the gun were returned with higher energy, indicating that a cloud of electrons may be circling the shuttle.

An atmospheric emissions photometric imager was designed to operate with SEPAC, but a problem with its clamps dictated that it stay in locked-down rather than pointed position. "Significant diagnostic data" were recorded in conjunction with SEPAC, a double-layer dayglow was detected, and tenuous magnesium ion clouds formed by meteors burning up in the atmosphere were observed. Additional SEPAC tests were made in conjunction with the phenomena induced by charged particle beams (PICPAB), thus demonstrating the flexibility of a manned system. A low-energy electron spectrometer detected auroral electrons. Data on cosmic ray were expected to come from an isotope of special plastics.

Some of the most valuable environmental data came from two instruments that tested designs never before flown in space. The imaging spectrometric observatory recorded the first broadband spectrum of the upper atmosphere dayglow in the 300 to 1,280 nanometer band (visible light covers 380 to 730 nanometers), obtained limb scans over the same range of the nightglow phenomena, and gathered an extensive database for assessing the shuttle environment. The observatory's success also verified the Spacelab concept, because it represented a significant technological risk and a major advance in instrument design at low cost, both of which could not be attempted with a conventional satellite.

The grille spectrometer, used in space for the first time, detected many gases not previously thought to be at certain levels of the atmosphere—carbon dioxide in the thermosphere and water vapor and methane in the mesosphere—measured ozone distribution in the upper atmosphere, and observed other gases such as carbon monoxide and hydrochloric acid. Because of sunlight problems, only 16 percent of the planned observations could be made, but the principal investigator expected several major scientific papers to come from his data.

A dual spectrophotometer measured the ratio of hydrogen to heavy hydrogen in the atmosphere. This will allow calculation of how gases migrate upward through the atmosphere and make environmental models more accurate. The spectrophotometers also observed auroras caused by protons hitting the dayside atmosphere, and saw lyman alpha (a common "color" emitted by hydrogen) in interplanetary space. (This now becomes a tool for measuring the solar wind.) And an infrared camera photographed large waves in the atmosphere around 53 miles altitude by the light emitted by hydroxyl, an oxygen-hydrogen compound.

The Earth itself was photographed with the metric mapping camera, a standard aerial mapping camera modified for space. About 900 frames were shot with it, including 80 given to the Spacelab crew to shoot "tourist" photos as thanks for Parker's having fixed a jam. The pictures were spectacular in their subjects and exquisite in their detail. The microwave remote sensing experiment suffered "considerable losses in science," because the traveling-wave tube in the radar portion apparently burned out, eliminating radar mapping of the ground. Passive observations were made as a radiometer observing natural radio wave emissions by the Earth and the atmosphere.

Most of the materials processing experiments were conducted in the materials science double rack, which had four facilities. In the rack's mirror heating facility, three samples were processed and two partially processed. On one, the scientist was able to direct the crew in processing a silicon rod, thus saving the sample. In its fluid physics module, all thirteen experiments were completed plus several additional experiments based on results seen by the scientists on TV or described by the Spacelab crew. Most striking in this set of experiments was confirmation of the Marangoni convection effects, which will have implications for the design of future materials processing in space. The isothermal heating facility processed eleven of twenty-two samples before its power system failed, and the gradient heating facility processed all five samples. Other materials experiments were equally successful, including growth of protein crystals several hundred times larger than are possible on Earth (thus making X-ray crystallography possible with them), and the use of bearing models to shed light on friction effects in zero gravity.

The bulk of the life science experiments was oriented toward how the human body adapts to space flight. Although they addressed the problem in different ways, they sometimes used the same equipment or studied the same parts of the body. Further, the results from each will be of use to the others because the human organism is complex and interactive, one in which no one organ works independently of all others.

One set of vestibular experiments used a rotating dome with a camera to watch eye reactions. Tests found that there is an increased reliance on vision for determining one's orientation in space. As the body adjusted to space, the motion of the dome increasingly gave the subject the impression that he was moving in the opposite direction as revealed by reflexive eye jerks (nystagmus). Only moderate problems with space adaptation syndrome (space sickness, apparently in Lichtenberg, who soon adapted) show that "mildly provocative testing appears practical," Lichtenberg said. Other tests measured reflexes when pulled down by an elastic cord (called "hop-and-drop") and the accuracy of the subject's knowledge about the position of his limbs and the location of other objects after keeping his eyes closed for a while. Failure of the power supply for the camera flash jeopardized the eye-motion experiments; but adapting a TV camera not only made the experiments 90 percent successful but allowed results to be analyzed as they happened.

A more complex experiment rig was a headset to study the "effects of rectilinear accelerations, optokinetic and caloric stimulation." This used a TV camera to watch one eye while the other viewed a TV screen showing images indicating motion. At the same time, warm or cold air could be blown onto the eardrums to confuse the vestibular system. A joystick was also provided for the test subjects to indicate which way they thought they were being moved by the crew member running the test. The most striking result here was the overturning of a Nobel prize–winning 1904 theory of the function of the inner ear system. Failure of a videotape player (for giving the eye the illusion of motion) cut one set of investigations in half, but 90 percent of the other tests were successful.

The hop-and-drop station was also used in other tests to measure vestibulospinal reflexes. The results indicate that the brain modifies a knee reflex that braces for a fall. This modification corresponds with whether a crew member becomes space sick, but after the fact. The principal investigator hopes that it may lead to a way to predict space sickness before flight.

A 3-D ballistocardiography harness measured body motion while the heart beats were made. And a mass discrimination was run by having a crew member "weigh" outwardly identical epoxy balls with different amounts of lead inside. These showed a "significant degree of adaptation" after one day.

Additionally, a wide array of blood, hormone, and enzyme tests were conducted with samples drawn during the flight.

Two non-human tests were conducted. The circadian rhythms in neurospora fungus was shown to persist in space in the absence of day-night and tidal cues, indicating that it has an internal clock. Photography of the nutating motion of dwarf sunflower seedlings was inconclusive, so that more flight tests are planned.

Almost unnoticed amidst the science results were the engineering tests run on Spacelab itself, ranging from baking it in the Sun to having the crew scramble out in a fire drill. These tests were the official purpose for Spacelab 1, but that they went so smoothly underscored how well the lab was designed and built.

The mission was not without its moments of concern, however. During re-entry preparations, one of the primary flight computers failed when there was a thruster firing, at 6:13 A.M. on Saturday, 9 December; a second followed 4 minutes later after another thruster firing. The de-orbit burn was postponed while engineers on the ground studied the problem. The first computer had a hardware failure and was left off, but the second one was revived and worked with address errors that were acceptable for the time being. After the flight the failures would be traced to foreign-object contamination that shorted the computer circuits.

Retrofire was rescheduled for 5:52 P.M. (EST) leading to a landing at Edwards at 6:47 P.M. (3:47 P.M. Pacific time), 10 days, 7 hours, 48 minutes after liftoff. The landing was the first made on a descending node, i.e., as the spacecraft traveled south toward the equator, taking it across the north Pacific coast of Asia and of Alaska and Canada.

Two more failures came before the mission ended. The second computer quit when the nose wheel touched down, and a fire broke out a few minutes later in the aft compartment when a seal leaked hydrazine into the compartment. Damage was modest and NASA said that the crew was never in any danger.

With its tenth mission, the shuttle program made one of its most daring moves: two space walks with the crew members untethered, flying their own rocket-propelled backpacks. It was the first "personal" spaceship since the cramped days of the one-man Mercury space capsules almost 20 years earlier. It was also to become the most frustrating mission to date.

Again the vehicle flown was SS Challenger, Flight 41-B. The crew members were commander Vance D. Brand, making his second shuttle mission; pilot Robert L. ("Hoot") Gibson; and mission specialists Bruce McCandless II, Robert L. Stewart, and Ronald E. McNair. The mission was the first not to receive an STS designation—e.g., STS-1, STS-2—but rather was designated 41 B. The first digit of this alphanumeric scheme refers to the last digit of the government's fiscal year (which begins on 1 October); the second digit indicates the launch site (1 = Kennedy Space Center at Cape Canaveral, 2 = Vandenberg Air Force Base in California); and the letter is the serial flight in the fiscal year. Launch dates are sometimes changed after mission designation has been assigned; usually the original designation is retained.

Launch of 41-B at 8 A.M. on 3 February 1984, was normal. And while the shuttle mission itself went well, its payloads soon ran into trouble. Two almost-identical satellites were the primary cargo—Westar VI and Palapa B2, each sitting atop identical PAM-D stages. Also on board was SPAS-01A, a reflight of the German satellite carried on STS-7.

Unfortunately, both Westar and Palapa were dropped in low orbits (about 190 × 750 miles each) rather than geostationary transfer orbit (190 × 22,300 miles) when the perigee kick motors in their PAM-D stages failed early in their burns. Deployment of Palapa was delayed 2 days while engineers tried to determine the cause of the Westar failure and estimate their chances of a duplicate failure. Watching via a TV camera on the robot arm that was crooked outboard while the shuttle's belly was to the rocket, the crew watched Palapa's PAM-D motor ignite and then apparently flame out a few seconds later. Stewart said they felt that they knew what had happened but tried to be optimistic because of the failure of the first motor. "We didn't want to let ourselves believe that it happened twice in a row," Stewart said.

A transfer to geostationary orbit like that made by TDRS-1 was not possible, because the satellites were placed in too low an orbit and did not have enough station-keeping propellant to make the journey. But a rescue later was proposed wherein the station-keeping propellant would lower the orbit to match the shuttle's and the satellites would be mounted on a special post on a Spacelab pallet. While technically feasible, the plan was held up because of uncertainties on the parts of the owners and their insurance carriers. Approval did not come until that August for a rescue attempt in November. Subsequent analysis showed that there were manufacturing flaws in the carbon-carbon cloth of the PAM-D's rocket nozzle that allowed it to burn through after a few seconds.

The third failure of the mission came on deployment of the integrated radar target—an aluminum-coated balloon—to be used in two rendezvous tests simulating approaches that Challenger would take in the Solar Max rescue. Apparently the stays in the balloon's canister failed to release and it ruptured as it inflated after deployment. Despite that loss, the target verified the radar's response out to a range of about 52,000 feet.

The two spectacular space walks clearly were the best part of the mission. For the first time in manned space flight history, an astronaut released the tether between him and his orbiting spacecraft and moved out to become a miniature spacecraft unto himself.

Making the walk possible was the 330-pound Manned Maneuvering Unit (MMU), which latches to the back of the astronaut's life-support pack. It has two large bottles of compressed nitrogen gas to power sixteen 1-pound thrusters and a gyro package to hold the MMU in a position or course set by the astronaut. Both the thruster system and the electronics were built in identical halves, so that any failure could be isolated and allow the crew member to return to the shuttle. Overall, the MMU is 49 inches tall, 32.6 inches wide, and 48 inches deep.

The first space walk started at 5:25 A.M., 7 February. "That may have been one small step for Neil [Armstrong]," McCandless said as he flew the MMU

*Astronaut Bruce McCandless became the first spaceman to float free of his spacecraft when, during STS-41-B, he tested the manned maneuvering unit, a gas-powered "Buck Rogers" backpack. He flew as much as 300 feet away from SS Challenger.* (NASA)

away from the flight support station, "but it was one heck of a big step for me." After checking the handling characteristics of the MMU in the area of the payload bay, McCandless moved out 150 feet from Challenger and back in about 12 minutes. He then moved out to 320 feet, 20 feet farther than planned.

"The view that you get out here is like the difference between the view you get flying a heavy aircraft and looking out little windows compared to flying a helicopter at Mach 25," he said, "It really is a beautiful panorama." For him the flight was especially rewarding, because most of his astronaut career had been spent in developing the MMU's design and helping put it through ground tests.

Stewart, meanwhile, removed the manipulator foot restraint (MFR) from its storage box and secured it in the grasp of the robot arm end effector. The MFR gives an astronaut a place to stand just as a cherry picker puts a utility worker next to a job above the ground. The MFR has no controls, though, only tool holders. The arm was operated by McNair at the aft flight deck controls. Because he had difficulty in getting his feet into the rings, Stewart's operations on the MFR were canceled so he could make his MMU excursion on schedule when McCandless returned from his flight.

Stewart made only one run straight to 306 feet with a brief stop at 150 feet. While Stewart did that, McCandless stood in the MFR and simulated repair of the Solar Max electronics box with a mockup mounted on SPAS-01A and used a wrench to measure how much force was required to move the arm out of position (about twice as much force as predicted). "It went pretty much as expected," McCandless said of the repair simulation. Before going inside, McCandless fixed a mass spectrometer switch on SPAS-01A and retrieved a TV camera that had failed 2 days earlier. The first EVA lasted almost 6 hours.

After a day of rest, necessary to let blood gases reach equilibrium, the second EVA started at 3:40 A.M., 9 February. McCandless donned MMU No. 2 and flew a simulation of the approach to the trunnion pin on Solar Max. (Because of a minor robot arm wrist problem, SPAS-01A was not picked up and rotated for this sequence as planned.) He and Stewart then traded places. In both cases, the astronaut without the MMU was operating a mockup of a hydrazine refueling station for tests to be conducted during an EVA on a later mission.

During the second half of this EVA McCandless found that the MFR broke loose and floated away. Brand vetoed having Stewart retrieve it and instead maneuvered Challenger into position so that McCandless could grab it with one hand while the other held a slidewire. This showed that the shuttle can be used to recover an astronaut who might become separated from the shuttle or have an MMU failure.

McCandless reported during the flight that the only unusual aspect of the MMU flights was vibration and noise encountered during some maneuvers forward. He said that it was caused by valves for the nitrogen thrusters popping open and shut in rapid fire in order to maintain certain motions, a phenomenon that had not been adequately modeled in ground simulators.

Both laser ranging and radar worked well when they did work—there were problems that prevented both from working at the same time—as did McCandless's "solid state ranging device." That was a modified forest ranger's tool that gave distances when viewed against the shuttle in the distance. It was consistently 25 feet off and thus usable only for approximate ranges. Within 50 to 70 feet it was easy to judge speed and direction by watching the shuttle, Stewart added.

And McCandless, in addition to the two repairs mentioned above, reattached a thermal cover on the Cinema 360 camera. "I think it showed a lot of flexibility for the man-machine system," Brand said.

Re-entry and landing were nominal except for the first-ever landing at Kennedy's shuttle landing facility. Ground fog was a worry early in the morning, but predictions that it would clear and not be a problem proved correct as Challenger cut across the Florida peninsula at Mach 5. Brand said he initially thought he was too high and fast, but instruments showed that the glide was perfect. Touchdown was at 218.5 miles per hour 1,930 feet from the runway's north threshold. Challenger rolled 10,770 feet and stopped with 2,300 feet of main runway left.

The most outstanding shuttle mission, coming just 3 years after the first launch, was the rescue of the Solar Maximum Mission satellite. It did more to attract attention to and prove the capabilities of the shuttle than had any flight since the first launch.

Solar Max, as it is called, was the first satellite to use the multimission modular spacecraft concept that groups most of a satellite's "bus" components in three modules, each 4 feet square and 1.5 feet thick. They are held to the spacecraft frame by two large bolts, top and bottom, and a blind electric connector on the rear side. The bolts and connector were designed so that an astronaut could replace the module in space without any special effort. All that was needed was a module servicing tool that resembled a large power drill.

Solar Max was built in this manner with three modules in the bus and its seven solar telescopes mounted in a large module atop that. It was launched on 14 February 1980 and worked well for 9 months when three of four fuses in the attitude control system blew because their helium coolant gradually leaked away and they overheated. This left the satellite without the use of its reaction control wheels and it was left in a slow spin that aimed its instruments slightly away from the Sun. Solar Max and other space officials started formulating a rescue plan and, emboldened by the success of the STS-1 mission, started pushing it within NASA.

When the repair mission was launched, the cost of replacing Solar Max was estimated at $235 million. The bill for the repair mission was less than $60 million. Equally important, it gave solar scientists a renewed observatory for continued studies of the Sun and NASA an excellent tool for advertising the shuttle.

Basically, the plan was for the shuttle to rendezvous with Solar Max and for an astronaut using an MMU to fly over and stabilize the satellite so that it could be picked up and repaired or brought back to Earth. While this sounds simple, many details had to be worked out, including developing software for the rendezvous radar and training crew members to fly up to the rotating satellite.

The thirteenth shuttle mission, to be made by Challenger, was selected to attempt the repair. Robert L. Crippen was named to command it. The pilot would be Francis R. Scobee. The repair astronauts would be George D. ("Pinky") Nelson and James D. A. ("Ox") van Hoften. Operating the robot arm would be Terry J. Hart. With changes in payloads because of the IUS problem, the repair flight became the eleventh shuttle mission. It was designated 41-C.

With the success of the MMU test flight on the 41-B mission, launch was set for 6 April. Also making the flight was the 21,400-pound Long Duration Exposure Facility, a metal frame 14 feet wide and 30 feet long carrying 80 trays of materials and test hardware to be left in space for a year and then retrieved by another shuttle mission. Such tests are considered crucial to developing hardware for advanced long-term missions, including the space station.

From the ground the launch of 41-C was normal, but it followed a special "direct ascent" trajectory that placed it in a higher orbit than that of past missions in order to conserve propellant for the rendezvous with Solar Max, now 309 miles high. One result was that the shuttle's external tank, which normally re-entered over the Indian Ocean, was hurled all the way to an area southeast of Hawaii, where it provided a spectacular cometlike display in the predawn hours as it broke up and was incinerated in the atmosphere.

Once in orbit the crew began setting up for the rescue. LDEF was deployed on the second day by Hart using the robot arm with such precision that there was no perceptible motion in the unpowered spacecraft as the shuttle backed away.

On the 8th, Challenger completed its final rendezvous and was less than 300 feet away from Solar Max. Nelson and van Hoften donned their space suits and went into the cargo hold, where Nelson climbed into the MMU that would take him 300 feet across the void of space to the slowly rotating Solar Max.

His approach was perfect, and in less than 10 minutes he had arrived. He waited a moment until the satellite had rotated into the right position and then flew between the solar arrays to dock with the trunnion pin at the satellite's waist at 10:33 A.M. A device called the trunnion pin adapter device, or TPAD, mounted between the arms of his MMU was to close rubber-coated steel jaws around the pin to hold him in place while he used his thrusters to stop the satellite's roll.

"OK, the jaws didn't fire that time," Nelson called out. "The jaws didn't fire." They didn't fire two more times as Nelson flew into the satellite harder each time. Even before Challenger returned to Earth, engineers decided that the most likely cause was a small blanket pin, which stood out exactly far enough to keep the TPAD from being triggered. Somehow, it had escaped being included in engineering drawings but was spotted (too late) in prelaunch photos of the satellite.

Giving up on the TPAD, Nelson decided to try an alternative approach that had been rehearsed in simulators. He flew over to the end of one solar array, grabbed it with his hand, and then tried to stabilize the satellite. But that and the repeated ramming earlier only left Solar Max rotating faster than it had been before, about 1 to 2 degrees per second on all three axes instead of the ½ degree per second before rendezvous.

Mission control here offered Crippen the option of having Nelson try again with the MMU, but he declined because Challenger had used up too much of its propellant in the forward thrusters. Subtracting propellant that would be needed for later maneuvers, Challenger was below the minimum required for rescuing a crew member in case his MMU had a "runaway" thruster. Nelson also had used more than half of the nitrogen charge in his MMU tanks.

For a while it appeared that Solar Max was lost because the solar arrays were not in sunlight long enough to keep the satellite working and the batteries were gradually being used. At one point the satellite had less than 30 minutes to live. But the solar arrays kept getting exactly enough sunlight to work and engineers at NASA's Goddard Space Flight Center in Greenbelt, Maryland, where Solar Max is controlled, radioed up a special attitude control program. Using magnetic torque bars and a program called B-dot (from the physicist's shorthand for electromagnetic flux), the bars worked as electromagnets dragging against the Earth's magnetic field line and gradually slowed the satellite. By 2 A.M. on the 9th, the roll rates had been reduced to 0.1 degree per second in all axes.

And this rate turned out to be too slow and could require that Challenger fly in formation for several minutes until the grapple fixture rotated into the right position. Because the forward thruster propellant, the pacing item, was down to about 25 percent (with 3 percent saved for getting away from the satellite), flight directors decided to postpone the grapple attempt until the 10th while Goddard engineers started Solar Max rolling at ½ degree per second, fast enough to bring the grapple around in short order but slow enough that the arm should be able to handle it without any problem.

The crew spent their third mission day using the large-format IMAX movie camera, running a student experiment on bees, and making preparations for rendezvous. By being stingy on the attitude control jets, they were able to go into the maneuvers with 21 percent propellant in the forward tanks. Engineers at Mission Control also found that there actually is 1.5 percent left when the tank gauges read zero and decided that this would be sufficient for moving away from the satellite, thus allowing them to burn to zero rather than 3 percent.

The maneuvers started with a thruster burn at 3:46 A.M. on 10 April. As Challenger started to close on Solar Max, it appeared on TV screens as an indistinct blob in the night. The crew operated in near-radio silence, saying little except to note that burns were made on schedule. By 8 A.M. Challenger was less than 2 miles away; by 8:27 it was less than 1,000 feet.

While out of TV contact the final gap was closed and Hart, moving the arm like a giant cobra, waited until the grapple fixture rotated into position and shot forward to grab the satellite.

"We've got it," Crippen reported. Surprisingly, there was 13 percent fuel remaining in the forward tanks, considered more than an adequate margin.

By 10 A.M., Solar Max was secure in the berthing ring, and TV from Challenger showed the satellite being tilted forward and rotated so that the attitude control system module faced forward and then tilted back into position. On TV from various angles, Solar Max appeared to be in good shape and looked like the training model at Goddard.

Originally the repairs were to have been made with two separate EVA's because of the time required to capture Solar Max. With the satellite now on board, flight directors were able to combine the two repair procedures into one EVA. Soon after the crew was awakened, Nelson and van Hoften started suiting up (at 1:58 A.M.). They were eager to start their work this morning with Nelson saying that he wanted to go early—without his suit if necessary. Airlock depressurization started at 3:58 A.M. At 4:19 the hatch was opened and Nelson and van Hoften, clad in space suits, emerged from the airlock of Challenger and went to work as high-flying maintenance men. An informal mission logo drawn on the daily plan for ground controllers expressed best the hopes for the day: a toolbox with wings and an arm grasping a satellite, with the motto, "Ace Satellite Repair Co."

By 4:30 the robot arm had grappled the manipulator foot restraint, and van Hoften was standing in it like a utility worker on a cherry picker. Replacement of the attitude control module had started at 3:50 when van Hoften went to work on it with the module servicing tool. With little initial comment or TV coverage, he disconnected the upper and lower bolts, removed the module, and installed the new module by 5:23 A.M. Nine minutes later, engineers at Goddard were reading temperature data from inside the new module.

The two repairmen then mounted a "butter dish" cap over a vent for the X-ray polychromator instrument to protect it from gases being rammed in and creating noise in its data.

At 5:54 the TV showed the two huddling together, van Hoften on the robot arm and Nelson hanging on the tool caddy at the front of the foot restraint, as they prepared to cut into the gold-coated mylar that covers the instrument module. This second repair was intended for the second day and was far more challenging. The main electronics box for the satellite's coronograph telescope had failed 2 months be-

*James "Ox" van Hoften prepares to install a new electronics module for a telescope on the Solar Max satellite. This was the first time that a satellite was retrieved and repaired in orbit.* (NASA)

fore the fuses. It was mounted behind an access panel that was designed for servicing on the ground, not by astronauts wearing bulky pressure gloves. But NASA again hoped to stretch beyond conventional wisdom and make it "doable," as they often said.

A Valentine's heart was clearly visible on the side of the module where its launch crew had drawn it. Van Hoften commented that the blanket is "in better shape than the stuff we've been using before" in the water tank. "Nothing wrong with it at all."

Using surgical scissors, van Hoften cut through the plastic blankets to expose the access plate. Pulling out a power drill, he installed a piano hinge and then loosened the screws that held the panel in place. At 6:12, the panel-turned-door swung open. "All right," Nelson said. "Looks good," van Hoften replied. Helmet-mounted TV cameras showed a jumble of wires snaking from inside the instrument module to the main electronics box mounted inside the door. "I've never had so many people looking over my shoulder," van Hoften said.

A blanket was placed over the opening to protect it from sunlight and van Hoften used the drill to back out the screws that held the box in place and removed the electric connectors from the box. Then he and Nelson traded places, and Nelson installed the replacement box. Rather than holding it in place with screws, he secured the box with special clips. Finally, by 7:30 A.M., all the screws holding the door were back in place and the blankets were being taped to cover the wound.

The repairs and propellant conservation had gone so well that van Hoften was given the go-ahead for a

*The shuttle carried its first array of space science instruments (in shadow) during STS-3. Here, the Canadian-built robot arm holds up a can-like plasma diagnostics package.* (NASA)

*A technician is dwarfed by the cargo hold of SS Discovery as it is prepared for its first mission (STS-41-D). From the bottom are: the large Syncom IV, two PAM-D satellites, and a solar array test model.* (NASA)

check ride on the MMU, a procedure that had been dropped from the timetable earlier in the flight.

As Solar Max entered its 23,000th orbit of the Earth, engineers at Goddard started checking out their newly activated satellite and announced that the repairs appeared to be good and the reaction control wheels were spinning.

At 10:59 A.M., the robot arm started lifting Solar Max up and away from the shuttle for systems checks. When the high-gain antenna was lowered from the bottom of Solar Max, it could transmit through the Tracking and Data Relay Satellite System rather than intermittently through ground stations. Early in the morning of the 12th, Solar Max was declared ready and released from the robot arm. In a month-long checkout that followed, the satellite was returned to full operating status, proving the repairs were successful.

Crippen, meanwhile, was to be frustrated once more in trying to land at Kennedy. There was too much ground fog and the landing was quickly rescheduled for Edwards. It went without incident and ended the mission after 144 hours.

After the Solar Max rescue, SS Discovery, the third shuttle orbiter, was to make its debut. But it was postponed by problems in the main engine that led to the first on-the-pad engine shutdown since Gemini 6 in 1965.

To avoid delays to other missions, NASA decided to cancel the 41-F mission and combine its payload with 41-D. The new flight carried the Westar VI and Telstar 3 communications satellites (using PAM-D stages fresh from intense redesign work), and the Syncom 4 leased communications satellite for the Navy. The latter used a unique "Frisbee-style" deployment that popped the satellite up from the payload bay. Also carried was a 102-foot-tall model of a lightweight solar array such as those that might be used by a space station. Only a few of the eighty-four panels on the array actually had solar cells, because the purpose of the test was not to generate electricity but to understand how a large structure vibrates in space when a spacecraft maneuvers. Three payloads were postponed to later missions—a second Syncom 4 (a total of four is in the series), a large-format camera for mapping the Earth, and the Spartan 1 X-ray astronomy satellite to be deployed and retrieved on the same mission.

The 41-D mission was commanded by Henry Hartsfield, the pilot from STS-4. His pilot this time was Michael Coates. The mission specialists were Judith A. Resnik, Richard M. ("Mike") Mullane, and Steven A. Hawley (husband of the first American woman astronaut, Sally Ride). One payload specialist, McDonnell Douglas engineer Charles Walker, was also carried. The continuous-flow electrophoresis system was aboard for a nonstop 80-hour run, and McDonnell

Douglas had requested that Walker fly with it because, as its chief engineer and a participant in its design, he knew more about the unit than could readily be taught to an astronaut. By contrast, it would be easier to teach him how to live in space.

Other minor problems seem to have dogged the mission, including a computer timing problem discovered only 2 days before the planned 29 August launch. Although NASA thought it unlikely, the problem posed the possibility that the boosters and tank could be jettisoned late, thus causing the shuttle to burn up precious propellant and go into a low orbit, forcing an early end to the mission.

But once liftoff came on 30 August, the mission went almost exactly by the timetable. All three satellites were deployed on schedule and had perfect burns that put them in geostationary orbit and restored no small degree of confidence in the shuttle as a satellite launcher. Work with the solar array went fast enough so that several tests that had to be dropped to a "shopping list" were accomplished. And the electrophoresis system had a few minor problems requiring Walker's skills, thus justifying his presence.

The one problem of note was an icicle several inches long that grew on the waste-water vent on Discovery's port side. This forced the crews to close the bathroom and resort to Apollo-era methods. There was also worry that the icicle might damage heatshield tiles when it broke off during re-entry. When baking the port side in the Sun failed to evaporate all the icicle, the robot arm was used to snap it loose.

Re-entry and landing were as clean as had been any in the past, and Discovery was pronounced to be in the best shape of any new orbiter.

On its thirteenth mission—which also was the 100th manned space flight—the space shuttle flew upgraded versions of the Earth observations carried on the STS-2 mission in 1981. Mission 41-G, flown between 5 and 13 October 1984, was commanded by Robert Crippen, making it a record fourth shuttle mission for him. His pilot was Jon A. McBride. Also aboard were mission specialists David C. Leestma, Sally K. Ride, Kathryn D. Sullivan, and payload specialists Paul Scully-Power (the first oceanographer in space) and Marc Garneau (the first Canadian in space).

The principal cargo was OSTA-3, occupying a single Spacelab pallet and almost identical to the OSTA-1 payload. The radar was improved to operate on a higher frequency, and the flat panel antenna was replaced by a slightly longer version that operated at a higher frequency and at different "look angles" for better viewing. It also folded up to make room for other cargoes in the shuttle. Also on the pallet were upgraded versions of the measurement of air pollution from space (MAPS) monitor and the feature identification and landmark experiment (FILE) from OSTA-1.

Other major cargoes on this flight were a large-format camera (using 9 × 18-inch film), the Earth Radiation Budget Satellite (ERBS), and an orbital refueling simulator. SS Challenger went into an orbit inclined 57 degrees to the equator, taking it over more than half of the Earth's surface.

Major Earth observation instruments were Scully-Power's eyes—and his experience of years of briefing astronauts on what to watch for from space. He was flown just to stare out the window and observe and photograph the oceans. In particular, he was to look for "internal waves," eddylike activities within ocean currents that are believed to play a major role in how energy is transferred in the oceans. Until seen by John Young, by chance, on STS-1, internal waves were unknown.

Scully-Power observed by "lying down" on the cockpit overhead and turning the cabin and its forward windows into a gondola, like those on research submarines. "The fact that these spiral eddies, which are new in oceanography, were interconnected over the whole length and breadth of the Mediterranean was just a tremendous observation," he said in a postflight interview. "The other thing that has never been heard of before involved the Gulf Stream, which has been studied for literally hundreds of years. But no one has ever made mention before of spiral eddies associated with the Gulf Stream."

Halfway through the mission Sullivan and Leestma made a 3½-hour space walk to try out a working model of a refueling station that might be used on future space missions to extend the lives of satellites.

The mission ended with a landing at Kennedy Space Center, Florida, the second for the shuttle and the third such attempt for Crippen. Problems were found in the heatshield tiles afterward—apparently waterproofing chemicals had damaged the bonding strips—leading to the replacement of about 4,000 tiles and the postponement of several missions from December 1984 through 1985 (including the one scheduled to be the first all-military flight).

NASA closed its third year of shuttle flights with a pair of spectacular rescues, which proved the versatility of man and shuttle in space. Although the Palapa B2 and Westar VI satellites launched by mission 41-B were not designed for retrieval, engineers at NASA and Hughes Aircraft (their builder) soon started thinking about how they might be brought back to Earth for reuse. The plan that emerged was to use the apogee kick motor to raise the satellites' orbits to more than 600 miles high so that Westar was ahead of Palapa (previously the two orbits had planes that intersected rather than coincided). Then the small station-keeping thrusters would be used in a series of 20 firings on each satellite to lower the orbit to 224 miles high, one the shuttle easily could reach.

The orbital maneuvering was tricky and precise. No less so was the legal maneuvering to obtain permission from the insurance carriers to perform the rescue. Merritt Syndicates and International Technology Underwriters wrangled with the firms that had paid more than $100 million total for the two satellites' loss (the $35 million value of each satellite plus launch fees, revenues, etc.). They paid NASA $2.75 million for each rescue.

In order to bring the satellites aboard, NASA and Hughes designed a unique set of hardware: a "stinger" to insert, harpoonlike, in the satellites' apogee rocket nozzle, an A-frame to place over the top of each so the robot arm could grapple it, and a berthing adapter and platform, all carried on a Spacelab pallet.

After a day's delay for bad weather, mission 51-A was launched on 8 November 1984 with commander Frederick H. Hauck, pilot David M. Walker, and mission specialists Joseph P. Allen IV, Dale A. Gardner, and Anna L. Fisher. The first task of SS Discovery was to deploy two satellites: Anik H for Canada and Syncom 4 for the United States Navy.

After successfully launching their satellite cargo Discovery rendezvoused with Palapa B2 on 12 November; Westar was about 700 miles ahead in the same orbit.

"Dale, it is right there, my friend," Allen said as he stuck his head out the airlock during final EVA preparations and saw Palapa. 'We could go out and get it just as it stands." Allen used the Manned Maneuvering Unit (MMU) to fly over to Palapa and inserted the stinger in the satellite's rocket nozzle to stop its spin. This allowed Fisher to grapple it with the robot arm.

"OK, entry, toggles pulled," Allen called out. "Soft dock, soft dock."

With Palapa rolled over so its top faced down into the cargo hold, Gardner was supposed to attach the A-frame to the top of the satellite, thus giving the robot arm a second grapple fixture by which to hold it. This would free the base to allow Allen to withdraw the stinger and let Gardner attach the berthing adapter that would hold Palapa to the pallet.

It did not work that way. Final adjustments to part of the feedhorn assembly that aimed the radio signals at Palapa's large dish antenna (not deployed because it never reached geostationary orbit) placed it ⅛ inch farther out than drawings had shown, and that blocked the A-frame from going into place. Capcom Jerry Ross, at Mission Control, quickly switched the crew to "Plan B," an improvisation he had helped work out less than a month before launch.

Allen stood on a foot restraint mounted on the sill of the pallet and held Palapa by a barlike omni-antenna atop its dish antenna. Gardner removed and stowed the stinger, then attached the berthing adapter to Palapa's base. Nine latches were torqued into position to clamp the adapter to the lip of the nozzle. All the while, for almost 2 hours, Allen held Palapa over his head. Although weightless, it still had more than 1,200 pounds of mass and all the inertia that goes with it.

"Boy, I'll tell you one thing," Allen said, "I'm sure glad this isn't Syncom"—which is twice as wide.

Despite that, he maneuvered the satellite with great precision. Once the latches were secure, the two men wrestled Palapa into the berthing latches. The whole procedure took less than 5 hours, about as much time as the original plan would have taken.

The crew had a day of rest while Discovery played a slow game of catch-up with Westar. With the experience gained in salvaging Palapa, they and officials at mission control back on Earth decided for a modified rescue plan for Westar. They would go directly to the handheld routine, but with a slight change: the foot restraint would be attached to the end of the robot arm (as it had been on the Solar Max rescue), permitting a crewman to hold the satellite there, acting as a substitute for the A-frame.

Gardner flew the MMU this time, making a dramatic, televised flight over to Westar and capturing the satellite.

Gardner then maneuvered it into position for Allen on the robot arm to hold. After stowing the MMU and stinger, he stood on the pallet to attach the berthing adapter. Fisher then used the robot arm to swing the satellite onto the latches, and Allen drove it home.

"We have two satellites retrieved and locked down in the cargo bay," Hauck announced at midmorning.

After another day to rest and stow equipment, Discovery flew its crew and treasure home to Kennedy Space Center. Only five missions had been flown in 1984, about half the total NASA had intended at the outset. But they were historic, setting the stage for satellite repairs and servicing, a new field in the space business.

The fourth and last shuttle orbiter, Atlantis, was delivered in 1985. More than 100 missions are planned for the space shuttle by the end of the decade, when the flight rate is expected to reach 24 a year.

The missions described so far will be typical of most of the flights. Each will attempt to expand man's capabilities and senses in space, pushing the frontier back a little each time. Many payloads will be seen in space repeatedly. The radar used in the OSTA-1 payload on STS-2 was reflown in the autumn of 1984 with several improvements and at a higher inclination than before.

It is expected to evolve into even more advanced forms and to be used in polar orbits in later years. So the shuttle program and its payloads are expected to progress. Some Spacelab payloads will be "dedicated discipline labs" assembled to tackle specific problems in space science and, between missions, will be left intact rather than incurring disassembly and reassembly expenses.

Smaller payloads are likely to multiply as well. Two NASA centers are working on a family known as Hitchhiker, a carrier midway (in size and capacity) between a Getaway Special and Spacelab and designed to fill the inevitable gaps that occur when some payloads are delayed (as happened on STS-8). A new concept called Spartan, blending the low cost of sounding rockets with the shuttle's ability to retrieve, has also been initiated.

New kinds of satellites are evolving to accommodate man—the fixer. Long before Solar Max failed, the Hubble Space Telescope's design was incorporating major space servicing features so that it can be kept operating in space rather than brought back to Earth or require paying exorbitant prices for reliability. Its solar arrays, five science instruments, three fine guidance sensors, and most of its electronics can be replaced in orbit—about everything except its 92-inch main mirror, the largest ever placed in space. Other astronomy payloads are following this trend toward space serviceability.

The first satellite for hire is being developed by Fairchild Space Co., using designs from the multimission modular spacecraft program. Leasecraft, as it is called, will be a satellite bus carrying payloads, such as materials processing plants, weighing up to 32,000 pounds. These payloads will be exchanged in orbit without having to return Leasecraft to the Earth, and even Leasecraft itself will be repairable in space. A smaller version called Proteus is being studied by NASA as a way of reviving the Explorer satellite program which has suffered from rising costs and declining launch rates. With Proteus, a handful of buses will be built and Explorer-type instrument packages will be placed on them for 1- or 2-year stays.

Low-cost measures are also being planned for shuttle-based planetary exploration. After this field nearly died in the dry budget spell of the 1970's, NASA commissioned a Solar System Exploration Committee to find out how to revive planetary research. Their key recommendations include using "production-line" satellites, such as the small spinners launched on STS-5 and -7, as the basis of a low-cost planetary observer series. The first of these, the Mars Geochemical/Climatology Observer, is to be ready late in the 1980's. More ambitious missions would be tackled with

*Although modest in size—its main mirror is only 8 feet wide—the Hubble Space Telescope is expected to perform better than any terrestrial telescope, because it will be above the Earth's atmosphere. It is to be launched and serviced by the shuttle.* (PERKIN-ELMER)

*A scale model of the Soviet "space plane" re-entry vehicle has been flown at least three times in orbital tests ending in splashdowns. One recovery was seen by an Australian patrol plane. The "conehead" pilot is a retrieval aid.* (ROYAL AUSTRALIAN NAVY)

a modular series of Mariner Mark II spacecraft based on the old Mariner series but using standardized parts. The first mission anticipated for these would be a comet flyby/asteroid rendezvous in the 1990's.

Despite its potential to carry out a wide array of missions, all is not well with the shuttle. Worried about the on-pad shutdown of the 41-D shuttle mission, mishaps with upper stages, and the fragility of the shuttle system, in 1984 the Air Force announced plans for a new expendable launcher. It, the Franco-European Ariane, and private-industry-financed expendable launchers will inevitably take business away from the shuttle.

And what of the Soviets? Because they are a closed society, an outline of the future of their manned space effort must be deduced. This is done principally by a small band of western enthusiasts who through years of close observation have worked out the general purpose of a mission soon after it takes off, and by intelligence analyses released by the U.S. Department of Defense. Recent summaries of Soviet progress are "Salyut—Soviet Steps Toward Permanent Human Presence in Space," released in December 1983 by the Congressional Office of Technology Assessment, and the annual "Soviet Military Power" report issued by the Department of Defense.

The Soviets appear to be building their equivalent of the American space shuttle. Actually, two reusable spacecraft are expected to begin operations within a few years. The first is a space plane resembling the HL-10 lifting body flown by the United States as an atmospheric test vehicle in the 1970's. The Soviet space plane may have started its atmospheric tests as early as 1976. Its existence did not become public knowledge until an Australian patrol plane captured on film the recovery of Kosmos 1445 on 15 March 1983. The photographs clearly showed a winged re-entry vehicle being hoisted onto the recovery ship in the Indian Ocean. It had been launched 2 hours earlier from the Kapustan Yar facility and flew once around the Earth to an ocean landing.

Kosmos 1374 on 3 June 1982 followed a similar trajectory but was not photographed. At least three such tests of a scale model space plane have been carried out to date. The full-size space plane is expected to have a crew of three persons and carry only small amounts of priority cargo, if any. The 33,000-pound craft is to be boosted into orbit by a new medium-lift launcher with a gross liftoff weight of 2,860,000 pounds. Despite its relatively small size, the space plane will give the Soviets greater flexibility in exchanging Salyut crews.

The size limitation is expected to be reduced when two larger vehicles become available. One is a reusable heavy-lift shuttle and the other is a heavy-lift expendable launcher of the United States Saturn 5 class. The heavy shuttle appears to have the same general configuration as the United States space shuttle with two major differences: the boosters are liquid propellant (their recoverability is unknown), and the orbiter's main engines are on the external tank and are expended. This is expected to cost the Soviets dearly in the price of each mission. But they have been known to sacrifice economics in order to achieve a desired goal. In this case, it may be double the United States shuttle's payload. Gross liftoff weight is estimated to be 3,300,000 pounds, considerably less than that of the shuttle because of the booster propellant difference.

For heavy payloads, the Soviets are expected to try once again with a booster reputed to be larger than the Saturn 5. It has failed, spectacularly, at least three times during the period 1969–1972. Redesigned and using six or more strap-on boosters, it will have a payload weighing about 400,000 pounds, more than six times that of the shuttle and almost double that of the Saturn 5, according to analysts. The key payload for this super launcher would be a 200,000-pound space station manned by up to a dozen persons. Such a station could become operational by the end of this decade.

In 1984 the Department of Defense predicted that

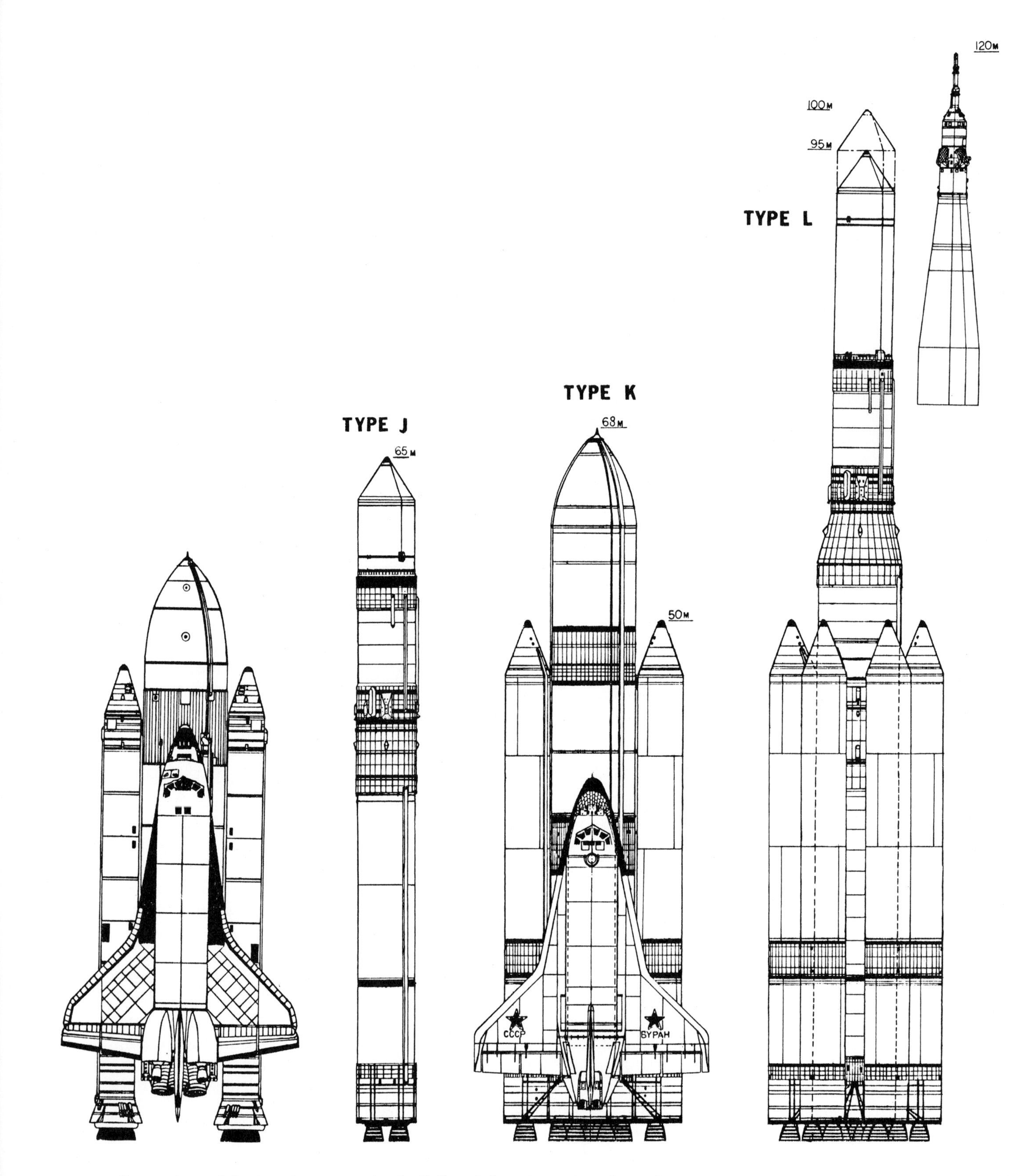


*Comparison of the American space shuttle (left) and three Soviet launchers: the type J standard expendable launch vehicle, the projected type K shuttle-like configuration, and the projected type L heavy launch configuration, according to studies by Charles P. Vick.* (© 1984 BY CHARLES P. VICK)

the Soviets would orbit "A larger, permanently manned space station. . . . [that] could be used as a stepping stone to interplanetary exploration and the establishment of bases on other planetary bodies. The Soviets, however, are more likely to use such a station to perform command and control, reconnaissance and targeting functions." DOD predicted that a "permanently manned Skylab-size space station" with a crew of six to twelve persons would be orbited by 1987, and that a "very large modular space station" with a crew of up to 100 would be orbited by the mid-1990's. The Defense Department report continued:

> The Soviets have embarked upon a long-term, broad-based effort to expand their operational military capability in space. A major Soviet objective is to expand warfighting capability in space and achieve a measure of superiority in that arena. Their technological base is strengthening and is being enhanced by technology transfer from the West. Their launch capability is increasing with the development of new facilities and booster systems. They continue to operate the world's only operational antisatellite system, while they test and develop more sophisticated space weaponry. It is clear the Soviets are striving to integrate their space systems with the rest of their Armed Forces to ensure superior military capabilities in all arenas.

Other nations are following suit. The Japanese are designing a small shuttlecraft of their own, as is France, with its proposed Hermes winged vehicle. Eureca, or European Retrieval Carrier, is to follow Spacelab and stay in orbit up to 9 months before being brought back to Earth. China is also making its presence in space known and launched a communications satellite in the spring of 1985.

And the United States? After two decades of frustration, it finally seemed to be embarking on a space station program. Although it has a presidential commitment comparable to what the space shuttle had in 1972, the station will have to leap the same hurdles in getting funding from the Congress when the expensive development years begin.

*At least two other nations have studied small shuttlecraft, Japan (as illustrated here) and France. The craft would be tied to a space station or would service small satellites that produced high-value materials in orbit.* (SPACE LIBRARY)

## Comparative Performance of United States Space Shuttle and Several Soviet Configurations

| | *United States Space Shuttle Configuration with Expendable Tank, Two Boosters, and Orbiter* | *Soviet Type J Medium-Lift Expendable Launch Vehicle* | *Soviet Type K Heavy-Lift Configuration with Ram-R Shuttle and Two Strap-On Boosters* | *Soviet Type L Heavy-Lift Expendable Configuration with Six Strap-On Boosters* |
|---|---|---|---|---|
| Liftoff weight, pounds (approx.) | 4.5 million | 880,000 | 3.3 million | 6–6.7 million |
| Liftoff thrust, pounds (approx.) | 6.5 million | 1.3 million | 4–6 million | 8–9 million |
| Payload weight to nominal 115-mile altitude, pounds (approx.) | 65,000 | 33,000 | 55,000–66,000 | 330,000–400,000; 150,000+ along an escape trajectory |

*EURECA, the European Retrievable Carrier, was designed to be launched by the shuttle and retrieved six months later, after materials and life science experiments had been run by remote control.* (MBB)

*China's first experimental communications satellite, Long March 3, heads for geostationary orbit on 8 April 1984. The craft successfully relayed radio, television, and other services throughout much of China and Tibet.* (COURTESY EMBASSY OF THE PEOPLE'S REPUBLIC OF CHINA IN THE UNITED STATES, WASHINGTON, D.C.; PHOTO BY YUNG WUMIN)

There had always been some level space station effort at NASA, but it was within the "noise level" of the agency's budget and barely detectable amidst other approved and adequately funded programs. It started to resurface in 1977 with studies of a space platform (then called a 25-kilowatt power module for extending the shuttle's stay in orbit) at the Marshall Space Flight Center, followed by studies of a space operations center (SOC) at the Johnson Space Center. But whereas the space platform would start out unmanned and grow into a manned capability, the SOC would be fully manned from the start. The space platform was viewed as a sort of supersatellite that would keep shuttle's Spacelab payloads in orbit for several months of operations rather than a few days or weeks. The SOC would be a service center for spacecraft heading to higher orbits and for science and applications satellites in lower orbits.

Although the two NASA field centers seemed to be in competition, their concepts were complementary. With the success of the space shuttle's first launch, the space station program office was formed at NASA headquarters in early 1982. That August, eight aerospace firms were awarded mission analysis contracts worth about $800,000 each. They were detailed to study what a space station would do and to outline how it would be developed rather than to investigate the nuts and bolts of a particular design.

Philip E. Culbertson, associate administrator for the space station, wrote in the September 1982 issue of *Astronautics and Aeronautics* that this focus on "architecture" rather than "configuration" will let NASA "be fully responsive to user thinking." A possible architecture would be a manned station with unmanned platforms in nearby orbits and robot spacecraft shuttling between them. The results, in May 1983, were not startling but gave NASA a firm idea of what user requirements would be. The agency then worked to lay out the presentation it gave to President Reagan and his senior advisors after Thanksgiving 1983.

His response came on 25 January 1984 in the State of the Union address when he directed NASA to take what it had touted as the "next logical step" in space.

"Our second great goal is to build America's pioneer spirit and develop our next frontier," the President announced. "A sparkling economy spurs initiative and ingenuity to create sunrise industries and make older ones more competitive," he said as he led into his brief passage on space.

Washington and the news services had been abuzz for several days with speculation that Reagan was going to commit to a space station program. Such rumors had been heard before—when space shuttle Columbia made its last test flight in 1982 and again when NASA celebrated its twenty-fifth anniversary in October 1983—

but this time they were stronger and came on the heels of extensive studies by NASA and briefings with Reagan and his cabinet.

"America has always been greatest when we dared to be great," he said. "We can reach for greatness again. We can follow our dreams to distant stars, living and working in space for peaceful, economic, and scientific gain.

"Tonight, I am directing NASA to develop a permanently manned space station, and to do it within a decade."

He cited the potentials for "quantum leaps" in science, communications and manufacturing, and invited foreign participation.

Details of the program were discussed by three top NASA officials—James M. Beggs, the administrator, and John D. Hodge and Louis J. Evans, Jr., directors of the space station and space commercialization task forces—the next day in a press briefing held at the White House.

Hodge said that NASA was then in the middle of the first year of "an extended development period" that would last at least 2 years. One reason for taking so long is to make sure that there are no technical challenges that might cause the same kinds of delays—and political problems—that put the space shuttle 3 years behind schedule.

Because everything making up the station would be brought up by the shuttle, and having the station would make United States shuttle operations more efficient, Hodge called the two space vehicles "a matched pair."

Four major functions have been set for the space station, Hodge explained: "first and foremost" as a laboratory for science, technology, and commercial ventures; as a service center for satellites; as a transportation node where satellites and upper stages will be handled; and, "perhaps most exciting of all," as a site for assembling vehicles too large for the shuttle to carry.

Included in the design studies will be selection of the "appropriate level of technology" to be used in the station's various systems, and how much of it will be manned versus robotic.

"It's important to make sure that we get a proper mix of things that man does best and those which are best done in an automated fashion," Hodge added.

The cost of what is viewed as an "initial capability" space station would be $8 billion. "We don't expect the program to stop at that point," Beggs said. "We expect that it will go on." Adding capabilities to the station and expanding its size would raise the price to around $20 billion by the end of the century. The down payment was $150 million in the fiscal 1985 budget plan.

One criticism leveled at NASA for planning a space station is that it will drain money from space science programs.

"I think it's a bum rap," Beggs replied. He said that the space science budget had declined and risen over the years, but that its share of the NASA budget remained much the same. "If the budget goes up for this space station," he said, "they [space scientists] will get their share—they always do."

Although a National Research Council study concluded that a space station is not needed for most space science missions, it admitted that most can be supported by a space station and the Council's space science board offered to help in designing one.

Two other potential customers are the military and industry.

Beggs said there is "very little military interest in the station. There's no direct military involvement in the station," although they are welcome to become users. Evans called the space station "a new place to work in space," one that will help put business in orbit "to benefit people." Admitting that the space environment is a tough, expensive one in which to work, he said there is great potential for manufacturing high-value drugs and other materials that cannot now be made on Earth. "The challenges will be great," he predicted, "but so will the opportunities."

By August 1984 NASA had settled on a "reference" design, called the Power Tower, to be used by all contractors as their starting point as they conduct advanced phase B definition studies.

The Power Tower uses a trick called gravity gradient stabilization to handle the bulk of its attitude control: if an object is long enough, the slight decrease in gravity across its length will keep the heavy end aimed at the Earth. This has been used to stabilize the shuttle and the long duration exposure facility (deployed on mission 41-C). This layout also has advantages for assembly and for accommodating shuttles that approach and dock with the station.

In one "typical" configuration, the Power Tower is 400 feet long with the pressure modules at the Earth end. An 8-foot-wide cubical truss forms the backbone of the Power Tower. Two-thirds of the way up from the base a cross arm sticks out for the solar panels that provide 75 kilowatts of electricity for the station. In the initial configuration there are eight 40 × 90 foot panels sticking 133 feet to each side of the centerline. They are on rotating joints so that as the station orbits the Earth the panels are kept flat to the Sun. Also on rotating joints are the radiators.

At the Earth end of the initial station are five pressure modules—two lab and two habitability command modules arranged in a square, and a supply module sticking to one side. A crew of at least six persons is expected, one for housekeeping and five or more for payload operations. They will work for 90 days at a stretch, although shuttle visits will probably be more frequent.

Weights will range from almost 9,000 pounds for the truss to more than 55,000 pounds for one of the lab modules. Total weight will be nearly 450,000 pounds. It will take at least five shuttle missions to build up the station so that it is large enough to be occupied.

But that will not be the stopping point. The station is expected to grow for at least 20 years, and the major challenge now facing NASA is how to design it today so that technologies of the twenty-first century can be added to the space station. Payloads will be attached to the sky end, such as the large solar optical telescope for studying details as small as 42 miles across on the Sun. This will grow into the advanced solar observatory to subject our closest star to intense scrutiny. At the Earth end active plasma physics instruments, such as those on Spacelab 1, may be installed.

Operational gear will be added also. An orbital maneuvering vehicle (OMV) will be based at the space station, being stored in a hangar between missions to service nearby satellites and platforms. The orbital transfer vehicle, once referred to as the space tug, may be based at the space station as well, turning it into a true space port. Tethers may be used to store materials such as propellants at a safe distance from the station. Construction of large payloads will be tried at the station before major projects such as a 66-foot-wide infrared telescope.

At least two unmanned platforms, using hardware copied from the space station, are expected to be built as well. One will be in the same orbit as the space station, only 100 or so miles ahead so the OMV can reach it for maintenance. The other will be in polar orbit and serviced by the shuttle.

Each will be about 195 feet long and have four solar panels delivering 20 kilowatts of electric power. Payloads such as Astro and the large Starlab ultraviolet telescope and an X-ray pinhole camera may be on the co-orbiting platform, and advanced atmospheric and oceanic monitors will be on the polar platform.

These, of course, are candidate payloads. The actual makeup will change as the space station design is refined and as scientists learn how to use it. The station design given here is somewhat conjectural, even though certainly possible. One purpose of the phase B studies is for industry to offer a variety of "best" solutions to this challenge. The cubical truss might be replaced with a larger tetrahedral truss taking more room and offering greater stiffness. The solar cell panels might be replaced with parabolic dishes focusing sunlight on the space equivalent of steam-driven turbogenerators.

*The "Power Tower" configuration was chosen in mid-1984 as the starting point for NASA's space station definition era. It has a six-man crew and hangars for space tugs, plus room for future additions.* (NASA)

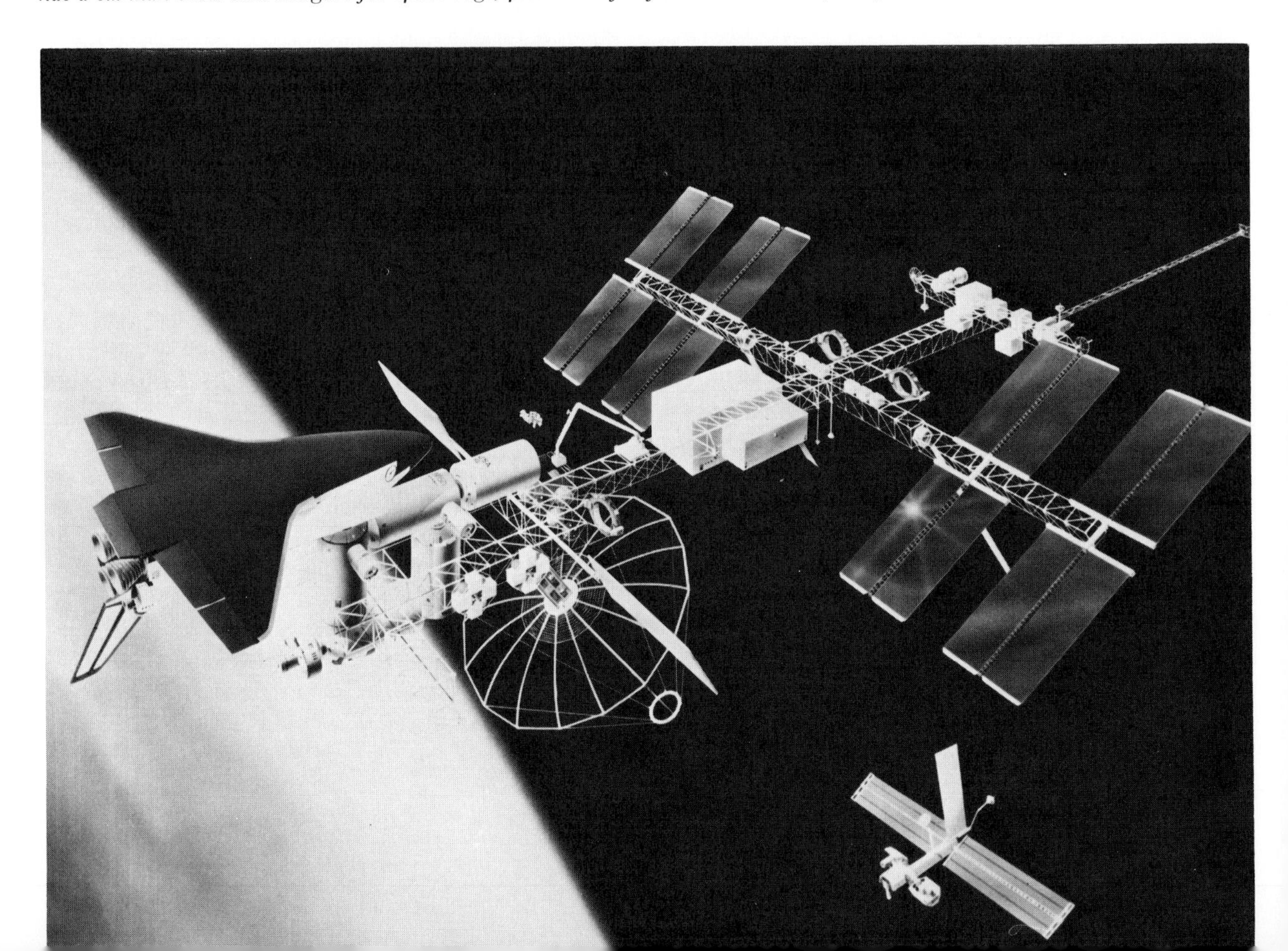

*An unmanned co-orbiting platform to support astronomy and other studies is a key part of NASA's space station "infrastructure."* (TRW/NASA)

Several test beds on Earth and in flight (such as the large solar panel model on the 41-D shuttle mission) are planned to help find the best answers.

With the space station in orbit, the gate will open a little more for returning to the Moon and setting off to the planets. In one scheme, unmanned rovers are sent to the Moon to sample the geology in several areas. A handful of orbital transfer vehicles will then land crews and a small factory to "mine" oxygen from the soil, liquefy it, and ship it to Earth orbit. The energy required to "drop" oxygen from the surface of the Moon is less than that required to bring it up from Earth, so such a deep-space version of the 1600's "rum route" may be more economical than bringing the oxygen up from Earth. In time, this mining operation would be expanded to refine aluminum, iron, silicon, and other materials for use as raw stock in building larger space stations and even orbiting power plants that, operating in constant sunlight, would beam microwaves to Earth for conversion to electricity.

All this would benefit the scientists wanting to study Earth's sister planet and gradually would build the "infrastructure" that would make possible manned missions to Mars and beyond in the twenty-first century.

During the Renaissance, Prince Henry the Navigator of Portugal established in his seaside castle of Sagres the closest precedent to what the space community is trying to accomplish in our time. He systematically collected maps, ship designs, and navigational instruments from all over the world. He attracted Portugal's most experienced mariners. He laid out a step-by-step program aimed at the exploration of Africa's Atlantic coast as well as the discovery of the continent's southernmost tip, which he knew had to be circumnavigated if India were to be reached by the sea. With equal determination he pushed for the possibly shorter westbound route to the Far East. Prince Henry trained the astronauts of his time—men such as Bartholomeu Diaz, Ferdinand Magellan, and Vasco de Gama, and he created the exploratory environment that launched Christopher Columbus from neighboring Spain on his historic voyage.

Henry the Navigator would have been hard put had he been requested to justify his actions on a rational basis, or to predict the payoff or cost-effectiveness of his program of exploration. He committed an act of faith and the world became richer for it. Exploration of space is the challenge of our day. If we continue to put our faith in it and pursue it, it will reward us handsomely.

# Bibliography

The following bibliography includes selected works that bear on historical or potentially historical aspects of the subject. Most of the bibliography is limited to book references, but occasionally an important journal article is cited, particularly if it treats pre-1900 events not otherwise recorded, or even cited, in books. Since most post-1900 events in one way or another have been recorded at least superficially in the book literature, and since many of the books listed cite original periodical sources, only an occasional listing of the latter is made.

## 1 The Lure of Other Worlds

Selected books on the history of astronomy and of man's ideas of the universe around him are first presented:


Abetti, Giorgio. *History of Astronomy.* New York: Abelard-Schuman, 1952.

Bentley, John. *A Historical View of the Hindu Astronomy.* London: Smith, Elder, 1825.

Berry, Arthur. *A Short History of Astronomy; From Earliest Times Through the Nineteenth Century.* New York: Dover, 1961.

Brewster, Sir David. *Memoires of the Life, Writings, and Discoveries of Sir Isaac Newton.* 2 vols. Edinburgh: Thomas Constable, 1855.

Carlos, Edward Stafford. *The Sidereal Messenger of Galileo Galilei.* London: Rivingtons, 1880.

Carmody, Francis J. *Arabic Astronomical & Astrological Sources in Latin Translation.* Berkeley: University of California Press, 1956. (Extensive bibliography.)

Christianson, Gale E. *In the Presence of the Creator: Isaac Newton and His Times.* New York: Free Press/Macmillan, 1984.

Clerke, Agnes M. *A Popular History of Astronomy During the Nineteenth Century.* London: A. & C. Black, 1902.

Copernicus, Nicolaus. *De Revolutionibus Orbium Coelestium.* New York: Johnson Reprint, 1965. (Facsimile reprint.)

Delambre, Jean Baptiste Joseph. *Histoire de l'astronomie ancienne; Histoire de l'astronomie du moyen age;* and *Histoire de l'astronomie moderne.* Paris: V. Courcier, 1817, 1819, and 1821, respectively.

Derham, William. *Physico-Teology: Or, A Demonstration of the Being and Attributes of God, from his works of Creation.* London: A. Strahan, 1798. (Critique of Huygens, possibility of habitation of planets, and so on.)

Dick, Steven J. *The Plurality of Worlds: The Origins of the Extraterrestrial Life Debate from Democritus to Kant.* Cambridge: Cambridge University Press, 1982.

Doig, Peter. *A Concise History of Astronomy.* London: Chapman & Hall, 1950.

Flammarion, Camille. *La Pluralité des mondes habités.* Paris: Didier & Cie, 1864.

Fontenelle, Bernard le Bouvier de. *Conversations on the Plurality of Worlds.* London: A. Bettesworth, 1715. Also: *A Discovery of New Worlds.* London: Will. Canning, 1688. (Translations of the original French *Entretiens sur la pluralité des mondes,* the first edition of which appeared in Paris in 1686; it went through four editions in the seventeenth century and many in the eighteenth, e.g., La Haye: Isaac Vander Kloot, 1733, in which a seventh *Entretien* was added.)

Galilei, Galileo. *Dialogue Concerning the Two Chief World Systems—Ptolemaic and Copernican.* Berkeley: University of California Press, 1953.

Grant, Robert. *History of Physical Astronomy from the Earliest Ages to the Middle of the Nineteenth Century.* London: Henry & Bohn, 1852.

Heath, Thomas Little. *Aristarchus of Samos*. Oxford: Clarenden, 1913; *Greek Astronomy*. London: Dent, 1932.
Huygens, Christian. *The Celestial Worlds Discover'd: Or Conjectures Concerning the Inhabitants, Plants, and Productions of the Worlds in the Planets*. London: Printed for Timothy Childe, 1698. (English edition of *Kosmotheoros*.)
Kesten, Hermann. *Copernicus and his World*. London: Secker and Warburg, 1945.
Learner, Richard. *Astronomy Through the Telescope: The 500-Year Story of the Instruments, the Inventors and their Discoveries*. New York: Van Nostrand, 1981.
Lewis, George Cornewall. *An Historical Survey of the Astronomy of the Ancients*. London: Parker, Son and Bourn, 1862.
Ley, Willy. *Watchers of the Sky*. New York: Viking, 1963.
Lowell, Percival. *Mars*. Boston: Houghton Mifflin, 1895; also his *Mars and its Canals*. New York: Macmillan, 1906; and *Mars as the Abode of Life*. New York: Macmillan, 1910.
Man, John, ed. *The Encyclopedia of Space Travel & Astronomy*. London: Octopus, 1979.
Martin, Th. H. *Aristarque de Samos*. Paris: Imprimerie des Sciences Mathématiques et Physiques, 1871.
Meadows, A. J. *The High Firmament: A Survey of Astronomy in English Literature*. Leicester: Leicester University Press, 1969.
Neugebauer, Otto Eduard, ed. *Astronomical Cuneiform Texts: Babylonian Ephemerides of the Seleucid Period for the Motion of the Sun, the Moon and the Planets*. London: Lund, Humphries, 1955. Also his *Egyptian Astronomical Texts*. London: Lund, Humphries, 1960.
Orr, Mary Ackworth. *Dante and the Early Astronomers*. London: Gall & Inglis, 1913.
Pannekoek, A. *History of Astronomy*. New York: Interscience, 1961.
Plutarch. *De Facie in Orbe Lunare*. Cambridge: Harvard University Press, 1957. (Found in his *Moralia* as *The Face on the Moon* in Harold Cherniss's translation in the Loeb Classical Library XII.)
Rosen, Edward. *Kepler's Conversation with Galileo's Sidereal Messenger*. New York: Johnson Reprint, 1965.
Rusinek, Michal. *Land of Nicholas Copernicus*. New York: Twayne, 1973.
Sagan, Carl. *Cosmos*. New York: Random House, 1980.
Sambursky, S. *The Physical World of the Greeks*. London: Routledge & Kegan Paul, 1956.
Seyyed Hossein Nasr. *An Introduction to Islamic Cosmological Doctrines*. Cambridge, Mass.: Harvard University Press, 1964.
Shapley, Harlow, and Howarth, Helen E., eds. *A Source Book in Astronomy*. New York: McGraw-Hill, 1929. Also Shapley's *Source Book in Astronomy 1900–1950*. Cambridge, Mass.: Harvard University Press, 1960.
Singer, Charles, ed. *Studies in the History and Method of Science*. 2 vols. Oxford: Oxford University Press, 1917 (Vol. 1) and 1921 (Vol. 2).
Singer, Dorothea Waley. *Giordano Bruno*. London: Abelard-Schuman, 1950.
Small, Robert. *An Account of the Astronomical Discoveries of Kepler*. London: J. Mawman, 1804. (Reprinted Madison: University of Wisconsin Press, 1963.)
Struve, O., and Zeberts, V. *Astronomy of the 20th Century*. New York: Macmillan, 1962.
Swedenborg, Emanuel. *Earths in our Solar System which are Called Planets and Earths in the Starry Heaven, their Inhabitants, and the Spirits and Angels there from Things Heard and Seen*. London: Swedenborg Society, 1962.
Tannery, Paul. *Recherches sur l'histoire de l'astronomie ancienne*. Paris: Gauthier-Villars, 1893.
Thorndike, Lynn. *The Sphere of Sacrobosco and its Commentators*. Chicago: University of Chicago Press, 1949.
Westfall, Richard S. *Never at Rest*. Cambridge: Cambridge University Press, 1980. (Biography of Isaac Newton.)
Whewell, William. *Of the Plurality of Worlds*. London: John W. Parker & Son, 1859.
Wilkins, John. *The Discovery of a World in the Moone; Or, A Discourse Tending to Prove 'tis Probable There May be Another Habitable World in that Planet*. London: E. G. for Michael Sparke and Edward Forrest, 1638.
Young, L. B., ed. *Exploring the Universe*. New York: McGraw-Hill, 1963.
Zeilik, M., and J. Gaustad. *Astronomy: The Cosmic Perspective*. New York: Harper & Row, 1983.

A number of surveys of the fictional literature dealing with space travel have been published. Among them are the following:

Aldriss, Brian W. *Billion Year Spree: The True History of Science Fiction*. New York: Doubleday, 1973.
Ash, Brian. *The Visual Encyclopedia of Science Fiction*. New York: Harmony, 1977.

Ashley, Michael, ed. *The History of the Science Fiction Magazine Part 1: 1926–1935.* London: New English Library, 1974. *Part 2: 1936–1945.* London: New English Library, 1975. *Part 3: 1946–1955.* Chicago: Contemporary Books, 1977.

Bailey, J. O. *Pilgrims Through Space and Time.* New York: Argus, 1947.

Berneri, Maria Louise. *Journey Through Utopia.* London: Routledge & Kegan Paul, 1950.

Bleiler, E. F., ed. *The Checklist of Fantastic Literature.* Chicago: Shasta, 1948.

———. *Science Fiction Writers: Critical Studies of the Major Authors from the Early Nineteenth Century to the Present Day.* New York: Scribner, 1982.

Bretnor, Reginald, ed. *Modern Science Fiction, Its Meaning and Its Future.* New York: Coward-McCann, 1953.

Carter, Paul A. *The Creation of Tomorrow: Fifty Years of Magazine Science Fiction.* New York: Columbia University Press, 1977.

Costello, Peter. *Jules Verne: Inventor of Science Fiction.* New York: Scribner, 1978.

Crawford, Joseph H. Jr., Donahue, James J., and Grant, Donald M. *A Bibliography of the Science-Fantasy Novel "333."* Providence: Grandon, 1953.

Davenport, Basil. *Inquiry into Science Fiction.* New York: Longmans, 1955.

Day, Bradford M. *The Supplemental Checklist of Fantastic Literature.* New York: Science-Fiction & Fantasy Publications, 1963.

Day, Donald B. *Index to the Science-Fiction Magazines 1926–1950.* Portland, Oreg.; Perri, 1952.

De Camp, L. Sprague. *Science-Fiction Handbook.* New York: Hermitage House, 1953.

———, and Ley, Willy. *Lands Beyond.* New York: Rinehart, 1952.

Derleth, August, ed. *Beyond Time and Space.* New York: Pellegrini and Cudahy, 1950.

Emme, Eugene M., ed. *Science Fiction and Space Futures: Past and Present.* San Diego: Univelt, 1982.

Flammarion, Camille. *Les mondes imaginaires et les mondes réels.* Paris: Didier & Cie, 1876. (Reviews many fictional works.)

Franklin, H. Bruce. *Future Perfect.* New York: Oxford University Press, 1966.

Freedman, R. *2000 Years of Space Travel.* New York: Holiday House, 1963.

Gaul, A. *Complete Book of Space Travel.* New York: World, 1956.

Gove, Philip Babcock. *The Imaginary Voyage in Prose Fiction: A History of its Criticism and a Guide for its Study, with an Annotated Check List of 215 Imaginary Voyages from 1700 to 1800.* New York: Columbia University Press, 1941.

Green, R. L. *Into Other Worlds.* New York: Abelard-Schuman, 1958.

Gunn, James. *Alternate Worlds: The Illustrated History of Science Fiction.* Englewood Cliffs, N.J.: Prentice-Hall, 1975.

Haining, Peter, ed. *The Fantastic Pulps.* London: Victor Gollancz, 1975. See also his *The H. G. Wells Scrapbook.* New York: Potter, 1978.

Heins, Henry Hardy. *A Golden Anniversary Bibliography of Edgar Rice Burroughs.* West Kingston, R.I.: Donald M. Grant, 1964.

Jules-Verne, Jean. *Jules Verne: A Biography.* New York: Taplinger, 1976.

Kyle, David. *The Illustrated Book of Science Fiction Ideas and Dreams.* London: Hamlyn, 1977. And his *A Pictorial History of Science Fiction.* London: Hamlyn, 1977.

Leighton, Peter. *Moon Travellers.* London: Oldbourne, 1960.

Locke, George. *Voyages in Space: A Bibliography of Interplanetary Fiction 1801–1914.* London: Ferret Fantasy, 1975.

Lundwall, Sam J. *Science Fiction: An Illustrated History.* New York: Grosset & Dunlap, 1977.

Lupoff, Richard A. *Edgar Rice Burroughs: Master of Adventure.* New York: Canaveral, 1965.

McHugh, Joseph, and Harris, Latif. *Journey to the Moon.* Millbrae, Calif.: Celestial Arts, 1974.

Moskowitz, Sam. *Explorers of the Infinite.* New York: World, 1963; *Seekers of Tomorrow.* Cleveland: World, 1966; ed. *Under the Moons of Mars.* New York: Holt, 1970; and ed. *Masterpieces of Science Fiction.* Westport, Conn.: Hyperion, 1974.

Nicholls, Peter, ed. *The Science Fiction Encyclopedia.* New York: Dolphin-Doubleday, 1979.

Nicholson, Marjorie. *Voyages to the Moon.* New York: Macmillan, 1948.

Philmus, Robert M. *Into the Unknown: The Evolution of Science Fiction from Francis Godwin to H. G. Wells.* Berkeley: University of California Press, 1970.

Pizor, Faith K., and Comp, T. Allan, eds. *The Man in the Moon and Other Lunar Fantasies.* New York: Praeger, 1971.

Porges, Irwin. *Edgar Rice Burroughs: The Man Who Created Tarzan.* Provo, Utah: Brigham Young University Press, 1975.

Rottensteiner, Franz. *The Science Fiction Book.* New York: Seabury, 1975.

Scholes, Robert, and Rabkin, Eric S. *Science Fiction: History, Science, Vision.* London: Oxford University Press, 1977.

Tuck, Donald H. *The Encyclopedia Of Science Fiction and Fantasy.* 3 vols. Chicago: Advent, 1974, 1978, and 1982.

Tymn, Marshall B., ed. *The Science Fiction Reference Book.* Mercer Island, Wash.: Starmount, 1981.

Versins, Pierre. *Encyclopédie de l'utopie des voyages extraordinaires et de la science fiction.* Lausanne: Éditions l'Age d'Homme, 1972.

Wright, Hamilton and Helen, and Rapport, Samuel, eds. *To the Moon: A Distillation of the Great Writings from Ancient Legend to Space Exploration.* New York: Meredith, 1968.

A selection of tales dealing with lunar and planetary voyages follows, grouped first from antiquity to the end of the eighteenth century and then from 1800 to the end of the nineteenth century.

*A. Antiquity Through End of Eighteenth Century*

*An Account of Count d'Artois and his Friend's Passage to the Moon, in a Flying Machine, Called, An Air Balloon, Which was Constructed in France.* Litchfield: Printed for Collier and Copp, 1785.

Behn, A[phra]. *The Emperor of the Moon: A Farce.* London: Holt, 1687.

[Bethune] *La relation du monde de mercure.* Found in Charles G. T. Garnier's *Voyages imaginaires, songes, visions et romans cabalistiques.* Vol. 16. Amsterdam and Paris: Rue et Hôtel Serpente, 1777. Also Geneva: Chez Barillot & Fils, 1750.

Brunt, Samuel. *A Voyage to Cacklogallinia.* London: J. Watson, 1727. See New York: Facsimile Text Society, Columbia University Press, 1940.

Cicero. *Somnium Scipionis.* New York: Loeb Classical Library, 1928.

Cyrano de Bergerac [Savinien]. *Histoire comique des états et empires de la lune et du soleil.* Paris: Chez Charles le Sercy, 1656. See *Voyages to the Sun and Moon.* London: Oxford University Press, 1965; and *The Comical History of the States and Empires of the Worlds of the Moon and Sun.* London: Henry Rhodes, 1687.

Daniel, Gabriel P. *Voiage du monde de Descartes.* Paris: Chez la Veuve de S. Bénard, 1691. See *A Voyage to the World of Cartesius.* London: T. Bennet, 1694.

Defoe, Daniel. *The Consolidator: Or Memoirs of Sundry Transaction from the World in the Moon.* London: Benj. Bragg at the Blue Ball, 1705; *A Journey to the World in the Moon.* Edinburgh: James Watson in Craig's Cross, 1705.

Firdausī. *The Shāh-Nāma of Firdausī,* translated into English by A. G. Warner and E. Warner. London: Kegan Paul, 1905.

*Furetiriana, ou les bons mots et les remarques d'histoire, de morale, de critique, de plaisanterie, et d'érudition de M. Furetière, Abbé de Chalivoy, de l'Académie Française.* Brussels: François Foppens, 1698.

Gonsales, Domingo (pseud. for Francis Godwin, Bishop of Hereford). *The Man in the Moone: Or a Discourse of a Voyage Thither.* London: John Norton for Ioshua Kirton and Thomas Warren, 1638.

Kepler, Johannes. See Lear, John. *Kepler's Dream, with Full Text and Notes of Somnium, Sive Astronomia Lunaris.* Berkeley: University of California Press, 1965.

La Folie, Louis Guillaume de. *Le philosophe sans prétention. . . .* Paris: Clusier, 1775.

Lodovico, Ariosto. *Orlando Furioso.* London: J. Harington, 1607. George Bell and Sons, 1876. In English heroical verse.

Lucianus, Samosatensis. *The Works of Lucian of Samosata.* Vol. 2. Oxford: Clarendon, 1905. (Contains Icaro-Menippus); *True History.* London: Bullen, 1902. (*Note:* the first English edition was entitled *Certaine Select Dialogues of Lucian, Together with his True Historie, translated from the Greeke into English by Mr. Francis Hickes. Whereunto is added the Life of Lucian gathered out of his owne Writings, with briefe Notes and Illustrations upon each Dialogue and Booke.* By T. H. Mr. of Arts of Christ-Church in Oxford. Oxford: 1634). See also Lucian. *Trips to the Moon.* London: Cassell, 1887.

Morris, Ralph. *A Narrative of the Life and Astonishing Adventure of John Daniel a Smith at Ryston in Herefordshire, for a Course of Seventy Years Containing . . . a Description of a Most Surprising Engine Invented by his Son Jacob, on which he Flew to the Moon, with some Account of its Inhabitants . . . Taken by his Own Mouth.* London: M. Cooper, 1751. Also see "The Life and Astonishing Adventures of John Daniel" in Penzer, N. M., ed. *The Library of Imposters,* Vol. 1. London: Robert Holden, 1926.

*A New Journey to the World in the Moon. Containing, I. A Full Description of the Manner of the Author's Performing his Journey; and his Reasons Why Former Lunarian Travellers Could not Find their Way Thither . . .* London: C. Corbett, 1741.

Roumier, Marie-Anne de. *Voyages de Milord Céton dans les sept planettes ou le nouveau mentor.* In Garnier's *Voyages imaginaires,* Vols. 17 and 18.

Russen, David. *Iter Lunare: or, A Voyage to the Moon.* London: J. Nutt, 1703. Reprinted Boston: Gregg, 1976.

[Vasse, Cornelie (Wouters), Baronne de]. *Le char vo-*

*lant, ou voyage dans la lune.* London and Paris: La Veuve Ballard & Fils, 1783.

Voltaire [Francois Marie Arouet de] *Le Micromégas.* London: J. Robinson et W. Meyer, 1752. Also *Micromegas: A Comic Romance.* London: D. Wilson and T. Durham, 1753.

Wilson, Miles. *The History of Israel Jobson, the Wandering Jew, Giving a Description of his Pedigree, Travels in this Lower World, and his Assumption Thro' the Starry Regions, Conducted by a Guardian Angel, Exhibiting in a Curious Manner the Shapes, Lives, and Customs of the Inhabitants of the Moon and Planets . . .* London: J. Nickolson, 1757. See Anderson, George K. *The Legend of the Wandering Jew.* Providence: Brown University Press, 1965.

*B. Nineteenth Century*

Astor, John Jacob. *Journey to Other Worlds.* New York: Appleton, 1894.

Atterley, Joseph (pseud. for George Tucker). *A Voyage to the Moon with Some Account of the Manners and Customs, Science and Philosophy of the People of Morosofia and Other Lunarians.* New York: Elam Bliss, 1827. Reprinted as Tucker, George. *A Voyage to the Moon.* Boston: Gregg, 1975.

Cromie, Robert. *A Plunge into Space.* London and New York: Frederick Warne, 1891.

Douglass, Ellsworth. *Pharaoh's Broker.* London: C. A. Pearson, 1899. Reprinted Boston: Gregg, 1976.

[Dulaure, Jacques Antoine]. *Le retour de mon pauvre oncle, ou rélation de son voyage dans la lune.* Paris: Chez Lejay, 1784.

Eyraud, Achille. *Voyage à Venus.* Paris: Michel Levy Frères, 1863.

Fowler, George. *A Flight to the Moon: or, the Vision of Randalthus.* Baltimore: A. Miltenberger, 1813.

Greg, Percy. *Across the Zodiac: The Story of a Wrecked Record.* 2 vols. London: Trubner, 1880. Reprinted Westport, Conn.: Hyperion, 1974.

Hale, Edward Everett. *His Level Best and Other Stories.* Boston: Robert Brothers, 1872. (Contains "The Brick Moon".)

Hodgson, John Edmund. *The History of Aeronautics in Great Britain from the Nineteenth Century.* London: Oxford University Press, 1924.

*A Journey to the Moon, and Interesting Conversations with the Inhabitants Respecting the Condition of Man. By the Author of Worlds Displayed.* London: Howar[d] and Evans, c. 1811.

Lasswitz, Kurd. *Auf zwei Planeten.* Leipzig: Elischer Nachfolger, 1897. Reprinted as *Two Planets,* abridged by Erich Lasswitz. Carbondale, Ill.: Southern Illinois Press, 1971.

Locke, Richard Adams. *The Celebrated "Moon Story," its Origin and Incidents.* New York: Bunnell and Price, 1852; and *The Moon Hoax: Or A Discovery that the Moon has a Vast Population of Human Beings.* New York: W. Gowans, 1859. Reprinted Boston: Gregg, 1975.

Poe, Edgar Allan. "The Unparalleled Adventure of One Hans Pfaall" from *Works of Edgar Allen Poe* edited by J. H. Ingram. Edinburgh: A. & C. Black, 1875.

Pope, Gustavus W. *Journey to Mars.* New York: Dillingham, 1894. Reprinted Westport, Conn.: Hyperion, 1974; *Journey to Venus.* Boston: Arena, 1895.

Serviss, Garrett P. *Edison's Conquest of Mars.* Los Angeles: Carcosa, 1947. (Originally published in New York *Evening Journal,* 1898.)

Tumble, Timothy, Quixote, Richard, and Telltruth, John (!). *A Trip to the Man in the Moon, from Terra Firma; in an Air Balloon.* London: D. Carvalho, c. 1810.

Verne, Jules. *De la terre à la lune.* Paris: J. Hetzel, 1865. And *Autour de la lune.* Paris: J. Hetzel, 1870. See also *From the Earth to the Moon . . . and a Trip Round it.* London: Sampson Low, Marston, Low, and Searle, 1873; and Miller, Walter James. *The Annotated Jules Verne: From the Earth to the Moon.* New York: Crowell, 1978.

Wells, H. G. *The War of the Worlds.* London: Heinemann, 1898; also, *The First Men in the Moon.* London: Newnes, 1901.

**2 A Thousand Years of Rocketry**

*General Works*

Books on the history of gunpowder and artillery often include rockets; and even though rockets may sometimes be given scant attention, the books provide a feeling for the environment in which rocketry evolved.

Bertholot, Marcelin. *La Chimie au Moyen Age.* 3 vols. Paris: Imprimerie Nationale, 1893.

Bethell, Colonel H. A. *Use of Rockets in Artillery in the Field.* London: Macmillan, 1911.

Brock, Alan St. H. *A History of Fireworks.* London: Harrap, 1949; also, *Pyrotechnics: The History and Art of Firework Making.* London: O'Connor, 1922.

Canby, Courtlandt. *A History of Rockets and Space.* New York: Hawthorn, 1963.

Clark, John D. *Ignition! An Informal History of Rocket Propellants.* New Brunswick, N.J.: Rutgers University Press, 1972.

Dickson, Katherine M. *History of Aeronautics and*

*Astronautics—A Preliminary Bibliography.* Washington, D.C.: NASA, 1968.
Durant, Frederick C., III, and George S. James, eds. *First Steps Toward Space.* Washington, D.C.: Smithsonian Institution Press, 1974.
Faber, Henry B. *Military Pyrotechnics.* 3 vols. Washington, D.C.: Government Printing Office, 1919.
Guttmann, Oscar. *Monumenta Pulveris Pyrii.* Text in English, French, and German. London: Artists Press, 1906.
Held, Robert. *The Age of Firearms.* New York: Harper, 1957.
Hime, Lieutenant Colonel Henry W. L. *Gunpowder and Ammunition: Their Origin and Progress.* London: Longman, 1904; also, *The Origin of Artillery.* London: Longman, 1915.
Hoeffer, Ferdinand. *Histoire de la chimie depuis les temps les plus reculées jusqu'à notre époque.* 2 vols. Paris: Fortin, Masson, 1842.
Hogg, O. F. G. *The Royal Arsenal: Its Background, Origin and Subsequent History.* London: Oxford University Press, 1963.
Jocelyn, Colonel Julian R. J. *The History of the Royal Artillery.* London: Murray, 1911.
Ley, Willy. *Rockets, Missiles and Men in Space.* New York: Viking, 1968.
Magne, Émile. *Les Fêtes en Europe au XVII^e siècle.* Paris: Martin-Dupuis, 1930.
Mourey, Gabriel. *Le Livre des fêtes françaises.* Paris: Librairie de France, 1930.
Partington, J. R. *A History of Greek Fire and Gunpowder.* Cambridge: Heffer, 1960.
Pollard, Major A. B. C. *A History of Firearms.* London: Geoffrey Bles, 1926.
Reinaud, Joseph Toussaint, and Ildephonse Favé. *Histoire de l'artillerie: feu grégeois, des feux de guerre et des origines de la poudre à cannon.* 2 vols. Paris: Dumaine, 1845.
Riling, Ray. *Guns and Shooting: A Selected Chronological Bibliography.* New York: Greenberg, 1953.
Romocki, S. J. von. *Geschichte des Explosivstoffe.* Vol. 1, Berlin. Oppenheim, 1895. Vol. 2, Hannover: Jänecke, 1896.
Schmidt, Rodolphe. *Développement des armes à feu.* Schaffhausen: Brodtmann, 1870.
Wilkinson, Henry. *Engines of War.* London: Longman, 1841.

*Use of Rockets for Military, Recreational, and Other Purposes*

For centuries, the rocket has been used for a variety of purposes—for military engagements, fireworks, lifesaving, and signaling. Typical books covering these and other uses follow:

Alberti, G. A. *La Pirotechnia, o sia Trattado die Fuochi d'Artificio.* Venice: Recurti, 1749.
Anderson, Robert. *The Making of Rockets.* London: Morden, 1696.
Antonj, Domenico. *Trattato Teórico-Pratico.* Trieste: Sambo, 1893.
*L'Arte de Fare i Fuochi d'Artifizio con poca Spesa.* Napoli: Tasso, 1834.
Babington, John. *Pyrotechnia, or, A Discourse of Artificiall Fire-works.* London: Harper & Mab, 1635.
Bate, John. *The Mysteries of Nature and Art.* London: Mabb, 1635.
Bem, Joseph. *Erfahrungen über Congrev'schen Brand-Raketan . . .* Weimar: Landes-Industrie Comptoirs, 1820.
Biringuccio, Vanoccio. *De la Pirotechnia.* New York: American Institute of Mining and Metallurgical Engineers reprint, 1942. (Originally appeared in 1540.)
Blümel, Johann Daniel. *Luft-Feuerwerkerey.* Strasbourg: König, 1771.
Boillot, Joseph. *Modelles artifices de feu et divers instrumens de guerre.* Chaumont-en-Bassigny: Mareschal, 1598.
Browne, William H. *The Art of Pyrotechny.* London: "The Bazaar" Office, 1883.
Chertier, François-Marie. *Nouvelles recherches sur les feux d'artifice.* Paris: Chertier, 1854.
Colliado, Luigi. *Prattica Manuale dell'Artiglieria . . .* Milano: Chisolfi, 1641.
Congreve, William. *A Concise Account of the Origin and Progress of the Rocket System.* London: Whiting, 1807; Dublin: O'Neil, 1817; *The Details of the Rocket System . . .* London: Whiting, 1814. (Reprinted in 1970 by the Museum Restoration Service, Ottawa, Ontario.); *The Different Modes of Use and Exercises of Rockets.* London: Whiting, 1808; *Memoir on the Possibility, the Means and the Importance of the Destruction of the Boulogne Flotilla in the Present Crisis . . .* London: Whiting, 1806; *A Treatise on the General Principles, Powers, and Facility of Application of the Congreve Rocket System . . .* London: Longman, 1827.
Corréard, Joseph. *Histoire des fusées de guerre.* Paris: Corréard, 1841.
Cutbush, James. *A System of Pyrotechny.* Philadelphia: Cutbush, 1825.
Denisse, Amédée. *Traité pratique complet des feux d'artifice.* Paris: Denisse, 1882.
Dennett, John. *A Concise Description of a Powerful Species of War Rockets.* London: Dennett, 1832.
D.M. *Pyrotechnia of Konstige Vuurwerken . . .* Rotterdam: Ryckhals, 1672.

Dollecsek, Anton. *Geschichte der Österreichischen Artillerie.* Vienna: Kreisel & Gröger, 1887.

Ellena, Giuseppe. *Corso di Materiale d'Artiglieria.* Turin: Scuola d'Applicazione delle Armi d'Artiglieria e Genio, 1877.

Ferré Vallvé, Juan Bautista. *La Pirotecnia Moderna.* Barcelona: Soler, 1904.

*Die Feuerwerkerei.* Leipzig; Paul, n.d.

Frezier, Amédée François. *Traité des feux d'artifice.* Paris: Nyon, 1747; rev. ed., Paris: Jombert, 1747.

Furttenbach, Joseph. *Architectura Navalis.* Frankfurt-on-Main: Clemens Sleichen, 1629.

[Grignon]. *La Pyrotechnie pratique.* Paris: Cellot & Jombert, 1780.

Hale, William. *A Treatise on the Comparative Merits of a Rifle Gun and Rotary Rocket.* London: Mitchell, 1863.

Hassenstein, W., ed. and trans. *Das Feuerwerkbuch von 1420.* Munich: Deutsche Technik, 1941.

[Jombert, Charles-Antoine] ed. *Manuel de l'artificier.* Paris: Jombert, 1757.

Jones, Captain Robert. *A New Treatise on Artificial Fireworks.* London: Millar, 1765; also, *Artificial Fireworks.* Chelmsford: Meggy & Chalk, 1801.

Kentish, Thomas. *The Pyrotechnists' Treasury.* London: Chatto & Windus, 1878; also, *The Complete Art of Firework-Making.* London: Chatto & Windus, 1905.

Kostantinov, Major General Konstantin I. *On Fighting Rockets.* St. Petersburg: Tipographia Eduard Veimar, 1864. (Principal text in French, title page in Russian.)

L——e, L. von. *Vollständiges Taschenbuch für Kunst und Lustfeuerwerker . . .* Budapest: Hartlebens, 1820.

[Leurechon, Jean (or Henry van Etten)]. *Mathematicall Recreations.* London: Hawkins, 1633.

Lorrain, Hanzelet [Jean Appier]. *La Pyrotechnie.* Pont-à-Mousson: Gaspard, 1630.

———, and François Thybourel, Français. *Recueil de plusieurs machines militaires, et feux artificiels pour la guerre, & recréation.* Pont-à-Mousson: Marchant, 1620.

*Livre de cannonerie et artifice de feu.* Paris: Sertenas, 1561.

Majendie, Sir Vivian Dering. *Ammunition.* London: Mitchell, 1867.

Malthe, François de. *Traité des feux artificiels pour la guerre, et pour la recréation.* Paris: Guillemot, 1632.

Malthus, François. *Pratique de la guerre.* Paris: Clovsier, 1650.

Malthus, Francis. *Treatise of Artificiall Fireworks Both for Warres and Recreation.* London: Hawkins, 1629.

Meyer, Franz Sales. *Die Feuerwerkerei.* Leipzig: Seemann, 1898.

Meyer, Moritz. *Traité de pyrotechnie.* Liège: Oudart, 1844.

Mongéry, Merignon de. *Des fusées de guerre, maintenant fusées à la Congreve.* Paris: Bachelier, 1825.

Moore, William. *A Treatise on the Motion of Rockets.* London: Robinson, 1813.

Morel, A. M. Th. *Traité pratique des feux d'artifice et pour la guerre.* Paris: Didot, 1800.

Mortimer, G. W. *A Manual of Pyrotechny.* London: Simpkin & Marshall, 1824.

Nye, Nathaneal. *The Art of Gunnery.* London: Leak, 1647. (Includes "A Treatise of Artificiall Fireworks for Warre and Recreation.")

[d'Orval, Perrinet]. *Essay sur les feux d'artifice pour le spectacle et pour la guerre.* Paris: Coustelier, 1745; also *Traité des feux d'artifice pour le spectacle et pour la guerre.* Bern: Wagner & Muller, 1750.

Ossorio, Marcello Calà. *Instituzioni di Pirotecnia per Istruzione di Coloro che Vogliono Apprendere a Lavorare i Fuochi d'Artifizio.* Naples: Stamperia Reale, 1819.

Porta, John Baptista. *Natural Magick.* London: Young & Speed, 1658.

Pralon, A. *Les Fusées de guerre en France.* Paris: Berger-Levrault, 1883.

*Pyrotechny, or, The Art of Making Fireworks at Little Cost and with Complete Safety and Cleanliness.* London: Ward, Lock and Tyler, 1873.

Rogier, Charles. *A Word for My King and Country: A Treatise on the Utility of a Rocket Armament . . .* Macclesfield: Wilson, 1818.

Ruggieri, Claude-Fortuné. *Elemens de pyrotechnie.* Paris: Barba & Magimel, 1802; also, *Pyrotechnie militaire.* Paris: Patris, 1812.

Ruggieri, Gaetano, and Giuseppi Sarti. *A Description of the Machine for the Fireworks . . .* London: Bowyer, 1749.

Scoffern, John. *Projectile Weapons of War and Explosive Compounds.* London: Cook & Whitley, 1852.

Simienowicz, Casimir. *The Great Art of Artillery.* London: Tonson, 1729. (English edition of Kazimierz Siemienowicz, *Artis Magnae Artilleriae.)*

Smith, Thomas. *The Art of Gunnery.* London: 1643; to which is added "Certain Additions to the Book of Gunnery with a Supply of Fireworks."

Sonzogno, Cesare. *L'Arte di Fare i Fouchi d'Artifizio . . .* Milan: Sonzogno, 1819.

Susane, M. *Les Fusées de guerre.* Metz: Blanc, 1865.

Thompson, C. R. *The Biography of Wm. Schermuly and the History of the Schermuly Pistol Rocket Apparatus Ltd.* London: Victoria House, 1946.

Todericiu, Doru. *Preistoria Rachetei Moderne Ma-*

*nuscrisul de la Sibiu (1400–1569)*. Bucharest: Editura Academiei, 1969.
Valle di Venafro, Battista de. *Vallo Libro Continente appertinentie a Capitanii* . . . Venice: Nicolo d'Aristotile detto Zoppino, 1529.
Vegetius. *Feuerwerkbuch, 1420*. Augsburg: Steiner, 1529.
Venn, Thomas. *The Compleat Gunner*. London: Pawlet, 1672.
Vergnaud, M. *Manuel de l'artificier*. Paris: Roret, 1826.

*Development of Rocketry in Asia*

Source material is not widely available in western book literature. Reference is thus made herein to key articles as well as a relatively few books.

GENERAL

Reinaud, Joseph Toussaint. "De l'Art militaire chez les Arabes au moyen age," *Journal Asiatique*, 12, No. 9 (Sept. 1848), 193–237.
———, and Ildephonse Favé. "Du Feu grégeois, des feux de guerre, et des origines de la poudre à canon chez les Arabes, les Persans, et les Chinois," *Journal Asiatique*, 14, No. 10 (Oct. 1849), 257–327.

INDIA

Beatson, Alexander. *A View of the Origin and Conduct of the War with Tippoo Sultaun* . . . London: Nicol, 1900.
Bowring, Lewin B. *Haidar Ali and Tipu Sultan and the Struggle with the Musalman Powers of the South*. Oxford: Clarendon, 1899.
Diron, Alexander. *A Narrative of the Campaign in India, Which Terminated the War with Tipoo Sultan*. London: Bulmer, 1793.
Egerton of Tatton, Lord. *A Description of Indian and Oriental Armour* . . . London: Allen, 1896.
Gode, P. K. "The History of Fireworks in India Between A.D. 1400 and 1900," *Transaction*, No. 17 (Indian Institute of Culture), May 1953.
Hammick, Murray, ed. *Historical Sketches of the South of India, in an Attempt to Trace the History of Mysoor* . . . Mysore: Government Branch Press, 1930.
Hook, Theodore E. *The Life of General the Right Honourable Sir David Baird*. 2 vols. London: Bentley, 1832.
Katre, S. M., and Gode, P. K. *Use of Guns and Gunpowder in India from A.D. 1400 Onwards*. Bombay: Karnatak, 1939. (Indian and Iranian studies presented to Sir E. Denison Ross on his sixty-eighth birthday.)
Lane-Poole, Stanley. *Medieval India Under Mohammedan Rule*, 712–1764. New York: Putnam, 1903.
Mackenzie, Lt. Roderick. *A Sketch of the War with Tippoo Sultaun*. . . . 2 vols. London: J. Sewell, Cornhill, T. Egerton, Whitehall, and J. Debrett, Piccadilly, 1799.
Mir Hussain, Ali Khan Kirmānī. *The History of the Reign of Tipú Sultán, Being a Continuation of the Neshani Hyduri*. Calcutta: Susil Gupta, 1858. (Translated from the Persian manuscript by Colonel W. Miles.)
Munro, Innes. *A Narrative of the Military Operations on the Coromandel Coast* . . . London: Bentley, 1789.
Oppert, Gustav. *On the Weapons, Army Organization, and Political Maxims of the Ancient Hindus*. London: Truebner, 1880.

CHINA

Amiot, Joseph-Maria. *Art militaire des Chinois*. . . . Paris: Didot, 1772.
Davis, Tenney L., and Ware, James R. "Early Chinese Military Pyrotechnics," *Journal of Chemical Education*, 24, November 1947, pp. 522-37.
Gaubil, Antoine. *Histoire de Gentchiscan et de toute la dynastie de Mongous, ses successeurs, conquérants de la Chine*. Paris: Nyon, 1739.
Goodrich, L. C., and Fêng Chia-Shêng. "The Early Development of Firearms in China," *Isis*, 36, No. 104, Pt. 2, 1946, pp. 114–23, and "Addendum," 36, Nos. 105–6, Pts. 3–4, pp. 250–51.
Mailla, Joseph Anne-Marie Moriac de. *Histoire genérále de la Chine*. Paris: Pierres, 1777.
Ohsson, Constantine Mouradgea d'. *Histoire des Mongols* . . . The Hague: Van Cleef, 1834.
Pan, Jixing. "On the Origin of Rockets," *Exploration of Nature*, No. 3, 1984, pp. 173–84. (Author at the Institute for History of Science, Chinese Academy of Science, Beijing.)
Schlege, Gustave. "On the Invention and Use of Fire-Arms and Gunpowder in China," *T'oung Pao*, 3, 1902, pp. 1–11.
Sun, Fang-Toh. "Rockets and Rocket Propulsion Devices in Ancient China," *Journal of Astronautical Sciences*, 29, No. 3, 1981, pp. 289–305. (Author provided further details on Chinese rocketry at 35th Congress of the International Astronautical Federation, Lausanne, Switzerland, October 1984; paper IAA-84-258, "On the Early Rocket Weapons in China.")
Wang Ling. "On the Invention, and Use of Gunpowder and Firearms in China," *Isis*, 37, Nos. 109–10, Pts. 3–4, July 1947, pp. 160–78.
Winter, Frank H. "On the Origin of Rockets," *Chem-*

*istry*, 49, No. 2, 1976, pp. 8–12; also, "The Genesis of the Rocket in China and its Spread to the East and West," 30th Congress of the International Astronautical Federation, Munich, 1979.

### 3 Pioneers of Space Travel

The works of the three principal pioneers of rocketry are included in this section: Konstantin Eduardovitch Tsiolkovsky of Russia, Robert Hutchings Goddard of the United States, and Hermann Oberth of German-speaking Transylvania.

Deutherty, Charles Michael. *Robert Goddard: Trail Blazer to the Stars.* New York: Macmillan, 1964.

Dewey, Anne Perkins. *Robert Goddard.* Boston: Little, Brown, 1962.

Gartmann, Heinz. *The Men Behind the Space Rockets.* New York: David McKay, 1956.

Goddard, Esther C., and Pendray, G. Edward, eds. *The Papers of Robert H. Goddard.* 3 vols. New York: McGraw-Hill, 1968.

Goddard, Robert H. *A Method of Reaching Extreme Altitudes.* Washington, D.C.: Smithsonian Institution, 1919; *Liquid-Propellant Rocket Development.* Washington, D.C.: Smithsonian Institution, 1936; *Rocket Development.* Englewood Cliffs, N.J.: Prentice-Hall, 1948.

Kosmodemyansky, A. *Konstantin Tsiolkovsky.* Moscow: Foreign Languages Publishing House, 1956.

Lehman, M. *This High Man.* New York: Farrar, Straus, 1963. (Biography of Goddard.)

Oberth, Hermann. *Die Rakete zu den Planetenräumen.* Munich: R. Oldenbourg, 1923. (Reprinted: Nürnberg: Reproduktionsdruck von Uni-Verlag Feucht, 1960, 1964, and 1984); also *Wege zur Raumschiffahrt.* Munich: R. Oldenbourg, 1929; *Man into Space.* New York: Harper, 1957; *The Moon Car.* New York: Harper, 1959.

Thomas, Shirley. *Men of Space.* Vol. 1. Philadelphia: Chilton, 1960. (Contains biographies of Goddard and Tsiolkovsky as well as other individuals.)

Tsiolkovsky, Konstantin Eduardovitch. *Sobranie Sochinenie (Collected Works).* Moscow: Izd. Akademii Nauk U.S.S.R., 1951, 1954, and 1959. (Note: These were translated and released by the National Aeronautics and Space Administration in Washington, D.C., 1965.) Also, *Works on Rocket Technology*, 1965, containing the 1903 article "A Rocket into Cosmic Space," the 1911 article "The Investigation of Universal Space by Means of Reactive Devices," and the 1926 "Investigation of Universal Space by Reactive Devices."

Verral, C. S. *Robert Goddard: Father of the Space Age.* Englewood Cliffs, N.J.: Prentice-Hall, 1963.

Walters, H. B. *Hermann Oberth: Father of Space Travel.* New York: Macmillan, 1962.

Williams, B., and Epstein, S. *The Rocket Pioneers.* New York: Julian Messner, 1955.

Winders, G. H. *Robert H. Goddard: Father of Rocketry.* New York: Day, 1963.

### 4 The Legacy of the Pioneers

Adams, Carsbie C., and others. *Space Flight.* New York: McGraw-Hill, 1958.

Ananoff, Alexandre. *L'Astronautique.* Paris: Librairie Arthème Fayard, 1950; also, *Les Mémoires d'un astronaute, ou l'astronautique Française.* Paris: Librairie Scientifique et Technique Albert Blanchard, 1978.

Barré, J. J. "Des fusées de guerre," *Revue Historique de l'Armée,* Vol. 12, No. 4, 1956, pp. 157–162; also "Historique des études Françaises sur les fusées à oxygene liquide," *Memorial de l'Artillerie Française,* 1er fasc., 1961. (Reprinted Paris: Imprimerie Nationale, 1961.)

Baumgarten-Crusius, Artur von. *Die Rakete als Weltfriedenstaube.* Leipzig: Rossberg'sche Buchdruckerei, 1931.

Bergman, A. "The Use of Rockets and Illuminating Shells in the Present War," *Journal of Acetylene Lighting,* Vol. 20, No. 1, July 1918, pp. 12–14.

Bloom, Ursula. *He Lit the Lamp.* London: Burke, 1958. (Contains details of A. M. Low's missile work.)

Brügel, Werner. *Männer der Rakete.* Leipzig: Hachmeister und Thal, 1933.

Butlerov, A. A., ed., and others. *Reaktivnoe Dvizhenie (Jet Propulsion).* Leningrad and Moscow: ONTI, Glavnaya Redaksiya Obshcheteknicheskoi Literatury, 1935. (A second volume was released the following year.)

Canby, C. *History of Rockets and Space.* Manhasset, N.Y.: New Illustrated Library of Science and Invention, 1963.

Cleater, P. E. *Rockets Through Space.* New York: Simon and Schuster, 1936.

De Leeuw, H. *From Flying Horse to Man in the Moon.* New York: St. Martin's, 1963.

Emme, Eugene. *A History of Space Flight.* New York: Holt, 1965.

Esnault-Pelterie, Robert. *L'Astronautique.* Paris: A. Lahure, 1930; *L'Astronautique-complément.* Paris: Sociétés des Ingénieurs Civils de France, 1935; *L'exploration par fusées de la très haute atmosphère et la possibilité des voyages interplanétaires.* Paris: Société Astronomique de France, 1928.

Essers, I. *Max Valier: A Pioneer of Space Travel.* Washington, D.C.: National Aeronautics and Space Administration, 1976. (Translation of "Max Valier: Ein Vorkamfer der Weltraumfahrt 1895–1930," Düsseldorf: VDI-Verlag Gmbh, 1968, in *Technikgeschichte in Einzeldarstellungen* NR. 5, pp. 1–314.)

Faber, Henry B. *Military Pyrotechnics.* 3 vols. Washington, D.C.: U.S. Government Printing Office, 1919.

Haley, Andrew G. *Rocketry and Space Exploration.* Princeton, N.J.: Van Nostrand, 1958.

Harper, Harry. *Dawn of the Space Age.* London: Sampson Low, Marson, 1946.

Hohmann, Walter. *Die Erreichbarkeit der Himmelskörper.* Munich: R. Oldenbourg, 1925.

Kaiser, H. K. *Kleine Raketenkunde.* Stuttgart: Mundus-Verlag, 1949.

Kleimenov, I.T., ed., and others. *Raketnaya Tekhnika (Rocket Technology).* Moscow and Leningrad: ONTI, 1936.

Korolov, Sergey Pavlovich. *Raketnyi Polet v Stratosfere (Rocket Flight in the Stratosphere).* Moscow: Voenizdat, 1934.

Kronstein, Max, and Dellenbag, J. *Rocket Mail Catalogue and Historical Survey of First Experiments in Rocketry.* Jamaica, N.Y.: F. Billig, 1955.

Langemak, G. E., and Glushko, V. P. *Rakety, Ikh Ustroistvo i Primeneniye (Rockets, Their Construction and Utilization).* Moscow: ONTI, 1935.

Lasser, David. *Conquest of Space.* New York: Penguin, 1931.

Lehmann, Ernst A. *Zeppelin.* New York: Longman, 1937. (Contains a description of French use of rockets against German dirigibles during World War I.)

Ley, Willy. *Die Fahrt ins Weltall.* Leipzig: Hachmeister und Thal, 1926; *Die Möglichkeit der Weltraumfahrt.* Leipzig: Hachmeister und Thal, 1928; *Rockets, Missiles and Men in Space.* New York: Viking, 1968.

Moore, Patrick. *Space: The Story of Man's Greatest Feat of Exploration.* New York: Macmillan, 1968.

Noordung, Hermann. *Das Problem der Befahrung des Weltraums.* Berlin: R. C. Schmidt, 1929.

Pendray, G. Edward. *The Coming Age of Rocket Power.* New York: Harper, 1945.

Perelman, Jakov Isidorovich. *Mezhplanetnye Puteshestviya (Interplanetary Travels).* Leningrad and Moscow: ONTI, 1935.

Philp, C. C. *Stratosphere and Rocket Flight.* London: Sir Isaac Pitman & Sons, 1937.

Rynin, Nikolai Aleksevich. *Mezhplanetnye Soobshcheniya (Interplanetary Communications).* This great nine-volume encyclopedic work was released between 1928 and 1932 by several different publishers in Leningrad and in numbers ranging from less than a thousand to fifteen thousand. Vol. 7 is a biography of Tsiolkovsky and Vol. 9 is a comprehensive bibliography. See *Interplanetary Flight and Communication.* Jerusalem: Israel Program for Scientific Translations, 1970. (Translation of complete encyclopedia.)

Sänger, Eugen. *Raketenflugtechnik.* Munich: R. Oldenbourg, 1933. See *Rocket Flight Engineering.* Washington, D.C.: National Aeronautics and Space Administration, 1965.

Scherschevsky, A. B. *Die Rakete für Fahrt und Flug.* Berlin: Volckmann, 1929.

Tikhonravov, M. K. *Raketnaya Tekhniya (Rocket Technology).* Moscow: ONTI, 1935.

Tsander, Fridrikh Arturovich. *Probleme Poleta Pri Pomoshchi Raketnykh Apparatov.* Moscow: ONTI, 1932. See *Problems of Flight by Jet Propulsion.* Jerusalem: Israel Program for Scientific Translation, 1964.

Valier, Max. *Der Vorstoss in den Weltenraum.* Munich and Berlin: Oldenbourg, 1925; *Raketenfahrt.* Munich: Oldenbourg, 1930.

Wilcox, A. *Moon Rocket.* London: Thomas Nelson, 1946.

Winter, Frank H. *Prelude to the Space Age: The Rocket Societies—1924–1940.* Washington, D.C.: Smithsonian Institution Press, 1983.

### 5 The Rocket Returns to War

In this section of the bibliography, works principally or exclusively concerned with World War II rocketry are recorded. Note that many books listed in Chapter 6 contain information on wartime rocketry, although they concentrate on postwar activities.

*U.S. Rocket Ordnance: Development and Use in World War II.* Washington, D.C.: Government Printing Office, 1946.

Arct, Bohdan. *Poles Against the "V" Weapons.* Warsaw: Interpress, 1972.

Benecke, Th., and Quick, A. W. *History of German Guided Missile Development.* Brunswick: Appelhaus, 1957.

Bolster, C. M. *Assisted Take-off of Aircraft.* Northfield, Vt.: Norwich University, 1960.

Bowman, Norman J. *Handbook of Rockets and Guided Missiles.* Chicago: Perastadion, 1957.

Boyce, Joseph C., ed. *New Weapons for Air Warfare.* Boston: Little, Brown, 1947. (Covers United States guided-missile developments.)

Burchard, John E., ed. *Rockets, Guns and Targets.* Boston: Little, Brown, 1948. (Part of the series cited above.)

Christman, Albert. *Sailors, Scientists and Rockets.* Washington, D.C.: Naval History Division, 1971.

Collier, Basil. *The Battle of the V-Weapons 1944–45.* Morey, Yorkshire: Elmfield, 1976; *The Defence of the United Kingdom.* London: Her Majesty's Stationary Office, 1957.

Cooksley, Peter G. *Flying Bomb: The Story of Hitler's V-Weapons in World War II.* New York: Scribner, 1979.

Crow, Sir Alwyn D. "The Rocket as a Weapon of War in the British Forces," *Institution of Mechanical Engineers Journal and Proceedings*, Vol. 158, No. 1, June 1948, pp. 15–21.

Dornberger, Walter. *V-2.* New York: Viking, 1954; *Peenemünde: Die Geschichte der V-Waffen.* Esslinger: Bechtle, 1981. (Expanded version of German edition of V-2.)

Ethell, Jeffrey L. *Komet: The Messerschmidt 163.* London: Sky Books, 1978.

Feist, Uwe, and Maloney, Edward T. *Messerschmitt Me 163.* Fallbrook, Calif.: Aero, 1967.

Ford, Brian. *German Secret Weapons: Blueprint for Mars.* New York: Ballantine, 1969.

Gatland, Kenneth W. *Development of the Guided Missile.* New York: Philosophical Library, 1952.

Gerrard-Gouch, J. D., and Christman, Albert B. *The Grand Experiment at Inyokern.* Washington, D.C.: Naval History Division, 1978.

Huzel, Dieter K. *Peenemünde to Canaveral.* Englewood Cliffs, N.J.: Prentice-Hall, 1962.

Irving, David. *The Mare's Nest.* London: William Kimber, 1964. (Efforts of British intelligence to discover nature and capabilities of German surface-to-surface missiles; countermeasures taken.)

Johnson, Brian. *The Secret War.* New York: Methuen, 1978.

Johnson, David. *V for Vengeance: The Second Battle of London.* London: William Kimber, 1983.

Jones, R. V. *Most Secret War: British Scientific Intelligence 1939–1945.* London: Hamish Hamilton, 1978. (Includes detailed information on intelligence activities applied to German V-weapons.)

Joubert de la Ferte, Sir Philip B. *Rocket.* New York: Philosophical Library, 1957. (Story of V-1 and V-2 seen from the British wartime viewpoint.)

Kennedy, Gregory P. *Vengeance Weapon: The V-2 Guided Missile.* Washington, D.C.: Smithsonian Institution Press, 1983.

Klee, Ernst, and Merk, Otto. *The Birth of the Missile.* New York: Dutton, 1965. (Deals with Peenemünde and the missiles developed there.)

Kooy, J. M. J., and Uytenbogaart, J. W. H. *Ballistics of the Future.* Haarlem, Netherlands: N.V. de Technische Vitgeverij H. Stam, 1946. (Contains information on V-1 and V-2 missiles and their combat firings from Holland.)

Lasby, Clarence G. *Project Paperclip.* New York: Atheneum, 1971. (Deals with roundup and movement of German scientists and engineers to the west following World War II.)

Longmate, Norman. *Hitler's Rockets: The Story of the U2s.* London: Hutchinson, 1985.

Lusar, Rudolf. *German Secret Weapons of the Second World War.* New York: Philosophical Library, 1959.

McGovern, James. *Crossbow and Overcast.* New York: Morrow, 1964. (Covers allied intelligence efforts to discover the capabilities of the V-weapons and the rounding up of leading German rocket experts and their documentation and materiel at the close of the war.)

Middlebrook, Martin. *The Peenemünde Raid: The Night of 17–18 August 1943.* London: Allen Lane, 1982.

Morey, Loren. *The Powder Rockets.* Manhatten, Kan.: Military Affairs/Aerospace Historian, 1982.

Napier, A. F. S. "British Rockets in the World War," *British Royal Artillery Journal*, Vol. 73, No. 1, January 1946, pp. 11–20.

Ordway, Frederick I., III, and Wakeford, Ronald C. *International Missile and Spacecraft Guide.* New York: McGraw-Hill, 1960. (Part 1 contains a detailed survey of World War II rockets and missiles.)

———, and Sharpe, Mitchell R. *The Rocket Team: From the V-2 to the Saturn Moon Rocket.* New York: Crowell; London: Heinemann, 1979; Cambridge, Mass.: M.I.T. Press, 1982. (The story of the Wernher von Braun rocket team from 1930's through Apollo expeditions to the Moon.)

Pile, General Sir Frederick. *Ack-Ack: Britain's Defence Against Air Attack During the Second World War.* London: Harrap, 1949.

Ross, Frank, Jr. *Guided Missiles: Rockets and Torpedoes.* New York: Lothrop, 1951.

Rosser, R. B., Newton, R. R., and Gross, G. L. *Mathematical Theory of Rocket Flight.* New York: McGraw-Hill, 1949. (Official final report on Section H, Division 3, of wartime National Defense Research Committee on exterior ballistics of fin-stabilized rockets.)

Sänger, E., and Bredt, I. *Über einen Raketenantrieb für Fernbomber.* Ainring: Deutsche Luftfahrtforschung, August 1944. (Translated and published by the Technical Information Branch, Bureau of Aeronautics, U.S. Navy, and distributed as *Rocket*

*Drive for Long Range Bombers*. Whittier, Calif.: Robert Cornog, 1952.)

Searby, J. *The Great Raids: Peenemünde*. Chippenham: Nutshell, 1978.

Walters, Helen B. *Wernher von Braun: Rocket Engineer*. New York: Macmillan, 1964.

Weyl, A. R. *Guided Missiles*. London: Temple, 1949.

Ziegler, M. *Rocket Fighter*. London: MacDonald, 1963. (Covers the Me 163 rocket airplane.)

Zim, H. S. *Rockets and Jets*. New York: Harcourt, 1945.

### 6 Postwar Military Rocketry

Chapters 6 to 10 cover the four decades from the end of World War II to 1985. Many if not most books dealing with the period are concerned with more than one subject, so for ease of reference, we simply group the works in these chapters in accordance with degrees of emphasis. For further reading and research, the reader is directed to the ample bibliographic material that appears at the end of many books.

Abel, Elie. *The Missiles of October: The Cuban Missile Crisis 1962*. London: MacGibbon & Key, 1966.

Akens, David S. *A Pictorial History of Rockets and Rocketry*. Huntsville, Ala.: Strode, 1966.

Altman, D., and others. *Liquid Propellant Rockets*. Princeton, N.J.: Princeton University Press, 1960.

Armacost, Michael H. *The Politics of Weapons Innovation: The Thor-Jupiter Controversy*. New York: Columbia University Press, 1969.

Baar, J., and Howard, W. E. *Combat Missileman*. New York: Harcourt, 1961; *Polaris: The Concept and Creation of a New and Mighty Weapon*. New York: Harcourt, 1960.

Baker, David. *The Rocket: The History and Development of Rocket and Missile Technology*. New York: Crown, 1978.

Barrère, N., and others. *Rocket Propulsion*. Amsterdam: Elsevier, 1960.

Beard, Edmund. *Developing the ICBM: A Study in Bureaucratic Politics*. New York: Columbia University Press, 1976.

Bergaust, Erik. *Reaching for the Stars*. New York: Doubleday, 1960. (Biography of Wernher von Braun; gives considerable attention to events in the 1945–1960 period and the U.S. Army ballistic missile program.)

Bottome, Edgar M. *The Missile Gap: A Study of the Formulation of Military and Political Policy*. Rutherford, N.J.: Fairleigh Dickinson University Press, 1971.

Brodie, B. *Strategy in the Missile Age*. Princeton, N.J.: Princeton University Press, 1959.

Burgess, Eric. *Guided Weapons*. New York: Macmillan, 1957; *Long-range Ballistic Missiles*. New York: Macmillan, 1962.

Caiden, Martin. *Rockets and Missiles—Past and Future*. New York: McBride, 1954.

Carter, Ashton B., and Schwartz, David N. *Ballistic Missile Defense*. Washington, D.C.: Brookings Institution, 1984.

Chapman, J. L. *Atlas: The Story of a Missile*. New York: Harper, 1960.

Chayes, Abram. *The Cuban Missile Crisis*. New York: Oxford University Press, 1974.

Clemow, J. *Short-range Guided Weapons*. London: Temple, 1961.

Detzer, David. *The Brink: Cuban Missile Crisis, 1962*. New York: Crowell, 1979.

Emme, Eugene M., ed. *History of Rocket Technology*. Detroit: Wayne State University Press, 1964. (Covers not only ballistic missile developments but launch vehicles and spacecraft.)

Gantz, K. F. *United States Air Force Report on the Ballistic Missile: Its Technology, Logistics, and Strategy*. Garden City, N.Y.: Doubleday, 1958.

Gavin, James M. *War and Peace in the Space Age*. New York: Harper, 1958. (By former Army general, book is concerned with tactical and strategic implications of missiles, spacecraft.)

Graham, Daniel O. *High Frontier: A New National Strategy*. Washington, D.C.: Heritage Foundation, 1982. (The concept led to the Strategic Defense Initiative of 1983.)

Gröttrup, Irmgard. *Rocket Wife*. London: Deutsche, 1959. (Story of the Germans who worked on Russia's missile program under Helmut Gröttrup from mid-1940's until early 1950's.)

*Guided Missiles—Operations, Design and Theory*. New York: McGraw-Hill, 1958.

Gunston, Bill. *Rockets & Missiles*. New York: Salamander-Crescent, 1979.

Hartt, J. *Mighty Thor*. New York: Duell, Sloan & Pearce, 1961.

Holst, Johan, and Schneider, William, Jr., eds. *Why ABM? Policy Issues in the Missile Defense Controversy*. New York: Pergamon, 1969.

Huggett, C., Bartley, C. F., and Mills, M. M. *Solid Propellant Rockets*. Princeton, N.J.: Princeton University Press, 1960.

Humphries, John. *Rockets and Guided Missiles*. New York: Macmillan, 1956.

Karas, Thomas. *The New High Ground: Systems and Weapons of Space Age War*. New York: Simon and Schuster, 1983.

Kit, B., and Evered, D. S. *Rocket Propellant Handbook*. New York: Macmillan, 1960.

Lee, A., ed. *Soviet Air and Rocket Forces.* New York: Praeger, 1959.
Loosbrock, J. F., and others, eds. *Space Weapons: A Handbook of Military Astronautics.* New York: Praeger, 1959.
Medaris, J. B. *Countdown for Decision.* New York: Putnam, 1960. (By former head of the Army Ballistic Missile Agency.)
Merrill, G., and others, eds. *Dictionary of Guided Missiles and Space Flight.* Princeton, N.J.: Van Nostrand, 1959.
Neal, R. *Ace in the Hole.* New York: Doubleday, 1962. (Story of the Minuteman ICBM.)
Ordway, Frederick I., III, and Wakeford, Ronald C. *International Missile and Spacecraft Guide.* New York: McGraw-Hill, 1960. (Part 2 describes post World War II missiles and spacecraft.)
Parry, Albert. *Russia's Rockets and Missiles.* Garden City, N.Y.: Doubleday, 1960.
Schwiebert, Ernest G. *A History of the U.S. Air Force Ballistic Missiles.* New York: Praeger, 1965.
Skelton, W. R. *Countdown: The Story of Cape Canaveral.* Boston: Little, Brown, 1960.
*Soviet Military Power 1984*, 4th ed. Washington, D.C.: Government Printing Office, 1985. (First edition, 1981, second edition, 1983, third edition, 1984.)
Sutton, George P. *Rocket Propulsion Elements.* New York: Wiley, 1963.
Taylor, Michael J. H., and Taylor, John W. R. *Missiles of the World.* London: Ian Allan, 1976.
Von Braun, Wernher, and Ordway, Frederick I., III. *The Rockets' Red Glare: An Illustrated History of Rocketry Through the Ages.* New York: Anchor/Doubleday, 1976.
Zucrow, Maurice. *Aircraft and Missile Propulsion.* 2 vols. New York: Wiley, 1958.

## 7 Probing the Fringe of Space

Relatively few books have been written that treat sounding rockets and launch vehicles exclusively, which is the subject of this chapter. The reader is also directed to the bibliographies of Chapters 8, 9, and 10 that also contain in part material covered in this chapter.

Angelo, Joseph A., Jr. *The Dictionary of Space Technology.* New York: Facts-on-File, 1982.
*Assessment of Candidate Expendable Launch Vehicles for Large Payloads.* Washington, D.C.: National Academy Press, 1984.
Benson, Charles D., and Faherty, William Barnaby. *Moonport: A History of Apollo Launch Facilities and Operations.* Washington, D.C.: National Aeronautics and Space Administration, 1978.
Berkner, Lloyd V., ed. *Manual of Rockets and Satellites.* New York: Pergamon, 1958.
Bilstein, Roger E. *Stages to Saturn: A Technological History of the Apollo/Saturn Launch Vehicles.* Washington, D.C.: National Aeronautics and Space Administration, 1980.
Boyd, R. L. F., and Seaton, M. J., eds. *Rocket Exploration of the Upper Atmosphere.* New York: Pergamon, 1954.
Buedeler, Werner. *Operation Vanguard.* London: Burke, 1957.
Cohan, Christopher J., Olstad, Walter B., Patterson, Donald W., and Salkeld, Robert. *Space Transportation Systems 1980–2000.* New York: American Institute of Aeronautics and Astronautics, 1979.
Corliss, William R. *NASA Sounding Rockets, 1958–1968.* Washington, D.C.: National Aeronautics and Space Administration, 1971.
Gatland, Kenneth W. *Spacecraft and Boosters.* London: Iliffe, 1964.
Green, Constance McLaughin, and Lomask, Milton. *Vanguard: A History.* Washington, D.C.: Smithsonian Institution Press, 1971.
Holder, W. G. *Saturn Five: The Moon Rocket.* New York: Messner, 1968.
Lange, O.H., and Stein, R. J. *Space Carrier Vehicles: Design, Development and Testing of Launching Rockets.* New York: Academic, 1963. (Supplement to Frederick I. Ordway, III, ed. *Advances in Space Science and Technology.*)
Newell, Homer E., Jr. *High Altitude Rocket Research.* New York: Academic, 1953; *Sounding Rockets.* New York: McGraw-Hill, 1959; *Beyond the Atmosphere: Early Years of Space Science.* Washington, D.C.: National Aeronautics and Space Administration, 1980.
Ordway, Frederick I., III, ed. *Advances in Space Science and Technology.* Vols. 6 and 7. New York: Academic, 1964 and 1965. (Contains two-part monograph on "Rocket, Missile and Carrier Vehicle Testing, Launching and Tracking Technology" within and outside the United States, by Mitchell R. Sharpe, Jr., and John M. Lowther.)
Rosen, Milton W. *Viking Rocket Story.* New York: Harper, 1955.
Samson, D. R., ed. *Development of the Blue Streak Satellite Launcher.* New York: Pergamon, 1963.
Southall, Ivan. *Woomera.* Sydney: Angus and Robertson, 1962. (Describes the large Australian launching site for sounding rockets, launch vehicles, and missiles.)
Stehling, Kurt R. *Project Vanguard.* Garden City, N.Y.: Doubleday, 1961.
Vaeth, J. Gordon. *200 Miles Up.* New York: Ronald, 1955.

Wexler, H., and Caskey, J. E., Jr., eds. *Rocket and Satellite Meteorology.* Amsterdam: North-Holland, 1963.

Wilson, Andres. *The Eagle Has Wings: The Story of American Space Exploration 1945–1975.* London: British Interplanetary Society, 1982.

Young, R. E., *Telemetry.* London: Temple, 1963. (Describes how information is sent by radio from sounding rockets and probes to receiving stations on the ground.)

## 8 The Remote Explorers

This section lists books that deal principally with unmanned automated artificial Earth satellites and planetary and interplanetary spacecraft. Some works also cover the general field of astronautics.

*Above and Beyond: The Encyclopedia of Aviation and Space Sciences.* 14 vols. Chicago: New Horizons, 1968.

Adams, Carsbie C., von Braun, Wernher, and Ordway, Frederick I., III. *Careers in Astronautics and Rocketry.* New York: McGraw-Hill, 1962.

Baar, James, and Howard, William E. *Spacecraft and Missiles of the World.* New York: Harcourt, 1966.

Beatty, J. Kelly, O'Leary, Brian, and Chaikin, Andrew, eds. *The New Solar System.* Cambridge, Mass.: Sky, 1981.

Berkner, Lloyd V., and Odishaw, Hugh, eds. *Science in Space.* New York: McGraw-Hill, 1961.

Berman, A. I. *Physical Principles of Astronautics.* New York: Wiley, 1961.

Bester, Alfred. *The Life and Death of a Satellite.* Boston: Little, Brown, 1966.

Blasingame, B. P. *Astronautics.* New York: McGraw-Hill, 1964.

Bova, Ben, with Bell, Trudy E. *Closeup: New Worlds.* New York: St. Martin's, 1977.

Boyd, R. L. F. *Space Research by Rocket and Satellite.* New York: Harper, 1960.

Bradbury, Ray, Clarke, Arthur C., Murray, Bruce, Sagan, Carl, and Sullivan, Walter. *Mars and the Mind of Man.* New York: Harper & Row, 1973.

Briggs, Geoffrey, and Taylor, Frederick. *The Cambridge Photographic Atlas of the Planets.* Cambridge: Cambridge University Press, 1982.

Bucheim, R. W., and Rand Corp. staff. *Space Handbook: Astronautics and Its Applications.* New York: Random House, 1959.

Burgess, Eric. *To the Red Planet.* New York: Columbia University Press, 1978; *By Jupiter: Odysseys to a Giant.* New York: Columbia University Press, 1982.

Carr, Michael H. *The Surface of Mars.* New Haven, Conn.: Yale University Press, 1981.

Carter, L. J. *Artificial Satellite.* London: British Interplanetary Society, 1951.

Chapman, Clark R. *Planets of Rock and Ice: From Mercury to the Moons of Saturn.* New York: Scribner, 1982.

Chester, M., and Kramer, S. B. *Discoverer: The Story of a Satellite.* New York: Putnam, 1960.

Chipman, Ralph, ed. *The World in Space: A Survey of Space Activities and Issues.* Englewood Cliffs, N.J.: Prentice-Hall, 1982. (Prepared for United Nations–sponsored Unispace 82.)

Clarke, Arthur C., ed. *The Coming of the Space Age.* New York: Meredith, 1967; *Interplanetary Flight.* New York: Harper, 1951; *The Making of a Moon.* New York: Harper, 1957; *Voices from the Sky.* New York: Harper, 1965; with paintings by Chesley Bonestell. *Beyond Jupiter: The Worlds of Tomorrow.* Boston: Little, Brown, 1973; *Ascent to Orbit.* New York: John Wiley, 1984.

Corliss, William R. *Propulsion Systems for Space Flight.* New York: McGraw-Hill, 1960; *Scientific Satellites.* Washington, D.C.: National Aeronautics and Space Administration, 1967; *Space Probes and Planetary Exploration.* Princeton, N.J.: Van Nostrand, 1965; *The Interplanetary Pioneers.* 3 vols. Washington, D.C.: National Aeronautics and Space Administration, 1972.

Cross, Charles A., and Moore, Patrick. *The Atlas of Mercury.* New York: Crown, 1977.

Davies, Merton E., and Murray, Bruce C. *The View from Space: Photographic Exploration of the Planets.* New York: Columbia University Press, 1971.

De Galiana, T. *Concise Encyclopedia of Astronautics.* London: Collins, 1968.

Deutsch, A. J., and Klemperer, W. B., eds. *Space Age Astronomy.* New York: Academic, 1962.

Dickson, Paul. *Out of This World: American Space Photography.* New York: Delta, 1977.

Durant, Frederick C., III, ed. *Between Sputnik and the Shuttle: New Perspectives on American Astronautics.* San Diego: American Astronautical Society, 1981.

———, and Miller, Ron. *Worlds Beyond: The Art of Chesley Bonestell.* Norfolk, Vir.: Donning, 1983.

Ehricke, Krafft A. *Space Flight.* 2 vols. Princeton, N.J.: Van Nostrand, 1960, 1962.

Fimmel, Richard O., Van Allen, James, and Burgess, Eric. *Pioneer: First to Jupiter, Saturn and Beyond.* Washington, D.C.: National Aeronautics and Space Administration, 1980.

———, Colin, Lawrence, and Burgess, Eric. *Pioneer*

*Venus.* Washington, D.C.: National Aeronautics and Space Administration, 1983.

French, Bevan, and Maran, Stephen P., eds. *A Meeting with the Universe: Science Discoveries from the Space Program.* Washington, D.C.: National Aeronautics and Space Administration, 1981.

Frye, William E., ed. *Impact of Space Exploration on Society.* Tarzana, Calif.: American Astronautical Society, 1966.

Gatland, Kenneth W. *Astronautics in the 60s.* New York: Wiley, 1962; *Spaceflight Today.* Los Angeles: Aero, 1964; ed. *Spaceflight Technology.* New York: Academic, 1977; *The Illustrated Encyclopedia of Space Technology: A Comprehensive History of Space Exploration.* New York: Harmony, 1981.

Giacconi, Riccardo, ed. *X-Ray Astronomy with the Einstein Satellite.* Boston: Reidel, 1981.

Gilmer, J. R., and others, eds. *Commercial Utilization of Space.* Tarzana, Calif.: American Astronautical Society, 1968.

Glass, Billy P. *Introduction to Planetary Geology.* Cambridge: Cambridge University Press, 1982.

Glasstone, Samuel. *Sourcebook of the Space Sciences.* Princeton, N.J.: Van Nostrand, 1965.

Grey, J., and Grey, V., eds. *Space Flight Report to the Nation.* New York: Basic Books, 1962.

Haber, Heinz. *Space Science.* New York: Golden, 1967.

Hall, R. Cargill. *Lunar Impact: A History of Project Ranger.* Washington, D.C.: National Aeronautics and Space Administration, 1977.

Hanly, Paul A., and Chamberlain, Von del, eds. *Space Science Comes of Age: Perspectives in the History of the Space Sciences.* Washington, D.C.: National Air and Space Museum-Smithsonian Institution, 1981.

Haviland, Robert P., and House, C. M., eds. *Handbook of Satellites and Space Vehicles.* Princeton, N.J.: Van Nostrand, 1965.

Hess, H. H., and others, eds. *Review of Space Research.* Washington, D.C.: National Academy of Sciences, 1962.

Hobbs, M. *Basics of Missile Guidance and Space Techniques.* New York: Rider, 1959. (Two volumes on control, guidance, telemetry, tracking, optics, etc.)

Howard, William E., and Barr, James. *Spacecraft and Missiles of the World.* New York: Harcourt, 1966.

Hubert, Lester F., and Lehr, Paul E. *Weather Satellites.* Waltham, Mass.: Blaisdell, 1967.

Hunten, D. M., Colin, L., Donahue, T. M., and Moroz, V. I. *Venus.* Tucson: University of Arizona Press, 1983.

Hunter, Maxwell W., II. *Thrust into Space.* New York: Holt, 1966.

Jacobs, H., and Whitney, E. E. *Missile and Space Projects Guide.* New York: Plenum, 1962.

Jacobs, Horace, ed. *Exploitation of Space for Experimental Research.* Tarzana, Calif.: American Astronautical Society, 1968.

Jaffe, Leonard. *Communications in Space.* New York: Holt, 1966.

Johnson, Nicholas L. *Handbook of Soviet Lunar and Planetary Exploration.* San Diego, Calif.: Univelt, 1979.

Killian, James R., Jr. *Sputnik, Scientists, and Eisenhower.* Cambridge, Mass.: M.I.T. Press, 1977.

King-Hele, D. *Satellites and Scientific Research.* New York: Dover, 1960; *The RAE Table of Earth Satellites,* 1957–1982. New York: Wiley-Interscience, 1983.

Klass, Philip J., *Secret Sentries in Space.* New York: Random House, 1971.

Koelle, H. H., ed. *Handbook of Astronautical Engineering.* New York: McGraw-Hill, 1961.

Koppes, Clayton R. *JPL and the American Space Program: A History of the Jet Propulsion Laboratory.* New Haven, Conn.: Yale University Press, 1982.

Krieger, F. J. *Behind the Sputniks: A Survey of Soviet Space Science.* Washington, D.C.: Public Affairs Press, 1958.

Lawton, A. T. *A Window in the Sky.* New York: Pergamon, 1979.

LeGalley, Donald P., ed. *Space Science.* New York: Wiley, 1963.

Leondes, C. T., and Vance, R. W., eds. *Lunar Missions and Exploration.* New York: Wiley, 1964.

Lewis, John S., and Prinn, Ronald G. *Planets and their Atmospheres.* Orlando: Academic, 1984.

Lewis, Richard S. *The Illustrated Encyclopedia of Space Exploration.* London: Salamander, 1984.

Ley, Willy, ed. *Harnessing Space.* New York: Macmillan, 1963; *Ranger to the Moon.* New York: New American Library, 1965; *Mariner to Mars.* New York: New American Library, 1966.

Liller, W. *Space Astrophysics.* New York: McGraw-Hill, 1961.

Lundquist, Charles A. *Space Science.* New York: McGraw-Hill, 1966.

McCauley, John F. *Moon Probes.* Morristown, N.J.: Silver Burdett, 1969.

*The McGraw-Hill Encyclopedia of Space.* New York: McGraw-Hill, 1968.

*The Martian Landscape.* Washington, D.C.: National Aeronautics and Space Administration, 1978.

Mesmer, G., and Stuhlinger, Ernst, eds. *Space Science and Engineering.* New York: McGraw-Hill, 1964.

McDougall, Walter A. *The Heavens and the Earth: A*

*Political History of the Space Age.* New York: Basic Books, 1985.

Miller, Ron, and Hartmann, William K. *The Grand Tour: A Traveler's Guide to the Solar System.* New York: Workman, 1981.

Mirabito, Michael M. *The Exploration of Outer Space with Cameras: A History of the NASA Unmanned Spacecraft Missions.* Jefferson, N.C.: McFarland, 1983.

Moore, Patrick, Hunt, Gally, Nicolson, Iain, and Cattermole. *The Atlas of the Solar System.* London: Beazley, 1983.

Morganthaler, George W., and Morra, Robert G., eds. *Unmanned Exploration of the Solar System.* North Hollywood, Calif.: Western Periodicals, 1965.

Morrison, David. *Voyages to Saturn.* Washington, D.C.: National Aeronautics and Space Administration, 1982.

Mueller, George E., and Spangle, Eugene R. *Communication Satellites.* New York: Wiley, 1964.

Murray, Bruce, Malin, Michael C., and Greeley, Ronald. *Earthlike Planets: Surfaces of Mercury, Venus, Earth, Moon, Mars.* San Francisco: Freeman, 1981.

———, and Burgess, Eric. *Flight to Mercury.* New York: Columbia University Press, 1977.

Naugle, John E. *Unmanned Space Flight.* New York: Holt, 1965.

Naylor, J. L., ed. *Advances in Space Technology.* London: George Newnes, 1962.

Needell, Allan, ed. *The First 25 Years in Space: A Symposium.* Washington, D.C.: Smithsonian Institution Press, 1983.

Newlan, Irl. *First to Venus: The Story of Mariner II.* New York: McGraw-Hill, 1963.

Nicks, Oran W., ed. *This Island Earth.* Washington, D.C.: National Aeronautics and Space Administration, 1970.

Ordway, Frederick I., III, ed. *Advances in Space Science and Technology.* New York: Academic, annual review volumes from 1959 to 1972.

———. *Pictorial Guide to Planet Earth.* New York: Crowell, 1975.

———, Gardner, James Patrick, and Sharpe, Mitchell R., Jr. *Basic Astronautics.* Englewood Cliffs. N.J.: Prentice-Hall, 1962. Also, with Wakeford, Ronald C.. *Applied Astronautics.* Englewood Cliffs, N.J.: Prentice-Hall, 1963.

———, and Wakeford, Ronald C. *Conquering the Sun's Empire.* New York: Dutton, 1963; and Adams, Carsbie C., and Sharpe, Mitchell R. *Dividends from Space.* New York: Crowell, 1971.

Ovenden, M. W. *Artificial Satellites.* Harmondsworth: Penguin, 1960.

Pardoe, G. K. C. *The Challenge of Space.* London: Chatto and Windus, 1964; *The Future of Space Technology.* London: Pinter, 1984.

Petrov, G. I., ed. *Conquest of Outer Space in the U.S.S.R.* New Delhi: Amerind, 1973.

Pierce, J. R. *The Beginnings of Satellite Communications.* San Francisco: San Francisco Press, 1968.

*Planetary Exploration Through Year 2000.* Washington, D.C.: National Aeronautics and Space Administration, 1983. (By the Solar System Exploration Committee.)

Popescu, Julian. *Russian Space Exploration: The First 21 Years.* Henley-on-Thames: Gothard House, 1979.

Poynter, Margaret and Lane, Arthur L. *Voyager: The Story of a Space Mission.* New York: Atheneum, 1981.

Ramo, S., ed. *Peacetime Uses of Outer Space.* New York: McGraw-Hill, 1961.

Ruppe, Harry O. *Introduction to Astronautics.* 2 vols. New York: Academic, 1966, 1967.

Seifert, Howard S., ed. *Space Technology.* New York: Wiley, 1959.

Sharpe, Mitchell R. *Satellites and Probes.* London: Aldus, 1970.

Sheffield, Charles. *Earth Watch: A Survey of the World from Space.* New York: Macmillan, 1981; *Man on Earth.* London: Sidgewick & Jackson, 1983.

Sheldon, Charles, S., II. *Review of the Soviet Space Program.* New York: McGraw-Hill, 1968.

Shelton, William R. *American Space Exploration.* Boston: Little, Brown, 1967.

Sherman, Madeline W., ed. *The Silver Anniversary of the Age of Space Exploration 1957–1982.* Redondo Beach, Calif.: TRW Electronics and Defense, 1983. (Record of spacecraft missions during first 25 years of the Space Age.)

Short, Nicholas M. *The Landsat Tutorial Workbook: Basics of Satellite Remote Sensing.* Washington, D.C.: National Aeronautics and Space Administration, 1982.

———, Lowman, Paul D., Jr., Freden, Stanley C., and Finch, William A., Jr. *Mission to Earth: Landsat Views the World.* Washington, D.C.: National Aeronautics and Space Administration, 1976.

Shternfeld, A. *Soviet Space Science.* New York: Basic Books, 1959.

Singer, S. F., ed. *Progress in the Astronautical Sciences.* Amsterdam: North-Holland, 1962.

Smith, Marcia S., et al. *Soviet Space Programs—1976–80.* Vol. 1. Washington, D.C.: U.S. Government Printing Office, 1983. Vol. 2 (1984) includes supplementary material through 1983 and data on manned flights.

———. *Space Activities of the United States, Soviet Union, and other Launching Countries.* Washing-

ton, D.C.: U.S. Government Printing Office, 1984. (Updated annually.)
Solomon, L. *Telstar.* New York: McGraw-Hill, 1962.
*Soviet Writings on Earth Satellites and Space Travel.* New York: Citadel, 1958.
*The Space Industry: America's Newest Giant.* Englewood Cliffs, N.J.: Prentice-Hall, 1962. (By the editors of *Fortune.*)
Spitzer, Cary R., ed. *Viking Orbiter Views of Mars.* Washington, D.C.: National Aeronautics and Space Administration, 1980.
Strong, J. *Search the Solar System.* London: David & Charles, 1973.
Stuhlinger, Ernst. *Ion Propulsion for Space Flight.* New York: McGraw-Hill, 1964.
———, Ordway, Frederick I., III, McCall, Jerry C., and Bucher, George C., eds. *Astronautical Engineering and Science.* New York: McGraw-Hill, 1963.
*The Teacher's Handbook of Astronautics.* London: British Interplanetary Society, 1963.
Trinklein, F. E., and Huffer, C. M. *Modern Space Science.* New York: Holt, 1961.
Tucker, Wallace H. *The Star Splitters: The High Energy Astronomy Laboratories.* Washington, D.C.: National Aeronautics and Space Administration, 1984.
——— and Riccardo Giaconni. *The X-Ray Universe.* Cambridge, Mass.: Harvard University Press, 1985.
Vaeth, J. Gordon. *Weather Eyes in the Sky: America's Meteorological Satellites.* New York: Ronald, 1965.
Van Allen, James A., ed. *Scientific Uses of Earth Satellites.* Ann Arbor, Mich.: University of Michigan Press, 1956; *Origins of Magnetospheric Physics.* Washington, D.C.: Smithsonian Institution Press, 1983.
Vassiliev, M., and Dobronravov, V. V. *Sputnik into Space.* London: Souvenir, 1958.
Vertregt, M. *Principles of Astronautics.* Amsterdam: Elsevier, 1960.
*Viking: The Exploration of Mars.* Pasadena: Jet Propulsion Laboratory, 1984.
Von Braun, Wernher, and Ordway, Frederick I., III. *New Worlds: Discoveries from Our Solar System.* Garden City, N.Y.: Anchor/Doubleday, 1979.
*Voyager Encounters Jupiter.* Washington, D.C.: National Aeronautics and Space Administration, 1979.
*The Voyager Flights to Jupiter and Saturn.* Pasadena: Jet Propulsion Laboratory, 1982.
*Voyager 1 Encounters Saturn.* Pasadena: Jet Propulsion Laboratory, 1980.
West, Richard M., ed. *Understanding the Universe: The Impact of Space Astronomy.* Boston: Reidel, 1983.
Wheelock, Harold J., ed. *Mariner Mission to Venus.* New York: McGraw-Hill, 1963.
Widger, William K., Jr. *Meteorological Satellites.* New York: Holt, 1966.

**9 Manned Space Flight**
**10 The Era of the Space Shuttle**

Books listed herein deal primarily or exclusively with manned space activities following launch from either expendable or recoverable vehicles.

Abbas, A. *Till We Reach the Stars: The Story of Yuri Gagarin.* Bombay: Asia Publishing House, 1961.
Ahn, Chung-Hae. *NASA's Biomedical Research Program.* Washington, D.C.: National Aeronautics and Space Administration, 1981.
Alexander, Tom. *Project Apollo: Man to the Moon.* New York: Harper, 1964.
Allen, Joseph P. *Entering Space: An Astronaut's Journey.* New York: Stewart, Tabori & Chang, 1984.
Anderson, Frank W., Jr. *Orders of Magnitude: A History of NACA and NASA 1915–1980.* Washington, D.C.: National Aeronautics and Space Administration, 1981.
Armstrong, Neil, Collins, Michael, and Aldrin, Edwin E., Jr. *First on the Moon.* Boston: Little, Brown, 1970.
Badgley, Peter C. ed. *Scientific Experiments for Manned Orbital Flight.* North Hollywood, Calif.: Western Periodicals, 1965.
Bainbridge, William Sims. *The Spaceflight Revolution: A Sociological Study.* New York: Wiley, 1976.
Baker, David. *The History of Manned Space Flight.* New York: Crown, 1981.
Bedini, Silvio A., von Braun, Wernher, and Whipple, Fred L. *Moon: Man's Greatest Adventure.* New York: Abrams, 1971.
Bell, J. N. *Seven into Space.* Chicago: Popular Mechanics, 1960. (About the seven original Mercury astronauts.)
Benson, O. O., and Strughold, H., eds. *Physics and Medicine of the Atmosphere and Space.* New York: Wiley, 1960.
Bergman, Jules. *Ninety Seconds to Space: The X-15 Story.* Garden City, N.Y.: Doubleday, 1960.
Bergwin, C. R., and Coleman, W. T. *Animal Astronauts: They Opened the Way to the Stars.* Englewood Cliffs, N.J.: Prentice-Hall, 1963.
Booker, P. J., Frewer, G. C., and Pardoe, G. K. C. *Project Apollo: The Way to the Moon.* London: Chatto & Windus, 1970.
Boston, Penelope J., ed. *The Case for Mars.* San Diego, Calif.: Univelt, 1984.
Bridgeman, W., and Hazard, J. *The Lonely Sky.* New

York: Holt, 1955. (On the D-558-2 rocket airplane.)
Calvin, Melvin, and Gazenko, Oleg, G., eds. *Foundations of Space Biology and Medicine.* 4 vols. Washington, D.C.: National Aeronautics and Space Administration, 1975.
Campbell, Paul A. *Earthman, Spaceman, Universal Man?* New York: Pageant, 1965; ed. *Medical and Biological Aspects of the Energies of Space.* New York: Columbia University Press, 1961.
Carpenter, M. S., and others. *We Seven.* New York: Simon and Schuster, 1962. (By the seven original Mercury astronauts.)
Carter, L. J., ed. *Realities of Space Travel.* New York: McGraw-Hill, 1957.
Clarke, Arthur C. *Exploration of Space.* New York: Harper, 1952; *Challenge of the Spaceship.* New York: Harper, 1959; *Exploration of the Moon.* New York: Harper, 1954; *Man and Space.* New York: Time, Inc., 1964; *The Promise of Space.* New York: Harper & Row, 1968.
Collins, Michael. *Carrying the Fire: An Astronaut's Journey.* New York: Farrar, Straus & Giroux, 1974. (Apollo 11 story by crew member.)
Compton, W. David, and Benson, Charles D. *Living and Working in Space: A History of Skylab.* Washington, D.C.: National Aeronautics and Space Administration, 1983.
Cooke, Hereward Lester, and Dean, James D. *Eyewitness to Space.* New York: Abrams, 1971.
Coombs, Charles. *Project Apollo Mission to the Moon.* New York: Morrow, 1965; *Skyrocketing into the Unknown.* New York: Morrow, 1954. (About rocket airplanes.)
Cooper, Henry S. F., Jr. *Thirteen: The Flight that Failed.* New York: Dial, 1973. (Apollo 13 story); *Apollo on the Moon.* New York: Dial, 1969.
Cortright, Edgar M., ed. *Apollo Expeditions to the Moon.* Washington, D.C.: National Aeronautics and Space Administration, 1975.
Cretien, Jean-Loup, and Baudry, Patrick, with the collaboration of Bernard Chabbert. *Spatiale première.* Paris: Plon, 1982. (French "spationauts" story.)
Crossfield, Alfred S., with Blair, Clay, Jr. *Always Another Dawn.* Cleveland: World, 1960. (On U.S. rocket-powered airplanes.)
Cummings, Clifford I., and Lawrence, H. R., eds. *Technology of Lunar Exploration.* New York: Academic, 1963.
Cunningham, Walter. *The All-American Boys.* New York: Macmillan, 1977. (Astronaut's story.)
Eddy, John A. *A New Sun: The Solar Results from Skylab.* Washington, D.C.: National Aeronautics and Space Administration, 1979.
El-Baz, Farouk, and Warner, D. M., eds. *Apollo-Soyuz Test Project.* Washington, D.C.: National Aeronautics and Space Administration, 1979.
Everest, F. K., Jr. *Fastest Man Alive.* New York: Dutton, 1958. (Story of X-2 rocket airplane.)
Ezell, Clinton, and Ezell, Linda Neuman. *The Partnership: A History of the Apollo-Soyuz Test Project.* Washington, D.C.: National Aeronautics and Space Administration, 1978.
Faget, Maxime. *Manned Space Flight.* New York: Holt, 1965.
Fallaci, Oriana. *If the Sun Dies.* New York: Atheneum, 1966.
Farmer, Gene, and Hamblin, Dora Jane. *First on the Moon: A Voyage with Neil Armstrong, Michael Collins, Edwin E. Aldrin, Jr.* Boston: Little, Brown, 1970.
French, Bevan M. *The Moon Book.* New York: Penguin, 1977.
Froehlich, Walter. *Spacelab: An International Short-Stay Orbiting Laboratory.* Washington, D.C.: National Aeronautics and Space Administration, 1983.
Gagarin, Yuri. *Road to the Stars.* Moscow: Foreign Languages Publishing House, 1962.
Gantz, K. F., ed. *Man in Space.* New York: Duell, Sloan & Pearce, 1959.
Gatland, Kenneth. *Manned Spacecraft.* New York: Macmillan, 1967.
———, and Kunesch, A. M. *Space Travel.* New York: Philosophical Library, 1953.
Gerathewohl, S. J. *Principles of Bioastronautics.* Englewood Cliffs, N.J.: Prentice-Hall, 1963.
Grey, Jerry L. *Beachheads in Space.* New York: Macmillan, 1983; *Enterprise.* New York: Morrow, 1979.
Grissom, Betty, and Still, Henry. *Starfall.* New York: Crowell, 1974.
Grissom, Virgil. *Gemini: A Personal Account of Man's Venture into Space.* New York: Macmillan, 1968.
Gurney, Gene. *Walk in Space: The Story of Project Gemini.* New York: Random House, 1967.
Hallion, Richard P. *Supersonic Flight: The Story of the Bell X-1 and Douglas D-588.* New York: Macmillan, 1972; and *Test Pilots: The Frontiersmen of Flight.* Garden City, N.Y.: Doubleday, 1981.
——— and Crouch, Tom D., eds. *Apollo: Ten Years Since Tranquility Base.* Washington, D.C.: Smithsonian Institution Press, 1979.
Heppenheimer, T. A. *Colonies in Space.* Harrisburg, Penn.: Stackpole, 1977.
Hilton, W. F. *Manned Satellites.* London: Hutchinson, 1965.
Hirsch, Richard, and Trento, Joseph John. *The National Aeronautics and Space Administration.* New York: Praeger, 1973.

Joels, Kerry. *Apollo to the Moon: A Dream of Centuries.* Washington, D.C.: Smithsonian Institution—National Air and Space Museum, 1982.

———, and Kennedy, Gregory P. *The Space Shuttle Operator's Manual.* New York: Ballantine, 1982.

Johnson, Nicholas L. *Handbook of Soviet Manned Space Flight.* San Diego, Calif.: Univelt, 1980.

Johnson, Richard S., and Dietlein, Lawrence F., eds. *Biomedical Results from Skylab.* Washington, D.C.: National Aeronautics and Space Administration, 1977.

Lansberg, M. P. *Primer of Space Medicine.* Amsterdam: Elsevier, 1960.

Larmore, Lewis, and Germais, Robert L., eds. *Space Stations.* Tarzana, Calif.: AAS Publications Office, 1970.

Levine, Arnold S. *Managing NASA in the Apollo Era.* Washington, D.C.: National Aeronautics and Space Administration, 1982.

Lewis, Richard S. *Appointment on the Moon.* New York: Viking, 1968.

Ley, Willy. *Conquest of Space.* New York: Viking, 1949.

———, and von Braun, Wernher. *The Exploration of Mars.* New York: Viking, 1956.

Logsdon, John M. *The Decision to Go to the Moon.* Cambridge, Mass.: M.I.T. Press, 1970.

Lundgren, William R. *Across the High Frontier.* New York: Morrow, 1955. (About the X-1A rocket-powered airplane.)

Lundquist, Charles A., ed. *Skylab's Astronomy and Space Sciences.* Washington, D.C.: National Aeronautics and Space Administration, 1979.

Maisak, Lawrence. *Survival on the Moon.* New York: Macmillan, 1966.

Mallan, Lloyd. *Men, Rockets and Space Rats.* New York: Messner, 1961.

Marbarger, J. P., ed. *Space Medicine.* Urbana, Ill.: University of Illinois Press, 1951.

Mason, Robert Grant, ed. *Life in Space.* Boston: Little, Brown, 1983.

Masursky, Harold, Colton, G. W., and El-Baz, Farouk, eds. *Apollo Over the Moon: A View from Orbit.* Washington, D.C.: National Aeronautics and Space Administration, 1978.

*NASA 1958–1983 Remembered—Images.* Washington, D.C.: National Aeronautics and Space Administration, 1983.

Oberg, James E. *Red Star in Orbit.* New York: Random House, 1981; *The New Race for Space.* Harrisburg, Penn.: Stackpole, 1984.

O'Leary, Brian. *Project Space Station.* Harrisburg, Penn.: Stackpole, 1984.

Olney, Ross. *Americans in Space.* Camden, N.J.: Nelson, 1967.

O'Neill, Gerald K. *The High Frontier: Human Colonies in Space.* New York: Morrow, 1977.

Parkinson, Bob. *High Road to the Moon.* London: British Interplanetary Society, 1979.

Pierce, P. N., and Schuon, Karl. *John H. Glenn: Astronaut.* New York: Watts, 1962.

Pirie, Norman W., ed. *Biology of Space Travel.* London: Institute of Biology, 1961.

Powers, Robert M. *Shuttle: The World's First Spaceplane.* Harrisburg, Penn.: Stackpole, 1979.

Rabinowitch, Eugene, and Lewis, Richard S., eds. *Man on the Moon.* New York: Basic Books, 1969.

Riabchikov, Evgeny. *Russians in Space.* New York: Doubleday, 1971.

Richardson, Robert S., ed. *Man and the Moon.* Cleveland: World Publishing, 1961.

Ruzic, Neil. *The Case for Going to the Moon.* New York: Putnam, 1965.

Ryan, Cornelius, ed., with Kaplan, Joseph, von Braun, Wernher, Haber, Heinz, Ley, Willy, Schachter, Oscar, and Whipple, Fred L. *Across the Space Frontier.* New York: Viking, 1952.

———, and von Braun, Wernher, Whipple, Fred L., and Ley, Willy. *Conquest of the Moon.* New York: Viking, 1953.

Seamans, Robert C., Jr., and Ordway, Frederick I., III. "The Apollo Tradition: An Object Lesson for the Management of Large-Scale Technological Endeavors," in Davidson, Frank P., and Meador, C. Lawrence, eds. *Macro-Engineering and the Future: A Management Perspective.* Boulder, Colo.: Westview, 1982.

Shapland, David, and Rycroft, Michael. *Spacelab: Research in Earth Orbit.* Cambridge: Cambridge University Press, 1984.

Sharpe, Mitchell R. *Living in Space: The Astronaut and his Environment.* Garden City, N.Y.: Doubleday, 1969; *Yuri Gagarin, First Man in Space.* Huntsville, Ala.: Strode, 1969.

Shelton, William R. *Flights of the Astronauts.* Boston: Little, Brown, 1963.

Simpson, Theodore R., ed. *The Space Station: An Idea Whose Time Has Come.* New York: IEEE Press, 1985.

*Skylab Explores the Earth.* Washington, D.C.: National Aeronautics and Space Administration, 1977.

Smith, Marcia S., Hellman, Alfred, and Dodge, Christopher H. *Manned Space Programs and Space Life Sciences.* Washington, D.C.: Government Printing Office, 1984.

Stambler, Irwin. *Project Gemini.* New York: Putnam, 1964.

Stockton, William, and Wilford, John Noble. *Spaceliner.* New York: Times Books, 1981.

Summerlin, Lee B., ed. *Skylab, Classroom in Space.* Washington, D.C.: National Aeronautics and Space Administration, 1977.

Swenson, Lloyd S., Jr., Grimwood, John M., and Alexander, Charles C. *This New Ocean: A History of Project Mercury.* Washington, D.C.: National Aeronautics and Space Administration, 1966.

Taylor, Stuart Ross. *A Post-Apollo View.* New York: Pergamon, 1975; *A Lunar Perspective.* Houston, Tex.: Lunar and Planetary Institute, 1982.

Thomas, Shirley. *Men of Space.* Vol. 1 to 8. Philadelphia: Chilton, 1960–1968.

Titov, G., and Caiden, M. *I Am Eagle!* Indianapolis: Bobbs-Merrill, 1962.

Tregaskis, Richard W. *X-15 Diary.* New York: Dutton, 1961.

Von Braun, Wernher. *The Mars Project.* Urbana, Ill.: University of Illinois Press, 1953; *Space Frontier.* New York: Holt, 1971.

*What Made Apollo a Success?* Washington, D.C.: National Aeronautics and Space Administration, 1971.

White, Clayton S., and Benson, O. O., Jr., eds. *Physics and Medicine of the Upper Atmosphere.* Albuquerque, N.M.: University of New Mexico Press, 1952.

Wolfe, Tom. *The Right Stuff.* New York: Farrar, Straus & Giroux, 1979.

Wunder, Charles C. *Life into Space: An Introduction to Space Biology.* Philadelphia: Davis, 1966.

Young, Hugo, Silcock, Bryan, and Dunn, Peter. *Journey to Tranquillity.* Garden City, N.Y.: Doubleday, 1970.

# Index

Page numbers in *italics* refer to material in illustrations or captions.